TAURINE 5

ADVANCES IN EXPERIMENTAL MEDICINE AND BIOLOGY

Recent Volumes in this Series

Volume 518
ADVANCES IN MALE MEDIATED DEVELOPMENTAL TOXICITY
Edited by Bernard Robaire and Barbara F. Hales

Volume 519
POLYMER DRUGS IN THE CLINICAL STAGE: Advantages and Prospects
Edited by Hiroshi Maeda, Alexander Kabanov, Kazurori Kataoka, and Teruo Okano

Volume 520
CYTOKINES AND AUTOIMMUNE DISEASE
Edited by Pere Santamaria

Volume 521
IMMUNE MECHANISMS IN PAIN AND ANALGESIA
Edited by Halina Machelska and Christoph Stein

Volume 522
NOVEL ANGIOGENIC MECHANISMS: Role of Circulating Progenitor Endothelial Cells
Edited by Nicanor I. Moldovan

Volume 523
ADVANCES IN MODELLING AND CLINICAL APPLICATION OF INTRAVENOUS
ANAESTHESIA
Edited by Jaap Vuyk and Stefan Schraag

Volume 524
DIPEPTIDYL AMINOPEPTIDASES IN HEALTH AND DISEASE
Edited by Martin Hildebrandt, Burghard F. Klapp, Torsten Hoffmann,
and Hans-Ulrich Demuth

Volume 525
ADVANCES IN PROSTAGLANDIN, LEUKOTRIENE, AND OTHER BIOACTIVE
LIPID RESEARCH: Basic Science and Clinical Applications
Edited by Zeliha Yazıcı, Giancarlo Folco, Jeffrey M. Drazen, Santosh Nigam,
and Takao Shimizu

Volume 526
TAURINE 5: Beginning the 21st Century
Edited by John B. Lombardini, Stephen W. Schaffer, and Junichi Azuma

TAURINE 5

Beginning the 21st Century

Edited by

John B. Lombardini

Texas Tech University Health Sciences Center
Lubbock, Texas

Stephen W. Schaffer

University of South Alabama School of Medicine
Mobile, Alabama

and

Junichi Azuma

Osaka University
Osaka, Japan

Kluwer Academic / Plenum Publishers
New York, Boston, Dordrecht, London, Moscow

Proceedings of the International Taurine Symposium 2002—Taurine: Beginning the 21st Century, held September 20–23, 2002, in Kauai, Hawaii

ISSN 0065-2598

ISBN 0-306-47769-6

©2003 Kluwer Academic/Plenum Publishers, New York
233 Spring Street, New York, New York 10013

http://www.wkap.nl/

10 9 8 7 6 5 4 3 2 1

A C.I.P. record for this book is available from the Library of Congress

Dr. Ralph Dawson, Jr.
1954 - 2002

It is with great sadness that the Editors of this Proceedings relate to the authors and the scientific community that Dr. Ralph Dawson Jr. was killed in an automobile accident on December 24, 2002. He was 48 years old. Ralph is survived by his parents, Ralph Sr. and Christine Dawson, and two children, a son Ralph Dawson III and a daughter Alecia Perkins. He also has a grandson, Ralph Dawson IV. Dr. Dawson was an expert in the role of taurine in aging and his studies indicated that a failure to maintain proper levels of nutrition could diminish taurine availability and elevate the risk to certain neurodegenerative conditions in the elderly. His studies had great future implications for the role of taurine as a dietary supplement in elderly individuals. His death is a great loss to his family, friends, and the scientific community. We extend our sincere condolences to his family.

PREFACE

The Taurine Symposium - *"Taurine: Beginning the 21st Century"* - was held September 20-23, 2002, on the beautiful island of Kauai in Hawaii. The headquarters of the meeting was the Radisson Kauai Beach Resort.

This international meeting was attending by approximately 80 individuals from 23 nations and 4 continents. Seventy-five papers were presented either as platform presentations or poster presentations. Taurine, first isolated from ox bile in 1827 by Tiedemann and Gmelin and named in 1838 by Demarcay, became of significant scientific interest in 1968 when the first extensive review article was published by Jacobsen and Smith. Interest in taurine grew exponentially after 1975 when the first taurine symposium was organized by Ryan Huxtable in Tucson, Arizona. Since that date, taurine symposia have been held approximately every two years held in various cities and resort areas around the world. Taurine investigators have had the privilege of attending these scientific meetings on three continents - Asia, Europe, and North America.

Since the initial meeting in 1975, a central question addressed during many of the symposia has been: *"What is physiological, pharmacological, nutritional, and pathological role of taurine?"*. Although taurine has been established as an important osmolyte, it appears to affect many other biological processes. However, the exact mechanism(s) by "which taurine acts" has not yet been definitively answered. In Kauai, the participants discussed many topics and asked many questions regarding the role and actions of taurine. For example: *"How is taurine synthesized in specific cells, where is taurine localized, is taurine involved in nutrition, does taurine have osmoregulatory properties, how is taurine transported into cells, what are the cardiovascular and renal effects of taurine, what is the role of taurine in diabetes, does taurine have an effect on growth, is taurine an antioxidant and does it have an effect on the inflammatory system?"*. Did we answer all of these question? No, but there was a great deal of discussion and the participants generated many new ideas. At this Kauai meeting, new friends were made and old friends became reacquainted. However, perhaps most important, these new ideas were taken home to be tested in the laboratory.

The organizers of the taurine symposium would like to thank Taisho Pharmaceutical Co., Ltd. of Tokyo, Japan, and Red Bull GmbH of Vienna, Austria, for their generous financial support. The organizers would also like to thank Mr. Craig Ricci, Dr. Viktor Pastukh, Ms. Kim Schaffer and Ms. Viktoriya Solodushko for their technical help in showing the slides and manning the registration desk. Mrs. Sandra Taylor is thanked for her expertise in Microsoft Word in helping to format the manuscripts. The staff of the Radisson Hotel and especially Mr. Bartlett were extremely helpful in making sure that the participants were most comfortable.

So as the participants left the island of Kauai where do we go from here. Future interest in taurine is strong. However, there is a lot of work yet to be done to determine *"the function of taurine"*. Finally, the organizers wish to thank all the participants for a very successful taurine symposium.

J. B. Lombardini
Stephen W. Schaffer
Junichi Azuma

PARTICIPANTS

Dr. C.S. Ahn
Department of Food and Nutrition
Ansan College
Ansan 425-701
KOREA

Dr. J. Albrecht
Department of Neurotoxicology
Medical Research Centre
Polish Academy of Sciences
Warsaw
POLAND

Dr. Hye Suk An
Department of Life Science
University of Seoul
Seoul 130-743
KOREA

Dr. Junichi Azuma
Clinical Evaluation of Medicines
    and Therapeutics
Graduate School of
    Pharmaceutical Sciences
Osaka University
Osaka 565-0871
JAPAN

Dr. Edith Baccichet
Centro de Salud Mental del Este
El Peñón
Caracas
VENEZUELA

Dr. Madhabi Barua
Laboratory of Molecular Cell
    Signaling
Department of Developmental
    Biochemistry
New York State Institute for Basic
    Research in Developmental
    Disabilities
Staten Island, NY
USA

Dr. Kathryn Boorer
Department of Physiology
The David Geffen School of
    Medicine at UCLA
Los Angeles, CA 90095
USA

Dr. Michele L. Brooks
Department of Family &
    Community Medicine
Texas Tech University Health
    Sciences Center
Lubbock, TX 79430
USA

Dr. Salvatore Caputo
Department of Internal and
    Geriatric Medicine
Catholic University
Rome
ITALY

Dr. Isabel Carreira
Centro de Salud Mental del Este El
Peñón
Caracas
VENEZUELA

Dr. Kyung Ja Chang
Department of Food and Nutrition
Inha University
Incheon
KOREA

Dr. A.S. Charu
Medical Hospital and Research
    Centre
School of Medicine
Moradabad
INDIA

Dr. Katherine B. Chauncey
Department of Family &
    Community Medicine
Texas Tech University Health
    Sciences Center
Lubbock, TX 79430
USA

Dr. Wen Chen
School of Food and Nutritional
    Sciences
The University of Shizuoka
Shizuoka 422-8525
JAPAN

Dr. Sun Hee Cheong
Department of Food and Nutrition
Inha University
Incheon
KOREA

Dr. Russell W. Chesney
Department of Pediatrics
University of Tennessee Health
    Science Center, and the
    Children's Foundation
Research Center at Le Bonheur
Children's Medical Center
Memphis, Tennessee
USA

Dr. Kwon H. Choi
Seoul National University
College of Pharmacy
San 56-1 Shinrim-Dong
Kwanak-Ku, Seoul
KOREA

Dr. Mi Ja Choi
Department of Food and Nutrition
Keimyung University
Taegu
KOREA

Dr. Sang-Keun Chon
PharmacoGenechips Inc.
Chunchon 200-160
KOREA

Dr. Julio Cortijo
Department of Pharmacology
Faculty of Medicine
University of Valencia
SPAIN

Dr. Suzana Cubillos
Laboratoria de Neuroquímica
Centro de Biofísica y Bioquímica
Instituto Venezolano de
    Investigaciones Científicas
Apdo. 21827
Caracas 1020-A
VENEZUELA

Dr. Randall L. Davis
Department of Family &
    Community Medicine
Texas Tech University Health
    Sciences Center
Lubbock, TX 79430
USA

Dr. Mauro A.S. Di Leo
Department of Emergency
    Medicine
Catholic University
Rome
ITALY

Dr. Xiaohui Ding
Department of Anatomy
Mie University School of
  Medicine
Tsu 51400001,
Mie
JAPAN

Dr. John Dominy, Jr.
Department of Pharmacodynamics
University of Florida
Gainesville, FL 32610
USA

Dr. Abdeslem El Idrissi
New York State Institute for Basic
  Research in Developmental
  Disabilities and the Center for
  Developmental Neuroscience at
  The City University of New
  York
Staten Island, NY 10314
USA

Dr. Fili Fazzino
Laboratoria de Neuroquímica
Centro de Biofísica y Bioquímica
Instituto Venezolano de
  Investigaciones Científicas
Apdo. 21827
Caracas 1020-A
VENEZUELA

Dr. Rodrigo Franco
Department of Biophysics
Cell Physiology Institute
National University of Mexico
Mexico City
MEXICO

Dr. Flavia Franconi
Department of Pharmacology
  and Center for Biotechnology
  Development and Biodiversity
  Research
University of Sassari
ITALY

Dr. Bruno Giardina
Institute of Biochemistry and
  Clinical Biochemistry
Catholic University
Rome
ITALY

Dr. Giovanni Ghirlanda
Internal and Geriatric Medicine
Catholic University
Rome
ITALY

Dr. Shri N. Giri
Department of Molecular
  Biosciences
School of Veterinary Medicine
University of California
Davis, California
USA

Dr. Ronald E. Gordon
Department of Pathology
Mount Sinai Medical School
New York, New York
USA

Dr. Ramesh C. Gupta
Department of Pharmacology
School of Dentistry
Kyung Hee University
Seoul 130-701
KOREA

Dr. Hee Chang Han
Department of Life Science
University of Seoul
Seoul 130-743
KOREA

Dr. Xiaobin Han
Department of Pediatrics
University of Tennessee Health
  Science Center, and the
  Children's Foundation
Research Center at Le Bonheur
Children's Medical Center
Memphis, Tennessee
USA

Dr. Mayo Hirata
Clinical Evaluation of Medicines
    and Therapeutics
Graduate School of
    Pharmaceutical Sciences
Osaka University
Osaka 565-0871
JAPAN

Dr. B. Hoffmann
Department of Pediatrics
University of Cologne
Cologne
GERMANY

Dr. Hoffman-Kuczynski
Department of Emergency
    Medicine
Wright State university School of
    Medicine
Dayton, Ohio
USA

Dr. Yu Hosokawa
Department of Food and Health
    Sciences
Jissen Women's University
JAPAN

Dr. M. Ikeda
Kyoto Industrial Health
Association
Kyoto 604-8472
JAPAN

Dr. Keisuke Imada
Pharmacological Evaluation
    Laboratory
Taisho Pharmaceutical, Co., Ltd.
JAPAN

Dr. Jun-Ichiro Inoue
Institute of Medical Science
University of Tokyo
Chiba and Tokyo
JAPAN

Dr. Takashi Itoh
Clinical Evaluation of Medicines
    and Therapeutics
Graduate School of
Pharmaceutical Sciences
Osaka University
Osaka 565-0871
JAPAN

Dr. Y. Jeong
Department of Food Science and
    Nutrition
Dankook University
Seoul 140-714
KOREA

Dr. Hong Jin
Department of Molecular
Biosciences
University of Kansas
Lawrence, KS 66045
USA

Dr. Betsy G. Jones
Department of Family &
    Community Medicine
Texas Tech University Health
    Sciences Center
Lubbock, TX 79430
USA

Dr. Eunhye Jung
PharacoGenechips Inc.
Chunchon 200-160
KOREA

Dr. Young S. Jung
Seoul National University
College of Pharmacy
San 56-1 Shinrim-Dong
Kwanak-Ku
Seoul
KOREA

Dr. Atsuhiro Kanayama
Department of Applied Biological
    Chemistry
University of Tokyo
Chiba and Tokyo
JAPAN

Dr. K. Kartikey
Medical Hospital and Research
    Centre
School of Medicine
Moradabad
INDIA

Dr. Robert T. Kennedy
Department of Chemistry
University of Michigan
Ann Arbor, Michigan
USA

Dr. E.S. Kim
Department of Food Science and
    Nutrition
Dankook University
Seoul 140-714
KOREA

Dr. Ha Won Kim
Department of Life Science
University of Seoul
Seoul 130-743
KOREA

Dr. Hyunjung Kim
Department of Food and Nutrition
Yonsei Univeresity
Seoul
KOREA

Dr. J.S. Kim
Department of Food Science and
    Nutrition
Dankook University
Seoul 140-714
KOREA

Dr. Seoyoun Kim
Department of Food and Nutrition
Yonsei Univeresity
Seoul
KOREA

Dr. Soon Ki Kim
Department of Pediatrics
Inha University Hospital
Incheon
KOREA

Dr. Sung-Jin Kim
Department of Pharmacology
School of Dentistry
Kyung Hee University
Seoul 130-701
KOREA

Dr. Young C. Kim
Seoul National University
College of Pharmacy
San 56-1 Shinrim-Dong
Kwanak-Ku
Seoul
KOREA

Dr. Kayoko Kinoshita
Clinical Evaluation of Medicines
    and Therapeutics
Graduate School of
Pharmaceutical Sciences
Osaka University
Osaka 565-0871
JAPAN

Dr. Y.S. Ko
Department of Food and Nutrition
Jeju National University
Jeju 690-756
KOREA

Dr. Yoshiko S. Konishi
Division of Microbiology
National Institute of Health
    Sciences
Tokyo
JAPAN

Dr. Ewa Kontny
Department of Pathophysiology &
    Immunology
Institute of Rheumatology
Warsaw
POLAND

Dr. Norman R. Kreisman
Department of Physiology
Tulane University Medical School
New Orleans, Louisiana
USA

Michio Kurachi
Pharmacological Evaluation
   Laboratory
Taisho Pharmaceutical, Co., Ltd.
JAPAN

Dr. Hye E. Kwak
Seoul National University
College of Pharmacy
San 56-1 Shinrim-Dong
Kwanak-Ku
Seoul
KOREA

Dr. Jinoh Kwak
Department of Food and Nutrition
Inha University
Incheon
KOREA

Dr. Ian H. Lambert
The August Krogh Institute
Universitetsparken 13
DK-2100
Copenhagen Ø
DENMARK

Dr. R.O. Law
Department of Preclinical
   Sciences
University of Leicester
Leicester
UNITED KINDGOM

Dr. James Leasure
Department of Emergency
   Medicine
Wright State University School of
   Medicine
Dayton, Ohio
USA

Dr. Haemi Lee
Department of Food and Nutrition
Yonsei University
Seoul 120-740
KOREA

Dr. Jeonghee Lee
Department of Food and Nutrition
Inha University
Incheon
KOREA

Dr. Mi Young Lee
Department of Food and Nutrition
Inha University
Incheon
KOREA

Dr. Yeojin Lee
Department of Food and Nutrition
Yonsei Univeresity
Seoul
KOREA

Dr. Young-Mi Lee
Department of Food and Nutrition
Keimyung University
Taegu
KOREA

Dr. Isabelle Lelong-Rebel
UPR 9003 du CNRS
Institut de Recherche Contre les
   Cancers de l'Appareil Digestif
Hôpitaux Universitaires
BP 426-F 67091
Strasbourg
FRANCE

Dr. R. Lezama
Department of Biophysics
Institute of Cell Physiology
National University of Mexico
Mexico City
MEXICO

Dr. Jenny Lim
Department of Emergency
   Medicine
Wright State University School of
   Medicine
Dayton, Ohio
USA

Dr. Lucimey Lima
Laboratoria de Neuroquímica
Centro de Biofísica y Bioquímica
Instituto Venezolano de
    Investigaciones Científicas
Apdo. 21827
Caracas 1020-A
VENEZUELA

Dr. Yong Liu
Laboratory of Molecular Cell
    Signaling
Department of Developmental
    Biochemistry
New York State Institute for Basic
    Research in Developmental
    Disabilities
Staten Island, NY
USA

John B. Lombardini
Department of Pharmacology
Texas Tech University Health
    Sciences Center
Lubbock, TX 79430
USA

Dr. Donald D. F. Loo
Department of Physiology
The David Geffen School of
    Medicine at UCLA
Los Angeles, CA 90095
USA

Dr. Ning Ma
Department of Anatomy
Mie University School of
Medicine
Tsu 5140-001,
Mie
JAPAN

Dr. Janusz Marcinkiewicz
Jagiellonian University Medical
    College
Cracow
POLAND

Dr. Włodzimierz Maśliński
Department of Pathophysiology &
    Immunology
Institute of Rheumatology
Warsaw
POLAND

Dr. Hideyasu Matsuda
Clinical Evaluation of Medicines
    and Therapeutics
Graduate School of
    Pharmaceutical Sciences
Osaka University
Osaka 565-0871
JAPAN

Dr. Takahisa Matsuda
Clinical Evaluation of Medicines
    and Therapeutics
Graduate School of
    Pharmaceutical Sciences
Osaka University
Osaka 565-0871
JAPAN

Dr. N. Matsuda-Inoguchi
Miyagi University
Taiwa-cho 981-3298
JAPAN

Dr. Norikazu Matsunaga
Clinical Evaluation of Medicines
    and Therapeutics
Graduate School of
    Pharmaceutical Sciences
Osaka University
Osaka 565-0871
JAPAN

Dr. Jeffrey Messing
New York State Institute for Basic
    Research in Developmental
    Disabilities and the Center for
    Developmental Neuroscience
    at The City University of New
    York
Staten Island, NY 10314
USA

Dr. D.V. Michalk
Department of Pediatrics
University of Cologne
Cologne
GERMANY

Dr. Th. Minor
Institute of Experimental
    Medicine
University of Bonn
Bonn
GERMANY

Toshifumi Mitani
Department of Biochemistry
Shimane Medical University
JAPAN

Dr. Tatsuo Miwa
Department of Anatomy
Mie University School of
    Medicine
Tsu 51400001,
Mie
JAPAN

Dr. Yusei Miyamoto
Department of Integrated
    Biosciences
University of Tokyo
Tokyo
JAPAN

Dr. Tetsunosuke Mochizuki
Department of Applied Biological
    Chemistry
The University of Tokyo
Bunkyo-ku, Tokyo 113-8657
JAPAN

Dr. Esteban Morcillo
Department of Pharmacology
Faculty of Medicine
University of Valencia
SPAIN

Dr. Mahmood S. Mozaffari
Department of Oral Biology &
    Maxillofacial Pathology
Medical College of Georgia
Augusta, Georgia
USA

Dr. Hironori Nakamura
Department of Clinical Nutrition
Kawasaki University of Medical
    Welfare
Kurashiki, Okayama
JAPAN

Dr. Kazumi Nakano
Pharmacological Evaluation
    Laboratory
Taisho Pharmaceutical. Co. Ltd.
JAPAN

Dr. H. Nakatsuka
Miyagi University
Taiwa-cho 981-3298
JAPAN

Dr. S. Nakatsuka
Miyagi University
Taiwa-cho 981-3298
JAPAN

Dr. Niaz
Medical Hospital and Research
    Centre
School of Medicine
Moradabad
INDIA

Dr. Naomichi Nishimura
Department of Human Life and
    Development
Nayoro City College
Nayoro 096-8641
JAPAN

Dr. Franciso Obregón
Laboratoria de Neuroquímica
Centro de Biofísica y Bioquímica
Instituto Venezolano de
    Investigaciones Científicas
Apdo. 21827
Caracas 1020-A
VENEZUELA

Dr. L.D. Ochoa-De La Paz
Department of Biophysics
Institute of Cell Physiology
National University of Mexico
Mexico City
MEXICO

Dr. Hiroaki Oda
Department of Applied Biological
    Sciences
Nagoya University
Nagoya 464-8601
JAPAN

Dr. Masanori Ohmoto
Clinical Evaluation of Medicines
    and Therapeutics
Graduate School of
    Pharmaceutical Sciences
Osaka University
Osaka 565-0871
JAPAN

Dr. M. Foster Olive
Ernest Gallo Clinic and Research
    Center
Department of Neurology
University of California at San
    Francisco, California
USA

Dr. Simo S. Oja
Tampere Brain Research Center
Medical School
FIN-33014
University of Tampere
FINLAND

Dr. James E. Olson
Department of Emergency
    Medicine and Department of
    Physiology and Biophysics
Wright State University School of
    Medicine
Dayton, Ohio
USA

Dr. B. Ordaz
Department of Biophysics
Institute of Cell Physiology
National University of Mexico
Mexico City
MEXICO

Dr. Michael H. O'Regan
Department of Biomedical
    Sciences
School of Dentistry
University of Detroit Mercy
8200 W. Outer Drive
P.O. Box 19900
Detroit, MI 48219
USA

Dr. Gregory L. Osterhaus
Department of Molecular
    Biosciences
University of Kansas
Lawrence, KS 66045
USA

Dr. Eunkyue Park
Department of Immunology
New York State Institute for Basic
    Research in Developmental
    Disabilities
Staten Island, New York
USA

Dr. Kun-Koo Park
PharmacoGenechips Inc.
Chunchon, 200-160
KOREA

Dr. Sungyoun Park
Department of Food and Nutrition
Yonsei University
Seoul, 120-749
KOREA

Dr. Taesun Park
Department of Food and Nutrition
Yonsei University
Seoul 120-749
KOREA

Dr. Nisha Patel
Department of Oral Biology &
    Maxillofacial Pathology
Medical College of Georgia
Augusta, Georgia
USA

Dr. Andrea Budreau Patters
Department of Pediatrics
University of Tennessee Health
    Science Center, and the
    Children's Foundation
Research Center at Le Bonheur
Children's Medical Center
Memphis, Tennessee
USA

Dr. Solisbella Peña
Centro de Salud Mental del Este
El Peñón
Caracas
VENEZUELA

Dr. Joanna Peris
Department of Chemistry
University of Michigan
Ann Arbor, Michigan
USA

Dr. John W. Phillis
Department of Physiology
Wayne State University School of
    Medicine
540 E. Canfield
Detroit, MI 48201
USA

Dr. O. Quesada
Department of Biophysics
Institute of Cell Physiology
National University of Mexico
Mexico City
MEXICO

Dr. Michael R. Quinn
Center for Developmental
    Neuroscience
Staten Island, NY
USA

Dr. R. Michael Ragain
Department of Family &
    Community Medicine
Texas Tech University Health
    Sciences Center
Lubbock, TX 79430
USA

Dr. Gérard Rebel
UPR 9003 du CNRS
Institut de Recherche Contre les
    Cancers de l'Appareil Digestif
Hôpitaux Universitaires
BP 426-F 67091
Strasbourg
FRANCE

Dr. Rocio Salceda
Department of Neurosciences
Cell Physiology Institute
National University of Mexico
Mexico City, C.P. 04510
MEXICO

Dr. Francesco Santangelo
Preclinical Development
R&D
Zambon Group
Bresso, Milan
ITALY

Dr. Stefano A. Santini
Institute of Biochemistry and
    Clinical Biochemistry
Catholic University
Rome
ITALY

Dr. Pirjo Saransaari
Tampere Brain Research Center
Medical School
FIN-33014
University of Tampere
FINLAND

Dr. Hemanta K. Sarkar
Department of Chemistry and
    Biochemistry
University of Massachusetts at
    Dartmouth
North Dartmouth, MA 02747
USA

Dr. Hiroyasu Satoh
Department or Pharmacology
Divison of Molecular and Cellular
    Biology
Nara Medical University
Kashihara
Nara 634-8521
JAPAN

Dr. Hideo Satsu
Department of Applied Biological
    Chemistry
The University of Tokyo
Bunkyo-ku, Tokyo 113-8657
JAPAN

Dr. Jason Scalia
New York State Institute for Basic
    Research in Developmental
    Disabilities and the Center for
    Developmental Neuroscience at
    The City University of New
    York
Staten Island, NY 10314
USA

Dr. Stephen W. Schaffer
Department of Pharmacology
University of South Alabama
    College of Medicine
Mobile, AL 3668
USA

Dr. Dana Schelble
Department of Emergency
    Medicine
Wright State University School of
    Medicine
Dayton, Ohio
USA

Dr. Georgia B. Schuller-Levis
Department of Immunology
New York State Institute for Basic
    Research in Developmental
    Disabilities
Staten Island, New York
USA

Dr. Reiji Semba
Department of Anatomy
Mie University School of
    Medicine
Tsu 5140-001,
Mie
JAPAN

Dr. Mooseok Seo
PharacoGenechips Inc.
Chunchon 200-160
KOREA

Dr. Valeria Serban
Center for Developmental
    Neuroscience
Staten Island, NY
USA

Dr. Di Sha
Biomedical Sciences
Florida Atlantic University
Boca Raton, FL 33431-0991
USA

Dr. S. Shimbo
Kyoto Women's University
Kyoto 604-8472
JAPAN

Dr. Makoto Shimizu
Department of Applied Biological
    Chemistry
University of Tokyo
Bunkyo-ku, Tokyo 113-8657
JAPAN

Dr. Ashfaq Shuaib
Department of Medicine
University of Alberta
Edmonton
CANADA

Dr. Nicolò Gentiloni Silveri
Department of Emergency
    Medicine
Catholic University
Rome
ITALY

Dr. R.B. Singh
Medical Hospital and Research
    Centre
School of Medicine
Moradabad
INDIA

Dr. Anthony Smith
Department of Pharmacodynamics
University of Florida
Gainesville, FL
USA

Dr. Miwon Son
Research Laboratories of Dong-A
    Pharmaceutical Co., Ltd
Kyunggi-do
KOREA

Dr. John Sturman
95 Fort Hill Circle
Staten Island, NY 10301
USA

Dr. Hyuni Sung
Department of Food and Nutrition
Inha University
Incheon
KOREA

Dr. Akihiko Sumita
Clinical Evaluation of Medicines
    and Therapeutics
Graduate School of
    Pharmaceutical Sciences
Osaka University
Osaka 565-0871
JAPAN

Dr. Koichi Takahashi
Department of Pharmaceutics
School of Pharmaceutical
Sciences
Mukogawa Women's University
Hyogo 663-8179
JAPAN

Dr. Kyoko Takahashi
Clinical Evaluation of Medicines
    and Therapeutics
Graduate School of
    Pharmaceutical Sciences
Osaka University
Osaka 565-0871
JAPAN

Dr. Tomoka Takatani
Clinical Evaluation of Medicines
    and Therapeutics
Graduate School of
Pharmaceutical Sciences
Osaka University
Osaka 565-0871
JAPAN

Dr. Takaaki Takenaga
Pharmacological Evaluation
    Laboratory
Taisho Pharmaceutical, Co., Ltd.
JAPAN

Dr. Yoshinori Tanigawa
Department of Biochemistry
Shimane Medical University
JAPAN

Dr. Thomas E. Tenner, Jr.
Department of Pharmacology
Texas Tech University Health
 Sciences Center
Lubbock, TX 79430
USA

Dr. Masaharu Terashima
Department of Biochemistry
Shimane Medical University
JAPAN

Dr. Ekkhart Trenkner
New York State Institute for Basic
 Research in Developmental
 Disabilities and the Center for
 Developmental Neuroscience at
 The City University of New
 York
Staten Island, NY 10314
USA

Dr. K.L. Tuz
Department of Biophysics
Institute of Cell Physiology
National University of Mexico
Mexico City
MEXICO

Dr. Toshihiko Ubuka
Department of Clinical Nutrition
Kawasaki University of Medical
 Welfare
Kurashiki, Okayama
JAPAN

Dr. Satoko Ueyama
Clinical Evaluation of Medicines
 and Therapeutics
Graduate School of
Pharmaceutical Sciences
Osaka University
Osaka 565-0871
JAPAN

Dr. M. Urbina
Laboratoria de Neuorquímica
Centro de Biofísica y Bioquímica
Instituto Venezolano de
 Investigaciones Científicas
Apdo. 21827
Caracas 1020-A
VENEZUELA

Dr. Shyamala Vinnakota
Department of Molecular
 Physiology & Biophysics
Baylor College of Medicine
Houston, TX 77054
USA

Dr. Chuanhua Wang
Department of Immunology
New York State Institute for Basic
 Research in Developmental
 Disabilities
Staten Island, New York
USA

Dr. Ronald D. Warner
Department of Family and
 Community Medicine
Texas Tech University Health
 Sciences Center
Lubbock, TX 79430
USA

Dr. T. Watanabe
Miyagi University of Education
Sendai 980-0845
JAPAN

Dr. Christopher J. Watson
Department of Chemistry
University of Michigan
Ann Arbor, Michigan
USA

Dr. Jianning Wei
Department of Molecular
 Biosciences
University of Kansas
Lawrence, KS 66045
USA

Dr. C. Wersinger
UPR 9003 du CNRS
Institut de Recherche Contre les
    Cancers de l'Appareil Digestif
Hôpitaux Universitaires
BP 426-F 67091
Strasbourg
FRANCE

Dr. Chen Wen
School of Food and Nutritional
    Sciences
The University of Shizuoka
Shizuoka 422-8526
JAPAN

Dr. Ernest M. Wright
Department of Physiology
The David Geffen School of
    Medicine at UCLA
Los Angeles, CA 90095
USA

Dr. Heng Wu
Biomedical Sciences
Florida Atlantic University
Boca Raton, FL 33431-0991
USA

Dr. Jang-Yen Wu
Biomedical Sciences
Florida Atlantic University
Boca Raton, FL 33431-0991
USA

Dr. Yasuhiro Yamamoto
Clinical Evaluation of Medicines
    and Therapeutics
Graduate School of
    Pharmaceutical Sciences
Osaka University
Osaka 565-0871
JAPAN

Dr. Yoko Yamauchi
Clinical Evaluation of Medicines
    and Therapeutics
Graduate School of
    Pharmaceutical Sciences
Osaka University
Osaka 565-0871
JAPAN

Dr.M.H. Yim
Department of Food Science and
    Nutrition
Dankook University
Seoul 140-714
KOREA

Dr. Hidehiko Yokogoshi
School of Food and Nutritional
    Sciences
The University of Shizuoka
Shizuoka 422-8526
JAPAN

Dr. Heaeun You
Department of Food and Nutrition
Inha University
Incheon
KOREA

Dr. Chai Hyeock Yu
Department of Biology
Inha University
Inchon
KOREA

Dr. Barbara Zablocka
Laboratory of Molecular Biology
Medical Research Centre
Polish Academy of Sciences
Warsaw
POLAND

Dr. X.J. Zhang
Department of Pharmacology
Texas Tech University Health
    Sciences Center
Lubbock, TX 79430
USA

Dr. M. Zielinska
Department of Neurotoxicology
Medical Research Centre
Polish Academy of Sciences
Warsaw
POLAND

CONTENTS

A Voice From the Past...1
  John Sturman

Part 1. Cardiovascular and Renal Effects of Taurine

1. Inhibitory Mechanism of Taurine on the Platelet-Derived
      Growth Factor BB-Mediated Proliferation in Aortic
      Vascular Smooth Muscle Cells...5
   Keisuke Imada, Yu Hosokawa, Masaharu Terashima, Toshifumi
      Mitani, Yoshinori Tanigawa, Kazumi Nakano,
      Takaaki Takenaga, and Michio Kurachi

2. Taurine on Sino-Atrial Nodal Cells:
   Ca^{2+}-Dependent Modulation...17
   Hiroyasu Satoh

3. Cellular Characterization of Taurine Transporter in Cultured
      Cardiac Myocytes and Nonmyocytes...25
   Tomoka Takatani, Kyoko Takahashi, Takashi Itoh,
      Koichi Takahashi, Mayo Hirata, Yasuhiro Yamamoto,
      Masanori Ohmoto, Stephen W. Schaffer, and Junichi Azuma

4. Taurine Transporter in Cultured Neonatal Cardiomyocytes:
    a Response to Cardiac Hypertrophy..33
    Takashi Itoh, Kyoko Takahashi, Yoko Yamauchi,
    Koichi Takahashi, Sato Ueyama, Stephen W. Schaffer,
    and Junichi Azuma

5. Effect of Taurine and Coenzyme Q10 in Patients
    with Acute Myocardial Infarction...·41
    R.B. Singh, K. Kartikey, A.S. Charu, M.A. Niaz, and S. Schaffer

6. Taurine Reduces Renal Ischemia/Reperfusion Injury
    in the Rat..49
    D.V. Michalk, B. Hoffmann, and Th. Minor

7. Taurine-Depleted Heart and Afterload Pressure..............................57
    Mahmood S. Mozaffari, Nisha Patel, and Stephen W. Schaffer

Part 2. Role of Taurine in Diabetes

8. Taurine Reduces Mortality in Diabetic Rats: Taurine
    and Experimental Diabetes Mellitus...67
    Flavia Franconi, Stefano A. Santini,
    Nicolò Gentiloni Silveri, Salvatore Caputo,
    Bruno Giardina, Giovanni Ghirlanda, and
    Mauro A.S. Di Leo.

9. The Effect of Dietary Taurine Supplementation on Plasma
    and Urinary Free Amino Acid Concentrations in
    Diabetic Rats..75
    Young-Mi Lee, Mi Ja Choi, and Kyung Ja Chang

10. Insulin-Stimulated Taurine Uptake in the Rat Retina.......................83
    Rocio Salceda

11. The Effects of Taurine Supplementation on Patients
    with Type 2 Diabetes Mellitus..91
    Katherine B. Chauncey, Thomas E. Tenner, Jr.,
    John B. Lombardini, Betsy G. Jones, Michele L. Brooks,
    Ronald D. Warner, Randall L. Davis, and R. Michael Ragain

12. Hypoglycemic Effects of Taurine in the Alloxan-Treated Rabbit:
    a Model for Type 1 Diabetes..97
    Thomas E. Tenner, Jr., X. J. Zhang, and John B. Lombardini

Part 3. Osmoregulatory Properties of Taurine and Taurine Transport

13. Taurine and Cellular Volume Regulation in the Hippocampus.......107
    James E. Olson, Norman R. Kreisman, Jenny Lim,
    Beth Hoffman-Kuczynski, Dana Schelble, and James Leasure

14. Regulation of the Volume-Sensitive Taurine Efflux
    Pathway in NIH3T3 Mouse Fibroblasts...................................115
    Ian Henry Lambert

15. Taurine Counteracts Cell Swelling in Rat Cerebrocortical
    Slices Exposed to Ammonia in Vitro and in Vivo...................123
    R.O. Law, M. Zielinska, and J. Albrecht

16. Cloning of Human Intestinal Taurine Transporter and
    Production of Polyclonal Antibody..131
    Hye Suk An, Hee Chang Han, Taesun Park, Kun Koo Park, and
    Ha Won Kim

17. Transactivation of TauT by p53 in MCF-7 Cells:
    The Role of Estrogen Receptors... 139
    Xiaobin Han, Andrea Budreau Patters, and Russell W. Chesney

18. Gating of Taurine Transport: Role of the
    Fourth Segment of the Taurine Transporter.............................149
    Xiaobin Han, Andrea Budreau Patters, and Russell W. Chesney

19. Finding of TRE (TPA Responsive Element) in the Sequence of
    Human Taurine Transporter Promoter.....................................159
    Kun-Koo Park, Eunhye Jung, Sang-Keun Chon,
    Mooseok Seo, Ha Won Kim, and Taesun Park

20. Protein Kinase C and cAMP Mediated Regulation of
    Taurine Transport in Human Colon Carcinoma
    Cell Lines (HT-29 & Caco-2).....................................167
    Sungyoun Park, Haemi Lee, Kun-Koo Park,
    Ha Won Kim, and Taesun Park

21. Characterization of Transcriptional Activity of Taurine
    Transporter Using Luciferase Reporter Constructs................175
    Kun-Koo Park, Sang-Keun Chon, Eunhye Jung,
    Mooseok Seo, Ha Won Kim, and Taesun Park

22. Isovolumetric Regulation in Mammal Cells: Role of Taurine.......183
    B. Ordaz, R. Franco, and K. Tuz

23. Osmosensitive Taurine Release: Does Taurine Share the Same
    Osmosensitive Efflux Pathway With Chloride and Other
    Amino Acid Osmolytes?..189
    Rodrigo Franco

24. Electrophysiological properties of the Mouse
    Na^+Cl^- - Dependent Taurine Transporter (mTauT-1):
    Steady-State Kinetics: Stoichiometry of
    Taurine Transport..197
    *Hemanta K. Sarkar, Ernest M. Wright, Kathryn J. Boorer,
    and Donald D.F. Loo*

25. Taurine Uptake and Release by the Pancreatic β-Cells:
    Taurine Transport in β-Cells...205
    Shyamala Vinnakota and Hemanta K. Sarkar

26. Regulation of Intestinal Taurine Transporter by Cytokines...........213
    Makoto Shimizu, Tetsunosuke Mochizuki, and Hideo Satsu

Part 4. Taurine Synthesis, Localization, Determination, and Nutrition

27. Determination of Taurine and Hypotaurine in Animal
    Tissues by Reversed-Phase High-Performance Liquid
    Chromatography after Derivatization with
    Dabsyl Chloride..221
    Hironori Nakamura and Toshihiko Ubuka

28. Immunohistochemical Localization of Taurine in the
    Rat Stomach...229
    Ning Ma, Xiaohui Ding, Tatsuo Miwa, and Reiji Semba

29. Modulation of Taurine on CYP3A4 Induction by Rifampicin
    in a HepG2 Cell Line..237
    *Kyoko Takahashi, Hideyasu Matsuda, Kayoko Kinoshita,
    Norikazu Matsunaga, Akihiko Sumita, Takahisa Matsuda,
    Koichi Takahashi, and Junichi Azuma*

30. Effect of Acute Ethanol Administration on S-Amino Acid
    Metabolism: Increased Utilization of Cysteine for
    Synthesis of Taurine Rather Than Glutathione........................245
    Young S. Jung, Hye E. Kwak, Kwon H. Choi, and Young C. Kim

31. Effects of Dietary Taurine Supplementation on Hepatic
    Morphological Changes of Rats in
    Diethynitrosamine-Induced Hepatocarcinogenesis...................253
    Kyung Ja Chang, Chai Hyeock Yu, and Miwon Son

32. Effect of Taurine on Cholesterol Degradation and Bile
    Acid Pool in Rats Fed a High-Cholesterol Diet.......................261
    *Wen Chen, Naomichi Nishimura, Hiroaki Oda, and
    Hidehiko Yokogoshi*

33. Effect of α-Tocopherol and Taurine Supplementation on
    Oxidized LDL Levels of Middle Aged Korean Women
    During Aerobic Exercise..269
    C.S. Ahn and E.S. Kim

34. Dietary Taurine Intake and Serum Taurine Levels of Women
    on Jeju Island..277
    *E.S. Kim, J.S. Kim, M.H. Yim, Y. Jeong, Y.S. Ko, T. Watanabe,
    H. Nakatsuka, S. Nakatsuka, N. Matsuda-Inoguchi,
    S. Shimbo, and M. Ikeda*

35. Effect of the Obesity Index on Plasma Taurine Levels in
    Korean Female Adolescents...285
    *Mi Young Lee, Sun Hee Cheong, Kyung Ja Chang,
    Mi Ja Choi, and Soon Ki Kim*

36. Regional Differences in the Dietary Taurine Intake in Korean
    College Students..291
    *Jeonghee Lee, Heaeun You, Hyuni Sung, Jinoh Kwak,
    Kyung Ja Chang, Seoyoun Kim, Hyunjung Kim,
    Yeojin Lee, and Taesun Park*

37. Taurine Concentration in Human Blood Peripheral
    Lymphocytes: Major Depression and Treatment with
    the Antidepressant Mirtazapine...297
    *Lucimey Lima, Francisco Obregón, Mary Urbina,
    Isabel Carreira, Edith Baccichet, and Solisbella Peña*

Part 5. TAURINE: GROWTH, ANTIOXIDANTS,
AND INFLAMMATION

38. Why Is Taurine Cytoprotective?...307
     Stephen Schaffer, Junichi Azuma, Kyoko Takahashi,
        and Mahmood Mozaffari

39. Taurine, Analogues and Bone: a Growing Relationship................323
     Ramesh C. Gupta and Sung-Jin Kim

40. Anti-Inflammatory Activities of Taurine Chloramine:
        Implication for Immunoregulation and Pathogenesis
        of Rheumatoid Arthritis..329
     Ewa Kontny, Włodzimierz Maśliński, and Janusz Marcinkiewicz

41. Taurine Chloramine Inhibits Production of Inflammatory
        Mediators and iNOS Gene Expression in
        Alveolar Macrophages; a Tale of Two Pathways:
        Part I, NF-κB Signaling..341
     Michael R. Quinn, Madhabi Barua, Yong Liu, and Valeria Serban

42. Taurine Chloramine Inhibits Production of Inflammatory
        Mediators and iNOS Gene Expression in
        Alveolar Macrophages; a Tale of Two Pathways:
        Part II, IFN-γ Signaling Through JAK/Stat............................349
     Michael R. Quinn, Yong Liu, Madhabi Barua, and Valeria Serban

43. Production of Nitric Oxide by Activated Microglial cells
        Is Inhibited by Taurine Chloramine.....................................357
     Valeria Serban, Yong Liu, and Michael R. Quinn

44. Production of Inflammatory Mediators by Activated C6 Cells
        Is Attenuated by Taurine Chloramine Inhibition of
        NF-κB Activation..365
     Yong Liu, Madhabi Barua, Valeria Serban, and Michael R. Quinn

45. Taurine Is Involved in Oxidation of IκBα at Met45:
        N-Halogenated Taurine and Anti-inflammatory Action..........373
     Yusei Miyamoto, Atsuhiro Kanayama, Jun-Ichiro Inoue,
        Yoshiko S. Konishi, and Makoto Shimizu

46. The Combined Treatment with Taurine and Niacin Blocks the
        Bleomycin-Induced Activation of Nuclear Factor-kB
        and Lung fibrosis..381
     Shri N. Giri

47. Taurine Reduces Lung Inflammation and Fibrosis
      Caused by Bleomycin..395
   *Georgia B. Schuller-Levis, Ronald E. Gordon, Chuanhua Wang,
      and Eunkyue Park*

48. Taurine and the Lung, Which Role in Asthma? 403
   Francesco Santangelo, Julio Cortijo, and Esteban Morcillo

49. Effect of Taurine and Other Antioxidants on the Growth
      of Colon Carcinoma Cells in the Presence of
      Doxorubicin or Vinblastine in Hypoxic or in
      Ambient Oxygen Conditions: Effect of Antioxidants
      on the Action of Antineoplastic Drugs in MDR and
      non-MDR Cells..411
   C. Wersinger, G. Rebel, and I. Lelong-Rebel

Part 6. TAURINE: RETINA AND THE BRAIN

50. The Role of Taurine in Cerebral Ischemia: Studies in
      Transient Forebrain Ischemia and Embolic Focal
      Ischemia in Rodents..421
   Ashfaq Shuaib

51. Studies on Taurine Efflux from the Rat Cerebral Cortex
      During Exposure to Hyposmotic, High K⁺ and
      Ouabain-Containing aCSF...433
   John W. Phillis and Michael H. O'Regan

52. Interactions of Taurine and Adenosine in the Mouse
      Hippocampus in Normoxia and Ischemia.................................445
   Pirjo Saransaari and Simo S. Oja

53. Involvement of Nitric Oxide in Ischemia-Evoked Taurine
      Release in the Mouse Hippocampus.......................................453
   Pirjo Saransaari and Simo S. Oja

54. Effect of Ammonia on Taurine Transport in C6 Glioma Cells......463
   Magdalena Zielinska, Barbara Zablocka, and Jan Albrecht

55. Taurine and Hypotaurine Dynamics in Activated C6 Glioma:
      The Effects of Lipopolysaccharide (LPS) and
      Taurine Administration on Intracellular Hypotaurine and
      Taurine Dynamics in C6 Glioma..471
   John Dominy, Jr. and Ralph Dawson, Jr.

56. The Anti-Craving Taurine Derivative Acamprosate: Failure
    to Extinguish Morphine Conditioned Place Preference............481
    M. Foster Olive

57. Ethanol-Induced Taurine Efflux: Low Dose Effects
    and High Temporal Resolution...................................485
    *Anthony Smith, Christopher J. Watson, Robert T. Kennedy,
    and Joanna Peris*

58. Tyrosine Kinases and Taurine Release: Signaling Events
    and Amino Acid Release Under Hyposmotic and
    Ischemic Conditions in the Chicken Retina.............................493
    L.D. Ochoa-De La Paz and R.A. Lezama

59. Effect of Taurine on Regulation of GABA and
    Acetylcholine Biosynthesis.....................................499
    *Di Sha, Jianning Wei, Hong Jin, Heng Wu,
    Gregory L. Osterhaus, and Jang-Yen Wu*

60. Taurine Effect on Neuritic Outgrowth from Goldfish
    Retinal Explants in the Absence and Presence
    of Fetal Calf Serum..507
    Lucimey Lima, Suzana Cubillos, and Fili Fazzino

61. Prevention of Epileptic Seizures by Taurine.......................515
    *Abdeslem El Idrissi, Jeffrey Messing, Jason Scalia,
    and Ekkhart Trenkner*

62. Taurine Regulates Mitochondrial Calcium Homeostasis.................527
    Abdeslem El Idrissi and Ekkhart Trenkner

63. Taurine in Aging and Models of Neurodegeneration.....................537
    Ralph Dawson, Jr.

64. Taurine Stimulation of Calcium Uptake in the Retina:
    Mechanism of Action...547
    Julius D. Militante and John B. Lombardini

65. The Nature of Taurine Binding to Retinal Membranes.................555
    Julius D. Militante and John B. Lombardini

Index...561

A VOICE FROM THE PAST: WHERE ARE THE CATS?

John Sturman

For more than half a century scientists have been studying the role of taurine in biology. A major dilemma in ascribing specific functions to taurine is its ubiquitous distribution in the body and its presence in high amounts. Frequently in science, insight into the function of a compound, metabolite, enzyme, or factor has been achieved by perturbing the agent in question and observing the consequences. The majority of studies into the function of taurine were, and still are, performed on rodents, which for a number of reasons proved difficult to perturb.

The scene changed in the mid 1970's when a landmark study showed that central retinal degeneration in cats resulted from decreased taurine levels attributed to insufficient taurine levels in the diet. Follow-up studies on the same animals showed that the taurine deficiency in the retina was shared by all the organs in the body, including the brain, indicating a major difference from rodents. Subsequent studies over the next 20 years described various abnormalities in cats which resulted simply from a dietary insufficiency of taurine. These abnormalities include degeneration of the tapetum lucidum, the mirror-like reflecting layer behind the retina, which accounts for the superior vision of cats in low-light situations; dilated cardiomyopathy exhibited by enlarged hearts with thinner muscle walls, which results in death unless the animals are rescued with dietary or IV taurine treatment; reduced reproductive performance with fewer surviving offspring, and increased abortions and stillbirths. Examination of surviving offspring showed a delayed migration of cerebellar granule cells from the outer layer to the inner layer, which resulted in fewer cerebellar granule cells reaching their final destination, the implications of which are profound. Other abnormalities observed in the cat offspring included failure to fully develop retinal and tapetal cells, and fully functioning kidneys. The specificity of these defects was firmly attributed to taurine by the observations that kittens born to taurine-deficient mothers, orally supplemented with taurine, largely overcame these abnormalities.

Taurine 5: Beginning the 21st Century
Edited by Lombardini *et al.*, Kluwer Academic/Plenum Publishers, New York 2003

These observations triggered a renewed interest in the functions of taurine. Rodents were used as models using dietary GES or β-alanine to deplete organ taurine levels, which led to some useful observations. There remain several problems, including the fact that rodents have an extensive synthetic capacity for taurine (cats and primates, including humans, do not), and the ability of the rodent brain to retain taurine is much greater than that of cats (and probably primates, although there are no data on humans).

The above discourse is by way of saying that I am disappointed that research on taurine is no longer being conducted with cats. It is a difficult proposition: it took me 5 years to set up my cat colony, which was rated by various research groups, zoo groups and humanitarians as the best they had seen. Over the following 20 years, although my own research interests (brain development) predominated, since they were N.I.H. supported, I felt it should, and was, used as a national resource for other investigators whenever this was possible. Sadly I had to terminate my research career due to illness, and there was nobody to take over the cat colony. I am convinced that the cat model, because of its simplicity, will yield further insight into the functions of taurine, and hope that someone else will take up the challenge.

Part 1:

Cardiovascular and Renal Effects of Taurine

Inhibitory Mechanism of Taurine on the Platelet-Derived Growth Factor BB-Mediated Proliferation in Aortic Vascular Smooth Muscle Cells

KEISUKE IMADA[1], YU HOSOKAWA[2], MASAHARU TERASHIMA[3], TOSHIFUMI MITANI[3], YOSHINORI TANIGAWA[3], KAZUMI NAKANO[1], TAKAAKI TAKENAGA[1], MICHIO KURACHI[1]

[1] *Pharmacological Evaluation Laboratory, Taisho Pharmaceutical, Co., Ltd.*
[2] *Department of Food and Health Sciences, Jissen Women's University*
[3] *Department of Biochemistry, Shimane Medical University*

1.   INTRODUCTION

Platelet-derived growth factor-BB (PDGF-BB) is involved in the development of atherosclerosis in conjunction with the migration and the proliferation of vascular smooth muscle cells. PDGF-BB is produced by activated macrophages, smooth muscle cells and endothelial cells, or released from platelets, and its expression is increased in atherosclerotic lesions[1,2]. Thus, PDGF-BB is thought to play a critical role for the intimal thickening in the lesions of atherosclerosis.

Taurine contributes to cholesterol catabolism by conjugation with bile acids[3]. The cholesterol-lowering effect of taurine has been confirmed in several hyperlipidemic animal models except for rabbits[4-7]. Since high serum cholesterol is one of the major risk factors for atherosclerosis and coronary heart diseases[8,9], taurine is thought to prevent the development of atherosclerosis. It has been reported, however, that taurine prevents the development of atherosclerosis in cholesterol-fed rabbits without a lipid-lowering effect[10,11], suggesting that taurine might directly prevent the development of atherosclerosis. We previously reported that taurine attenuates the induction of the PDGF-BB-mediated expression of immediate-early genes in Balb/3T12 cells and NIH/3T3 cells[12]. In this study we examined whether taurine prevents the proliferation of vascular smooth muscle cells induced by PDGF-BB, and also examined the effect of

Taurine 5: Beginning the 21st Century
Edited by Lombardini *et al.*, Kluwer Academic/Plenum Publishers, New York 2003

taurine on the expression of immediate-early genes induced by PDGF-BB to clarify the mechanism of the anti-atherosclerotic effects of taurine.

2.   MATERIALS AND METHODS

2.1   Cell Cultures

A7r5 cells (rat aortic vascular smooth muscle cell line) from the American Type Culture Collection (Rockville, MD) were cultured in Dulbecco's modified Eagle's medium (DMEM) containing 10% (v/v) fetal bovine serum (FBS) under 95% Air-5% CO_2. Rat aortic vascular smooth muscle cells (rVSMC) were isolated from Sprague-Dawley rats (male, 4 weeks of age) by collagenase digestion[13]. Isolated rVSMC were cultured in Dulbecco's modified Eagle's medium (DMEM) containing 10% (v/v) FBS under 95% Air-5% CO_2 . Human aortic vascular smooth muscle cells (hVSMC) from Clonetics Corporation (Walkersville, MD) were cultured using SmBM Bullet kit (Clonetics Corporation, Walkersville, MD) under 95% Air-5% CO_2.

For the experiments, each cell type was seeded in culture dishes or plates at a concentration of 5×10^4 cells/ml. After 48 h (50-60% confluent), culture medium was changed to a starvation medium (DMEM containing 0.5% (v/v) FBS) and the cells were further incubated for 48 h to attain the G_0 resting phase. The resultant quiescent cells were then treated with taurine, guanidinoethane sulfonic acid and PDGF-BB. All experiments were carried out in triplicate.

2.2   Determination of Immediate Early Gene Expression

Quiescent cells were treated with taurine, GES and/or PDGF-BB for 45 min (for c-fos and c-jun) or 3 h (for c-myc). The cells were then washed with phosphate buffered saline (PBS) and total RNA was extracted by the acid guanidinium-phenol-chloroform (AGPC) method[14]. Extracted total RNA was subjected to Northern blot analysis, RT-PCR and Real-time PCR as described below.

Northern blot analysis: A method of Northern blot analysis using ^{32}P-labeled c-fos cDNA probe was described previously[12].

RT-PCR: RT-PCR was performed using SuperScript™ One-Step RT-PCR system (Invitrogen). PCR primers are shown below, 5'-agctgacagatacgctccaa-3' (Forward) and 5'-taggtgaagacaaaggaagacg-3' (Reverse) for c-fos (GenBank Accession number: X06769), 5'-cgaccttctacgacgatgcc-3' (Forward) and 5'-tcggtgtagtggtgatgtgc-3' (Reverse) for c-jun (GenBank Accession number: X17163),   5'-ccacaaggaaggactatccagc-3'   (Forward)   and 5'-ttgtgctcatctgcttgaacgg-3' (Reverse) for c-myc (GenBank Accession number:

Y00396), 5'-cggtgtgaacggatttggcc-3' (Forward) and 5'-catgagcccttccacgatgc-3' (Reverse) for GAPDH (GenBank Accession number: AF106860), respectively. Thermal cycler conditions were comprised as a 3-step program consisting of 94 °C for 1 min, 60 °C (for c-fos and GAPDH) or 57 °C (for c-jun and c-myc) for 1 min, and 72 °C for 2 min for 24 or 25 cycles (for c-fos and c-jun), for 27 cycles (for c-myc), or 24 cycles (for GAPDH), respectively. These conditions were determined as optimal for semi-quantitative PCR. Each PCR product (c-fos; 556 bp, c-jun; 258 bp, c-myc; 295 bp and GAPDH; 508 bp, respectively) was analyzed in a 2% (w/v) agarose gel containing ethidium bromide; the DNA bands were photographed. The relative expression was quantified by densitometric scanning, and revised by GAPDH (NIH Image version 1.62).

Real-time PCR: Relative quantitation of gene expression was performed using the Applied Biosystems ABI Prism® 7000 sequence detection system (Applied Biosystems). Probes and primers are shown below: Probe; 5'-FAM-cggagacagatcaactt-MGB-3' for c-fos and 5'-FAM-accttgaaagcgcaaaactccgagc-TAMRA-3' for c-jun, Primers; 5'-gaggagggagctgacagatacact-3' (Forward) and 5'-tctgcaacgcagacttctcatc-3' (Reverse) for c-fos, and 5'-ccaagtgccggaaaaggaa-3' (Forward) and 5'-agcatgttggccgtggat-3' (Reverse) for c-jun. Thermal cycler conditions comprised an initial holding at 50 °C for 2 min, 60 °C for 30 min, and then 95 °C for 5 min. This was followed by a 2-step program consisting of 94 °C for 20 sec, and 60 °C for 1 min for 40 cycles.

2.3 Cell Number Counting and Measurement of Thymidine Incorporation

Cell number counting: Quiescent cells were treated with taurine, GES and/or PDGF-BB. After 72 h, the cells were harvested, and the viability was determined by the trypan blue dye exclusion test. The number of viable cells was counted by microscopy.

Measurement of thymidine incorporation: DNA synthesis was determined by [^{3}H]-thymidine incorporation into DNA. Cells were treated with taurine, GES and/or PDGF-BB. After 24 h, [^{3}H]-thymidine was added into each well and was incubated for 4 h. The cells were then washed with 10% (w/v) trichloroacetic acid, and the radioactivity of the acid insoluble fraction was counted in a liquid scintillation counter.

2.4 Measurement of Cytotoxicity

Cytoplasmic leakage due to cell membrane damage was examined by measuring lactate dehydrogenase (LDH) in the culture medium using a commercial LDH assay kit (Wako Purechemial Industries, Japan).

2.5   **Western Blot Analysis**

Equal protein amounts (20 µg) of the cell lysates were separated by SDS/PAGE (12% polyacrylamide), transferred onto a nitrocellulose membrane and probed with specific antibodies against phosphorylated ERK (p44/p42) (New England Biolabs, Ltd) or ERK (p44/p42) (New England Biolabs, Ltd). The results were visualized by the ECL detection system (Amersham Pharmacia Biotech), and the intensity of the band was quantified using imaging analysis (NIH Image version 1.62)

2.6   **Statistical Analysis**

All data were expressed as mean values ± SEM. Differences between groups were assessed by the Student's t-test or Dunnett's multiple comparison.

3.   **RESULTS**

3.1   **Effect of Taurine on the PDGF-BB-Mediated Proliferation in A7r5 Cells**

To examine the effect of taurine on the immediate-early gene expression induced by PDGF-BB, A7r5 cells were treated with PDGF-BB (10 ng/mL), taurine (1 and 10 mM) or guanidinoethane sulfonic acid (GES, 10 mM). Relative expression of c-fos, c-jun and c-myc was determined by Northern blot analysis and RT-PCR. Taurine suppressed the expressions of c-fos, c-jun and c-myc (Figs. 1A and B, 2A and 3A). By contrast, GES, which is a competitor of taurine transport, increased their expression   (Figs. 1C, 2B and 3B). Thus, taurine suppressed the PDGF-BB-mediated gene expression concerning the cell proliferation.

The number of cells treated with taurine, GES and/or PDGF-BB for 3 days were then counted. Taurine suppressed the PDGF-BB-mediated increase of the cell number in A7r5, but did not affect the cell number without PDGF-BB (Fig. 4). By contrast, GES suppressed not only the PDGF-BB-mediated increase of the cell number, but also decreased the cell number without PDGF-BB (Fig. 4). This phenomenon suggested that GES was toxic in these cells.

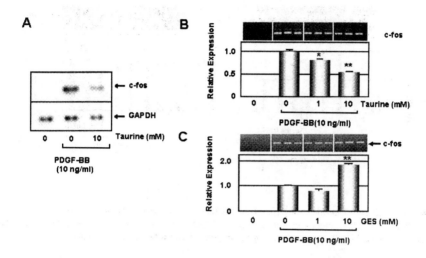

Figure 1. Effect of taurine and GES on the PDGF-BB-mediated expression of c-fos in A7r5 cells A: Northern blot analysis, B: RT-PCR, Effect of taurine, and C: RT-PCR, Effect of GES and **: Significantly different from the cells treated with PDGF-BB (Dunnett's test, p<0.05 and p<0.01, respectively).

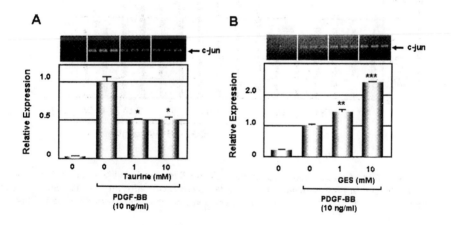

Figure 2. Effect of taurine and GES on the PDGF-BB-mediated expression of c-jun in A7r5 cells A: Effect of taurine, and B: Effect of GES *, ** and ***: Significantly different from the cells treated with PDGF-BB (Dunnett's test, p<0.05, p<0.01 and p<0.001, respectively).

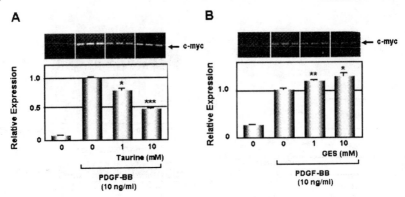

Figure 3. Effect of taurine and GES on the PDGF-BB-mediated expression of c-myc in A7r5 cells A: Effect of taurine, and B: Effect of GES *, ** and ***: Significantly different from the cells treated with PDGF-BB (Dunnett's test, p<0.05, p<0.01 and p<0.001, respectively).

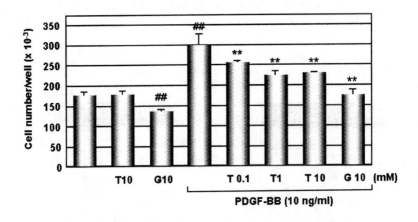

Figure 4. Effect of taurine and GES on the PDGF-BB-mediated proliferation in A7r5 cells T: treated with taurine, and G: treated with GES ##: Significantly different from the untreated cells (Dunnett's test, p<0.01) **: Significantly different from the cells treated with PDGF-BB (Dunnett's test, p<0.01).

To test the effects of taurine and GES on cell toxicity, lactate dehydrogenase which leaked from the cells was measured. As shown in Table 1, GES increased LDH leakage, indicating its cytotoxicity (Table 1). By contrast taurine significantly decreased LDH leakage.

Table 1.   Effect of taurine and GES on the LDH leakage from A7r5 cells

	LDH (IU/L)	
Control	71.5 ± 1.47	
Taurine	51.3 ± 3.72	***
GES	106.7 ± 8.23	*

* and ***: Significantly different from the cells treated with PDGF-BB (Dunnett's test, p<0.05 and p<0.001, respectively)

3.2   Effect of Taurine on the PDGF-BB-Mediated Proliferation in Primary Cultured Rat Aortic Vascular Smooth Muscle Cells

Functional differences between the cell lines such as A7r5 cells and native vascular smooth muscle cells were evident[15]. Thus, we demonstrated the effects of taurine on the cell proliferation of primary cultured rat aortic vascular smooth muscle cells (rVSMC).

The inhibitory effect of taurine on the PDGF-BB-mediated proliferation of rVSMC was determined by counting the cell number and the measurement of thymidine incorporation. Quiescent cells were treated with taurine (0.2 and 0.4 mM) and PDGF-BB (10 ng/mL). Taurine suppressed both cell number (Fig. 5A) and thymidine incorporation (Fig. 5B) increased by PDGF-BB. Thus, it is clear that taurine suppresses the PDGF-BB-mediated proliferation of rVSMC.

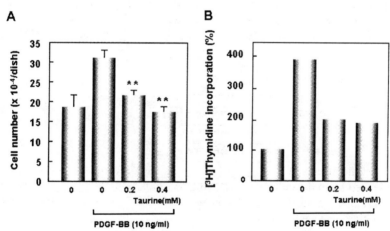

Figure 5. Effect of taurine on the PDGF-BB-mediated proliferation in rVSMC **: Significantly different from the cells treated with PDGF-BB (Student's t-test, p<0.01).

The effect of taurine on the PDGF-BB-mediated expression of immediate-early genes, c-fos and c-jun, was confirmed using Real-time PCR analysis. Taurine (0.4 mM) completely inhibited the expressions of c-fos (Fig. 6A) and c-jun (Fig. 6B) induced by PDGF-BB.

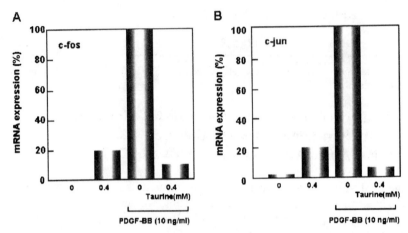

Figure. 6. Effect of taurine on the PDGF-BB-mediated expressions of c-fos and c-jun in rVSMC A: Result of real-time PCR analysis for c-fos, and B: Result of Real-time PCR analysis for c-jun.

3.3 Taurine Inhibited the Activation of MAP Kinase, ERK (p44/p42)

Since intracellular signaling by PDGF-BB is primarily due to the MAP kinase/ERK (p44/p42) cascade, we examined the effect of taurine on ERK (p44/p42) phosphorylation by Western blot analysis. This analysis revealed that taurine inhibited the phosphorylation of ERK (p44/p42) induced by PDGF-BB (Fig.7).

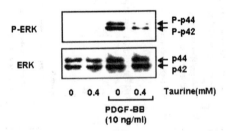

Figure 7. Effect of taurine on the PDGF-BB-mediated phosphorylation of ERK (p44/p42) in rVSMC P-ERK: Phosphorylated ERK.

3.4 Effect of Taurine on PDGF-BB-Mediated Proliferation of Primary Cultured Human Aortic Vascular Smooth Muscle Cells

The inhibitory effects of taurine on the PDGF-BB-mediated proliferation in humans was demonstrated. Primary cultured human aortic vascular smooth muscle cells (hVSMC) were used for this test. Taurine suppressed the cell proliferation by PDGF-BB as well as in rVSMC (Fig. 8). Thus, the inhibitory effect of taurine on the PDGF-BB-mediated proliferation of rVSMC is highly reproducible in humans, suggesting that it is not species specific.

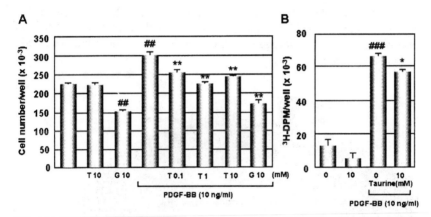

Figure 8. Effect of taurine and GES on the PDGF-BB-mediated proliferation in hVSMC. A: Cell number, and B: Thymidine incorporation. T: Cells treated with taurine, and G: Cells treated with GES ##: Significantly different from the untreated cells (Dunnett's test, p<0.01), **: Significantly different from the cells treated with PDGF-BB (Dunnett's test, p<0.01), ###: Significantly different from the untreated cells (Student's t-test, p<0.01), *: Significantly different from the cells treated with PDGF-BB (Student's t-test, p<0.05).

4. DISCUSSION

In this study, we focused on the proliferation of VSMC to clarify the mechanism of the anti-atherosclerotic effect of taurine. Suppressive effects of taurine on the PDGF-BB-mediated proliferation of VSMC might partially explain a mechanism for taurine on the prevention of atherosclerosis. Taurine is also thought to affect the development of atherosclerosis as taurine chloramines, which are produced by taurine and hypochlorous acid in activated neutrophils and inhibit the activation of NF-kappa B[16]. Thus, taurine also acts as an anti-inflammatory factor. Several actions of taurine including a lipid lowering action are thought to work in concert and thus are responsible for the prevention of atherosclerosis.

It was also demonstrated in this study that taurine suppressed the PDGF-BB-mediated activation of ERK. It is known that taurine suppresses angiotensin II-mediated expression of c-fos and c-jun in cardiac myoblasts[17]. Thus, taurine is thought to act on ERK signaling. However, taurine does not suppress c-fos expression induced by serum and phorbol 12-myristate 13-acetate (PMA), which activate ERK in Balb/3T12 and NIH/3T3 cells[12]. Further studies are needed to reveal the action of taurine.

5.   CONCLUSION

This study revealed that taurine suppresses the PDGF-BB-mediated proliferation of aortic vascular smooth muscle cells through the inhibition of ERK phosphorylation and subsequently the suppression of the immediate-early genes expression (Fig. 9).

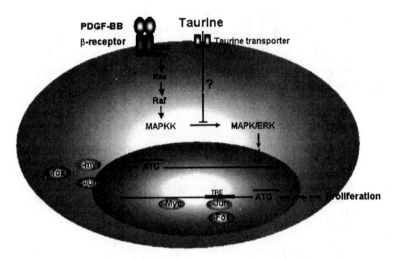

Figure 9. Scheme: Inhibitory mechanism of taurine on the PDGF-BB-mediated proliferation in aortic VSMC.

REFERENCES

1. Ross, R., 1993, The pathogenesis of atherosclerosis: a perspective for the 1990s, *Nature* 362:801-809.
2. Heldin, C.-H. and Westermark, B., 1999, Mechanism of action and in vivo role of platelet-derived growth factor, *Physiol. Rev.* 79:1283-1316.
3. Huxtable, R.J., 1992, Physiological actions of taurine, *Physiol. Rev.* 72:101-163.

4. Herrmann, R.G., 1959, Effect of taurine, glycine and β-sitosterols on serum and tissue cholesterol in the rat and rabbit, *Circulation Res.* 7:224-227.
5. Mann, G.V., 1960, Experimental atherosclerosis: Effects of sulfur compounds on hypercholesteremia and growth in cysteine-deficient monkeys, *Am. J. Clin. Nutr.* 8:491-498.
6. Yokogoshi, H., Mochizuki, H., Nanami, K., Hida, Y., Miyachi, F. and Oda, H., 1999, Dietary taurine enhances cholesterol degradation and reduces serum and liver cholesterol concentrations in rats fed a high-cholesterol diet, *J. Nutr.* 129:1705-1712.
7. Takenaga, T., Imada, K. and Otomo, S., 2000, Hypolipidemic effect of taurine in golden Syrian hamsters, *Adv. Exp. Med. Biol.* 483:187-92.
8. Castelli, W.P., Garrison, R.J., Wilson, P.W.F., Abbott, R.D., Kalousdian, S. and Kannel, W.B., 1986, Incidence of coronary heart disease and lipoprotein cholesterol levels: the Framingham Study, *JAMA* 256:2835-2838.
9. Austin, M.A., 1991, Plasma triglyceride and coronary heart disease, *Arterioscler. Thromb.* 11:2-14.
10. Petty, M.A., Kintz, J. and DiFrancesco, G.F., 1990, The effect of taurine on atherosclerosis development in cholesterol-fed rabbits, *Eur.J.Pharmacol.* 180:119-127.
11. Murakami, S., Kondo, Y., Sakurai, T., Kitajima, H. and Nagate, T., 2002, Taurine suppresses development of atherosclerosis in Watanabe heritable hyperlipidemic (WHHL) rabbits, *therosclerosis* 163:79-87.
12. Imada, K., Takenaga, T., Otomo, S., Hosokawa, Y. and Totani, M., 2000, Taurine attenuates the induction of immediate-early gene expression by PDGF-BB, *Adv. Exp. Med. Biol.* 483:589-594.
13. Thommes, K.B., Hoppe, J., Vetter, H. and Sachinidis, A., 1996, The synergistic effect of PDGF-AA and IGF-1 on VSMC proliferation might be explained by the differential activation of their intracellular signaling pathways. *Exp. Cell. Res.* 226:59-66.
14. Chomczynski, P. and Sacchi, N., 1987, Single-step method of RNA isolation by acid guanidinium thiocyanate-phenol-chloroform extraction, *Anal. Biochem.* 162:156-159.
15. Firulli, A.B., Han, D., Kelly-Roloff, L., Koteliansky, V.E., Schwartz, S.M., Olson, E.N. and Miano, J.M., 1998, A comparative molecular analysis of four rat smooth muscle cell lines, *In Vitro Cell Dev. Biol. Anim.* 34:217-226.
16. Kanayama, A., Inoue, J., Sugita-Konishi, Y., Shimizu, M. and Miyamoto, Y., 2002, Oxidation of I-kappa B alpha at methionine 45 is one cause of taurine chloramine-induced inhibition of NF-kappa B activation. *J. Biol. Chem.* 277:24049-24056.
17. Takahashi, K., Azuma, M., Taira, K., Baba, A., Yamamoto, I., Schaffer, S.W. and Azuma, J., 1997, Effect of taurine on angiotensin II-induced hypertrophy of neonatal rat cardiac cells. *J. Cardiovasc. Pharmacol.* 30:725-730.

Taurine on Sino-Atrial Nodal Cells
Ca^{2+}-Dependent Modulation

HIROYASU SATOH
Department of Pharmacology, Division of Molecular and Cellular Biology, Nara Medical University, Kashihara, Nara 634-8521, Japan

1. INTRODUCTION

Taurine, 2-aminoethanesulfonic acid, is a sulfur amino acid that is widely distributed in the body and is involved in many physiological processes[1,2]. Although taurine is classified as an α-amino acid, it is not incorporated into proteins or membranes. Taurine is present in high concentration (around 10 mM) within myocardial cells, but is found in relatively low levels in the plasma[3]. It has been found to exert many electrical and mechanical actions on cardiac muscle cells[4-7], including a positive inotropic effect[8,9] and protective action against cellular calcium overload and cardiomyopathy[10]. Thus, there are many electropharmacological effects of taurine on cardiac muscle cells, but the underlying mechanisms still remain unclear. In the sino-atrial (SA) node, taurine might be a potent modulator of spontaneous activity. The actions of taurine on cardiac pacemaker currents of the SA node are discussed here.

1.1 Initial Comments

Spontaneous beating is regulated by alterations in the slow diastolic potential (phase 4 depolarization) and by action potential duration (APD) of the SA nodal action potential. Major contributors to the spontaneously beating SA nodal cells are the following; (1) L-type Ca^{2+} current (I_{CaL}), (2) gK decay, (3) sustained inward current (I_{st}), (4) rapidly activated K^+ current (I_{Kr}), (5) hyperpolarization-activated inward current (I_f), and (6) T-type Ca^{2+} current (I_{CaT})[11,12].

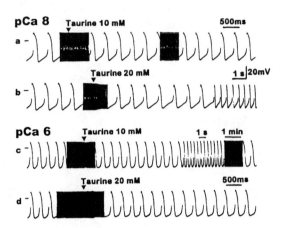

Figure 1. Modulation by taurine at low and high pCa levels in spontaneously beating rabbit SA nodal cells. Short line before the action potential recordings indicates 0 mV level.

Recently, we have demonstrated that the pacemaker mechanisms in rat SA nodal cells are different from those of rabbit and other cell types[13]. Taurine modulates these pacemaker currents in order to maintain regular rhythm.

2.   SPONTANEOUS BEATING

In rabbit SA nodal multicellular preparations exposed to normal extracellular Ca^{2+} concentration ($[Ca^{2+}]_o$), taurine produces a biphasic effect; an initial positive chronotropic effect is followed by a negative chronotropic effect[14]. By contrast, in spontaneously beating embryonic chick cardiomyocytes, taurine mediates a positive chronotropic effect at pCa 10, and a negative chronotropic effect at pCa 6[15].

Modulation of pacemaking activity and the underlying ionic currents of SA nodal cells by taurine was investigated at low and high intracellular Ca^{2+} concentrations ($[Ca^{2+}]_i$) using a patch-clamp technique. In rat SA nodal cells, no positive chronotropic effect of taurine occurred at both low and high pCa levels. However, a potent negative effect was seen which was greater at pCa 6 than at pCa 8. In rabbit SA nodal cells, taurine had similar actions at pCa 8 and 6 (Fig. 1).

It is generally accepted that APD is dependent on pCa levels while the spontaneous rate may be partly regulated by alterations in APD. In rabbit SA nodal multicellular preparations, taurine was found to prolong APD and increase the sinus rate at low $[Ca^{2+}]_o$, but at high $[Ca^{2+}]_o$ it shortened APD and decreased sinus rate. In SA nodal cells, taurine shortened APD at pCa 8, but prolonged it at pCa 6. Although the discrepancy is difficult to understand, these results indicate that spontaneous beating of SA nodal cells are affected by the actions of taurine on APD, presumably resulting from the modulation of the delayed rectifier K^+ current (I_K)[16].

3.   RABBIT SINO-ATRIAL NODAL CELLS

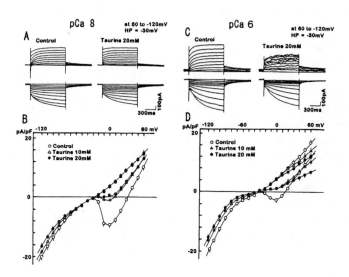

Figure 2. Voltage-dependency of taurine for the ionic currents in rabbit SA nodal cells. The values are represented as means ± S.E.M.

Taurine modulates ionic channels to maintain normal $[Ca^{2+}]_i$ levels; it inhibits I_{CaL} when $[Ca^{2+}]_i$ is high, but enhances I_{CaL} when $[Ca^{2+}]_i$ is low. Similarly, taurine modulates I_K, which is comprised of the rapidly activated current (I_{Kr}) and the slowly activated current (I_{Ks}).   While I_{Kr} is altered by taurine, I_{Ks} is not[17].   Also affected by taurine is spontaneous activity and APD. In rabbit SA nodal cells, the rate of spontaneous beating is regulated by taurine's action on I_{Kr}. The modulation of APD by taurine is consistent with the observed effects on I_K at high pCa, but not at low pCa. Although the reason underlying the discrepancy is unclear, it is likely that I_K is not the only current contributing to APD in SA nodal cells.

As shown in Fig. 2, taurine (20 mM) inhibited I_{CaL} by 56.9 ± 2.8% (n = 6, P<0.001) at pCa 8, and by 97.6 ± 3.8% (n = 7, P<0.001) at pCa 6. By contrast, taurine (20 mM) decreased I_K by 26.8 ± 2.6% (n = 6, P<0.01) at pCa 6, but had little or no effect at pCa 8.   In the presence of taurine (20 mM), I_f was reduced at pCa 8 and 6 by 18.3 ± 1.3% (n = 8, P<0.05) and by 20.8 ± 3.3% (n = 8, P<0.05), respectively. Thus, taurine exerted more potent inhibitory actions at the higher pCa level. Interestingly, dysrhythmias were observed in 3 of 17 cells at pCa 8 and in 12 of 16 cells at pCa 6 in the presence of taurine. During washout, there was a further increase in the incidence of dysrhythmias or arrest.   Presumably, cellular Ca^{2+} overload occurred when taurine exposure was discontinued. These results indicate that taurine exerts more potent inhibitory actions on ionic currents after Ca^{2+} overload in rabbit SA nodal cells, representative of a cardioprotective action.

4.   RAT SINO-ATRIAL NODAL CELLS

Modulation by taurine of pacemaking activity and the underlying ionic currents, in particular I_f current and the sustained inward current (I_{st}), in rat SA nodal cells was investigated at different pCa levels. Although increasing pCa levels stimulated spontaneous activity, spontaneous activity was more effectively inhibited by taurine at high pCa levels. At pCa 7, taurine (20 mM) inhibited I_{CaL} by $35.8 \pm 2.5\%$ (n = 8, P<0.01) and decreased I_K by $26.8 \pm 2.6\%$ (n = 6, P<0.01). Similarily, taurine inhibited I_{Kr}, but not I_{Ks} (Fig. 3). By comparison, taurine (20 mM) enhanced I_{CaT} by $29.3 \pm 2.9\%$ (n = 8, P<0.05).

In addition, taurine (20 mM) increased I_f by $60.1 \pm 1.7\%$ (n = 8, P<0.001) at pCa 10, and by $48.0 \pm 1.4\%$ (n = 8, P<0.01) at pCa 7. I_{st} was decreased by $13.3 \pm 1.1\%$ (n = 5, P<0.05) at 10 mM taurine and by $38.1 \pm 2.4\%$ (n = 8, P<0.01) at 20 mM taurine. Depending on the concentration of taurine and Ca^{2+}, dysrhythmias could also occur. These results indicate that taurine causes a negative chronotropic effect by inhibiting pacemaker ionic currents, such as I_{CaL}, I_{Kr} and I_{st}. Moreover, I_f and I_{CaT} are minor contributors to the pacemaker activity in rat SA nodal cells.

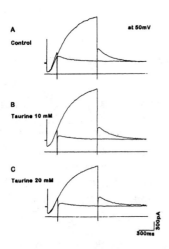

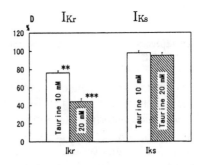

Figure 3. Depressant effects of taurine on the I_{Kr} and I_{Ks} currents. **: P<0.01, P<0.001, with respect to control value.

5.   CARDIOPROTECTION

The intracellular taurine level of myocardial cells is reduced in ischemic heart failure and hypoxia[18-20]. Under those conditions, $[Ca^{2+}]_i$ increases due to the inhibition of the sarcolemmal Ca-ATPase. The inhibition of ATP production and the increase in $[Ca^{2+}]_i$ activate the ATP-sensitive K^+ (K_{ATP}) channels and the Ca^{2+}-activated K^+ (K_{Ca}) channels, resulting in a shortening of APD. Simultaneously, calcium overload often produces triggered activity and elicits arrhythmias.

Because of the decline in taurine levels during ischemia and cardiac failure, taurine release from heart cells should be cytoprotective. If external application of taurine inhibits the ionic channels, especially I_K, taurine should improve the abnormally shortened APD. Thus, APD regulation would represent a key cardioprotective action of taurine.

In our experiments, the $[Ca^{2+}]_i$ level is easily altered by adding different concentrations of EGTA to the cytosol. When 100 μM EGTA is present in the pipette (intracellular) solution, cellular calcium overload readily occurs. Under those conditions, the application of taurine helps to normalize cellular responses (unpublished data).

Thus, it has recently been demonstrated that the physiological and pharmacological effects of taurine are greatly altered by $[Ca^{2+}]_i$ and $[Ca^{2+}]_o$[19,21,22]. At high $[Ca^{2+}]_o$, taurine reduces contractile force, while at low $[Ca^{2+}]_o$ it enhances force[23,24]. These findings suggest that taurine may exert a normalizing action on $[Ca^{2+}]_i$ and contractile force, which protects the cell against cardiac failure and arrhythmias during Ca^{2+} overload.

6. CONCLUSION

As compared with other cells of the heart, the pharmacological actions of taurine on SA nodal cells are less pCa-dependent. However, maintaining Ca^{2+} homeostasis

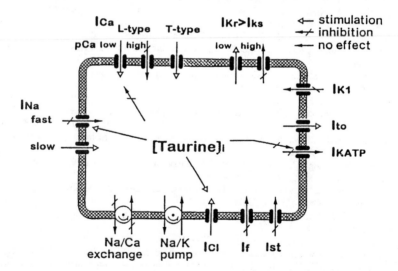

Figure 4. Summary of taurine's actions on the ionic channel currents in cardiomyocytes. Usually taurine was extracellularly administrated. [Taurine]$_i$ indicates an intracellular application of taurine. I_{K1}: Inwardly rectifying K$^+$ current current. I_{to}: Transient outward current. I_{KATP}: ATP-sensitive K$^+$ current. I_{Cl}: Chloride current.

appears to be the physiological function of taurine in SA nodal cells. The SA nodal cells beat spontaneously because of tight regulation of Ca^{2+}. Pharmacological actions of taurine on ion channels and transport systems in the cell membrane and intracellular organelles help to maintain Ca^{2+} homeostasis. In both rabbit and rat SA nodal cells, I_{Kr} plays a more important role as pacemaker than I_{Ks}. Therefore, the author concludes that the I_{Kr} current may be a key site for taurine action in the SA node. However, the effect of taurine on I_f in rabbit and rat SA nodal cells is quite different. This is because the pacemaker mechanisms of rat are different from those of rabbit and other SA nodal cells[13]. Figure 4 summarizes the effects of taurine on various ionic channel currents of the cardiac myocyte.

REFERENCES

1.   Huxtable, R.J., and Sebring, L.A, 1983, Cardiovascular actions of taurine. In *Sulfur Amino Acids: Biochemical and Clinical spects* (K. Kuriyama, R.J. Iwata and A.R. Liss, eds.), Clin. Aspec., New York, pp. 5-37.
2.   Pasantes-Morales, H., and Gamboa, A., 1980, Effect of taurine on ^{45}Ca accumulation in rat brain synaptosomes. *J. Neurochem.* 34: 244-246.
3.   Huxtable, R.J., and Sebring, L.A., 1980, Cardiovascular actions of taurine. In *Sulfur Amino Acids: Biochemical and Clinical Aspects* (K. Kuriyama, R.J. Huxtable, and H. Iwata, eds.), Alan R Liss, New York, pp. 5-37.
4.   Huxtable, R.J., 1992, The physiological actions of taurine. *Physiol. Rev.* 72: 101-163.
5.   Schaffer, S.W., Kramer, J., and Chovan, J.P., 1980, Regulation of calcium homeostasis in the heart by taurine. *Fed. Proc.* 39: 2691-2694.
6.   Sperelakis, N., and Satoh, H., 1993, Taurine effects on ion channels of cardiac muscle. In *Ionic Channels and Effect of Taurine on the Heart* (D. Noble, and Y.E. Earm, eds.), Kluwer Academic Publisher, Boston, pp. 93-118.
7.   Satoh, H., and Sperelakis, N., 1998, Review of some actions of taurine on ion channels of cardiac muscle cells and others. *Gen. Pharmacol.* 30: 451-463.
8.   Franconi, F., Martini, F., Stendari, I., Matucci, R., Zilleti, L., and Giotti, A., 1982, Effect of taurine on calcium level and contractility in guinea pig ventricular strips. *Biochem. Pharmacol.* 31: 3181-3185.
9.   Takihara, K., Azuma, J., Awata, N., Ohta, H., Hamaguchi, T., Sawamura, A., Tanaka, Y., Kishimoto, S., and Sperelakis, N., 1986, Beneficial effect of taurine in rabbits with chronic congestive heart failure. *Am. Heart J.* 112: 1278-1284.
10. Azari, J., Brumbaugh, P., and Huxtable, R., 1980, Prophylaxis by taurine in the hearts of cardiomyopathic hamsters. *J. Mol. Cell. Cardiol.* 12: 1353-13663.
11. Guo, J., and Noma, A., 1995, A sustained inward current activated at the diastolic potential range in rabbit sino-atrial node cells. *J. Physiol. (Lond.)* 483: 1-13.
12. Noble, D., 1984, The surprising heart: a review of recent progress in cardiac electrophysiology. *J. Physiol. (Lond.)* 353: 1-50.
13. Shinagawa, Y., Satoh, H., and Noma., A., 2000, The sustained inward current and inward rectifier K^+ current in pacemaker cells dissociated from rat sinoatrial node. *J. Physiol. (Lond.)* 523: 593-605.
14. Satoh, H., 1995, Electrophysiological actions of taurine on spontaneously beating rabbit sino-atrial nodal cells. *Jpn. J. Pharmacol.* 67: 29-34.
15. Satoh, H., 1995, Regulation of automaticity in spontaneously beating embryonic chick cardiomyocytes. *J. Cardiovasc. Pharmacol.* 25: 3-8.
16. Satoh, H., 1995, A dual action of taurine on the delayed rectifier K^+ current in young embryonic chick cardiomyocytres. *Amino Acids* 9: 235-246.
17. Satoh, H., 1999, Taurine modulates I_{Kr} but not I_{Ks} in guinea pig ventricular cardiomyocytes. *Br. J. Pharmacol.* 126: 87-92.

18. Lombardini, J.B., 1980, Effect of ischemia on taurine levels. In *Natural Sulfur Compounds* (D. Cavallini, G.E. Gaull, and V. Zappia, eds.), Plenum Press, New York, pp. 255-306.

19. Kramer, J.H., Chovan, J.P., Schaffer, S.W., 1981, The effect of taurine on calcium paradox and ischemic heart failure. *Am. J. Physiol.* 240: H238-H246.

20. Satoh, H., Nakatani, T., Tanaka, T., and Haga, S., 2002, Cardiac functions and taurine's actions at different extracellular calcium concentrations in forced swimming stress-loaded rats. *Biol. Trac Ele. Res.* 87: 171-182.

21. Sperelakis, N., Yamamoto, T., Bkaily, G., Sada, H., Sawamura, A., and Azuma, J., 1989, Taurine effects on action potentials and ionic currents in chick myocardial cells. In *Taurine and the Heart* (H. Iwata, J.B. Lombardini and T. Sagawa, eds.), Kluwer Academic Publisher, Boston, pp. 1-19.

22. Satoh, H., 1994, Cardioprotective actions of taurine against intracellular and extracellular Ca^{2+}-induced effects. In *Taurine in Health and Disease* (R.J. Huxtable, and D. Michalk, eds.), Plenum Press, New York, pp. 181-196.

23. Dietrich, J., and Diacono, J., 1971, Comparison between ouabain and taurine effects on isolated rat and guinea pig hearts in low calcium medium. *Life Sci.* 10: 499-507.

24. Dolara, P., Ledda, F., Mugelli, A., Mantelli, L., Zilletti, L. , Franconi, F., and Giotti, A., 1978, Effect of taurine on calcium, inotropism, and electrical activity of the heart. In *Taurine and Neurological Disorders* (A. Barbeau, R.J. Huxtable, eds.), Raven Press, New York, pp. 151-159.

Cellular Characterization of Taurine Transporter in Cultured Cardiac Myocytes and Nonmyocytes

TOMOKA TAKATANI, KYOKO TAKAHASHI, TAKASHI ITOH, KOICHI TAKAHASHI[1], MAYO HIRATA, YASUHIRO YAMAMOTO, MASANORI OHMOTO, STEPHEN W. SCHAFFER[2], and JUNICHI AZUMA
Clinical Evaluation of Medicines and Therapeutics, Graduate School of Pharmaceutical Sciences, Osaka University, Osaka 565-0871 JAPAN: [1]Department of Pharmaceutics, School of Pharmaceutical Sciences, Mukogawa Women's University, Hyogo 663-8179 JAPAN: [2]Department of Pharmacology, University of South Alabama, School of Medicine, Alabama, USA

1. INTRODUCTION

Taurine is by far the most abundant free amino acid in the mammalian heart, comprising in excess of 50% of the total free amino acid pool[1, 2]. Although its physiological function remains undefined, taurine exhibits an extensive cardiovascular pharmacology. Carnivores depend to a large extent on taurine obtained through the diet, which must be transported across cell membranes to accumulate in the heart. The taurine transporter, which belongs to a gene family that encodes Na^+- and Cl^--coupled transporters[2,3], has been cloned from mammalian tissues and cells[3-10]. Although numerous cell types are present in the heart, cardiomyocytes and cardiac fibroblasts are the predominant cell type in the neonatal rat heart. There are no known studies showing the detailed characterization of the taurine transporter in cardiac cells. The present study was undertaken to clarify the physiological characteristics of the cardiac taurine transporter using cultured myocytes and nonmyocytes prepared from neonatal rats.

2. METHODS

2.1 Preparation of Cardiac Myocytes and Nonmyocytes

Preparation of primary cardiac myocyte and nonmyocyte cultures from 1-day-old Wistar rats was according to the method described by Sadoshima et al.[11]. This procedure yielded cultures with 90-95% myocytes, as assessed by microscopic observation of cell beating. The nonmyocyte preparation contained mostly cardiac fibroblasts, although there were some neurons, endothelial cells and vascular smooth muscle cells. The quality of the nonmyocyte preparation was evaluated using three different chemical stains, one for acetylated-low density lipoprotein, the second for α-smooth muscle actin, and the final for neuron-specific nuclear protein. Based on the chemical staining pattern, we estimated that the preparation generally contained less than 10 % of endothelial cells, neurons and smooth muscle cells (data not shown). All experiments were performed using highly purified nonmyocytes that had been incubated in serum-free medium for 24 hr.

2.2 Intracellular Taurine Content

The intracellular taurine content of cardiocytes was measured by the procedure of Jones and Gilligan[12] using a high-performance liquid chromatography (HPLC) system (JASCO880-PU) equipped with a HITACHI F1000 fluorescence detector and HITACHI D-2500 integrator.

2.3 Isolation and Northern Blot Analysis of RNA

Total RNA of the cardiac cells (myocytes and nonmyocytes) was isolated using the guanidinium thiocyanate-phenol-chloroform extraction method[13]. Northern blot analysis was performed according to the procedure described by Kim et al.[14]. The cDNA probes used were as follows: canine taurine transporter cDNA, a 7-kb fragment (from Dr. S. Uchida, Medical School, Tokyo Medical and Dental University) and rat glyceraldehyde-3-phosphate dehydrogenase (GAPDH), a 1.3-kb fragment.

2.4 Western Blot Analysis

The taurine transporter protein was determined by Western blot analysis[15]. The membranes were incubated with rat taurine transporter primary antibody (ALPHA DIAGNOSTIC, TX, USA) 1:1000, and with a secondary antibody, goat anti-rabbit immunoglobulin G (IgG). The bands were detected by the enhanced chemiluminescence reaction (Amersham pharmacia biotech, NJ, USA).

2.5 Taurine Uptake Measurements

The uptake reaction was performed in medium containing [1,2-^{3}H] taurine (0.1 µCi /µl, 1 mL). After uptake medium was added to the dish, the cells were incubated at 37°C for 30 min. The medium was then removed and the reaction was terminated by the addition of cold tritium-free medium and 0.5 ml of 0.1 M NaOH. After 30 min, the content of each dish was transferred to a scintillation vial, and their radioactivity was determined by liquid scintillation spectrometry.

2.6 Statistics

Statistical significance was determined by the Student's t-test or analysis of variance (ANOVA) (Bonferroni's method was used to compare individual data when a significant F value was shown) and χ^2-test, depending on the design of the experiment. Each value was expressed as the mean ± S.E.M. Differences were considered statistically significant when the calculated p value was less than 0.05.

3.   RESULTS AND DISCUSSION

3.1 The Comparison of Intracellular Taurine Content Between Myocytes and Nonmyocytes

Total protein content of the myocyte and the nonmyocyte was 136 ± 15 and 86 ± 10 µg/10^6 cells, respectively. Taurine content of both cell types was normalized relative to protein content and cell number (Fig 1). Based on these calculations, the intracellular taurine content of myocytes was 2-fold to 3-fold higher than that of nonmyocytes, a difference that could account for alterations in both cellular function and morphology. Myocytes in culture contain contraction-relaxation proteins that allow them to contract spontaneously; nonmyocytes lack this characteristic. Moreover it is possible that the contracting cell may become more susceptible to the Ca^{2+}-modulating actions of taurine[16-18].

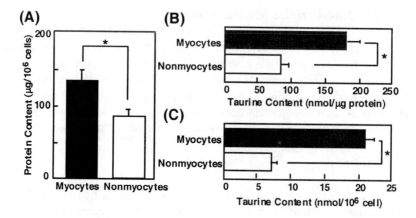

Figure 1. The intracellular taurine concentration in neonatal rat heart cells. (A) Total protein contents in neonatal rat heart cells. (B) Intracellular taurine content per μg protein in neonatal rat heart cells. (C) Intracellular taurine content per 10^6 cells protein in neonatal rat heart cells. Mean ± SEM, *p<0.01, n=6-15.

3.2   Identity and Characteristics of Taurine Transporter on Myocytes and Nonmyocytes in Culture

This represents the first evidence that the taurine transport system was present in cardiac fibroblasts obtained from neonatal rat hearts. It was found that both myocytes and nonmyocytes contain high levels of taurine transporter mRNA. Nonetheless, the myocyte content of transporter mRNA (normalized to GAPDH mRNA) was lower than that found in the nonmyocyte (Fig 2A). Although the levels of transporter protein in the two cell types were related to the levels of mRNA (Fig 2B), total protein content per cell was 1.6-fold higher in the myocyte than in the nonmyocyte. Thus, the difference in the content of transporter protein between the two cell types could merely reflect the difference in total protein per cell.

Uptake of radiolabeled taurine into monolayer cultures of heart cells was markedly stimulated by the presence of Na^+ in the medium, with the uptake almost abolished in the absence of Na^+ (Fig 3A). When the Na^+-dependent uptake of taurine was plotted against Na^+ concentration, a sigmoidal curve was obtained, suggesting that more than one Na^+ ion was associated with the transfer of one taurine molecule. Based on the Hill equation, the stoichiometry for Na^+/taurine was 2 : 1 in both myocytes and nonmyocytes (Fig 3B, C). The present findings in cultured heart cells are in good agreement with previous studies using other cell lines, such as Caco-2 cells and HRPE[15, 19, 20]. Competition between taurine and the taurine analog, β-alanine, was examined over a β-alanine concentration range of 0-250 μM. β-alanine competitively inhibited the uptake of taurine by the heart

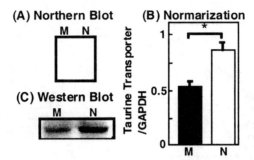

(A) Northern Blot (B) Normarization

(C) Western Blot

Figure 2. The levels of taurine transporter mRNA and protein in neonatal rat heart cells. (A) shows typical Northern blot autoradiograms of myocytes and nonmyocytes mRNA for taurine transporter. (B) Bar shows mean ± SEM related to GAPDH, obtained by densitometric analysis. (C) represesnts Western blots of taurine transporter protein in myocytes (M) and nonmyocytes (N). Equal protein of 24 μg/lane was loaded onto a 10 % SDS-polyacrylamide gel.

cells, with a Ki value of about 30 μM for β-alanine observed in both the myocyte and the nonmyocyte (data not shown). Kinetic analysis revealed that the uptake of taurine over a concentration range of 10–100 μM involved a single saturable, high affinity system in both the myocyte and the nonmyocyte (Fig 4A). Based on Lineweaver-Burk and Eadie-Hofstee plots, the Km for the taurine transporter was about 21-26 μM in both cell types (Fig. 4B,C). However, Vmax was approximately 2-fold higher in the myocyte. Since total protein content per cell was 1.6-fold higher in the myocyte than in the nonmyocyte, the difference in cellular protein may account for the observed difference in the content of transporter protein.

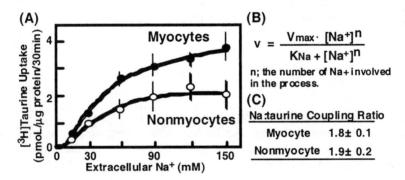

(A)

(B)

$$v = \frac{V_{max} \cdot [Na^+]^n}{K_{Na} + [Na^+]^n}$$

n; the number of Na+ involved in the process.

(C) Na:taurine Coupling Ratio

Myocyte 1.8± 0.1

Nonmyocyte 1.9± 0.2

Figure 3. Sodium-dependent taurine transport by neonatal rat heart cells. (A) Taurine transport of neonatal rat heart cells was evaluated as a function of extracellular sodium concentration. Mean ± SEM, n=6. (B) Hill-type equation. (C) Na⁺:taurine coupling ratio.

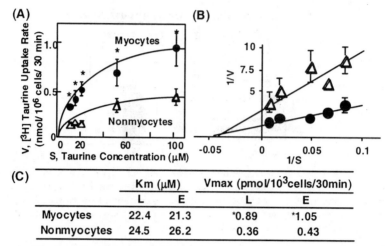

Figure 4. Kinetic analysis of taurine uptake in neonatal rat heart cells. (A) Taurine uptake was incubated for 30 min with NaCl-containing buffer. Mean ± SEM, n=6-12, *p<0.01 versus nonmyocytes at each concentration. (B) Graph shows Lineweaver-Burk plots. (C) Kinetic parameters. L; Lineweaver-Burk plot, E; Eadie-Hofstee plot. n=6-12, *p<0.01 versus nonmyocytes.

It is plausible that the higher Vmax in the myocyte may lead to a greater accumulation of taurine. Nonetheless, there are other factors to consider, such as the possible existence of storage pools in the myocyte. Further studies are needed to fully understand the role of the taurine transporter in the heart. Interestingly, based on the plasma levels of taurine, which range from 30 – 500 μM depending upon the species[2] and Km for the transporter (about 20 μM), the taurine carrier should be operating close to Vmax. Thus, like other cell types with Km values varying from 0.9 μM (in Caco-2 cells) to 11 μM (in HT-29 cells[20]), the transporter is unlikely to be substrate limited.

4. CONCLUSION

These findings indicate that both myocytes and nonmyocytes express nearly identical taurine transporters with similar Km values. Nonetheless, each cell type exhibits different transport activity. As predicted, the cell with the higher taurine content (namely, myocytes) also exhibited the higher Vmax.

REFERENCES

1. Pisarenko OI, 1996, Mechanisms of myocrdial protection by amino acids : Facts and hypothesis. *Clin. Exp. Pharmacol. Physiol.* 23: 627-633.
2. Huxtable R J, Sebring L, 1982, Cardiovascular actions of taurine. *Amino Acids* 5:29-51.
3. Uchida S, Kwon HM, Yamauchi A, Preson AS, Marumo F, Handler JS, 1992, Molecular cloning of the cDNA for an MDCK cell Na^+-and Cl^--dependent taurine transporter that is regulated by hypertonicity. *Proc. Natl. Acad. Sci.* 89 : 8230-8234.
4. Smith KE, Borden LA, Wang CD, Hartig PR, Branchek TA, Weinshank RL, 1992, Cloning and expression of a high affinity taurine transporter from rat brain. *Mol. Pharmacol.* 42 : 563 - 569.
5. Liu O, L-Corcuera B, Nelson H, Mandiyan S, Nelson N, 1992, Cloning and expression of a cDNA encoding the transporter of taurine and β-alanine in mouse brain. *Proc. Natl. Acad. Sci.* 89 : 12145-12149.
6. Jhiang SM, Fithian L, Smanik P, McGill J, Tong Q, Mazzaferri EL, 1993, Cloning of the human taurine transporter and characterization of taurine uptake in thyroid cells. *FEBS Lett.* 318 : 139 - 144.
7. Ramamoorthy S, Leibach FH, Mahesh VB, Han H, Yang-Feng T, Blakely RD, Ganapathy V, 1994, Functional characterization and chromosomal localization of a cloned taurine transporter from human placenta. *Biochem. J.* 300 : 893-900.
8. Jayanthi LD, Ramamoorthy S, Mahesh VB, Leibach FH, Ganapathy V, 1995, Substrate-specific regulation of the taurine transporter in human placental choriocarcinoma cells (JAR). *Biochim. Biophys. Acta.* 1235 : 351 - 360.
9. Miyamoto Y, Liou GI, Sprinkle TJ, 1996, Isolation of a cDNA encoding a taurine transporter in the human retinal pigment epithelium. *Curr. Eye Res.* 15 : 345 - 349.
10. Leibach JW, Cool DR, Monte MAD, Ganapathy V, Leibach FH, Miyamoto Y, 1993, Properties of taurine transport in a human retinal pigment epithelial cell line. *Curr. Eye Res.* 12 : 29 - 36.
11. Sadoshima J, Izumo S, 1993, Molecular characterization of angiotensin II-induced hypertrophy of cardiac myocytes and hyperplasia of cardiac fibroblasts. *Circ. Res.* 73 : 416 - 423.
12. Jones BN, Gilligan JP, 1983, O-phthaladehyde precolumn derivatization and reversed-phase high-performance liquid chromatography of polypeptide hydrolysates and physiological fluids. *J. Chromatogr.* 266 : 471 - 482.
13. Chomczynski P, Sacchi N, 1987, Single-step method of RNA isolation by acid guanidium thiocyanate-phenol-chloroform extraction. *Anal. Biochem.* 162 : 156 - 159.
14. Kim S, Ohta k, Hamaguchi A, Yukimura T, Miura K, Iwao H, 1995, Angiotensin II induces cardiac phenotypic modulation and remodeling in vivo in rats. *Hypertension* 25 : 1252 - 1259.
15. Mollerup J, Lambert IH, 1998, Calyculin A modulates the kinetic constants for $Na+$-coupled taurine transport in Ehrlich ascites tumour cells. *Biochim. Biophys. Acta.* 1371 : 335-344.
16. Takahashi K, Azuma M, Taira K, Baba A, Yamamoto Y, Schaffer SW, Azuma J, 1997, Effect of taurine on anigotensin II-induced hypertrophy of neonatal rat cardiac cells. *J. Cardiovasc. Pharmacol.* 30 : 725 - 730.
17. Takahashi K, Schaffer SW, Azuma J, 1997, Taurine prevents intracellular calcium overload during calcium paradox of cultured cardiomyocytes. *Amino Acids* 13 : 1 -11.
18. Schaffer SW, Azuma J, Madura JD, 1995, Mechanisms underlying taurine-mediated alterations in membrane function. *Amino Acids* 8 : 231 - 46.
19. Kundaiker S, Hussain AA, Marshall J, 1996, Component characteristics of the vectorial transport system for taurine in isolated bovine retinal pigment epithelium. *J. Physiol.* 492 : 505 - 516.
20. Ganapathy V, Leibach FH, 1994, Expression and regulation of the taurine transporter in cultured cell lines of human origin. In: Taurine in Health and Disease, (Eds. R. Huxtable and DV Michalk), pp 51-57. Plenum Press, New York.

Taurine Transporter in Cultured Neonatal Cardiomyocytes:
A Response to Cardiac Hypertrophy

TAKASHI ITOH, KYOKO TAKAHASHI, YOKO YAMAUCHI, KOICHI
TAKAHASHI[1], SATO UEYAMA, STEPHEN W. SCHAFFER[2], and
JUNICHI AZUMA
*Clinical Evaluation of Medicines and Therapeutics, Graduate School of Pharmaceutical Sciences,
Osaka University, Osaka JAPAN*
[1]*Department of Pharmaceutics, School of Pharmaceutical Sciences, Mukogawa Women's University,
Hyogo JAPAN*
[2]*Department of pharmacology, University of South Alabama, School of Medicine, Alabama, USA*

1.   INTRODUCTION

Hemodynamic overload results in cardiac hypertrophy, and may ultimately
lead to congestive heart failure.   Two factors are thought to contribute to the
development of cardiac hypertrophy, mechanical stress and neurohormonal stress
from angiotensin II (Ang II), endothelin-1, and the cytokines.

Taurine is one of the most abundant free amino acids in animals, and is found
in very high concentration in cardiac tissue[1].   Myocardial taurine content has
been reported to increase in patients suffering from congestive heart failure[2].
Because taurine is beneficial in treating patients with congestive heart failure[3],
the increase in taurine content may represent a physiological adaptation.

Taurine transporter (TAUT) plays a key role in the modulation of intracellular
taurine content by stresses such as hypertonic stress[4].   Recently, we identified
TAUT in neonatal rat cardiac myocytes[5].   Although cardiac levels of taurine are
altered in response to heart disease, the role of the taurine transport system
remains undefined.   Therefore, this study was undertaken to clarify the role of
the cardiac taurine transporter(s) in the modulation of cellular taurine content by
hypertrophic stimuli.   In this study, cultured myocytes were stimulated by either
mechanical stretching or by exposure to angiotensin II, both of which trigger
myocyte hypertrophy.

2.   METHODS

2.1   Preparation of Cardiomyocytes Culture

Preparation of primary cardiac myocyte cultures from 1-day-old Wister rats was as previously described[5]. For selective enrichment of cardiac myocytes, dissociated cells were preplated for 1 hr. Non-adherent cells were seeded onto silicon rubber bottom plates at a density of 2-5 x 10^6 cells/ml/dish. Cells were kept in serum-containing culture medium for 48 hr and then incubated with serum-free medium. After 24 hr in serum-free medium, the cells were stimulated for 24 hr by either mechanical stretching them or exposing them to Ang II (100 nM).

2.2   Cyclical Stretch on Cultured Cardiomyocytes

The FX-2000 stretch unit (Flexcell)[6] consists of a vacuum unit linked to a value controlled by a computer program. Cardiomyocytes cultured on silicon rubber bottom plates were subjected to cyclical stretch produced by a computer-controlled application of sinusoidal negative pressure. The silicon rubber membranes were extended by 120% of the cell length at a frequency of 60 cycles/min. This degree of mechanical stretching was shown not to damage the cardiomyocytes, as evidenced by a lack of CPK leakage into the culture medium.

2.3   Measurement of Intracellular Taurine Content

The intracellular taurine content of cardiomyocytes was measured using an HPLC system as previously described[5].

2.4   Measurement of Taurine Uptake

Uptake of taurine into cardiomyocytes was measured at 37°C as previously described. The uptake medium containing [1,2-^{3}H] taurine (0.1 µCi / µL) and cold taurine (10~1000µM) was added to the dish. After a 30 min incubation, the medium was removed, and the cells were quickly washed 3 times with cold PBS (-). The cells were then lysed with 0.1M NaOH. The lysate was transferred to a scintillation vial, and the radioactivity was determined by liquid scintillation spectrometry.

2.5   Isolation and Northern Blot Analysis of RNA

Total RNA of cardiomyocytes was isolated using the guanidinium thiocyanate-phenol-chloroform extraction method[7]. The RNA concentration was determined spectrophotometrically by absorbance at 260nm. Northern blot analysis was performed according to the procedure described by Kim et al[8]. The cDNA probes used were as follows: canine taurine transporter cDNA, a 7-kb fragment (from Dr. S. Uchida, Medical School, Tokyo Medical and Dental University[4]); and rat glyceraldehyde- 3-phosphate dehydrogenase (GAPDH), a 1.3-kb fragment. Results were normalized relative to the mRNA levels of GAPDH.

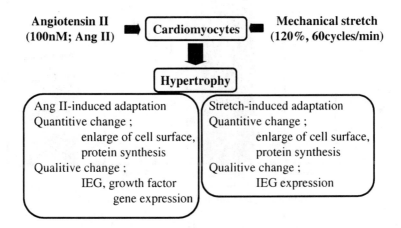

Figure 1. Hypertrophic responses mediated by Angiotensin II and mechanical stretch. We reconfirmed these changes in cultured neonatal rat cardiomyocytes.

2.6 Statistics

Statistical significance was determined by the Student's t-test. Each value was expressed as the mean ± S.E.M. Differences were considered statistically significant when the calculated p value was less than 0.05.

3. RESULTS

3.1 Intracellular Taurine Content in Hypertrophic Cardiomyocytes

We, as well as other investigators, have demonstrated that stimulation by either mechanical stretching (120%, 60 cycles/min) or Ang II (100 nM) leads to hypertrophy in cultured neonatal rat cardiomyocytes[6,9]. The type of quantitative and qualitative hypertrophic changes mediated by angiotensin II and mechanical stress are depicted in Figure 1.

As shown in Figure 2, we measured the intracellular taurine content of

(A)   (B)

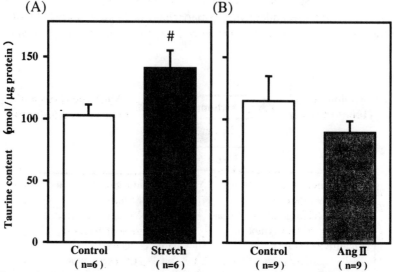

Figure 2. Change in intracellular taurine content of hypertrophic cardiomyocytes. Neonatal rat cardiomyocytes were stimulated for 24 hr by mechanical stretching (120%, 60 cycles/min) (A) or with Ang II (100 nM) (B). Intracellular taurine content was measured by HPLC. Results were obtained from 3-4 separate experiments. #P<0.05 vs. control. Data were mean ± SE.

hypertrophic cardiomyocytes induced by either mechanical stretching or Ang II exposure. After 24 hr. of mechanical stretching the cells, taurine content was increased 1.3-fold compared with the control. However, Ang II treatment for 24 hr did not influence taurine content.

3.2   Kinetics of Taurine Uptake in Hypertrophic Cardiomyocytes

To determine whether changes in taurine uptake accounted for the alteration in taurine content of the hypertrophic cardiomyocyte, taurine uptake activity was measured after stimulation by either mechanical stretching or with Ang II. Figure 3 shows that the velocity of taurine uptake at each extracellular taurine concentration examined was increased after mechanically stretching the cells, but not after stimulation with Ang II. This result indicates that the effect of hypertrophic stimuli on taurine content is related to the modulation of taurine uptake.

The maximum velocity (Vmax) and Michaelis Menten constant (Km) of taurine uptake were calculated from Eadie-Hoffstee plots generated from control and stretch-induced hypertrophic cardiomyocytes. The Eadie-Hofstee plot demonstrated that cardiomyocyte taurine uptake involves two-transport systems. The plot indicated that only the high-affinity transport system contributed to the uptake of taurine in the low concentration range (~about 50 μM taurine). However, the low-affinity transport system contributed to the uptake of taurine in the high concentration range (about 100 μM taurine). Data obtained from control cardiomyocytes yielded a Km = 7.4 μM and a Vmax = 3.0 pmol/μg protein/30min for the high-affinity transport system, while a Km = 316 μM and a Vmax = 11 pmol/μg protein/30min was obtained for the low-affinity transport system. The Vmax value for the high-affinity taurine transport system in the mechanically stretched, hypertrophic cardiomyocyte was increased 1.5-fold compared to the control, whereas the Km values were unaffected by stretching. By comparison, relative to the control, the values of Vmax and Km of the low affinity transport system in the mechanically stretched, hypertrophic cardiomyocyte were increased by 1.9-fold and 1.3-fold, respectively.

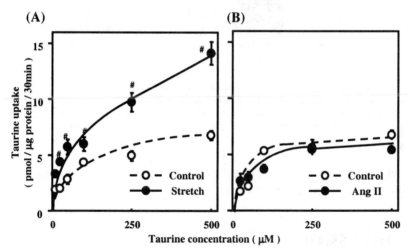

Figure 3. Changes in taurine uptake activity of hypertrophic cardiomyocytes stimulated for 24 hr by mechanical stretching (120%, 60 cycles/min) (A) or Ang II (100 nM) (B). Cells were incubated for 30 min with uptake buffer containing different concentrations of taurine (10~500 μM) and ^{3}H-taurine (0.1 μCi) for 30 min. Results were obtained from at least 3 separate experiments. #P<0.05 vs. control. Data were mean ± SE.

3.3   Expression Level of Taurine Transporter mRNA in Hypertrophic Cardiomyocytes

As shown in Figure 4, the levels of TAUT mRNA in cardiomyocytes subjected to 24 hr of mechanical stretching increased 2-fold relative to the control cells, but not after stimulation with Ang II for 24 hr.   This result indicates that

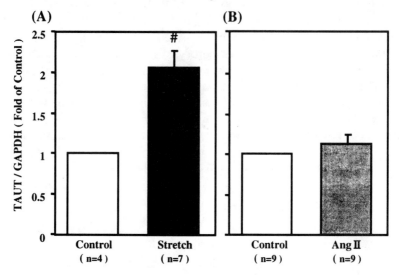

Figure 4.   Change in TAUT mRNA levels in hypertrophic cardiomyocytes.   After stimulation by mechanical stretching (120%, 60 cycles/min) (A) or with Ang II (100 nM) (B) for 24 hr, total RNA was isolated, and Northern blots were performed.   Results were obtained from 3-4 separate experiments.   #P<0.05 vs. control.   Data were mean ± SE.

this activation of taurine uptake by cell stretching involves an upregulation of the high-affinity taurine transporter at the transcriptional level.

4.   DISCUSSION

Taurine content is increased in hypertrophic left ventricles of patients suffering from congestive heart failure[2].   However, the cause of the taurine increase in the failing heart has not been determined.   Reasoning that it might be associated with the development of myocyte hypertrophy, taurine content was determined in cells either subjected to mechanical stretching or exposed to medium containing angiotensin II.   Mechanical stretch of cultured myocytes leads to hypertrophic responses[10] while Ang II has been implicated in the development of cardiac hypertrophy associated with hemodynamic overload[11]. In this report, we found that the movement of taurine was different in the two hypertrophic stimuli. While intracellular taurine content increased in hypertrophic neonatal rat cardiomyocytes subjected to mechanical stretching, the hypertrophy associated with Ang II treatment did not lead to an elevation in taurine levels. This is interesting because mechanical stretching initiates the secretion of Ang II, which in turn plays a critical role in many stretch-induced hypertrophic responses[12,13], but obviously not in the elevation of cellular taurine content. Cell stretching is a complex process that results in the transcriptional activation of a number of proteins, including TAUT. These results suggest that in the intact heart, pressure overload-induced cell stretching is the likely cause of increased intracellular taurine content.

Our findings illustrate that mechanical stretching increases the maximum velocity for taurine uptake by both the high- and low-affinity taurine transporters. In the mechanical stretch-induced hypertrophic cardiomyocyte, the low- affinity taurine transport system has a 4.5-fold greater transport capacity for taurine than the high-affinity taurine transport system. Therefore, we suggest that the low-affinity taurine transport system may play an important role in controlling the intracellular taurine content during mechanical stress.

From these data we conclude that the myocardial taurine transporters are responsive to mechanical stress. This is important because taurine exerts several critical actions in cardiac tissue, including the modulation of calcium movement, the stabilization of the cell membrane, and the regulation of cellular osmolality[14]. Indeed, the activation of myocardial taurine transport systems may be a physiological adaptation, resisting dysfunction caused by mechanical stress.

REFERENCES

1. Huxtable, R. J., Sebring, L., 1982, Cardiovascular actions of taurine. *Sulfur Amino Acids* 5:29-51.
2. Huxtable, R. J., and Bressler, R., 1974, Taurine concentrations in congestive heart failure. *Science* 184:1187-1188.
3. Azuma, J., Sawamura, A., and Awata, N., 1992, Usefulness of taurine in chronic congestive heart failure and its prospective application. *Jpn. Circ. J* 56:95-99.
4. Uchida, S., Kwon, H.M., Preston, A.S., and Handler, J.S., 1991, Expression of Madin-Darby canine kidney cell Na$^+$-and Cl$^-$-dependent taurine transporter in Xenopus laevis oocytes. *J. Biol. Chem* 266:9605-9609.
5. Takahashi, K., Azuma, M., Yamada, T., Ohyabu, Y., Takahashi, K., Schaffer, S., and Azuma, J., Taurine transporter in primary cultured neonatal rat heart cells. *J. Cardiovasc. Pharmcol* in press.
6. Kijima, K., 1996, Mechanical stretch induces enhanced expression of angiotensin II receptor subtypes in neonatal rat cardiac myocytes. *Circ. Res.* 79:887-897.

7. Chomczynski, P., and Sacchi, N., 1987, Single-step method of RNA isolation by acid guanidium thiocyanate-phenol-chloroform extraction. *Anal. Biochem* 162: 156 - 159.
8. Kim, S., Ohta, K., Hamaguchi, A., Yukimura, T., Miura, K., and Iwao, H., 1995, Angiotensin II induces cardiac phenotypic modulation and remodeling in vivo in rats. *Hypertension* 25: 1252-1259.
9. Takahashi, K., Azuma, M., Taira, K., Baba, A., Yamamoto, Y., Schaffer, S. W., and Azuma, J., 1997, Effect of taurine on anigotensin II-induced hypertrophy of neonatal rat cardiac cells. *J. Cardiovasc. Pharmacol* 30 : 725 - 730.
10. Sadoshima, J., 1992, Molecular characterization of the stretch-induced adaptation of cultured cardiac cells. An in vitro model of load-induced cardiac hypertrophy. *J. Biol. Chem* 267:10551-10560.
11. Baker, K. M., Chernin, M. I., Wixson, S. K., and Aceto, I. F,. 1990, Renin-angiotensin system involvement in pressure-overload cardiac hypertrophy in rats. *Am. J. Physiol* 259: H324-H332.
12. Sadoshima, J., Xu, Y., Slayter, H. S., and Izumo, S., 1993, Autocrine release of angiotensin II mediates stretch-induced hypertrophy of cardiac myocytes in vitro. *Cell* 75:977-984.
13. Yamazaki, T., Komuro, I., Kudoh, S., Zou, Y., Shiojima, I., Mizuno, T., Takano, H., Hiroi, Y., Ueki, K., Tobe, K., Kadowaki, T., Nagai, R., and Yazaki, Y., 1995, Angiotensin II partly mediates mechanical stress-induced cardiac hypertrophy. *Circ. Res* 77:258-265.
14. Schaffer, S. W., Solodushko, V., and Azuma, J., 2000, Taurine-deficient cardiomyopathy: role of phospholipids, calcium and osmotic stress. *Adv. Exp. Med. Biol* 483:57-69.

Effect of Taurine and Coenzyme Q10 in Patients with Acute Myocardial Infarction

R.B. SINGH[*], K. KARTIKEY[*], A.S. CHARU[*], M.A. NIAZ[*], and
S. SCHAFFER[#]

[*]Medical Hospital and Research Centre, Moradabad, India, School of Medicine, [#]University of South Alabama, College of Medicine, Mobile, AL, USA

1.   INTRODUCTION

Reperfusion induced free radical stress, lipid peroxidation and a deficiency in antioxidant vitamins, minerals and coenzyme Q10 are common during acute myocardial infarction (AMI)[1-3]. These alterations are associated with cardiac necrosis, resulting in arrhythmias, myocardial dysfunction and coronary thrombosis which continue as a chain reaction for several days after the infarction[3,4]. A fall in the amino acid content of the heart, with a consequent rise in the blood, is seen in experimental studies after ligation of the coronary artery and during both hypoxia and drug-induced cardiac necrosis[5-7]. Due to its loss from the cardiac cell, the concentration of taurine in blood is also raised in patients with AMI, unstable angina and after cardiac surgery[8-10]. Experimental studies in cats indicate that feeding a taurine deficient diet can predispose the animal to a cardiomyopathy[11]. Nonetheless, taurine depletion during ischemia appears to be a novel mechanism for cardioprotection from regional ischemia[12].

Taurine administration exerts a positive ionotropic effect which is independent of adenosine 3,5-cyclic monophosphate and Na^+-K^+ adenosine triphosphatase[13]. Treatment with taurine protects against injury in several models of heart failure, including the calcium paradox, cardiomyopathic hamster, isoproterenol cardiotoxicity and doxorubicin induced cardiac damage[12-15]. However, the most useful effect of taurine has been observed in heart failure, in which taurine administration can reduce cardiac dysfunction and mortality[16].

Coenzyme Q has been demonstrated to enhance cell membrane stabilization in vitro and to exert bioenergetic and antioxidant effects by acting as a free radical scavenger[4,17]. Coenzyme Q is a rate limiting factor in mitochondrial respiratory activity. Its deficiency may result in an intracellular decrease in ATP whereas its supplementation is useful in AMI and heart failure[4,17]. Since taurine and coenzyme Q10 have independent beneficial effects in heart failure, angina pectoris and AMI, it is possible that combined therapy with both the agents may have synergistic effects. In the present study, we examine the effects of such a combination in the prevention of cardiac necrosis and dysfunction in patients with AMI.

2.   SUBJECTS AND METHODS

All the patients with proven diagnosis of AMI, possible AMI and unstable angina admitted in the cardiac care unit of the Medical Hospital and Research Centre, Moradabad, India were included in this study. AMI was diagnosed in the presence of symmetric ST-segment elevation of 1 mm from baseline in limb leads or of 2 mm from baseline in chest leads, or T wave inversions with or without Q waves in a 12-lead electrocardiogram in association with an increase in creatinine phosphokinase of at least twice the upper limit of the normal value on serial examination. Possible AMI was diagnosed by the presence of convincing history of cardiac chest pain accompanied by an increase of less than twice the upper limit of the normal value of cardiac enzymes and electrocardiographic abnormalities that were suggestive of AMI. Unstable angina was diagnosed by cardiac chest pain lasting for >30 minutes without a significant increase in enymes, the absence of diagnostic, electocardiographic changes, or a combination of the two. Patients were excluded if they had cardiogenic shock, blood urea >40 mg/dl, death before randomization, stable angina and noncardiac chest pain.

All the patients were supplied test agents in identical bottles from our laboratory in both the groups. The placebo contained a route powder which was unlikely to provide any benefit or adverse effect to AMI patients. The active agents taurine and coenzyme Q10 were mixed and filled in sachets to provide 2g thrice daily (6.0g/day) of taurine and 40mg thrice daily of coenzyme Q10 (120mg/day).

After a written informed consent, all the patients with AMI and unstable angina were randomized by the pharmacist to receive either taurine+coenzyme Q10 or placebo. The study was blinded for physicians examining the patients and technicians analyzing the blood. All patients were stratified into anterior or posterior wall infarction before randomization with the help of computer generated numbers. All other treatments concerned with the clinical situation were similar in the two groups during 28 days of follow up. Clinical, electrocardiographic, radiological and laboratory data were recorded during hospitalization. All of the patients were monitored on lead II of the electrocardiogram for 48-72 hours and remained in the hospital for 7 days and then followed-up weekly for 28 days. Clinical data, complications and drug intake were recorded by an interviewer at the end of follow-up of 28 days; the interviewer was blind to the groupings. A fasting venous blood sample was taken at entry and later on in the morning and was analyzed for blood urea, glucose, vitamins E and C and beta-carotene, thiobarbituric acid reactive substances (TBARS), malondialdehyde (MDA) and conjugated dienes, creatine phosphokinase, creatine phosphokinase-MB, lactate dehydrogenase (cardiac isoenzyme) by standard methods followed in our laboratory[4].

Statistical analyses were performed by two sample t-test using one-way analysis of variance and by Z score test for proportions. Only p values less than 0.05 and a two tailed t-test were considered significant.

3. RESULTS

We randomized 51 patients with AMI or unstable angina to an intervention or control group as shown in Table 1. There was no significant difference in mean age, body mass index and sex between the two groups. Approximately 22% of the subjects were females. History of previous AMI, angina pectoris and hypertension, current and ex-smokers and concerned drug therapy showed no significant difference between the groups. Delay from onset of symptoms of AMI to the beginning of treatment was also comparable in the two groups. All remaining characteristics, such as final diagnosis, extent of myocardial disease, complications and drug therapy administered on admission showed no significant difference between the two groups (Table 1).

Table 2 reveals that mean infarct size was significantly smaller in the taurine group compared to the placebo group, as calculated from levels of gram-equivalent creatine phosphokinase and creatine phosphokinase-MB in the two groups. Electrocardiographic assessment of infarct size also showed smaller infarcts in the intervention group than in control group. The sum of Q or R waves, the site anterior or inferior to the necrosis, the amplitude and

Table 1. Characteristics of randomized subjects at entry to the study.

Data	Taurine+Coenzyme Q10 (n=25)	Placebo (n=26)
Men	19 (76.0) n(%)	21 (80.7)
Body mass index (Kg/m2)	23.6±1.4	23.4±1.4
Previous acute myocardial infarction.	2 (8.0)	2 (7.7)
Previous angina pectoris	5 (20.0)	1 (3.8)
Known systemic hypertension (>140/90mmHg)	6 (24.0)	7 (26.7)
Current smokers	8 (32.0)	10 (38.4)
Ex-smokers	7 (28.0)	6 (23.1)
Drug intake		
Metaprolol (50-300mg/day)	2 (8..0)	3 (11.5)
Nifedipiñe (10-30mg/day)	3 (12.0)	2 (7.7)
Aspirin (75-150mg/day)	10 (40.0)	8 (30.7)
Final diagnosis.		
Q-wave myocardial infarction	18 (72.0)	15 (57.7)
Non Q-wave myocardial infarction.	4 (16.0)	7 (26.7)
Unstable angina	3 (12.0)	4 (15.4)
Extent of disease.		
Anterior and universal	14 (56.0)	16 (61.4)
Posterior/inferior	6 (24.0)	7 (26.7)
Elapsed time from symptom onset to Intervention (hrs)	11.2±3.5	10.7±3.3
Complications.		
Ventricular ectopic beats	4 (16.0)	3 (11.5)
Left ventricular hypertrophy	2 (8.0)	2 (7.7)
Peak creatinine phosphokinase (IU)	710±82	725±90
Drug therapy given on admission.		
Streptokinase (0.75-1.5million IU)	1.22±0.2	1.27±0.3
Metaprolol (50- 300mg/day)	150±28.6	200±39
Diltiazem (60-160mg/day)	90±15.5	85±12
Nitrates (20-60mg/day)	45±5.8	50±6.5
Aspirin (75-150mg/day)	100±15	110±18

Values are number (%) and mean±standard deviation.

changes in the curves during the surveillance were significantly fewer in the taurine group than the placebo group.

Baseline levels of lactate dehydrogenase showed no significant difference between the two groups, however, the enzyme level on the 6th day was significantly lower in the taurine group compared to the placebo group as shown in Table 3. Thiobarbituric acid reactive substances (TBARS), malondialdehyde (MDA), and conjugated dienes, which are indicators of free radical stress, were comparable at entry to the study in the two groups. After treatment, these parameters showed a significant decline in the intervention group compared to the control group. Plasma levels of the antioxidant vitamins E and C and beta-carotene showed a significantly greater increase in the taurine group than the control group.

Table 2. Cardiac enzyme and electrocardiographic data showing infarct size.

Data	Taurine (n=25)	Placebo (n=26)
Creatine phosphokinase (CPK) size of necrosis (G equivalent)	80.6±15.5	115.5±25.6
Maximal time (Minutes latent period before enzyme peak)	1075±225*	1450±250
Maximal height of enzyme peak concentration (IU/L)	1.30±0.32*	1.87±0.85
Area under the curve	2855±600*	4150±688
CPK-MB size of necrosis.	52.2±8.2*	75.5±11.5
Maximal time (Minutes, latent period before enzyme peak)	950±198*	1208±205
Maximal height of enzyme peak concentration (IU/L)	1.22±0.31*	1.57±0.42
Area under the curve	2312±505*	3150±606
QRS score on electrocardiogram	7.0±1.1**	12.8±1.9

Values are means±standard deviation; * =P<0.05 **=0.02; P values were obtained by comparison of taurine and placebo group by using analysis of variance.

Table 3. Plasma levels of biochemical data at entry and changes after treatment.

Data	Taurine (n=25)		Placebo (n=26)	
	At entry	After 28 days	At entry	After 28days
LDH (IU) (6th day)	105±10.6	198±20.5*	102±8.8	278±31.5
TBARS (pmol/L)	2.56±0.52	1.26±0.30**	2.45±0.56	2.25±0.41
MDA (pmol/L)	1.82±0.32	0.95±0.21**	1.91±0.33	1.56±0.31
Conjugated dienes (units)	29.0±4.1	24.2±3.6*	29.5±4.2	28.6±4.0
Vitamin E (umol/L)	15.5±2.6	25.6±3.5**	14.6±2.5	17.0±3.0
Vitamin C (umol/L)	5.1±0.72	17.6±3.6**	5.4±0.75	9.6±2.1
Beta-carotene (umol/L)	0.19±0.03	0.49±0.07**	0.18±0.3	0.25±0.08

Values are mean±standard deviation, *=P<0.05, **=P<0.02; P values were obtained by comparison of taurine group with placebo group after treatment by analysis of variance. LDH=Lactate dehydrogenase, MDA=Malondialdehyde; TBARS=Thiobarbituric acid reactive substances.

Table 4. Complications in the intervention and control groups.

Complications	Taurine (n=25)	Placebo (n=26)
Prolonged Q-T interval	2 (8.0)	8 (30.7)
Angina pectoris	2 (8.0)	6 (23.1)
NYHA class III and IV heart failure	1 (4.0)	4 (15.3)
Ventricular ectopics (>8/minute)	1 (4.0)	4 (15.3)
Total complications	6 (24.0)*	21 (80.7)
Cardiac events.(Deaths+reinfarction)	4 (12%)	8 (30.8)

Values are number (%). P values were obtained by comparison of taurine group with placebo group by Z score test for proportions.* P =<0.05

Prolonged QT-interval, angina pectoris, NYHA class III and IV heart failure and ventricular ectopics were significantly lower in the taurine group compared to the control group. Cardiac events (12 vs 31%), including cardiac deaths (1 vs 3) and reinfarctions (2 vs 5), were not significantly different in the two groups (Table 4).

4.   DISCUSSION

This study shows for the first time that treatment with taurine (6g/day) + coenzyme Q10 (120mg/day) was associated with a significant reduction in complications, including angina pectoris, left ventricular dysfunction, arrhythmias and prolonged QT interval in the intervention group compared to the control group (Table 4). Cardiac events were slightly less in the intervention group than the placebo group (12 vs 31%). These benefits appear to be due to the cardioprotective effects of taurine and coenzyme Q10 against myocardial dysfunction as a result of the loss of both taurine and coenzyme Q10 from the heart. Recent clinical and experimental studies indicate a protective role of taurine and coenzyme Q10 in congestive heart failure, hypertension and ischemia-induced reperfusion injury[15-24]. A comprehensive review of the literature reveals that the first published clinical studies with taurine were Italian-language reports indicating that taurine supplementation might be beneficial in angina pectoris, intermittent claudication and psychological dysfunction associated with cerebral atherosclerosis[22]. In one randomized, double blind crossover trial among 62 patients with heart failure, treatment with taurine (6g/day) for 4 weeks was associated with a highly significant improvement in dyspnea, palpitation, crackles, edema, cardiothoracic ratio and NYHA class III and IV heart failure[13].

Treatment with the test agents was administered within a mean of 11 hours after the onset of clinical manifestation of AMI, indicating that taurine and coenzyme Q10 must have provided rapid protection against ischemia-induced cardiac damage due to the antioxidant and free radical scavenging effects of taurine and coenzyme Q10. Our study showed that there was a significant decrease in the CPK size of necrosis, as well as CPK-MB size of necrosis, in the taurine group compared to the control group. The QRS score for infarct size also showed a significant decline in the intervention group compared to the control group (Table 2).

Taurine regulates a number of functions, including cardiac rhythm, cardiac contractility, blood pressure, platelet aggregation, neuronal excitability, memory function, motor behaviour, food consumption, vision, cell proliferation, viability, energy metabolism and neutrophil adhesion and activation. The majority of these actions are associated with alterations in ion transport or protein phosphorylation[18,22,24]. Therefore the modulation of ion transport appears to be the most important action of taurine in providing cardioprotection. It is also possible that taurine inhibits hyperinsulinemia common in AMI patients and modulates the suprachiasmatic nucleus responsible for brain-heart interactions, leading to the inhibition of melatonin, catecholamines, cortisol, serotonin and other important factors involved in the pathogenesis of cardiovascular events[15,17]. The loss of cardiac taurine may be a mechanism of adaptation in an attempt to provide cardioprotection[24]. Taurine also exhibits antioxidant activity against hypochlorous acid and suppresses the activity of neutrophils, which may synergize with the antioxidant activity of coenzyme Q10. Interestingly, the levels of vitamins E and C and beta-carotene were significantly higher in the taurine group following the myocardial infarction than in the control group (Table 3). This may also contribute to the reduction in oxidative stress in the taurine group.

In brief, the findings indicate that treatment with taurine + coenzyme Q10 can provide cardioprotective benefits in patients with AMI. Larger, long-term follow-up studies would be necessary for confirmation of our results.

ACKNOWLEDGEMENTS

We appreciate the assistance of Mr Raj Chopra, president of Tishcon Corporation, Westbury, NY, USA, who provided us with coenzyme Q10 and taurine for the trial.

REFERENCES

1. Ceremuzynski, L., 1981, Hormonal and metabolic reactions evoked by acute myocardial infarction. *Circ. Res.* 48:765-76.
2. Singh, R.B., Niaz, M.A. ,Rastogi, S.S., Sharma, J.P., Kumar, Bishnoi, I., Beegum, R., 1994, Plasma levels of antioxidant vitamins and oxidative stress in patients with suspected acute myocardial infarction. *Acta Cardiol.* 49:441-52.
3. Grech, E.D., ,Jackson, M., Ramsdale, D.R., 1995, *Br. Med. J.* 310:477-78.
4. Singh, R.B., Wander, G.S., Rastogi, A., Shukla, P.K., Mittal, A., Sharma, J.P., Mehrotra, S.K., Kapoor, R., Chopra, R.K., 1998, Randomized,double blind,placebo controlled trial of coenzyme Q10 in patients with acute myocardial infarction. *Cardiovasc. Drug Ther.* 12:347-53.
5. Crass, M.F., Lombardini, J.B., 1978, Release of tissue taurine from the oxygen deficient perfused rat heart. *Proc. Soc. Exp. Ther.* 157:486-88.
6. Crass, M.F., Lombardini, J.B., 1977, Loss of cardiac muscle taurine after acute left ventricular ischaemia. *Life Science* 21:951-58.
7. Lombardini, J.B., 1980, Effects of isoproterenol and methoxamine on the contents of taurine in rat tissue. *J. Pharmacol. Exp. Ther.* 213:399-405.
8. Lombardini, J.B., Cooper, M.W., 1981, Elevated blood taurine levels in acute evolving myocardial infarction. *J. Lab. Clin. Med.* 98:849-59.

9.   Bhatanagar, S.K., Welty, J.D., Al-Ysuf, A.R., 1990, Significance of blood taurine levels in patients with first time acute ischemic cardiac pain. *Int. J. Cardiol.* 27:361-66.

10. Cooper, M.W., Lombardini, J.B., 1981, Elevated blood taurine levels after myocardial infarction or cardiovascular surgery. Is there any significance? *Adv. Exp. Med. Biol.* 139:191-205.

11. Pion, P.D., Kittleson, M.D., Rogers, Q.R., Morris, J.G., 1987, Myocardial failure in cats associated with low plasma taurine:a reversible cardiomyopathy. *Science* 237:764-68.

12. Allo S.N. Bagby, L,, Schaffer, S.W., 1997, Taurine depletion,a novel mechanism for cardioprotection from regional ischemia. *Am. J. Physiol.* 273: H1956-61.

13. Azumo, J., Sawamura, A., Awata, N., Hasegawa, H., Ogura, K., Harada, H., Ohta, H., Yamayuchi, K., Kishimoto, S., Yamagami, T., Ueda, T., Ishiyama, T., 1983, Double blind randomized crossover trial of taurine in congestive heart failure. *Curr. Ther. Res.* 34:543-58.

14. Kramer, J.H., Chovan, J.P., Schaffer, S.W., 1981, Effect of taurine on calcium paradox and ischemic heart failure. *Am. J. Physiol.* 240:H238-H246.

15. Raschake, P., Massoudy, P., Becker, B.F., 1995, Taurine protects the heart from neutrophil-induced reperfusion injury. *Free Rad. Biol. Med.* 19:461-71.

16. Takihara, K., Azuma, J., Awata, N., Ohta, H., Hamaguchi, T., Sawamura, A., Tanaka, Y., Kishimoto, S., Sperelakis, N., 1986, Beneficial effect of taurine in rabbits with chronic congestive heart failure. *Am. Heart J.* 112:1278-84.

17. Kaikonen, J., Toumainen, T.P., Nyyssonen, K., Salonen, J.T., 2002, Coenzyme Q10: absorption, antioxidative properties, determinants and plasma levels. *Free Rad. Res.* 36(Suppl):389-99.

18. Schaffer, S., Takahashi, K., Azuma, J., 2000, Role of osmoregulation in the actions of taurine. *Amino Acids* 19:527-46.

19. Lake, N., deRoode, M., Nattel, S., 1987, Effects of taurine depletion on rat cardiac electrophysiology:in vivo and in vitro studies. *Life Sci.* 40:997-1005.

20. Green, T.R., Fellman, J.H., Eicher, A.L., Pratt, K.L., 1991, Antioxidant role and subcellular location of hyporaurine and taurine in human neutrophils. *Biochem. Biophys. Acta* 1073:91-97.

21. Milei, J.,Ferreira, R., Llesuy, S., 1992, Reduction of reperfusion injury with preoperative rapid intravenous infusion of taurine during myocardial revascularization. *Am. Heart J.* 123:339-45.

22. McCarty, M.F., 1999, The reported clinical utility of taurine in ischemic disorders may reflect a down regulation of neutrophil activation and adhesion. *Med. Hypotheses* 53:290-99.

23. Azuma, M., Takahashi, K., Fukuda, T., Ohyabu, Y., Yamamoto, I., Kim, S., Iwao, H., Schaffer, S.W., and Azuma, J., 2000, Taurine attenuates hypertrophy induced by angiotensin II in cultured neonatal rat cardiac myocytes. *Eur. J. Pharmacol.* 403: 181-188,

24. Schaffer, S.W., Solodushko, V. and Kakhniashvili, D., 2002, Beneficial effect of taurine depletion on osmotic, sodium and calcium loading. *Am. J. Physiol.* 282: C1113-C1120.

Taurine Reduces Renal Ischemia/Reperfusion Injury in the Rat

D.V. MICHALK[1], B. HOFFMANN[1], and Th. MINOR[2]
[1]*Department of Pediatrics, University of Cologne*
[2]*Institute of Experimental Medicine, University of Bonn*

1.   INTRODUCTION

In renal transplantation, the deleterious effect of ischemia/reperfusion injury has a major impact on graft survival by triggering or influencing acute rejection as well as late chronic changes[12]. Thus protective maneuvers at the time of transplantation should ameliorate the fate of the organ.

In previous experiments with renal tubular cell cultures we found a positive effect of taurine on cell viability and survival during hypoxic preservation and after reperfusion due to a reduction of osmotic swelling and maintenance of the intracellular calcium homeostasis[15,16,23,24].

The positive effect of taurine on organ preservation was reassured in isolated perfused kidneys where taurine administered during cold ischemia maintained vascular permeability and ATP-content of the organ and reduced oxygen consumption[13].

The aim of the present study was to investigate the effect of taurine on ischemia reperfusion injury *in vivo.*

2.   MATERIAL AND METHODS

2.1.   Experimental Design

Wistar rats were anaesthetized with injections of ketamine and xylocain. A midline peritoneal incision was made and both renal pedicles were clamped for 60 minutes using non-traumatic vascular clamps. After 90 minutes of reperfusion the animal was sacrificed by blood aspiration from the aorta. Both kidneys were

Taurine 5: Beginning the 21st Century
Edited by Lombardini *et al.*, Kluwer Academic/Plenum Publishers, New York 2003

harvested, the right kidney stored in liquid nitrogen until further investigation and the left kidney fixed with 4% formalin solution (group IR). In the taurine treated group (IR + Tau) 40 mg/kg of taurine was administered intravenously 10 minutes before arterial clamping, leading to a taurine plasma level of about 8 mM/l. Sham-operated animals without pedicle clamping (group Sh-C) served as normal controls (Fig. 1). Each group consisted of 6-8 animals.

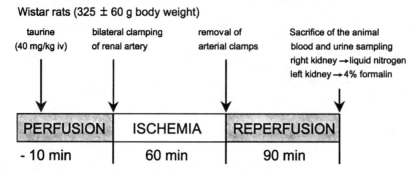

Figure 1. Design of the experiment.

2.2.   Measured parameters

Serum creatinine and urea levels, which are measuares of renal function were determined by standard methods. Serum lactate dehydrogenase (LDH) and γ-glutathione-S-transferase (γ-GST), measured enzymatically or by enzyme-immunoassay (EIA) served as parameters for tubular damage. ATP, ADP and AMP content of kidney extracts, which serve as measures of energy metabolism, were determined according to Warburg. As parameters for NO production and synthesis of prostaglandins, the mRNA of inducible nitric oxide synthase (iNOS) and cyclooxygenase 1 and 2 (Cox 1 and 2) were measured in kidney specimens by reverse transcriptase polymerase chain reaction (RT-PCR) using RNeasy mini kit (Qiagen GmbH) for RNA extraction and a light cycler (Roche) for cDNA determination. mRNA of glyceroaldehyd-3-phosphate-dehydrogenase (GAPDH) served as internal reference standard.

2.3.   Statistics

Statistical analysis was obtained by unidirectional variance analysis (ANOVA) and the least significant difference post hoc test. Data are given in mean $\pm$ standard deviation. Differences were considered to be significant at $p < 0.05$.

3.   RESULTS

3.1   Renal Function and Tubular Damage

Despite the short period of ischemia (60 min.) and reperfusion (90 min), serum creatinine was markedly increased in both the untreated and taurine treated animals (Fig.2). However, the increase was about 25% less in the taurine-treated animals, indicating less severe renal failure in the taurine group. There was a modest increase in serum urea after IR in both groups, but it was not significant, probably because the period of renal failure was too short to affect urea synthesis.

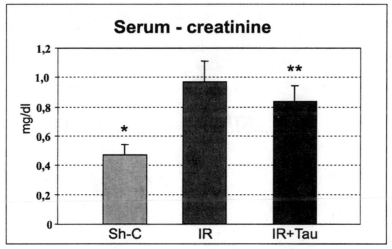

* p<0.001 Sh-C vs IR and IR+Tau; ** p<0.05 IR+Tau vs IR

Figure 2. Serum concentration of creatinine after 60 min. of   renal ischemia and 90 min.of reperfusion without (IR) or with (IR+Tau) taurine administration (40 mg/kg bodyweight i.v.). Mean ± SD. Sh-C = sham-operated control.

The serum levels of the renal-tubular enzymes LDH and   γ-GST were elevated more than twice in serum of the IR-group compared to sham-operated controls indicating significant damage of the renal tubular cells by ischemia and reperfusion. This damage was almost prevented by the administration of taurine, as there was no significant difference between the Sh-C and IR + Tau animals (Table 1).

Table 1. Serum concentration of the renal tubular enzymes
lactate dehydrogenase (LDH) and α-glutathione-S-transferase (α-GST)

Animal group	LDH U/l	α-GST µg/l
sham-operated control	524 ± 193	256 ± 109
ischemia/ reperfusion	1219[a] ± 84	590[a] ± 151
ischemia/ reperfusion + taurine	787[b] ± 384	374[c] ± 68

[a] $p < 0.001$ IR vs Sh-C; [b] $p < 0.01$ IR+Tau vs IR; [c] $p < 0.05$ IR vs IR+Tau

The ATP-content was reduced by about 50% in the kidneys of the animals subjected to IR, but there was no difference between the untreated and taurine treated groups (Fig. 3). The renal tissue content of ADP and AMP was similar in all groups (Table 2).

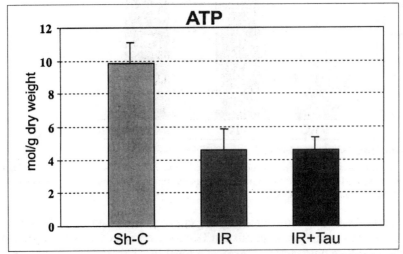

Figure 3. ATP-content of rat kidneys after 60 min. of renal ischemia and 90 min.of reperfusion without (IR) or with (IR+Tau) taurine administration (40 mg/kg bodyweight i.v.). Mean ± SD. Sh-C = sham-operated control.

Table 2. Kidney content of phospho-adenine nucleotides

Animal group	ATP µM/g dry wt.	ADP µM/g dry wt.	AMP µM/g dry wt.
shamoperated control	9.88[a] ± 1.26	2.79 ± 0.71	1.09 ± 1.24
ischemia/ reperfusion	4.58 ± 1.26	2.15 ± 0.97	0.86 ± 0.43
ischemia/ reperfusion + taurine	4.53 ± 0.82	2.22 ± 1.08	0.58 ± 0.36

[a] p< 0.01 Sh-C vs. IR and IR+ Tau

3.2   iNOS and COX 1 and 2

The mRNA of iNOS was significantly increased only in those animals who received taurine prior to ischemia and reperfusion.

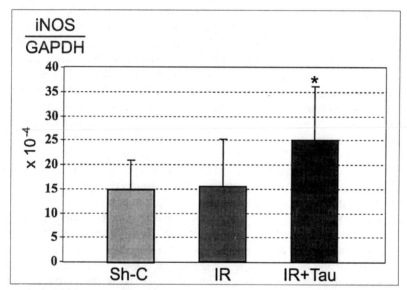

* p<0.05 IR+Tau vs Sh-C and IR

Figure 4. mRNA-expression of iNOS in rat kidneys after 60 min. of renal ischemia and 90 min. of reperfusion without (IR) or with (IR+Tau) taurine administration (40 mg/kg bodyweight i.v.). Mean ± SD. Sh-C = sham-operated control.

A similar trend was observed for the induction of COX 2 only in the taurine treated animals, but this increase did not reach significance because of a high SD. In contrast there was no change in COX 1 - mRNA expression after ischemia and reperfusion.

Table 3. mRNA-expression of iNOS, COX 1 and 2 in rat kidneys

Animal group	iNOS/GAPDH $\times 10^{-4}$	COX 1/GAPDH $\times 10^{-4}$	COX 2/GAPDH $\times 10^{-4}$
shamoperated control	15.23 ± 5.83	285 ± 174	348 ± 162
ischemia/ reperfusion	15.70 ± 9.44	190 ± 94	347 ± 231
ischemia/ reperfusion + taurine	26.56[a] ± 10.17	243 ± 190	504 ± 339

[a] p< 0,05 IR + Tau vs Sh-C and IR

4.   DISCUSSION

In addition to *ex vivo* experiments from our group[13,15,16,23,24] and others[1,17], the present data show for the first time that taurine is able to reduce cellular injury induced by ischemia and reperfusion under *in vivo* conditions. A similar protective effect of taurine was seen in rats in which renal tubular necrosis was induced by gentamicin[2] or cisplatin [19]. All routes of taurine application (orally, intraperitoneal or intravenously) seem to be equally effective. Moreover, chronic taurine treatment also improved renal function in uni-nephrectomized rats[14]. Although the mechanism underlying taurine mediated protection are only partly known   (osmoregulation, calcium homeostasis, membrane stabilizing, cell proliferation[7,15]), it is worth noting that taurine seems to counterbalance de-arrangements of cellular integrity irrespective of the underlying cause (ischemia/ reperfusion, toxicity, etc.). In this respect taurine behaves as an enantiostatic agent, which opposes the change in one physicochemical property by a change in another physiocochemical property in order to maintain physiologic function[8].

Since taurine positively influenced perfusion flow and vascular resistance in isolated perfused kidneys[13], it has been suggested to look for those enzymes which are responsible for the synthesis of the major regulators (NO and prostacyclin) of renal blood flow, *i.e.* nitric oxide synthase and cyclooxygenase.

The increase of iNOS and to a less degree of COX 2 in the taurine treated animals suggests that the protective effect of taurine was partly due to the induction of NO and prostaglandin synthesis. With respect to iNOS-expression our data are surprising; to date, NO has been associated mostly with the mechanisms of ischemia damage[4,9,22]. Therefore, one would have expected an increase in iNOS expression in the untreated rats after IR than in the taurine treated. But there is growing evidence about the role of NO as intrinsic cellular protector[5,18,20]. Besides its vascular effects NO induces two antioxidant and anti-apoptotic proteins, heat shock protein 70 and hemo-oxygenase 1, which attenuate the ischemia/reperfusion injury in the heart[10]. In addition NO downregulates connective tissue growth factors thereby reducing glomerular and interstitial fibrosis[3,11,21]. Whether NO exerts beneficial or detrimental effects after IR seems to be dependent upon the experimental conditions, the microenvironment and the time and extent of induction[4,20]. Also the interaction between NO and oxygen radicals (ROS) seems to be relevant[6,18]. A low continuous NO synthesis with rapid onset and low levels of ROS proved to be beneficial[18,20]. In agreement with those studies, in our relatively short experiment (150 min.) the rapid induction of iNOS in the taurine treated animals was associated with a beneficial effect in ischemia/reperfusion kidney. Whether taurine directly induced iNOS or positively influenced the conditions for a rapid induction remains to be investigated.

REFERENCES

1. Eppler, B., Dawson, R.Jr., 2002, Cytoprotective role of taurine in a renal epithelial cell culture model. *Biochem.Pharmacol* 63:1051-60.
2. Erdem, A., Gundogan, N.U., Usubutun, A., Kilinc, K., Erdem S.R., Kara A., and Bozkurt, A., 2000, The protective effect of taurine against gentamicin-induced acute tubular necrosis in rats. *Nephrol Dial Transplant* 15:1175-82.
3. Ferrini, M.G., Vernet, D., Magee, T.R., Shahed, A., Qian, A., Rajfer, J., and Gonzalez-Cadavid, N.F., 2002, Antifibrotic role of inducible nitric oxide synthase. *Nitric Oxide* 3:283-94.
4. Goligorsky, M.S., Brodsky, S.V., and Noiri, E., 2002, Nitric oxide in acute renal failure: NOS versus NOS. *Kidney Int.* 61:855-861.
5. Garcia-Criado, G., Eleno, N., and Santos-Benito, F., 1998, Protective effect of exogenous nitric oxide on the renal function and inflammatory response in a model of ischemia-reperfusion. *Transplantation* 27:982-990.
6. Halliwell, B., and Guttridge, J.M.C., 1999, *Free radicals in biology and medicine.* Oxford University press, Oxford. pp. 73-82.
7. Huxtable, R.J., 1992, Physiological actions of taurine. *Physiol Rev* 72:101-163.
8. Huxtable, R.J., 2000, Expanding the circle 1975-1999; Sulfur biochemistry and insights on the biological functions of taurine. *Adv. exp. med. biol.* 483:1-25.
9. Joles, J.A., Vos, I.H., Gröne, H.-J., and Rabelink T.J., 2002, Inducible nitric oxide synthase in renal transplantation. *Kidney Int.* 61: 872-875.
10. Katori, M., Tamaki, T., and Takahashi, T., 2000, Prior induction of heat shock proteins by a nitric oxide donor attenuates cardiac ischemia/reperfusion injury in the rat. *Transplantation* 69:2530-2537.
11. Keil, A., Blom, I.E., Goldschmeding, R., and Rupprecht, H.D., 2002, Nitric oxide down-regulates connective tissue growth factor in rat mesangial cells. *Kidney Int.* 62:401-411.
12. Land, W., and Messmer, K., 1996, The impact of ischemia/reperfusion injury on specific and non-specific early and late chronic events after organ transplantation. *Transplant Rev* 10:108-127.
13. Licht, Ch., Kriegesmann, E., Minor, T., Wingenfeld, P., Isselhard, W., and Michalk, D.V., 1998, Influence of Taurine Supplementation on Ischemic Preservation of the Isolated Rat Kidney. *Adv. exp. med. biol.* 442:201-206.

D.V. Michalk et al.

14. Mozaffari, M.S., and Schaffer St.W., 2002, Chronic taurine treatment ameliorates reduction in saline-induced diuresis and natriuresis. *Kidney Int.* 61: 1750-1759.
15. Michalk, D.V., Wingenfeld, P, and Licht Ch., 1997, Protection against cell damage due to hypoxia and reoxygenation: the role of taurine and the involved mechanisms, *Amino Acids* 13:337-346.
16. Minor T., Wingenfeld P., Gehrmann U., Strübind S., Isselhard, W., and Michalk D., 1994, Energetic status and viability of porcine kidney cells after hypoxic hypothermic preservation in UW solution and subsequent reoxygenation: the influence of taurine. *Pathophysiology* 1:247-250.
17. Oz, E., Erbas, D., Gelir, E., and Aricioglu, A., 1999, Taurine and calcium interaction in protection of myocadium exposted to ischemic reperfusion injury. *Gen Phamacol* 33:137-41.
18. Pautz, A., Franzen, R., Dorsch, S., Böddinghaus, B., Briner, V.A., Pfeilschifter, J. and Huwiler, A. 2002, Cross-talk between nitric oxide and superoxide determines ceramide formation and apoptosis in glomerular cells. *Kidney Int.* 61:790-796.
19. Saad, S.Y., and Al-Rikabi, A.C., 2002, Protection Effects of Taurine Supplementation against Cisplatin-Induced Nephrotoxicity in Rats. *Chemotherapy* 48: 42-8.
20. Torras, J., Herrero-Fresneda, I., Lloberas, N., Riera, M., Cruzado J.M., and Grinyó, J.M., 2002, Promising effects of ischemic preconditioning in renal transplantation. *Kidney Int.* 61:2218-2227.
21. Trachtman, H., Futterweit St., Pine, E., Mann, J., and Valderrama E., 2002, Chronic diabetic nephropathy: role of inducible nitric oxide synthase. *Pediatr. Nephrol.* 17:20-29.
22. Weight S.C., Furness P.N., and Nicholson, M.L., 1998, Nitric oxide generation is increased in experimental renal warm ischemia-reperfusion in jury. *Br. J. Surg.* 85:1663-1668.
23. Wingenfeld P., Minor, T., Gehrmann U., Strübind S., Isselhard, W., and Michalk D., 1994, Protecting effect of taurine against hypoxic cell damage in renal tubular cells cultured in different preservation solutions. *Adv. exp. med. biol.* 403:159-169.
24. Wingenfeld P., Minor, T., Gehrmann U., Strübind S., Isselhard, W., and Michalk D., 1995, Hypoxic cellular deterioration and its prevention by an amino acid taurine in a transplantation model with renal tubular cells (LLCPK1). *In Vitro Cell Dev Biol* 31.:483-486.

Taurine-Depleted Heart and Afterload Pressure

MAHMOOD S. MOZAFFARI, NISHA PATEL, and
STEPHEN W. SCHAFFER*
*Department of Oral Biology & Maxillofacial Pathology, Medical College of Georgia, Augusta, Georgia and * Department of Pharmacology, School of Medicine, University of South Alabama, Mobile, Alabama, USA*

1.   INTRODUCTION

Taurine is an ubiquitous amino acid with many putative actions, including the modulation of Na^+ and Ca^{++} homeostasis and the alteration in membrane structure and function[1-2]. These actions of taurine impact myocardial function, as evidenced by the development of heart failure in cats nutritionally depleted of taurine[3]. According to Novotny et al.[4], diastolic dysfunction is the initial myocardial defect observed in the taurine deficient cat. Indeed, isolated neonatal cardiomyocytes with reduced taurine levels show a delay in the decay phase of the calcium transient[1]. The resulting prolongation in the calcium transient correlates with the impairment in relaxation seen in the taurine deficient heart[1]. However, less clear is the mechanism underlying the progression from diastolic dysfunction to a prominent defect in systolic function. Several factors may contribute to the taurine-linked decline in systolic function. First, cell loss through apoptosis is considered an important event in the development of overt heart failure[5-6]. Second, impaired handling of calcium could contribute to the decline in systolic function[7]. Third, defects in muscle protein function could be a cause of contractile dysfunction[8]. Fourth, the presence of a coexisting disorder could contribute to heart failure.

Prominent among the disorders that can exacerbate heart failure is systemic hypertension. The decline in systolic function and cardiac output in the setting of coexisting hypertension and cardiac dysfunction can be attributed to both impaired contractile function and the impediment to ventricular outflow[6,9]. Since myocardial function is compromised in the taurine deficiency state, we reasoned that the taurine-depleted heart might show a diminished ability to cope with acute changes in afterload pressure. To test this hypothesis, functional parameters were determined in control and β-alanine-induced, taurine-depleted hearts following sequential changes in afterload pressure.

2.   METHODS

Two days after arrival in the vivarium, male Wistar-Kyoto rats were randomly assigned to two groups: the control group received tap water (n=5; body weight: 189 ± 2g) while the taurine-depleted group received water containing 3% β-alanine for 3 weeks (n=5; body weight: 188 ± 3g)[10-14]. Three weeks after arrival, each animal was heparinized (1000 U/kg, i.p., 30 minutes), decapitated and the heart quickly removed. The spontaneously beating heart was perfused on a Langendorff apparatus with Krebs-Henseleit buffer (37^0 C) containing 118 mM NaCl, 27.1 mM $NaHCO_3$, 2.8 mM KCl, 1 mM KH_2PO4, 1.2 mM $MgSO_4$, 2.5 mM $CaCl_2$ and 11 mM glucose and equilibrated with 95% O_2-5% CO_2[10]. After a 15-min stabilization period at an afterload pressure of 155 cm H_2O (~ 103 mm Hg), two coronary effluents were collected at 19-20 and 29-30 minutes. This was followed by an acute reduction in afterload pressure (< 30 seconds) to 55 cm H_2O (~ 37 mm Hg). Thereafter, additional coronary effluents were collected at 49-50 and 59-60 minutes. The average of the two samples was considered the coronary flow rate during each phase of the perfusion protocol; coronary flow data were normalized for heart weight (HW). Throughout the experiment, ventricular function was monitored via a 23-gauge needle inserted into the left ventricle. The needle was connected to a pressure transducer, which in turn was connected to a computerized heart performance analyzer (MicroMed, Louisville, KY). During each phase of the perfusion protocol, ventricular function data were recorded at five-minute intervals and the average of the measurements was calculated for each functional parameter.

All data were analyzed by the analysis of variance (ANOVA). Duncan's post-hoc test was used to establish the significance between the mean values ($p<0.05$). All data are reported as means ± SEM.

3.   RESULTS

The coronary flow rate was significantly lower in β-alanine-induced, taurine-depleted hearts than control hearts perfused at an afterload pressure of

155 cm H_2O (Figure 1A). An abrupt reduction in afterload pressure from 155 cm H_2O to 55 cm H_2O led to an immediate and marked decline in the coronary flow rate of hearts from both groups (Figure 1A). However, when the data were expressed as percent change, the taurine-depleted group displayed a more marked reduction in the coronary flow rate (25.5 ± 1.2 vs. $37.7 \pm 4.8\%$; p<0.05).

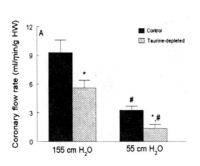 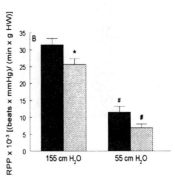

Figure 1. Panel A shows coronary flow rate while panel B shows rate-pressure-product (RPP) values for the control and the taurine-depleted hearts; data are normalized for heart weight (HW). Data are means ± SEM of 5 hearts/group.
* p<0.05 compared to the control group.
# p<0.05 compared to their counterparts at the afterload pressure of 155 cm H_2O.

Heart rate was similar between the two groups prior to the decrease in afterload (241 ± 8 vs. 225 ± 18 beats/min in the control vs. taurine-depleted group). However, at 155 cm H_2O, the maximum ventricular pressure was slightly lower in the taurine-depleted group than the control group (123.1 ± 2.6 vs. 111.6 ± 3.5 mm Hg; p<0.05). As a result, the rate-pressure-product, an index of heart performance, was lower in the taurine-depleted group prior to the change in afterload (Figure 1B). In both groups of hearts, the abrupt reduction in afterload pressure was associated with a marked reduction in heart rate and ventricular pressure, accounting for the dramatic decline in the rate-pressure-product (Figure 1B). However, it is noteworthy that while ventricular pressure was similar between the two groups at the lower afterload pressure (78.4 ± 7 vs. 76.3 ± 4.5 mm Hg), the abrupt decline in afterload resulted in a more marked decline in heart rate in the taurine-depleted heart than in the control group (56.3 ± 4.9 vs. $37.8 \pm 4.9\%$, p<0.05). Nonetheless, the differential in the rate-pressure-product between the two groups did not reach statistical significance at the lower perfusion pressure (p=0.056).

The effect of taurine depletion on positive and negative dP/dt, indices of myocardial contractility and relaxation, is illustrated in Figure 2. Prior to the afterload transition, the control hearts exhibited higher values of positive

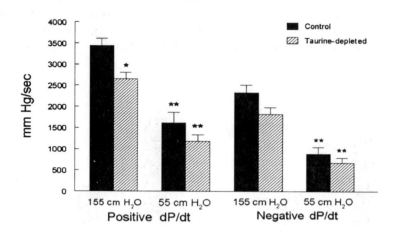

Figure 2. Bar graphs show positive and negative dP/dt values of the control and the taurine-depleted hearts. Data are means ± SEM of 5 hearts/ group.
* p<0.05 compared to the control group.
** p<0.05 compared to their counterparts at the afterload pressure of 155 cm H_2O.

dP/dt than the taurine-depleted hearts; no significant difference was noted in negative dP/dt between the two groups (p=0.0695). The abrupt decrease in afterload pressure to 55 cm H_2O resulted in a marked reduction in these parameters (p<0.05). Although both groups experienced a comparable percentage decrease in positive and negative dP/dt, a significant difference in positive dP/dt between the control and taurine-deficient hearts was not observed at the lower afterload pressure (Figure 2).

4.   DISCUSSION

This study supports the notion that the taurine-deficient heart is less capable of coping with the stress of high afterload pressure than the control heart. As seen in Figure 2, no significant difference in either positive or negative dP/dt was observed in hearts subjected to an afterload of 55 cm H_2O. Nonetheless, a defect in positive dP/dt was seen in the taurine deficient heart at 155 cm H_2O. Similarly, isolated trabeculae of taurine deficient hearts exhibit reduced contractile function[8]. These findings contrast with results obtained from taurine-depleted hearts perfused on a working heart apparatus at an afterload pressure of about 100 cm H_2O[10,12]. The reason for these seemingly variable results presumably relates to the preparation. The muscle fibers of both the working heart and isolated trabeculae undergo considerable degrees of stretching. By contrast, the muscle fibers of hearts perfused in the Langendorff mode at a fixed afterload pressure of 55 cm H_2O would be at a relatively reduced level of mechanical stress. Nonetheless, when the heart was subjected to an afterload pressure of 155 cm H_2O, myofibril stretching assumed an important role. In response to the stretching, myocardial function (i.e. rate-pressure-product, myocardial contractility) was elevated. However, the taurine-deficient heart was unable to adequately cope with the elevation in afterload pressure as indicated by the reduced contractile state.

The mechanism(s) by which taurine depletion affects the complex cascade of events triggered by myofibril stretching remains unclear. The "physical" effect of myofibril stretching includes increased cross-bridge formation and cooperativity, the latter resulting from enhanced affinity of troponin C for calcium. Aside from the "physical" effects on the contractile elements, stretching also augments the activity of stretch-activated ion channels that regulate cytosolic calcium content and promotes the release of autocrine and paracrine factors (*e.g.*, angiotensin II, endothelin, etc.)[15-18]. Of these factors, taurine is known to affect the activity of angiotensin II and modulate cellular calcium metabolism[1-2]. It is less likely to affect the "physical" effect of stretching.

Another noted change accompanying the abrupt reduction in afterload pressure is a decrease in heart rate, an effect that was more prominent in the taurine-depleted heart. Although the anti-arrhythmic activity of taurine is documented[19], this study provides the first evidence that taurine depletion can also affect heart rate. Both stretch and taurine depletion affect the electrical properties of the heart. Stretch activates stretch-dependent ion channels and increases beating frequency of the sino-atrial node, the primary pacemaker of the heart[15-18]. Thus, the bradycardia that occurs following an abrupt decline in afterload pressure involves a reduction in the degree of myofibril stretching. Taurine, on the other hand, modulates ion flux through cation transporters[1-2]. Therefore, it is plausible to suggest that the effect of taurine depletion on heart rate involves some interaction between taurine and stretch-mediated events.

Although contractile function at both high and low afterload pressures was only modestly affected by taurine depletion, taurine loss markedly reduced the coronary flow rate. The apparent uncoupling between the coronary flow rate and contractile function likely relates to the Langendorff preparation. Presumably, in this preparation, the decline in coronary flow rate can be compensated for by greater oxygen extraction from the perfusate, thereby minimizing the effect of reduced coronary circulation.

β-Alanine exerts certain actions on excitable tissues[20, see 2]. Therefore, the increase in coronary vascular resistance in the taurine-depleted heart could be caused by a direct effect of β-alanine on the vasculature rather than an effect of taurine depletion. In this regard it is important to note that β-alanine does not accumulate in the cell, as it is readily metabolized to malonic semialdehyde and eliminated as carbon dioxide[see11]. Moreover, acute exposure of rat aorta to medium containing 40 mM β-alanine does not affect contractile responsiveness to angiotensin II[13]. However, the effects of chronic β-alanine treatment are more closely linked to taurine depletion. In the β-alanine-induced, taurine-depleted rat abnormalities develop that are reversed by combination of β-alanine withdrawal and taurine repletion[10,14,21]. Therefore, the reduction in coronary flow in the β-alanine-treated rat is likely related to taurine depletion. This notion is supported by the demonstration that taurine affects vascular smooth muscle function. Acetylcholine-induced relaxation is augmented in the aorta of the taurine-treated rat while contractile responses to norepinephrine and potassium chloride are attenuated[22]. Therefore, it is likely that taurine depletion alters coronary vascular reactivity to vasoactive substances, such as adenosine, that play a pivotal role in the coronary circulation.

5.   CONCLUSION

The taurine-depleted heart displays a maladaptive response to the extremes of afterload pressure. This was manifested as a marked impairment in coronary flow rate and a modest reduction in contractile function. The functional deficit is likely caused by an interaction of taurine deficiency with the mechanisms that convert stretch-induced mechanical stimulation into excitation-contraction coupling events. The cause of the increase in coronary vascular resistance is unknown, although it is proposed that taurine deficiency alters vascular responses to vasoactive substances.

ACKNOWLEDGEMENTS

This study was supported, in part, by residual funds from a grant by Taisho Pharmaceutical Company. The authors thank Ms. Champa Patel for her expert technical assistance.

REFERENCES

1.   Schaffer, S., Solodushko, V., and Azuma, J., 2000, Taurine deficient cardiomyopathy role of phospholipids, calcium and osmotic stress. *In Taurine 4: Taurine and Excitable Tissues* (L. Della Corte, R.J. Huxtable, G. Sgaragli, and K.F. Tipton, eds), Kluwer Academic Publishers, New York/Boston. Dordrecht/London, Moscow, pp. 57-69.

2. *Taurine 3: Cellular and Regulatory Mechanisms.* 1998. (S. Schaffer, J.B. Lombardini and R.J. Huxtable, eds.) Plenum Press, New York/London.
3. Pion, P.D., Kittleson, M.D., Rogers, Q.R., and Morris, J.G., 1987, Myocardial failure in cats associated with low plasma taurine: a reversible cardiomyopathy. *Science* 237: 764-768.
4. Novotny, M.J., Hogan, P.M., Paley, D.M., and Adams, H.R., 1991, Systolic and diastolic dysfunction of the left ventricle induced by taurine deficiency in cats. *Am. J. Physiol.* 261: H121-H127.
5. Sabbah, H.N., 2000, Apoptotic cell death in heart failure. *Cardiovasc. Res.* 45: 704-712.
6. Ruffolo, R. R., Feuerstein, G., Z., 1998, Neurohormonal activation, oxygen free radicals, and apoptosis in the pathogenesis of congestive heart failure. *Journal of Cardiovascular Pharmacology* 32 (Suppl. 1): S22-S30.
7. Marks, A.R., 2001, Ryanodine receptors/calcium release channels in heart failure and sudden cardiac death. *J. Mol. Cell Cardiol.* 33: 615-624.
8. Eley, D.W., Lake, N., and ter Keurs, H.E.D.J., 1994, Taurine depletion and excitation-contraction coupling in rat myocardium. *Circulation Research* 74: 1210-1219.
9. Teerlink, J. R., 1994, The evolving role of angiotensin-converting enzyme inhibition in heart failure: Expanding the protective envelope. *Journal of Cardiovascular Pharmacology* 24 (Suppl. 3): S32-S37.
10. Allo, S.N., Bagby, L., and Schaffer, S. W., 1997, Taurine depletion, a novel mechanism for cardioprotection from regional ischemia. *Am. J. Physiol.* 273 (Heart Circ. Physiol.) 42: H1956-H1961.
11. Shaffer, J.E., Kocsis, J.J., 1981, Taurine mobilizing effects of beta alanine and other inhibitors of taurine transport. *Life Sciences*, 28: 2727-2736.
12. Mozaffari, M.S., Tan, B.H., Lucia, M.A., and Schafer, S.W., 1986, Effect of drug-induced taurine depletion on cardiac contractility and metabolism. *Biochem. Pharmacol.* 35:985-989.
13. Mozaffari, M.S., and Abebe, W., 2000, Cardiovascular responses of the taurine-depleted rat to vasoactive agents. *Amino Acids* 19: 625-634.
14. Mozaffari, M.S., and Schaffer, D., 2001, Taurine modulates arginine vasopressin-mediated regulation or renal function. *Journal of Cardiovascular Pharmacology* 37: 742-750.
15. Ruwhof, C., and van der Laarse, A., 2000, Mechanical stress-induced cardiac hypertrophy: mechanisms and signal transduction pathways. *Cardiovascular Research* 47: 23-37.
16. Crozatier, B., 1996, Stretch-induced modifications of myocardial performance: from ventricular function to cellular and molecular mechanisms. *Cardiovascular Research* 32: 25-37.
17. Calaghan, S.C., and White, E., 1999, The role of calcium in the response of cardiac muscle to stretch. *Progress in Biophysics & Molecular Biology* 71: 59-90.
18. Yamazaki, T., and Yazaki, Y., 2000, Molecular basis of cardiac hypertrophy. *Z. Kardiol.* 89: 1-6.
19. Lake, N., 1992, Effects of taurine deficiency on arrhythmogenesis and excitation-contraction coupling in cardiac tissue. Taurine, edited by J.B. Lombardini et al. New York: Plenum Press, 173-179.
20. Mori, M., Gahwiler, B.H., Gerber, U., 2002, Beta-alanine and taurine as endogenous agonists at glycine receptors in rat hippocampus in vitro. *J Physiol* 15; 539 (Pt 1): 191-200.
21. Lombardini, J.B., Young, R.S.L., and Props, C.L., 1996, Taurine depletion increases phosphorylation of a specific protein in the rat retina. *Amino Acids* 10: 153-165.
22. Abebe, W., and Mozaffari, M.S., 2000, Effects of chronic taurine treatment on reactivity of the rat aorta. *Amino Acids* 19: 615-623.

Part 2:

Role of Taurine in Diabetes

Taurine Reduces Mortality in Diabetic Rats
Taurine and Experimental Diabetes Mellitus

FLAVIA FRANCONI[§], STEFANO A. SANTINI[#],
NICOLÒ GENTILONI SILVERI[°], SALVATORE CAPUTO[+],
BRUNO GIARDINA[#], GIOVANNI GHIRLANDA[+], and MAURO A.S. DI LEO[°]
Department of Pharmacology and Center for Biotechnology Development and Biodiversity Research[§], University of Sassari, Italy; Departments of °Emergency Medicine, +Internal and Geriatric Medicine, and #Institute of Biochemistry and Clinical Biochemistry, Catholic University, Roma, Italy

1.   INTRODUCTION

Diabetes mellitus affects more that 6% of the US population and its prevalence is increased by 30% in the last decade, dramatically in younger individuals. In fact, it kills more people annually than AIDS and breast cancer combined and when the long-term complications and their costs are considered, the implications of these numbers are sobering[1]. Although, good glycemic control can delay the development and progression of microvascular complications (retinopathy, nephropathy), such metabolic control is often difficult to achieve and to maintain and it is not yet clear whether the same is true for macrovascular complications, such as coronary heart disease, stroke[2]. In this regard, it is important to note that clinical studies show that diabetic patients treated with phenformin and tolbutamide[3] have lower blood glucose levels, but also an increased overall mortality and cardiovascular mortality. Thus, diabetic therapy should lead to a lower incidence of diabetic complications and, ultimately to lower mortality. The logical corollary to this question is which of the new treatment options will affect long-term diabetic mortality. In this contest, the correction of oxidative stress may have important implications in preventing diabetes-induced alterations, since one of the consequences of chronic hyperglycemia is enhanced oxidative stress[3] and reactive oxygen species have been implicated in the pathogenesis of many diabetic complications[4]. In fact, diabetic patients appear to have significant defects in antioxidant protection compared to healthy non diabetic controls[5] and they produce more reactive oxygen species[4].

Therefore, we examined whether taurine and vitamin E plus selenium supplementations prolong survival in diabetic animals. Taurine has been proposed for therapy of diabetes mellitus[6,7] and it is beneficial in relatively short clinical and experimental studies which do not reflect the duration of illness[8-13]. In addition, clinically, both type 1 and type 2 diabetic patients have lower plasma taurine in comparison with healthy individuals[6,14]. Moreover, the administration (3 months) of taurine normalizes plasma and platelet levels and reduces platelet aggregation[6]. However, in a double–blind one-year duration trial taurine failed to delay or prevent kidney complications associated with type 2 diabetes[15]. Vitamin E, in the form of alpha-tocopherol, is quantitatively the most important lipophilic antioxidant of human circulation and vasculature[16]. It is a well known chain-branching antioxidant which in diabetes mellitus has been shown to be beneficial in short-term studies[17,18]. We selected a mixture of vitamin E + selenium because it delays early renal diabetic lesions more actively than other antioxidants[19].

2. MATERIALS AND METHODS

2.1 Animals

Eight week-old male Wistar rats were purchased from Harlan-Nossan (Milan, Italy). Animals were housed in the Catholic University College of Medicine Animal Facility in accordance with the Guidelines of the American Physiology Society. Water and food were provided ad libitum. Rats were allowed to acclimatize for one week before starting the project. After the induction of diabetes, they were randomly assigned to different experimental groups. Riefer (Bolzano, Italy) supplied all diets.

2.2 Experimental Design

Diabetes was induced by a single intraperitoneal injection of 60 mg/kg streptozotocin (STZ) (Sigma, St. Louis, MO, USA). 48 hours later and subsequently at the times indicated in table 1, blood samples were taken from the tail vein for the determination of glucose: rats with glucose > 15 mmol/L were considered diabetic. Glucose was measured by glucose oxidase reagent strips (Lifescan, Milpits, CA, USA).

The experimental groups included diabetic rats fed either standard (which includes 80 UI/Kg of alpha - tocopherol but not selenium) or enriched diets. 2 enriched diets were used: 5% w/w taurine and 500 IU vitamin E + 8 mg selenium/kg. Mortality rate, glycemia and body weight were measured.

The survival data are examined using Kaplan-Meier survival analysis and the difference in glycaemia and body weight is analyzed by ANOVA followed by Scheffè test. P was considered significant at a level of 0.05.

3.   RESULTS

As shown in Fig 1, 5% taurine supplementation is able to prolong survival in diabetic rats in very bad metabolic control. The statistical analysis shows a level of significance of 0.04. The taurine effect on mortality is not associated with a variation in growth (data not shown) since a statistically significant decrease in glycaemia is observed after 15 months to 18 months of illness (Table 1).

Table 1.   Time-course of glycaemia (mmol/L) in diabetic rats fed standard and supplemented diets (either taurine or Vitamin E + selenium).

Months of illness	Non supplemented diabetic rats	5% Taurine supplemented diabetic rats	500 UI Vitamin E + 8 mg selenium supplemented diabetic rats
0	5.3±0.7 (23)	5.3±0.7 (23)	5.3±0.7 (24)
2	22.6±1.8 (19)	20.7±1.6(23)	20.4±2.1(22)
4	26.4±2.3(19)	22.5±1.7 (22)	24.6±1.8(20)
12	17.4±9.8 (9)	14.2±7.9(18)	18.7±4.1(7)
15	15.5±5.7 (7)	7.0±1.1*(12)	15.1±3.9(5)
18	14.6±1.8 (3)	7.1±0.8* (8)	12.3±2.9(4)

Values are mean ±SD, in brackets the number of animals, * P<0.05

Vitamin E + selenium affects neither the survival of diabetic rats (Fig.1) nor either animal growth (data not shown) nor blood glycemia (Table 1).

4.   CONCLUSIONS

This is the first observation showing that taurine supplementation reduces the mortality rate in diabetic rats. The increase in survival is a very "difficult" end point which many antidiabetic drugs fail to achieve. In particular, phenformin and tolbutamide, despite their hypoglycemic effect, increase overall mortality and cardiovascular mortality[3]. Moreover, the UKPDS[2], designed to achieve good metabolic control, has shown that, despite the decrease in microvascular complications, pharmacological treatments do not reduce mortality in diabetic patients. In the case of sulfonylureas the hypothetical negative influence that these drugs may have under condition of relative myocardial ischemia will continue to be closely monitored[20]. Consequently, new therapeutic treatments should affect long-term diabetic mortality, also considering that certain agents have detrimental effects despite their ability to reduce glycemia.

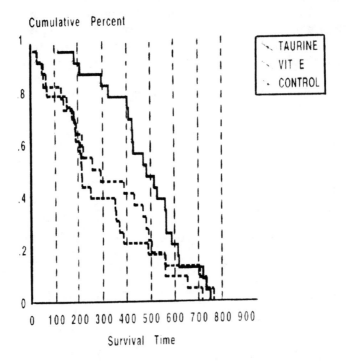

Figure 1. Kaplan-Maier survival curve.

Thus, the effect of chronic taurine supplementation on mortality is very interesting. After 15 months of diabetes, taurine decreased glycaemia while the effects on mortality are more precocious suggesting that some other factors contribute to the effect of taurine on survival. The data obtained with vitamin E + selenium suggest that the taurine effect can not be ascribed to the antioxidant activity of taurine because vitamin E + selenium do not modify any of the studied parameters. Classically, vitamin E is considered a chain-branching antioxidant. However, recent findings suggest that it may act as a promoter of oxidation[16].

This activity is not necessarily linked to its redox–active properties but could involve other metabolic pathways such protein kinase C system[16] and could be determinated by its localizations within biological systems[16]. Therefore, the protective effect of vitamin E could be observed only if other pro-oxidant factors are relatively weak and this is not the case of diabetes mellitus where oxidative stress is increased[4]. The lack of a beneficial effect of vitamin E is in line with the results of human interventional trials[21]. On the other hand, the effect of taurine could be due to its specificity in scavenging hypochlorous acid[8] and carbonyl groups[10]. Thus, in diabetes mellitus, the key to a successful antioxidant therapy could rely on the choice of the right antioxidants, which are not vitamin E + selenium. In this context, our findings encourage the use of taurine in diabetes mellitus while they discourage the use of vitamin E. The hypoglycemic effect of taurine is late in onset while, as previously shown[9,12], the amino acid does not modify glycaemia in the first two months of the disease. In normal rats, a hypoglycaemic effect of taurine has first been reported during the 1930s and then confirmed by later studies[9]. The mechanism of the hypoglycemic effect seems to involve an increase of insulin activity rather than enhanced insulin release[9].

Moreover, it is important to note that taurine is an osmotic agent[8] and osmotic stress is present in diabetes mellitus. However, the mortality in diabetes mellitus mainly depends on cardiac mortality[22], as cardiovascular disease is two to five times more common in patients with diabetes than in non diabetic patients of the same age[23]; taurine has been shown to be protective in the cardiovascular system including post-ischemical reperfusion[24-26].

Behind the mechanism of action of taurine, our results suggest that taurine is useful in the treatment of diabetes mellitus at least in an experimental model of the disease. Thus, these findings suggest that it is time to start appropriate clinical studies in humans. The clinical studies could also take advantage by the low toxicity of this supplement[8,9] and its low cost. Obviously, the effect of taurine in human diabetes mellitus should be carefully examined considering the slowly developing complications of the disease.

ACKNOWLEDGEMENTS

The work from our laboratory was supported by grants (60%) and MURST (40%).

REFERENCES

1. 1999 Diabetes Surveillance Report, 2000, Atlanta, Ga: Division of Diabetes Translation, National Center for Chronic Disease Prevention and Health Promotion, Center for Disease Control and Prevention. Available at www.cdc.gov/diabetes/statistics/index.htm. Accessed December 14, 2001.

72   *Flavia Franconi et al.*

2. UK Prospective Diabetes Study (UKPDS) Group, 1998, Intensive blood-glucose control with
    sulphonylureas or insulin compared with conventional treatment and risk of complications in
    patients with type 2 diabetes (UKPDS3). *Lancet* 352: 837-853.
3. Goldner. M.G., Knattrud, G.L., and Prout, T.E., 1971, Effect of hypoglycaemic agents on
    vascular complications in patients with adult onset diabetes. III Clinical Implications of UGDP
    results. *JAMA* 218: 1400-1410.
4. Baynes, J.W. and Thorpe, S.R., 1999, Role of oxidative stress in diabetic complications: a new
    perspective on an old paradigm. *Diabetes* 48: 1-9.
5. Santini, S.A., Marra, G., Giardina, B., Cotroneo, P., Mordente, A., Martorana, G.E., Manto,
    A., and Ghirlanda G., 1997, Defective plasma antioxidant defenses and enhanced susceptibility
    to lipid peroxidation in uncomplicated IDDM. *Diabetes* 46: 1853-1858.
6   Franconi, F., Bennardini, F., Mattana, A., Miceli, M., Ciuti, M. Mian, M., Gironi, A.,
    Bartolomei, G., Anichini, R., and Seghieri, G., 1995, Taurine deficiency in plasma and
    platelets of diabetic insulin-dependent patients: effects of taurine supplementation. *Am. J. Clin.
    Nutr.* 61: 1115-1119.
7. Kelly, G.S., 2000, Insulin resistance: lifestyle and nutritional interventions. *Altern. Med. Rev.*
    5: 109-132.
8. Huxtable, R.J., 1992, Physiological actions of taurine. *Physiol. Rev.* 72: 101-163.
9. Hansen, A.H., 2001, The role of taurine in diabetes and the development of diabetic
    complications. *Diab. Met. Res. Rev.* 17: 330-346.
10. Devanmanoharan, C., Ali, P.S., and Varna, A.H., 1997, Prevention of lens protein glycation by
    taurine *Mol. Cell Biochem.* **177**: 245-250.
11. Di Leo, M.A.S., Santini, S.A, Cercone, S., Marra, G., Lepore, D., Caputo, S., Antico, L.,
    Giardina, B., Pitocco, D., Franconi, F., and Ghirlanda, G., 1999, Dose-dependent effects of
    taurine in the prevention of Na, K-ATPase impairment and lipid peroxidation in the retina of
    streptozotocin-diabetic rats. *Diabetologia* 42 (Suppl. 1): A315.
12. Di Leo, M.A.S., Santini, S.A., Cercone, S., Lepore, D., Gentiloni Silveri, N., Caputo, S.,
    Greco, A.V., Giardina, B., Franconi, F., and Ghirlanda, G., 2002, Chronic taurine
    supplementation ameliorates oxidative stress and Na⁺K⁺ATPase impairment in the retina of
    diabetic rats. *Amino Acids* in press.
13. Obrosova, I., Minchenko, A.G., Marinescu, V., Fathallah, L., Kennedy, A., Stockert, C.M.,
    Frank, R.N., and Stevens, M.J., 2001, Antioxidants attenuate early up regulation of retinal
    vascular endothelial growth factor in streptozotocin-diabetic rat. *Diabetologia* 44:1102-1110.
14. De Luca, G., Calpona, P.R., Caponetti, A., Romano, G., Di Benedetto, A., Cucinotta, D., and
    Di Giorgio R.M., 2001, Taurine and osmoregulation: platelet taurine content, uptake, and
    release in type 2 diabetes. *Metabolism* 50: 60-64.
15. Nakamura, T., Ushiyama, C., Suzuki, S., Shimada, N., Ohmuro, H., Ebihara, I., and Koide, H.,
    1999, Effects of taurine and vitamin E on microalbuminuria, plasma metallo proteinase-9, and
    serum type IV collagen concentrations in patients with diabetic nephropathy. *Nephron* 83:
    361-362.
16. Neuzil, J., Weber, C., and Kontush A., 2001, The role of vitamin E in atherogenesis: linking
    the chemical, biological and clinical aspects of the disease. *Atherogenesis* 157: 257-283.
17. van Dam, P.S., Bravenboer, B., van Asbeck, BS, Marx, J.J.S., and Gispen, W.H., 1999, High
    rat food vitamin E content improves nerve function in streptozotocin-diabetes rat. *Eur. J.
    Pharmacol.* 376: 217-222.
18. Faure, P., Rossini, E., Lafond, J.L., Favier, R.M.J., and Halimi, S., 1977, Vitamin E improves
    the free radicals defense system potential and insulin sensitivity of rats fed high fructose diets.
    J. Nutr 127: 103-107.
19. Douillet, A., Tabib, M., Bost, M., Accominotti, F., Borson-Chazot R., and Ciavatti, M., 1996,
    A selenium supplement associated or not with vitamin E delay early renal lesions in
    experimental diabetes in rats. *Proc. Soc. Exp. Biol. Med.* 211: 323-331.
20. Matz, R., 1998, Sulfonylureas and ischemic heart disease *Arch. Intern. Med.* 158: 411-412.

21. Heart Protection Study Collaborative Group, 2002, MRC/BHF heart protection study of antioxidant vitamin supplementation in 20536 high-risk individuals: a randomised placebo controlled trial. *Lancet* 360: 23-33.
22. Wagner, A.M., Martinez-Rubio, A., Ordonez-Llanos, J., and Perez-Perez, A., 2002, Diabetes mellitus and cardiovascular disease. *Eur. J. Intern. Med.* **13**: 15-30.
23. Laasko, M., and Letho, S., 1997, Epidemiology of macrovascular disease in diabetes. *Diabetes Rev.* **5**, 294-310.
24. Militante, J.D., Lombardini, J.B., and Schaffer, S.W., 2000, The role of taurine in the pathogenesis of the cardiomyopathy of insulin-dependent diabetes. *Cardiovascular Res.* **393**: 393-400.
25. Franconi, F., Stendardi, I., Failli, P., Matucci, R., Baccaro, C., Montorsi, L., Bandinelli, R., and Giotti, A., 1985, The protective effects of taurine on hypoxia (performed in the absence of glucose) and on reoxygenation (in the presence of glucose) in guinea-pig heart. *Biochem. Pharmacol.* **34**: 2611-2614.
26. Takahashi, K., Ohyabu, Y., Schaffer, S.W., and Azuma J., 2000, Taurine prevents ischemia damage in cultured neonatal rat cardiomyocytes. *Adv. Exp. Med. Biol.* **483**:109-116.

The Effect of Dietary Taurine Supplementation on Plasma and Urinary Free Amino Acid Concentrations in Diabetic Rats

YOUNG-MI LEE, MI JA CHOI, and KYUNG JA CHANG[#]

Department of Food and Nutrition, Keimyung University, Taegu, Korea, [#]Department of Food and Nutrition, Inha University, Incheon, Korea

1.   INTRODUCTION

Taurine (2-aminoethanesulfonic acid) is a normal constituent of the human diet and is found in animal food sources. Taurine has many biological functions including bile acid conjugation, osmoregulation, membrane protection, antioxidant defense, and regulation of cellular calcium homeostasis[1-3]. It has been suggested that taurine might serve as a physiological regulator of insulin secretion and plasma glucose levels[4]. It has been reported that taurine acts as a potent antioxidant and exhibits anti-diabetic activity in streptozotocin-induced diabetes mellitus[5,6]. Recently an immunohistochemical study showed that taurine protects pancreatic β-cells of rats against destruction by streptozotocin injection in a dose-dependent way[7]. Also, it has been reported that the plasma and platelet taurine content are reduced in subjects with insulin-dependent diabetes mellitus, although this defect can be reversed by taurine supplementation[8]. Therefore, the purpose of this study was to investigate the effect of dietary taurine supplementation on the plasma and urinary free amino acid content in diabetic rats.

2.   MATERIALS AND METHODS

Forty-five male Sprague-Dawley rats were obtained from the Korea Life Engineering Corporation. The average body weight of the animals upon arrival was 110 ± 3.5g. Animals were housed individually in stainless steel cages in a room with controlled temperature and exposed to an alternating twelve hour period of light and dark. All animals were fed a chow diet and water *ad libitum*

Taurine 5: Beginning the 21st Century
Edited by Lombardini *et al.*, Kluwer Academic/Plenum Publishers, New York 2003   75

after arrival. Body weights were recorded weekly. When the animals attained a weight of from 190 to 210 grams, diabetes was induced in one-half of the animals by an intramuscular injection of streptozotocin (in 0.25 M citrate buffer, pH 4.5, Sigma, U.S.A) at a dose of 50 mg/kg body weight[9]. At the same time, 0.25 M citrate buffer, pH 4.5, was injected intramuscularly into the non-diabetic rats. Three days after streptozotocin treatment, the appearance of glucose in the urine was used to confirm the diabetic state (Multistix, U.S.A). Blood sample were taken from the tail vein of unanesthetized rats using heparinized, microhematocrit capillary tubes, and the glucose concentration of the plasma was analyzed using GLUCOPAT (Daiichi, Japan). Animals were diagnosed as being diabetic if they had a non-fasting blood glucose concentration which was greater than 300 mg/dl. Rats were then randomly assigned to two experimental dietary groups (Table 1) for both the diabetic and non-diabetic rats. At the end of 21 days on the experimental diet, rats were treated with light anesthesia (ethyl ether) and then sacrificed. To obtain plasma, blood samples were taken from the inferior vena

Table 1. Composition of experimental diets (g/100g diet).

Ingredient	Experimental Diets	
	Control diet	Taurine-supplemented diet
Corn starch	65.0	63.5
Casein	20.0	20.0
Corn oil	5.0	5.0
Cellulose	5.0	5.0
Mineral mixture	3.5	3.5
Vitamin mixture	1.0	1.0
Choline	0.2	0.2
DL-methionine	0.3	0.3
Taurine	-	1.5
Total	**100.0**	**100.0**

cava and were then centrifuged for 30 minutes at 3000 rpm at 4°C to obtain plasma. The plasma was stored at -70°C until glucose and amino acid assays could be performed. Blood glucose, total cholesterol, triglyceride and HDL-cholesterol concentration were determined by standard enzymatic methodology. LDL-cholesterol was calculated according to Friedewald[10]. Liver lipid concentration were measured by the method of Folch[11]. Plasma and urine free amino acid concentrations were determined using an automated amino acid analyzer based on ion-exchange chromatography (Biochrom 20, England). The data were analyzed by a computer using the Statistical Analysis System (SAS); the general linear model (GLM) procedure allowed for an unequal number of samples to be evaluated in each experimental group. Values were expressed as the mean ± SD. Comparisons relative to the two different dietary groups (control or taurine supplementation) and the animal's status (non-diabetic and diabetic) were performed using two-way analysis of variance. Statistical significance was evaluated according to the Duncan's test. Correlation coefficients for plasma taurine, glucose, and lipids were determined by the Pearson correlation coefficient.

3. RESULTS AND DISCUSSION

Diabetic rats showed significantly decreased body weight gain and FER but significantly increased food intake and urine excretion when compared to normal rats. Diabetic rats fed a taurine-supplemented diet showed significant improvement in lower growth, polyphagia and polyuria compared to diabetic rats fed the control diet (data are not shown). The results of the present study are in agreement with previous studies[5,12,13].

The effect of taurine on plasma glucose and lipid content is shown in Table 2. Diabetic groups showed significantly increased plasma glucose concentrations when compared to non-diabetic groups. The results of the present study are in agreement with the previous study[14]. Hypertriglyceridemia occurs frequently in diabetes mellitus[14]. In the present study, diabetic rats had significantly increased plasma triglyceride concentrations compared to non-diabetic rats. The results of this study confirm previous results[5]. The plasma triglyceride concentration of diabetic rats fed the control diet was significantly higher than those fed the taurine-supplemented diet. The results of this study support an earlier study[15] examining the hypolipidemic effect of taurine in diabetic rats[5,16].

Table 2. Effects of taurine supplementation on plasma glucose and lipid concentrations in normal and diabetic rats.

Variables	Normal		Diabetic	
	control	Taurine-supplemented	Control	Taurine-supplemented
Glucose(mg/dℓ)	11.9± 9.0[a]	98.8±11.8[b]	343.6±27.9[c]	295.5±21.6[d]
Triglyceride(mg/dℓ)	49.3±15.3[a]	34.2± 5.2[a]	65.9± 8.4[b]	52.3± 9.1[c]
T-cholesterol(mg/dℓ)	88.4±12.4[a]	73.4±10.6[b]	93.0±21.6[c]	79.7±13.4[b]
HDL-cholesterol(mg/dℓ)	44.1±11.4[a]	47.1±13.9[a]	44.9±10.7[a]	45.9±15.1[a]
LDL- cholesterol(mg/dℓ)	34.4±12.6[a]	19.4± 9.8[b]	34.9±12.8[a]	23.3±15.8[b]
Atherogenic index	1.09±0.45[a]	0.61± 0.3[a]	1.09± 0.26[a]	0.87±0.64[a]

Values are Mean±SD. Values with different superscripts within the same row are significantly different at $p<0.05$

Diabetic animals exhibit increased total plasma cholesterol concentrations[14,17]. The plasma total cholesterol concentrations of diabetic rats fed a control diet were significantly higher than those fed a taurine–supplemented diet. The results of this study support previous results examining the hypocholesterolemic effect of taurine in diabetic rats[15,16]. The concentration of LDL-cholesterol was significantly decreased (47%) in non diabetic rats fed a taurine-supplemented diet compared to rats fed the control diet. Moreover, the concentrations of LDL-cholesterol were significantly decreased (33%) in diabetic rats fed a taurine-supplemented diet compared to rats maintained on the control diet. The results of the present study are in agreement with a previous study[16]. Diabetic rats had significantly greater overall liver triglyceride concentrations, GOT and GPT than the non-diabetic rats.   Among the diabetic rats, the concentrations of liver triglyceride, cholesterol, GOT, and GPT were significantly lower in the taurine fed group (p<0.05) than in the group fed the control diet (Table 3, data are not shown).

Table 3. Effects of taurine supplementation on liver lipid concentrations in normal and diabetic rats.

Variables	Normal		Diabetic	
	Control	Taurine-supplemented	Control	Taurine-supplemented
Total cholesterol(mg/g)	2.52±0.15[a]	2.48±0.19[a]	2.26±0.46[a]	1.53±0.30[b]
Triglyceride(mg/g)	6.91±1.18[a]	5.85±0.35[b]	7.26±1.46[c]	5.58±0.94[b]

Values are Mean±SD. Values with different superscripts within the same row are significantly different at p<0.05

Plasma GOT activity was significantly decreased in the non-diabetic (17%) and diabetic (26%) rats fed the taurine-supplemented diet compared to those fed the control diet. GPT activity was significantly higher (22%) in the diabetic rats than the non-diabetic rats. However, GPT activity was significantly lower in diabetic rats fed the taurine-supplemented diet than in those fed the control diet (data are not shown).

The effects of taurine on plasma amino acid content of the diabetic and non-diabetic rats are given in Table 4-1 and 4-2. Diabetic rats fed the control diet contained normal plasma levels of the essential amino acids, but the levels of taurine (-24%) and serine (-27%) were significantly lower while those of proline (218%), ornithine (150%), ethanolamine (111%), phosphoserine (175%), and citrulline (10%) were significantly higher. Taurine-supplementation affected the plasma content of ethanolamine (-34%), citrulline (-15%), ornithine (+41%) and taurine (+448%). However, taurine supplementation in diabetic rats did not cause a characteristic change in the plasma aminogram pattern. Non-diabetic rats fed the taurine-supplemented diet had significantly higher concentrations of plasma taurine. In the non-diabetic rat, there was no significant effect of taurine supplementation on plasma amino acid content, the exception being

Table 4-1. Effects of taurine supplementation on plasma free essential amino acid concentration in normal and diabetic rats.

Variables	Normal		Diabetic	
	Control	Taurine-supplemented	Control	Taurine-supplemented
EAA (μmol/L)				
Arginine	109.9±33.9[a]	140.5±36.0[b]	105.8±26.4[a]	106.2±20.8[a]
Histidine	77.0±13.6[a]	82.9±12.4[a]	73.9±12.2[a]	82.1±10.5[a]
Isoleucine	57.8±6.5[a]	63.2±17.5[a]	81.5±21.3[a]	92.3±45.6[a]
Leucine	171.9±16.2[a]	173.1±32.3[a]	186.6±35.0[a]	204.1±79.2[a]
Lysine	500±106[a]	504±73[a]	573±59[a]	536±161[a]
Methionine	61.0±6.5[a]	67.2±6.0[a]	60.1±26.3[a]	59.5±10.7[a]
Phenylalanine	72.6±6.6[a]	68.9±7.4[a]	84.1±31.0[a]	68.0±17.6[a]
Valine	189.7±27.2[a]	196.9±48.2[a]	228.6±89.5[a]	189.4±56.7[a]

Values are Mean±SD. Values with different superscripts within the same row are significantly different at p<0.05.

Table 4-2. Effects of taurine supplementation on plasma free nonessential amino acid concentration in normal and diabetic rats.

Variables	Normal		Diabetic	
	Control	Taurine-supplemented	Control	Taurine-supplemented
NEAA (μmol/L)				
Alanine	459.1±68.2[a]	481.5±53.6[a]	484.0±135.2[a]	519.5±62.3[a]
α-aminobutyric acid	27.04±12.21[a]	44.09±9.87[b]	25.57±12.12[a]	26.79±1.54[a]
Citrulline	68.72±6.69[a]	65.62±9.90[a]	75.27±6.74[b]	63.88±10.05[a]
Ethanolamine	11.86±4.58[a]	18.90±1.50[a]	25.01±7.69[b]	16.30±3.60[a]
Glutamic acid	183.9±16.7[a]	195.8±27.7[a]	189.7±67.9[a]	201.9±28.2[a]
Glycine	275.4±42.1[a]	247.7±36.0[a]	289.6±75.1[a]	273.4±53.2[a]
1-Methyl histidine	8.44±1.31[a]	9.86±1.08[a]	7.56±1.25[a]	7.77±1.74[a]
Ornithine	72.3±13.9[a]	105.8±25.1[b]	108.6±44.8[b]	153.2±56.9[c]
Phosphoserine	5.33±1.36[a]	8.38±0.73[b]	14.65±2.01[c]	16.20±1.40[c]
Proline	74.8±6.4[a]	89.7±9.4[a]	163.0±20.4[b]	185.6±41.5[b]
Serine	244.2±33.9[a]	234.0±50.6[a]	158.0±35.0[b]	135.5±38.3[b]
Taurine	71.5±14.0[a]	218.1±25.6[b]	57.4±17.0[c]	257.7±33.1[b]
Tyrosine	93.6±12.3[a]	96.7±11.5[a]	93.1±41.3[a]	82.2±24.6[a]

Values are Mean±SD. Values with different superscripts within the same row are significantly different at p<0.05

Table 5-1. Effects of taurine supplementation on urinary free essential amino acid concentration in normal and diabetic rats.

Variables	Normal		Diabetic	
	Control	Taurine-supplemented	Control	Taurine-supplemented
EAA (µmol/day)				
Arginine	0.47±0.24[a]	0.68±0.17[a]	1.05±0.38[a]	1.07±0.55[a]
Histidine	0.61±0.12[a]	0.80±0.27[a]	2.30±0.72[b]	2.37±0.44[b]
Isoleucine	10.8±7.2[a]	10.5±6.9[a]	35.6±31.6[b]	44.5±13.1[b]
Leucine	7.85±8.45[a]	7.67±8.02[a]	62.2±39.5[b]	39.2±35.8[b]
Lysine	1.59±0.65[a]	2.55±0.68[a]	3.63±1.63[b]	3.60±1.04[b]
Methionine	0.58±0.27[a]	0.74±0.24[b]	1.93±0.66[c]	2.00±0.45[c]
Phenylalanine	66.2±6.6[a]	68.4±94.3[a]	67.3±62.5[a]	65.6±44.5[a]
Valine	3.7±6.7[a]	4.8±8.4[a]	29.9±40.1[b]	20.0±0.3[b]

Values are Mean±SD. Values with different superscripts within the same row are significantly different at p<0.05.

Table 5-2. Effects of taurine supplementation on urinary free nonessential amino acid concentrations in normal and diabetic rats.

Variables	Normal		Diabetic	
	Control	Taurine-supplemented	Control	Taurine-supplemented
NEAA (µmol/day)				
Alanine	2.24± 0.31[a]	3.23± 1.32[a]	8.21± 2.47[b]	7.73± 1.72[b]
α-aminobutyric acid	3.58± 0.67[a]	4.54± 8.43[a]	34.02±10.85[b]	33.10± 9.10[b]
Citrulline	0.16± 0.04[a]	0.23± 0.11[a]	0.75± 0.30[b]	0.68± 0.21[b]
Ethanolamine	2.59± 0.90[a]	3.13± 0.59[a]	4.82± 2.37[a]	4.19± 2.25[a]
Glutamic acid	0.21± 0.21[a]	0.15± 0.00[a]	0.61± 0.68[a]	0.89± 0.16[b]
Glycine	0.81± 1.06[a]	1.30± 1.42[a]	7.40± 4.40[b]	6.17± 5.29[b]
1-Methyl histidine	0.15± 0.02[a]	0.20± 0.06[a]	0.41± 0.22[b]	0.40± 0.11[b]
Ornithine	0.37± 0.12[a]	0.39± 0.09[a]	1.29± 0.44[b]	1.02± 0.15[b]
Proline	3.55± 0.26[a]	4.72± 1.86[a]	11.11± 3.42[b]	13.76± 4.93[b]
Serine	2.24± 0.32[a]	2.14± 1.04[a]	5.26± 1.89[b]	5.13± 1.71[b]
Taurine	68.1±11.5[a]	234.5±22.7[b]	140.6±45.2[b]	444.5±90.5[c]
Tyrosine	8.9± 8.6[a]	34.7±55.6[a]	58.5±51.5[b]	55.5±66.0[b]

Values are Mean±SD. Values with different superscripts within the same row are significantly different at p<0.05

α-aminobutyric acid (+63%), ornithine (+46%) and phosphoserine (+57%). The plasma levels of all of the nonessential amino acids, with the exception of citrulline and glycine, were increased, especially α-aminobutyric acid (63%), ornithine (46%)·and phosphoserine (57%). The diabetic rats had a significantly lower plasma taurine content than the nondiabetic rats. In addition, diabetic rats had significantly higher concentrations of plasma ethanolamine (111%), citrulline (10%), ornithine (50%), posphoserine (175%), and praline (118%), and a significantly lower taurine (20%) and serine (29%) content.

The effect of taurine on urinary amino acid excretion in the diabetic and non-diabetic rats is given in Table 5-1 and 5-2. The urinary excretion of the amino acids did not correlate with the changes in plasma levels. Diabetic rats fed the control diet had significantly higher overall urinary excretion of the essential amino acids, such as histidine, isoleucine, leucine, lysine, methioine, valine and of most of the nonessential amino acids, the exception being ethanolamine and glutamic acid than non-diabetic rats fed the control diet. However, diabetic rats fed the taurine-supplemented diet excreted normal levels of the essential amino acids but significantly elevated levels of glutamic acid and taurine. The pattern of excretion of most urinary free amino acids was very similar in the two dietary groups, the exception being taurine and glutamic acid in the diabetic rats.

Plasma taurine content of the non-diabetic rats exhibited a significantly negative correlation with respect to plasma glucose, triglyceride, total cholesterol, LDL-cholesterol. In the diabetic rats, plasma taurine concentrations exhibited a significantly negative correlation with respect to plasma glucose, LDL-cholesterol, liver cholesterol, and atherogenic index (Table 6).

Table 6. Correlation coefficient of plasma glucose, Plasma lipids and liver lipids with plasma taurine concentration in normal and diabetic rats.

Variables	Normal		Diabetic	
	R	P	r	P
Plasma Glucose	-0.65*[1]	0.03	-0.78**[2]	0.006
Plasma triglyceride	-0.65*	0.03	-0.56	0.08
Plasma total cholesterol	-0.87**	0.001	-0.58	0.07
Plasma HDL-cholesterol	0.06	0.86	0.07	0.84
Plasma LDL-cholesterol	-0.76**	0.009	-0.82**	0.003
Liver triglyceride	-0.49	0.15	-0.56	0.09
Liver cholesterol	-0.32	0.35	-0.63*	0.04
Atherogenic index	-0.60	0.06	-0.81**	0.004

1) *: significantly different at p<0.05
2) **: significantly different at p<0.001

In conclusion, taurine supplementation was hypolipidemic and hypoglyceridemic in streptozotocin-induced diabetic rats. This study demonstrates that taurine has marked effects on liver cholesterol and triglyceride levels. There was no effect of taurine supplementation on plasma free amino acid concentrations in the streptozotocin-induced diabetic rats, the exception being plasma taurine concentrations. Taurine supplementation caused a lowering in plasma glucose and triglyceride levels in the streptozotocin-induced diabetic rats. Both of these effects suggest that taurine supplementation may be of potential value in hypoglyceridemic and hypotriglycelidemic diets of diabetic individuals.

REFERENCES

1. Huxtable, R.J., 1992, Physiological Action of Taurine. *Physiological Reviews* 72(1): 101-163.
2. Redmond, H.P., Stepleton, P.P., Neary, P., Bouchier-Hayes, D., 1998, Immunoutrition: The role of taurine. *Nutriion* 14:599-604.
3. Monte, M.J., El-Mir, M.Y., Saninz, G.R., Bravo, P., Marin, J.J., 1997, Bile acid secretion during synchronized rat liver regeneration. *Biochim Biophys Acta* 1362: 56-66.
4. Tokanaga, H., Yoned, Y., Kuriyama, K, 1979, Protective action of taurine against streptozotocin-induced hyperglycemia. *Biochem Pharmacol* 28: 2807-2811.
5. You, J.S., Chang, K.J., 1998, Effects of taurine supplementation on lipid peroxidation, blood glucose and blood lipid metabolism in streptozotocin-induced diabetic rats. *Adv Exp Med Biol* 442: 163-168.
6. Lim, E., Park, S., Kim, H., 1998, Effect of taurine supplementation on the lipid peroxide formation and the activities of glutathione-related enzymes in the liver and islet of type and type II diabetic model mice. *Adv Exp Med Biol* 442: 99-103.
7. Chang, K.J., Kwon, W., 2000, Immunohistochemical localization of insulin in pancreatic beta-cells of taurine-supplemented or taurine-depleted diabetic rats. *Adv Exp Med Biol* 483: 579-587.
8. Franconi, F., Bennardini, F., Mattana, A., 1995, Plasma and platelet taurine are reduced in subject with insulin-dependent diabetes mellitus ; Effect of taurine supplementation. *Am J Clin Nutr* 61: 1115-1119.
9. Sauberlich, H.E., 1961, Studies on the toxiology and antagonism of amino acid for weanling rats. *J Nutr* 75: 61-72.
10. Fridewald, W.T., Lavy, R.I., Fredricson, D.S., 1972, Estimation of low density lipoprotein cholesterol in plasma, without use of the preparative ultracentrifuge. Clin Chem 18: 499-502.
11. Folch, J., Lees, M., Slane-Stanly, G.H., 1957, A simple method for the isolation and purification of total lipids from animal tissue. *J Biochem* 226: 497-509.
12. McNeill, J.H., Delgatty, H.L, Battell, M.L, 1991, Insulinlike effects of sodium selenate in streptozocin-induced diabetic rats. *Diabetes* 40(12): 1675-1678.
13. Malabu, U.H., Dryden, S., McCarthy, H.D., Kilpatrick, A., Williams, G, 1994, Effects of chronic vanadate administration in the STZ-induced diabetic rat. The antihyperglycemic action of vanadate is attributable entirely to its suppression of feeding. *Diabetes* 43: 9- 5.
14. Kahn, C.R., 1985, The molecular mechanism of insulin action. *Ann Rev Med* 36: 429-433.
15. Tamborlane, W.V., Sherwin, R.S., Genel, M., Felig, P., 1979, Restoration of normal lipid and aminoacid metabolism in diabetic patients treated with a portable insulin-infusion pump. *Lancet* 16: 1258-1261.
16. Goodman, H.O., Shihabi, Z.K., 1990, Supplemental taurine in diabetic rats: effects on plasma glucose and triglycerides. *Biochem Med Metab Biol* 43(1): 1-9.
17. Ressin, K., Gaull, G.E., Jarvenpaa, A.L., 1983, Feeding the low-birth-weight infant: Effect of taurine and cholesterol supplementation on amino acids and cholesterol. *Pediatrics* 71 179-186.

Insulin – Stimulated Taurine Uptake in the Rat Retina

ROCIO SALCEDA
Department of Neurociences, Cell Physiology Institute, National University of Mexico, Mexico City, C.P. O4510, MEXICO

1. INTRODUCTION

The sulfur amino acid taurine is found in millimolar concentrations in excitable tissues, especially those which generate oxidants. Taurine is involved in a variety of biological phenomena such as osmoregulation, antioxidation, detoxification, neuronal modulation, cell proliferation and cell viability [1,2]. Taurine is the most abundant amino acid in the retina of all species studied. In the rat retina, taurine accounts for more than 50% of the total free amino acid pool [3-5]. The intraretinal distribution of taurine is heterogeneous being largely concentrated in photoreceptors. A high taurine concentration is essential for maintenance of the structural and functional integrity of retina [6].

Retinal pigment epithelium (RPE) is primarily responsible for selective transport of substances between the choroidal and neural retina [7]. Passage of taurine through the RPE to the photoreceptor cells has been detected after injection of ³H taurine to mouse and frogs [8,9].

Although some degree of endogenous biosynthesis exists, much of the taurine present in mammalian tissue is exogenous in origin, requiring a very efficient transport system to achieve high taurine concentration. Disruption of the taurine transport gene was recently reported to cause retinal degeneration in mice [10]. More information on the regulation of taurine transport may provide insight into the role of taurine in retinal function.

2.   PROPERTIES OF THE TAURINE TRANSPORTER

Taurine is taken up by an energy dependent system, as it is temperature sensitive and inhibited by metabolic inhibitors such as ouabain and dinitrophenol[11]. Retinal transport processes for taurine are Na^+- and Cl^--dependent, as has been described in other tissues[1,2,11,12].

It has been reported that taurine is taken up into the retina and RPE by two transport systems[11,13] a low-affinity (K_M in the mM range) but high-capacity uptake system, and a high-affinity system (K_M 50 - 100 μM) having low capacity for uptake. Only the high-affinity system is present in isolated photoreceptor outer segments[13].

Isoforms of the taurine transport (taut) have been cloned and sequenced from different tissues including retina and RPE[14-17]. They belong to the family of Na^+- and Cl^--dependent cotransporters, which include neurotransmitters, osmolytes and amino acid transporters[12].

The substrate specificity of the taurine uptake system has been studied by investigating the ability of various unlabeled amino acids to compete with radiolabeled taurine for the uptake process. Substrates of the A, ASC or L amino acid transport systems, such as α-alanine, leucine or proline, do not inhibit taurine uptake, whereas hypotaurine and β-alanine can be substrates of the taurine transporter[14-17].

Immunohistochemical studies have been performed using antibodies against the taurine transporter cloned from rat brain[14] (taut-1) or from mouse brain[16] (taut-2). Intense labeling for taut-1 were observed in photoreceptors and populations of bipolar cells[18]. These results are in agreement with the distribution of taurine immunoreactivity and autoradiographic detection of taurine-accumulating cells[8,9,18]. Taut-1 labeling was absent from glial cell elements at the vitreal margin of the retina. By contrast, taut-2 which is predominantly associated with glial cells in different brain areas, was strongly expressed by glial elements on the vitreal surface of the retina[18]. Similarly, *in situ* hybridization experiments indicated that mouse taut mRNA is expressed at low levels in retinal cells and in the RPE[17]. These results suggest that retina may contain different taurine transporter isoforms.

3.   THE ROLE OF TAURINE IN DIABETES

Diabetes mellitus is known to produce a number of behavioral and pathological abnormalities, including retinopathy[19]. In diabetic patients, electroretinographic response and light and color sensitivity are severely reduced in comparison with the normal population[20,21]. RPE, which control the flow of molecules to the outer retina, plays an important role in a number of ocular lesions, including diabetic retinopathy[22].

The binding of insulin to its tyrosine kinase receptor, triggers the activation of signaling pathway whose function is first to stimulate the transport of nutrients[24]. Insulin is also known to stimulate the recruitment of glucose transporters to the cell surface membrane, thereby allowing an increase in glucose flux into the cell[24].

Dietary taurine supplementation counteracts taurine reduced content, oxidative stress, nerve blood flow and conduction deficits in experimental diabetic neuropathy[25,26].

In this regard it is relevant that taurine levels in retina and RPE are significantly reduced in streptozotocin-induced diabetes[5] and in RPE under hyperglycemic conditions[27]. In addition, an increase in taurine uptake was observed in retina and RPE from diabetic rats[5]. These results suggest that a regulatory mechanism controling taurine levels in the retina is modified during diabetes.

4.   EFFECT OF GLUCOSE AND INSULIN ON TAURINE TRANSPORT.

Endocrine effects in the transport of amino acids have been studied in a number of tissues, but only a few studies have investigated transport in retina. To better understand the regulation of taurine transport in diabetes we studied the effect of high glucose concentration and insulin on ³H taurine uptake into retina and RPE. A high glucose concentration (30 mM) in the incubation medium slightly stimulated taurine uptake (30%) in both retina and RPE. Lower glucose concentrations had no effect on taurine uptake[28].

On the other hand, taurine uptake by retina and RPE was significantly enhanced in a dose-dependent manner by physiological concentrations of insulin[28.] Addition of 10 ng/ml to the incubation medium increased taurine uptake by 50% in both retina and RPE.

Basal and insulin-stimulated taurine accumulation was strongly reduced (90%) in both retina and RPE when sodium chloride was replaced by either choline chloride or potassium gluconate. Taurine accumulation was significantly inhibited (60%) in the presence of the specific inhibitor of the taurine carrier, guanidinoethyl sulfonate (GES). These results indicate that insulin-stimulated taurine uptake involves the specific taurine transporter.

Kinetic analysis of the effects of insulin and high glucose concentration on taurine uptake revealed significant differences between the control and samples treated with insulin or high glucose concentrations[28]. Both insulin and high glucose concentrations increase Vmax and K_M of taurine uptake. The effect of glucose is a consequence of glucose concentration itself and not to variations of osmolarity[28].

In agreement with the *in vitro* experiments, the kinetic parameters of taurine uptake in retina from streptozotocin-treated rats were significantly different from those of the normal rat. A single transport system with low affinity (K_M = 3 mM) and high capacity was detected in diabetic rat retina. Similar results were seen in RPE[28]. These results strongly suggest a significant regulatory effect of insulin on taurine transport *in vivo*.

5. THE ROLE OF PROTEIN SYNTHESIS AND CYTOESKELETON ON INSULIN-INDUCED TAURINE UPTAKE.

To examine the involvement of protein synthesis and intracellular translocation of the taurine transporter, the actions of insulin, the effect of cycloheximide, an inhibitor of protein synthesis, and drugs that affects the cytoskeleton were studied.

The effect of insulin on taurine uptake is seen within a short time, suggesting that it is not acting at the level of taurine transporter synthesis. This idea is also supported by the observation that insulin-induced taurine uptake was unaffected by the treatment of the retina with 100 µM of cycloheximide, even after preincubation lasting up to 15 min.

The insulin stimulated taurine uptake was reduced to control levels in the presence of 20 µM of cytochalasin D, a drug which disrupts actin filaments, but was not affected by colchicine or nocodazol, inhibitors of the microtubular network. However, a ten minute preincubation with colchicine or nocodazol prior to initiating uptake measurements, significantly inhibited the insulin-stimulated taurine uptake (Table 1). Preincubation of the retinas with cytochalasin D reduced taurine uptake to values lower than the basal taurine accumulation (Table 1). These results indicate that insulin-stimulated taurine uptake is caused by an increase in the density of the plasma membrane transporter.

These results strongly suggest that taurine uptake is regulated by changes in the number of transporters in the membrane. This is in agreement with previous results in which activation of protein kinases induced reduction of Vmax of the taurine transporter concomitant to a decrease in the number of transporters in the plasma membrane of *Xenopus oocytes* expressing the mouse retinal transporter[29].

Table 1. Effect of cytoskeleton affecting drugs on insulin-stimulated taurine uptake in the rat retina.

Addition	Taurine uptake (nmol/g)	
	A	B
Insulin	50.0 ± 5.0	48.9 ± 5.4
Insulin + colchicine (50 µM)	47.2 ± 5.7	26.9 ± 1.2^b
Insulin + nocodazole (20 µM)	45.2 ± 4.9	29.0 ± 2.0^b
Insulin + cytochalasin (20 µM)	29.2 ± 6.6^c	15.2 ± 2.0^a

Uptake of ^{3}H-taurine (20 µM) was carried out at 20 min incubation in a Krebs medium in the presence of insulin (10 ng/µl) alone or coincubation with different drugs (A). In other experiments (B) retinas were preincubated for 10 min with the drugs before the addition of taurine and insulin. a, $p<0.0007$; b, $p<0.007$; c, $p<0.01$.

6. REGULATION OF TAURINE UPTAKE BY PROTEIN KINASES

The binding of insulin to its tyrosine kinase receptor on the outside surface of cells induces the self phosphorylation of several tyrosine residues located inside the cell. This triggers the recruitment of a lipid kinase, phosphatidylinositol 3 kinase (PI3K)[23,24,30]. PI3K has been shown to regulate several intracellular serine-threonine kinases, including protein kinase C (PKC)[31].

Taurine transporter cDNA cloned in different cell types contains putative consensus sites for protein phosphorylation within cytoplasmic domains, suggesting that its activity may be regulated by phosphorylation[14-17]. Effectively, there is evidence that kinases are involved in regulation of taurine transport[13,29].

Neither basal nor insulin-stimulated taurine uptake was affected when the PKC inhibitor staurosporine (25 nM) was added 10 min before or at the beginning of the taurine uptake measurements (Table 2). However, activation of PKC by 100 nM phorbol 12-myristate 13-acetate (PMA), resulted in a significant increase in the basal and insulin stimulated taurine uptake (Table 2).

PKC activation is known to down-regulate taurine uptake in rat astrocytes and human glioma cells but not in rat neuronal cells, in human neuroblastoma, nor in whole retinal preparations[13,32,33]. This may be explained to the fact that the action of protein kinases depends on the transporter isoform being expressed, as well as in the kinase isozyme present in the tissue.

Insulin-induced taurine accumulation was reduced (30%) by the presence of wortmannin, a specific inhibitor of the PI3K, and enhanced by okadaic acid, an inhibitor of protein phosphatases. The finding that inhibitors of PI3K inhibit most of the cellular responses to insulin indicates that the insulin-stimulated taurine transport is mediated by the activation of the insulin receptor, which in turn activates PI3K and leads to the recruitment of taurine transporter in the plasma membrane. These results do not exclude the action of other protein kinases. Whether activation of PKC leads to an increase in the number of taurine transporters into the plasma membrane or directly affects the transporter itself remains to be studied.

Table 2. Effect of drugs affecting protein kinases activity on taurine uptake.

Addition	^{3}H-taurine uptake (nmol/g)
Control	30.9 ± 3.4
Insulin (10 ng/ml)	$48.9 \pm 5.0^*$
PMA (100 nM)	$46.8 \pm 6.1^*$
Insulin + staurosporine (25 nM)	$42.1 \pm 3.5^*$
Insulin + wortmannin (25 nM)	$35.1 \pm 4.0^{**}$
Insulin + okadaic acid (100 nM)	$68.3 \pm 8.0^*$

Retinas were incubated 10 min with the drug before initiating uptake measurements in the absence (control) or presence of insulin. Significant difference from control $^*p<0.01$, $^{**}p<0.05$.

7.   CONCLUSION

A high taurine concentration is essential for maintenance of the structural and functional integrity of the retina. This high taurine concentration is most likely achieved by the activity of the taurine transporter. Insulin enhanced taurine uptake through a mechanism that involves protein kinases and the recruitment of the transporter into the plasma membrane. Further studies on the regulation of taurine transport will provide insight into the cellular mechanisms underlying taurine action and the role of taurine in retinal alterations in diabetes.

ACKNOWLEDGEMENTS

This work was partially supported by Conacyt (M0250-9911) and DGAPA-UNAM (IN-204000). The author thanks Mr. G. Sánchez-Chávez for technical assistance.

REFERENCES

1. Huxtable, R.J., 1989, Taurine in the central nervous system and the mammalian actions of taurine. *Prog. Neurobiol* 32: 471-533.
2. Wright, C.E., Tallan, H.H., Lin, Y.Y., and Gaull, G.E., 1986, Taurine: Biological Update. Ann. Rev. Biochem 55:427-453.
3. Pasantes-Morales, H., Klethi, J., Ledig, M., and Mandel, P., 1972, Free amino acids in chicken and rat retina. Brain Res 41:494-497.
4. Macaioni, S., Ruggeri, P., De Luca, F., and Tucci, G., 1974, Free amino acids in developing rat retina. J. Neurochem 22: 887-891.
5. Vilchis, C., and Salceda, R., 1996, Effect of diabetes on levels and uptake of putative amino acid neurotransmitters in rat retina and retinal pigment epithelium. Neurochem. Res 21: 1167-1171.
6. Hayes, K.C., Carey, R.E., and Schmidt, S.Y., 1975, Retinal degeneration associated with taurine deficiency in the cat. Science 188:949-951.
7. Cunha-Vaz, J.G., 1976, The blood-retinal barriers. Doc. Ophthalmol 41: 287-327.
8. Lake,N., Marshall, S., and Voaden, M.J., 1977, The entry of taurine into the neural retina and pigment epithelium of the frog. Brain Res 128:497-503.
9. Pourcho, R.G., 1977, Distribution of ^{35}S-taurine in mouse retina after intravitreal and intravascular injection. Exp. Eye Res 25: 119-127.
10. Heller-Still, B., Van Roeyen, C., Rascher, K., Hartwig, H.G., Huth, A., Seeliger, M.W., Warskulat, U., and Häussinger, D., 2002, Disruption of the taurine transporter gene (taut) leads to retinal degeneration in mice. FASEB J 16: 231-233.
11. Salceda, R., 1980, High-affinity taurine uptake in developing retina. Neurochem. Res 5: 561-572.
12. Amara, S.G., and Arriza, J.L., 1993, Neurotransmitter transporters: three distinct gene families. Curr. Opin. Neurobiol 3: 337-344.
13. Militante, J.D., and Lombardini, J.B., 1999, Taurine uptake activity in the rat retina: protein kinase C-independent inhibition by cherlerythrine. Brain Res 818: 368-374.
14. Smith, K., Borden, L.A., Wang, C.D., Hartig, P.R., Branchek, T.A. , and Weinshank, R.L., 1992, Cloning and expression of a high affinity taurine transporter from rat brain. Mol. Pharmacol 42: 563-569.
15. Uchida, S., Kwon, H.M., Yamamuchi, A., Preston, A.S., Marumo, F., and Handler, J.S., 1992, Molecular cloning of a cDNA for an MDCK cell Na$^+$ and Cl$^-$ dependent taurine transporter that is regulated by hypertonicity. Proc. Natl. Acad. Sci. USA 89:8230-8234.
16. Liu, Q.R., Lopez-Corcuera, B., Nelson, H., Mandiya, S., and Nelson, N., 1993, Cloning and expression of a cDNA encoding the transporter of taurine and β-alanine in mouse brain. Proc. Natl. Acad. Sci. USA 89: 12145-12149.
17. Vinnakota, S., Qian X., Egal, H., Sarthy, V., and Sarkar, H.K., 1997, Molecular characterization and in situ localization of a mouse retinal taurine transporter. J. Neurochem 69: 2238-2250.
18. Pow, D.V., Sullivan, R., Reye, P., and Hermanussen, S., 2002, Localization of taurine transporters, taurine, and ^{3}H-taurine accumulation in the rat retina, pituitary, and brain. Glia 37: 153-168.
19. Frank, R.N., 1995, Diabetic retinopathy. Progr. Retinal Eye Res. 14: 361-392.
20. Hood, C., and Birch, D.G., 1990, The a-wave of the human ERG and rod receptor function. Invest. Ophthalmol. Vis. Sci 31: 2070-2081.
21. Holopigian, K., Seiple, W., Lorenzo, M., and Carr, R., 1992, A comparison of photopic and scotopic electroretinographic changes in early diabetic retinopathy. Invest. Ophthalmol. Vis. Sci 33: 2773-2780.

22. Kirber, W.M., Nichols, C.W., Grimes, P.A., Winegard, A.I., and Laties, A.M., 1980, A permeability defect of the retinal pigment epithelium. Arch. Ophthalmol. 98: 725-728.

23. Lizcano, J.M., and Alessi, D.R., 2002, The insulin signaling pathway. Curr. Biology 12: R236-R238.

24. Zick, Y., 2001, Insulin resistance: a phosphorylation-based uncoupling of insulin signaling. Trends Cell Biol 11: 437-441.

25. Van Dam, P.S., Van Asbeck, B.S., Bravenboer, B., Van Oirschot, J.F., Marx, J.J., and Gispen, W.H., 1999, Nerve conduction and antioxidant levels in experimentally diabetic rats: effects of streptozotocin dose and diabetes duration. Metabolism 48:442-447.

26. Obrosova, I.G., Fathallah, L., and Stevens, M.J., 2001, Taurine counteracts oxidative stress and nerve growth factors deficit in early experimental diabetic neuropathy. Exp. Neurol 172: 211-219.

27. Reddy, V.N., 1992, Study of the polyol pathway and cell permeability in human lens and retinal pigment epithelium in tissue culture. Invest. Ophthalmol. Vis. Sci 33: 2334-2339.

28. Salceda, R., 1999, Insulin-stimulated taurine uptake in rat retina and retinal pigment epithelium. Neurochem. Int 35: 301-306.

29. Loo, D.D.F., Hirsch, J.R., Sarkar, H.K., and Wright, E.M., 1996, Regulation of the mouse retinal taurine transporter (Taut) by protein kinases in Xenopus oocytes. FEBS Lett 392: 250-254.

30. Vankaesebroeck, B., and Waterfield, M.D., 1999, Signaling by distinct class of phosphoinositide 3-kinases. Exp. Cell Res 253: 239-254.

31. Standaert M.L., Gallaway, L., Karnam, P., Bandyopadhyay, G., Moscat, J., and Farese, R.V., 1997, Protein kinase Cz as a downstream effector of phosphatidylinositol 3 kinase during insulin stimulation in rat adipocytes. Potential role in glucose transport. J. Biol. Chem 272: 30075-30082.

32. Tchoumkeu-Nzouessa, G.C., and Rebel, G., 1996, Activation of protein kinase C down-regulates glial but not neuronal taurine uptake. Neurosci. Lett 206: 61-64.

33. Tchoumken- Nzouessa, G.C., and Rebel, G., 1996, Characterization of taurine transport in human glioma GL15 cell line: regulation by protein kinase C. Neuropharmacol 35:37-44.

The Effect of Taurine Supplementation on Patients with Type 2 Diabetes Mellitus

KATHERINE B. CHAUNCEY*, THOMAS E. TENNER, JR#, JOHN B. LOMBARDINI#, BETSY GOEBEL JONES*, MICHELE L. BROOKS*, RONALD D. WARNER*, RANDALL L. DAVIS#, and R. MICHAEL RAGAIN*

*Department of Family & Community Medicine and #Department of Pharmacology, Texas Tech University Health Sciences Center, Lubbock, Texas, USA

1. INTRODUCTION AND HYPOTHESES

Diabetes Mellitus (DM) is the fifth leading cause of death in the US. Nearly 80% of diabetic mortality is secondary to cardiovascular disease resulting from atherosclerosis. Current therapy is based upon control of blood glucose, cholesterol and triglycerides, primarily through insulin replacement in Type 1 diabetes or oral hypoglycemic agents and/or insulin replacement in Type 2 diabetes. The more intensive the control, the lower the incidence of diabetic complications such as atherosclerosis. The principal investigators were interested in finding safe and effective nutritional supplements that would reduce the need for insulin replacement therapy, provide tighter glucose control, and protect against oxidative stress and the vascular pathology associated with diabetes mellitus. One such supplement is taurine.

Taurine was chosen as the initial compound for study because its effectiveness in controlling glucose levels has already been demonstrated in animal models[1-3]; moreover, it is relatively safe and has been administered routinely as an "over-the-counter" energy drink. Elizarova and Nedosugova[4] outlined a study in which taurine lowered the insulin requirement in human diabetic patients. Franconi[5] described decreased platelet aggregability in diabetic patients receiving taurine. These reports encouraged further investigation of taurine in human with diabetes mellitus.

The hypothesis that taurine, a nontoxic amino acid normally consumed in the human diet, can be beneficial in the treatment of patients with Type 2 DM, was tested through the following specific aims:

1. To determine whether taurine supplementation has a hypoglycemic effect in patients with Type 2 DM

2. To determine if taurine supplementation can reduce the oxidative stress normally observed in plasma of patients with Type 2 DM

2. METHODS

This study was a randomized, double-blind, placebo-controlled clinical trial.

2.1 Subject Inclusion Criteria:

*   Male or female
*   30-65 years of age
*   Type 2 DM as a primary diagnosis
* $HBA_{1C} > 7.0\%$
*   Not taking (or willing to stop taking) any dietary supplements; e.g., vitamins, minerals (particularly antioxidant vitamins/minerals, amino acids, fatty acids, herbs, fortified drinks, etc.)

2.2 Subject Exclusion Criteria:

*   Pregnancy
*   Receiving insulin therapy
*   Receiving lipid lowering medication
*   Recent or prospective surgical interventions
*   Complications of Type 2 DM, e.g., active cardiovascular disease, nephropathy, retinopathy, neuropathy

2.3 Subject Enrollment:

Subjects who met study criteria were assigned randomly to one of two groups:

- Placebo only (three capsules of an inert substance, twice per day)
- 3000 mg. of taurine (three 500 mg capsules, twice per day)

Projected enrollment was 30 experimental subjects and 15 control subjects. Over 15 months, 45 patients were enrolled in the study, but 13 failed to complete the study due to changes in medication, noncompliance or failure to follow-up, leaving 32 subjects: 10 control and 22 experimental.

Table 1. Procedures Required of Study Subjects.

Timeline	Activity
Baseline	1. Measured height and weight; calculated body mass index (BMI) 2. Evaluated blood pressure by standard measurements 3. Drew 3 tablespoons of blood for laboratory testing • insulin, taurine and TBARS measured in the Pharmacology labs • glucose, HbA_{1c}, and lipid profile measured in the hospital lab 4. Dispensed one month supply of capsules
1 month / 3 months	1. Returned unused capsules to study investigator 2. Received one month supply of new capsules 3. Answered questions concerning tolerance and compliance
2 months / 4 months	1. Measured weight; calculated body mass index (BMI) 2. Evaluated blood pressure by standard measurements 3. Drew 3 tablespoons of blood for laboratory testing • insulin, taurine and TBARS measured • glucose, HbA_{1c}, and lipid profile measured 4. Returned unused capsules to study investigator 5. Dispensed one month supply of capsules (2 months only) 6. Answered questions concerning tolerance and compliance

2.4 Statistical Analysis

All data are presented as means, plus or minus their standard deviation. These data were analyzed using the student's two-tailed t-test to evaluate differences of continuous outcome variables within and between groups ($p<0.05$).

3. RESULTS

At the conclusion of the 4-month study, subjects in both the taurine and control groups had higher levels of $HbgA_{1C}$ (Fig. 1) and fasting glucose (Fig. 2) than at baseline. At baseline, subjects in both groups had whole blood taurine levels within normal range. Whole blood taurine levels (Fig. 3) in the taurine group were significantly higher at 2 months and 4 months than at baseline. At the conclusion of the trial, the taurine supplemented group had a significantly higher level of whole blood taurine (407.3 umol/L) than at baseline (316.7 umol/L); p<0.0002.

Laboratory data revealed no significant differences in glucose, HgbA1c, lipids, insulin, or TBARS between the two groups at the conclusion of the study. As with other studies that have addressed taurine and Type 2 DM, this study failed to demonstrate a glucose-lowering effect of taurine in humans.

Some studies in Type 2 DM have demonstrated low plasma taurine levels and reduced taurine uptake by platelets. Our study indicated whole blood levels of taurine to be within the normal range; plasma and platelet taurine were not measured.

Supplementation of 3000 mg of taurine significantly increased whole blood levels of taurine; effect in other tissues was not measured. No change was found in insulin levels or oxidative stress in taurine supplemented Type 2 DM subjects, versus similar non-supplemented subjects.

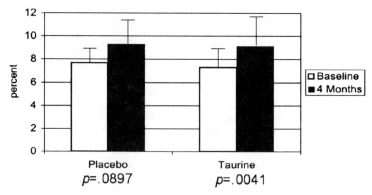

Figure 1. Hemoglobin A1C levels at baseline and four months.

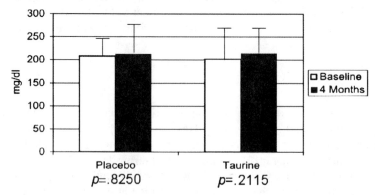

Figure 2. Fasting Glucose levels at baseline and four months.

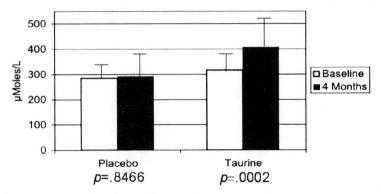

Figure 3. Taurine levels at baseline and four months.

3.1 Study Limitations

Study results were likely impacted by a number of factors. For example, excluding patients who were taking cholesterol-lowering medications severely limited the number of available subjects. Although study objectives included an assessment of the effect of taurine supplementation on patients' lipid levels, most potential subjects had already been placed on lipid-lowering medications and thus were excluded as participants. Moreover, all subjects were in poor glycemic control at the time of their enrollment in the study. Focusing on subjects with newly diagnosed Type 2 DM might have been more enlightening. Measures of plasma and platelet taurine levels would have been beneficial, but the lack of laboratory capacity prevented those specimens from being assayed in a timely manner.

4. CONCLUSIONS AND FUTURE DIRECTIONS

Taurine supplementation at the level of 3000 mg. per day significantly increased the whole blood taurine levels in patients with Type 2 DM. However, taurine supplementation at this level did not improve glucose, lipid, insulin or TBAR levels.

Future projects related to taurine and diabetes should analyze the following:

- Differences between Type 1 DM and Type 2 DM

- Measures of earlier (and possibly reversible) oxidative stress, such as reactivity to carbonyl groups[6,7]

- Measures of platelet and plasma taurine[8,9]

- Changes in taurine status among newly diagnosed DM, mid-stage DM, and end-stage DM[7]
- Earlier intervention with less complicated subjects[7,10]
- Larger doses of taurine supplementation
- Taurine as a complementary agent for the prevention of diabetic complications[10]
- Effect of study participation on subject behavior and compliance

ACKNOWLEDGEMENTS

The authors wish to acknowledge Charles A. Bradley, Ph.D., Professor of Pathology, Texas Tech University Health Sciences Center, for his guidance regarding laboratory measurements of HgA$_{1C}$. This study was supported by Grant #99151 of the Dean's Clinical Research Program at Texas Tech School of Medicine.

REFERENCES

1. Lampson, W.G., Kramer, J.H. and Schaffer, S.W., 1983, Potentiation of the actions of insulin by taurine. *Can. J. Physiol. Pharmacol.* 61: 457-463.
2. Kulakowski, E.C., Maturo, J. and Schaffer, S.W., 1978, Identification of taurine receptors from rat heart sarcolemma. *Biochem. Biophys. Res. Commun.* 80: 936-941.
3. Dokshina, G.A., Silaeva, T.Y. and Yartsev, E.I., 1976, Some insulin-like effects of taurine. *Vopr. Med. Khim.* 22: 503-507.
4. Elizarova, E.P., and Nedosugova, L.V., 1996, First experience in taurine administration for diabetes mellitus. *Adv. Exp. Med. Biol.* 403: 583-587.
5. Franconi, F., Miceli, M., Fazzini, A., Seghieri, G., Dileo, M., Lepore, D., and Ghirlanda, G., 1996, Taurine and Diabetes. *Adv. Exp. Med. Biol.* 403: 579-582.
6. Baynes, J.W. and Thorpe, S.R., 1999, Role of oxidative stress in diabetic complication. *Diabetes* 48:1-9.
7. Hansen, S.H., 2001, The role of taurine in diabetes and the development of diabetic complications. *Diabetes Metab. Res. Rev.* 17:330-346.
8. Mustard J.F., and Packman, M.A., 1984, Platelets and diabetes mellitus. *N. Engl. J. Med.* 311:665.
9. Franconi, F., Bennardini, F., Mattana, A., *et al.*, 1995, Plasma and platelet taurine are reduced in subjects with insulin-dependent diabetes mellitus: effects of taurine supplementation. *Am. J. Clin. Nutr.* 61:1115-1119.
10. McCarty, M.F., 1997, Exploiting complementary therapeutic strategies for the treatment of type II diabetes and prevention of its complications. *Med. Hypotheses* 49:143-152.

Hypoglycemic Effects of Taurine in the Alloxan-Treated Rabbit, a Model for Type 1 Diabetes

THOMAS E. TENNER, JR. [*], XIN JIAN ZHANG, [*] and JOHN B. LOMBARDINI[*,#]

*Department of Pharmacology and #Department of Ophthalmology & Visual Sciences, Texas Tech University Health Sciences Center, Lubbock, Texas USA

1. INTRODUCTION

Diabetes is the third leading cause of death in the US[1]. Prior to the discovery of insulin by Banting and Best in 1921, most diabetic mortality resulted from hyperglycemic ketoacidosis. Since that time, cardiovascular disease secondary to the diabetic state has become the leading cause of death in the diabetic population. While insulin replacement therapy appears sufficient to prevent severe hyperglycemic episodes, control is not sufficient to avert cardiovascular complications. Indeed, there is considerable evidence that insulin, itself, may play a deleterious role in the adverse cardiovascular sequelae associated with diabetes. Under normal conditions, the body releases only that amount of insulin required for control of blood glucose. In contrast, replacement therapy for the diabetic individual involves a bolus of insulin, resulting in early excessive levels followed by declining levels with time.

Consequently, many investigators have searched for compounds, which can produce insulin-like effects in the diabetic without the potential adverse sequelae. Indeed, a compound that could potentiate the hypoglycemic actions of insulin might allow a reduction in the amount of insulin-required daily.

Taurine has long been considered a non-essential but nutritionally important amino acid in certain animals including humans. There is now an increasing body of knowledge that taurine, at pharmacological concentrations, can alter carbohydrate metabolism. Several authors have demonstrated that taurine can produce a hypoglycemic effect[2-4]. Similarly, taurine has been shown to potentiate he actions of insulin in cardiac muscle[5], and, may bind to the insulin receptor[6].

Taurine 5: Beginning the 21st Century
Edited by Lombardini *et al.*, Kluwer Academic/Plenum Publishers, New York 2003

Indeed, taurine has been shown to reduce blood glucose levels acutely in diabetic animals[7,8,9]. Finally, taurine has been proposed to blunt the atherosclerotic actions of elevated cholesterol levels[10]. This latter study reported that while plasma cholesterol was not markedly altered, atherosclerotic lesions induced by an atherogenic, high cholesterol (2%), diet were significantly reduced.

The purpose of the current work was to reproduce the hypoglycemic action of taurine in control rabbits, and then extend this work to determine if taurine could chronically reduce the hyperglycemia and dyslipidemia associated with the alloxan-treated rabbit model for Type 1 diabetes. Rabbits received either normal drinking water or taurine-supplemented (0.5%) drinking water for six weeks. Blood chemistries were run on the rabbits to determine if taurine had an intrinsic hypoglycemic action in control and diabetic animals.

2.   METHODS

All studies were conducted in New Zealand White rabbits (male, 1.0 kg) allowed to adjust to the Animal Resource Facility for two weeks. At the end of this time, blood samples were drawn from the central ear vein and plasma glucose, cholesterol, triglycerides, and serum insulin levels were determined according to previously established methods[11-13]. Whole blood taurine levels were measured according to Lombardini[14]. For initial studies, control rabbits were placed on taurine supplementation (0.5% in drinking water) for six weeks. At the end of this time plasma glucose was again measured.

For studies involving diabetic rabbits, animals were fasted overnight one week following the initial blood draw, and subsequently received a single intravenous injection of alloxan monohydrate (100 mg/kg) in saline via the marginal ear vein[10,11]. In an effort to prevent the potentially fatal hypoglycemia that results from massive insulin release, all rabbits received dextrose in water (25% solution), subcutaneously every four to six hours after alloxan treatment for the first 24 hours. Thereafter, rabbits were maintained with free access to food and water for 6 weeks. At the end of this time period, blood samples were again drawn and blood chemistries and whole blood taurine levels measured. Subsequently, all rabbits received taurine (0.5% in drinking water). Six weeks later, blood samples were again drawn as described above and blood chemistries and taurine levels determined and compared. As such, each rabbit served as its own control and diabetic reference. Data are presented as means plus or minus the standard error of the means. Statistical significance was determined either by paired "t" test or by one-way analysis of variance followed by the Newman-Keuls test for multiple comparisons. Significance was determined as P <0.05.

3.   RESULTS

An initial study was performed to determine if taurine, administered as a dietary supplement in the drinking water, would result in a hypoglycemic effect. The initial concentration of taurine studied was 1%, however, several rabbits developed diarrhea at this concentration and would not drink. Consequently, the concentration was lowered to 0.5% taurine and this concentration was well tolerated by all animals subsequently studied. Six rabbits were initially studied with a mean plasma glucose level of 163 ± 22 mg/dl. Six weeks of taurine-supplementation at a concentration of 0.5% resulted in a significant reduction in plasma glucose (Fig. 1). Some degree of plasma glucose reduction was apparent in all animals studied.

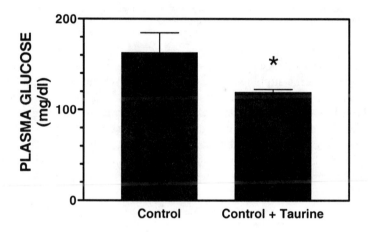

Figure 1. Comparison of plasma glucose before and six weeks after taurine supplementation in drinking water (0.5%). Each rabbit was its own control (n=6). Values are significantly (P< 0.05) different as determined by the paired "t" test.

Subsequent studies were designed to determine if chronic taurine administration could reduce the hyperglycemia and dyslipidemia associated with alloxan-induced Type 1 diabetes in the rabbit. Five rabbits were studied. Normal plasma glucose was 150 ± 17 mg/dl (Fig. 2) with serum insulin levels of 17.3 ± 7 µU/ml (Fig. 3).

Six weeks after alloxan-treatment, the average plasma glucose level was 565 ± 86 mg/dl (Fig. 2) and serum insulin levels were 2.42 ± 0.5 µU/ml (Fig. 3). Rabbits were then provided taurine supplementation in their drinking water as described above for six weeks. At the end of the study, plasma glucose levels were significantly reduced, 315 ± 88 mg/dl (Fig 2), from diabetic levels but serum insulin levels were unaffected (Fig. 3).

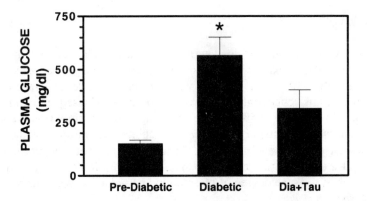

Figure 2. Plasma glucose levels measured in pre-diabetic, diabetic, and taurine-supplemented diabetic rabbits. Each rabbit was its own control (n=5). Diabetic plasma glucose values were statistically (P<0.05) greater than either pre-diabetic levels or those following taurine-supplementation.

Subjective observations consistent with the decrease in plasma glucose levels associated with taurine-supplementation included reversal of the polydipsia, polyphagia, and polyurea normally observed in diabetic animals. Indeed, one animal, whose blood glucose was not excessively elevated by alloxan-treatment, responded so well to taurine that it was initially assumed to have overcome the toxic effects of alloxan and reverted back to the non-diabetic state. This animal was switched back to normal drinking water. Within three weeks, plasma glucose levels returned to the diabetic levels observed prior to taurine-supplementation (Fig. 4).

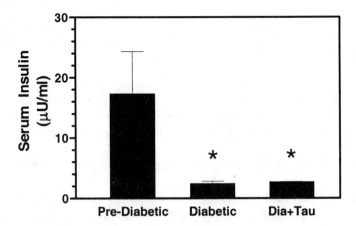

Figure 3. Serum insulin levels measured in pre-diabetic, diabetic, and taurine-supplemented diabetic rabbits. Each rabbit was its own control (n=5). Diabetic serum insulin levels as well as those following taurine-supplementation were statistically (P<0.05) less than pre-diabetic levels; and, indeed, were at the limit of detection of the assay.

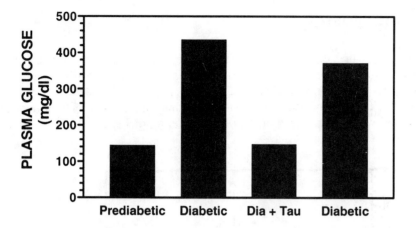

Figure 4. Plasma glucose levels measured in one rabbit during the pre-diabetic, diabetic, and taurine-supplemented treatment periods. After six weeks on taurine-supplementation, the rabbit was switched back to normal drinking water and hyperglycemia reappeared.

Measurements were also made to determine plasma cholesterol (Fig. 5) and triglyceride (Fig. 6) levels over the course of the study. While cholesterol levels were found not to be statistically different across the study, triglyceride levels were significantly increased following induction of diabetes with alloxan and were reduced following taurine-supplementation. Finally, whole blood taurine levels were determined to assure that taurine levels in rabbits could actually be altered through supplementation in the drinking water. Taurine levels appeared to drop in the diabetic relative to pre-diabetic levels but significance was not achieved. As expected, taurine supplementation (0.5% in the drinking water) significantly increased taurine levels relative to both the pre-diabetic and diabetic state.

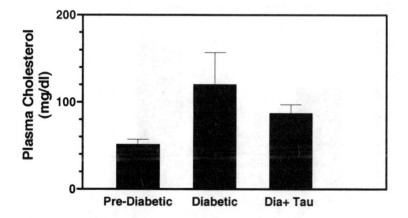

Figure 5. Plasma cholesterol levels measured in pre-diabetic, diabetic, and taurine-supplemented diabetic rabbits. Each rabbit was its own control (n=5). While diabetic cholesterol levels appeared elevated, no statistically significant differences were observed between the three measurements.

4.   DISCUSSION

Results of the current study demonstrate the effects of oral taurine supplementation in control and diabetic rabbits. The alloxan-treated rabbit represents an extensively studied and well characterized model of Type 1 diabetes[12,13]. The current study indicates that taurine-supplementation in drinking water (0.5%) was well tolerated in this rabbit model for diabetes.

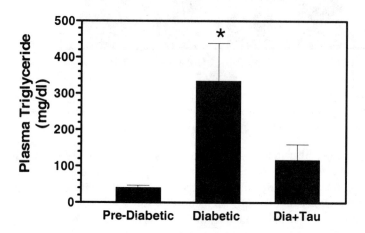

Figure 6. Plasma triglyceride levels measured in pre-diabetic, diabetic, and taurine-supplemented diabetic rabbits. Each rabbit was its own control (n=5). Diabetic plasma triglyceride values were statistically (P<0.05) greater than either pre-diabetic levels or those following taurine-supplementation.

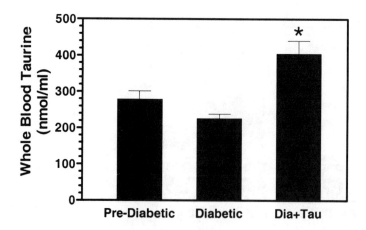

Figure 7. Whole blood taurine levels measured in pre-diabetic, diabetic, and taurine-supplemented diabetic rabbits. Each rabbit was its own control (n=5). Diabetic taurine levels appeared to be diminished, but did not reach the level of significance. In contrast, taurine supplemented rabbits demonstrated values statistically (P<0.05) greater than either pre-diabetic or diabetic levels.

Taurine levels in whole blood of treated animals were significantly increased over the period of treatment. Taurine (0.5%) produced a hypoglycemic action in both non-diabetic and diabetic animals that is not related to elevations in insulin levels. Similarly, taurine appears to possess an anti-dyslipidemic effect, at least relative to triglyceride levels.

Diabetes has been associated with taurine depletion in rats[15]. Similarly, human studies have reported increases in urinary taurine excretion[16] and reduced plasma and platelet concentrations in patients with Type 1 diabetes[17]. Indeed, taurine supplementation was shown to normalize or slightly raise plasma taurine levels[17]. In the current study, induction of Type 1 diabetes resulted in a decrease in whole blood taurine that did not achieve the level of significance. The lack of significance may be related to the low numbers of animals studied. In contrast, taurine supplementation resulted in a significant increase in whole blood taurine and was associated with decreases in plasma glucose. Consistent with the reduction in plasma glucose, taurine supplemented rabbits did not display the polydipsia, polyphagia, and polyurea normally observed in diabetic animals. Of note, taurine supplementation did not restore or improve serum insulin levels, indicating that the hypoglycemic effect of taurine was not related to recovery of the β islet cells of the pancreas. Furthermore, taurine supplementation resulted in a decrease in plasma triglyceride levels but not cholesterol. A similar lack of effect of taurine on plasma cholesterol levels has been reported [10]. Likewise, the ability of taurine to lower plasma triglycerides has been demonstrated in diabetic rats[15].

While preliminary in nature, these data and those reported in the literature, indicate that taurine supplementation might play a protective role in Type 1 diabetes. Not only might taurine reduce the insulin needed for adequate plasma glucose control as suggested by Elizarova and Nedosugova[18], but also the apparent anti-dyslipidemic effect exerted on plasma triglycerides may be protective in the atherosclerotic complications known to accompany diabetes. This speculation is strengthened by the observations that taurine decreases atherosclerotic lesions in rabbits fed atherogenic diets[10].

ACKNOWLEDGEMENTS

This work was supported by grants from the South Plains Foundation, American Heart Association, Texas Affiliate, and TTUHSC Seed Grant Program.

REFERENCES

1. Kannel, W.B. and McGee, D.L.,1979, Diabetes and cardiovascular disease. The Framingham study. *JAMA* 241: 2035-2038.
2. Ackerman, D. and Heinsen, H.A. ,1935, Uber die physiologische wirkung des asterubins and anderer, zum tiel neu dargestellter, schwefelhaltiger guanidinderivate. *Z. Physiol. Chem.* 235:115 – 121.
3. Macallum, A.B., and Silvertz, C., 1942, The Potentiation of insulin by sulfones. *Can. Chem. Process.* 26:669.
4. Kulakowski, E.C., Maturo, J., 1984, Hypoglycemic properties of taurine: not mediated by enhanced insulin release. *Biochem. Pharmacol.* 33:2835-2838.
5. Lampson, W.G., Kramer, J.H., and Schaffer, S.W. Potentiation of the actions of insulin by taurine. *Can. J. Physiol. Pharmacol.* 61: 457 – 463.

6. Kulakowski, E.C., Matsuro, J., and Schaffer, S.W., 1985, The low affinity taurine-binding protein may be related to the insulin receptor. In *Taurine: Biological Actions and Clinical Perspectives,* eds. Oja, S.S., Ahtee, L., Kontro, P., and Paasonen, M.K. Allan R. Liss, Inc., New York, pp. 126-136.

7. Shimuza, T., Shimizu, S., Nagami, S., and Wada, M., 1960, Effect of hypotaurine and cysteine on alloxan- and dehydroxyascorbic acid- diabetes. *Niigata Med. J.* 74:931-939.

8. Silaeva, T.Y., Dokshina, G.A., Yartsev, E.I., Yakoleva, V.V., and Arkhangelskaya, T.E., 1976, The effect of taurine on the carbohydrate metabolism of diabetic animals. *Prob. Endokrinol.* 22:99-103.

9. Tokunaga, H., Yoneda, Y., and Kuriyama, K., 1979, Protective actions of taurine against streptozotocin-induced hyperglycemia. *Biochem. Pharmacol.* 28:2807-2811.

10. Petty, M.A., Kintz, J., DiFrancesco, G.F., 1990, The effects of taurine on atherosclerosis development in cholesterol-fed rabbits. *Eur. J. Pharmacol.* 180:119-127.

11. Ramanadham, S. and Tenner, T.E., Jr., 1987, Alterations in the myocardial ß-adrenoceptor system of streptozotocin diabetic rats. *Europ. J. Pharmacol.* 136(3): 377-390.

12. Alipui, C., Ramos, K. and Tenner, T.E., Jr., 1993, Alterations of aortic smooth muscle cell proliferation in diabetes mellitus. *Cardiovascular Research* 27: 1229-1232.

13. Davis, R.L., Lavine, C.L., Arredondo, M.A., McMahon, P., and Tenner, T.E., Jr., 2002, Differential indicators of diabetes-induced oxidative stress in New Zealand White rabbits: Role of dietary vitamin E supplementation. *Experimental Diabetes Research* 3: 1-7.

14. Lombardini, J.B., 1986, Taurine levels in blood and urine of vitamin B6 deficient and estrogen-treated rats. *Biochem. Med.* 35:125-131.

15. Goodman, H.O., Shihabi, Z.K., 1990, Supplemental taurine in diabetic rats: effects on plasma glucose and triglycerides. *Biochem. Med. Metab. Biol.* 43:1 – 9.

16. Martensson, J. Hermansson, G., 1984, Sulfur amino acid metabolism in juvenile onset nonketotic and ketotic diabetic patients. *Metabolism* 33:425-428.

17. Franconi, F., Bennardini, F., Mattana, A., Miceli, M., Ciuti, M., Mian, M,. Gironi, A., Anichini, R. and Seghieri, G., 1995, Plasma and platelet taurine are reduced in subjects with insulin-dependent diabetes mellitus: effects of taurine supplementation. *Am. J. Clin. Nutr.* 61(5):1115-1119.

18. Elizarova, E.P. and Nedosugova, L.V., 1996, First experience in taurine administration for diabetes mellitus. *Adv. Exp. Med. Biol.* 403: 583-587.

Part 3:

Osmoregulatory Properties of Taurine and Taurine Transport

Taurine and Cellular Volume Regulation in the Hippocampus

JAMES E. OLSON*[#], NORMAN R. KREISMAN[†], JENNY LIM*, BETH HOFFMAN-KUCZYNSKI*, DANA SCHELBLE*, and JAMES LEASURE*

*Department of Emergency Medicine and [#]Department of Physiology and Biophysics, Wright State University School of Medicine, Dayton, Ohio, USA and [†]Department of Physiology, Tulane University Medical School, New Orleans, Louisiana, USA

1. INTRODUCTION

Taurine is an important osmolyte used for volume regulation in a variety of cell culture systems, including neurons and astroglial cells[1-2]. Taurine efflux from brain slices and taurine mobilization to the extracellular space of the intact brain appears during exposure to hypoosmotic conditions also suggesting this amino acid is used for brain volume regulation[3-4]. Taurine contents of isolated cells and brain slice preparations decrease within minutes of exposure to hypoosmolality and other conditions which cause cell swelling. However, little change is observed in brain taurine content of animals exposed to hypoosmotic hyponatremia for several hours[5-6]. Thus, taurine may be redistributed between different cell types of the brain in response to cell swelling. To obtain a better understanding of the role taurine plays in volume regulation of the brain *in situ*, we examined taurine contents and cellular volume in slices of rat hippocampus during hypoosmotic exposure.

Although substantial hypoosmotic volume regulation is observed in cultures or suspensions of neurons and glia, similar volume regulation in rat hippocampal slices has not been observed[6]. We reasoned that cellular volume regulation in brain slices is diminished because taurine and other important organic osmolytes are lost during the preparation and subsequent incubation of the slices in physiological saline solution. Because of the importance of taurine as an intracellular osmolyte, we examined the role of this amino acid for volume regulation in hippocampal slices swollen under hypoosmotic conditions. We hypothesized that slices containing normal amounts of taurine would show volume regulation during hypoosmotic swelling.

2.   MATERIALS AND METHODS

Slices of adult rat hippocampus were prepared as previously described[8]. Anesthetized rats were perfused with 4°C isoosmotic bicarbonate-buffered saline (290 mOsm, Iso-BBS) containing 124 mM NaCl, 3.5 mM KCl, 2.0 mM $CaCl_2$, 1.0 mM $MgCl_2$, 26 mM $NaHCO_3$, 1.0 mM NaH_2PO_4, and 10 mM glucose, via the left cardiac ventricle. The brain was removed and immediately chilled in a frozen slurry of Iso-BBS for 5-10 min. The hippocampus was dissected from the brain and the middle one-third was cut into 6-8, 400 μm-thick slices. Slices then were placed in Iso-BBS at room temperature (23°C) in an incubation chamber with a humidified atmosphere of 95% O_2 plus 5% CO_2. For some studies, the Iso-BBS contained 1 mM taurine.

Slices remained in the incubation chambers for 1.5-2 hr. Then they were placed on nylon mesh in a perfusion chamber with Iso-BBS at 35°C. For some slices, this solution contained 1 mM taurine for the first 30 min. After a total of 60 min in this chamber, the perfusion solution was changed to hypoosmotic bicarbonate-buffered saline (200 mOsm, Hypo-BBS) made by reducing the concentration of NaCl of Iso-BBS.

The intensity of transmitted light through the CA1 region of the slice was continuously measured as an index of cellular volume[9]. At various times before and during the hypoosmotic exposure, slices were removed for fixation in 4% paraformaldehyde plus 80 mM Na_2HPO_4 (pH=7.4) or 0.6 M $HClO_4$ for morphometric or chemical analyses, respectively. Paraformaldehyde-fixed slices were embedded in Durcupan, sectioned at 1 μm thickness, and stained with cresyl violet. The pyramidal cell layer in the CA1 region was digitally photographed. Diameters of pyramidal cell somas with a visible nucleolus were measured in the plane of the pyramidal cell layer using image analysis software. Tissue processing and morphometric measurements were done contemporaneously for taurine-treated and untreated slices by individuals unaware of the treatment group of the slices and the time of exposure to Hypo-BBS. Slices fixed in $HClO_4$ were sonicated and centrifuged at 10,000 × g for 1 min. Amino acids and potassium contents were determined in the supernatant by HPLC and atomic absorption spectroscopy, respectively[10]. Protein was determined in the pellet by the method of Lowry et al[11]. Other slices were removed from the perfusion chamber, blotted free of adherent water and placed on a kerosene-bromobenzene column to determine the slice specific gravity[12]. A separate group of slices were dried at 105°C for 24 hr to determine the specific gravity of the dry hippocampal material.

Data are expressed as the mean ± SEM. Results were analyzed by one-way or repeated measures ANOVA with *post hoc* two-tailed Student's t-test or Dunnett's test as appropriate. Significance was indicated for p < 0.05.

3. RESULTS

Slices fixed in $HClO_4$ immediately after dissection from the brain had taurine contents of 47.3 ± 4.5 nmol/mg protein (N=16), a mean value significantly lower that that reported for rat hippocampus *in situ*[13]. The taurine content of slices incubated 1.5-2 hr at room temperature and then 60 min at 35°C with Iso-BBS containing no taurine was further diminished (Table 1). However, slices incubated at room temperature and for 30 min at 35°C with 1 mM taurine as described above had taurine contents which were similar to those measured in the intact hippocampus. Exposure to Hypo-BBS for 60 min caused a reduction in taurine contents of both taurine-treated and untreated slices. Potassium content of freshly dissected slices was 623 ± 25 nmol/mg protein. Incubation without taurine caused an insignificant decrease in potassium content; however, ATP content of the slices was not altered by taurine treatment (data not shown), indicating the slices remained viable. Contents of potassium and amino acids other than taurine were not modified by taurine treatment during slice incubation in Iso-BBS or subsequent exposure to Hypo-BBS.

Exposure to Hypo-BBS caused an increase in the intensity of transmitted light through the stratum radiatum of the CA1 region of all slices (Fig. 1). Light transmission intensity increased to a maximum value after 16 ± 2 min of exposure to Hypo-BBS in taurine-treated slices and 18 ± 2 min of Hypo-BBS exposure in untreated slices (N=12). The maximum change in light intensity was similar for both taurine-treated and untreated slices (Table 2). Significant recovery of the light transmission signal was observed after 60 min of exposure to Hypo-BBS in taurine-treated but not in untreated slices.

In a separate series of experiments, we exposed taurine-treated slices to 100 μM NPPB (5-nitro-2-(3 phenylpropylamino)benzoic acid), an inhibitor of volume activated anion channels and osmotically induced taurine efflux[14], during exposure to Hypo-BBS. Other slices were exposed to 0.3% ethanol vehicle used to solubilize NPPB. No changes in transmitted light intensity were observed during NPPB exposure in Iso-BBS. Slices treated with NPPB or vehicle alone demonstrated a similar maximal increase in light transmission during Hypo-BBS exposure. Vehicle-exposed slices showed a similar recovery of the light transmission signal as slices without vehicle exposure. However, NPPB completely blocked the recovery of the light transmission signal.

The diameters of pyramidal cells in the CA1 layer were similar for taurine-treated and untreated cells at the end of exposure to Iso-BBS (Table 3). Cells in untreated slices demonstrated significant swelling during the first 15 min of Hypo-BBS exposure. In contrast, cells from taurine-treated slices demonstrated no change in diameter during this initial period of hypoosmotic exposure. After 60 min in Hypo-BBS, both groups of slices showed a significant increase in pyramidal cell diameter.

Table 1. Taurine and potassium contents of hippocampal slices.

	End of exposure to Iso-BBS	After 15 min in Hypo-BBS	After 60 min in Hypo-BBS
Taurine Contents			
Taurine-treated Slices	89.3 ± 7.2	61.2 ± 3.4*	41.1 ± 10.1*
Untreated Slices	27.6 ± 2.0	28.7 ± 2.8	13.9 ± 2.1*
Potassium Contents			
Taurine-treated Slices	647 ± 46	632 ± 42	633 ± 68
Untreated Slices	522 ± 49	516 ± 65	544 ± 61

Values are given in units of nmol/mg protein and are calculated from 7-9 (taurine-treated) and 9-10 (untreated) independent determinations. * indicates mean values significantly different from that measured at the end of exposure to Iso-BBS.

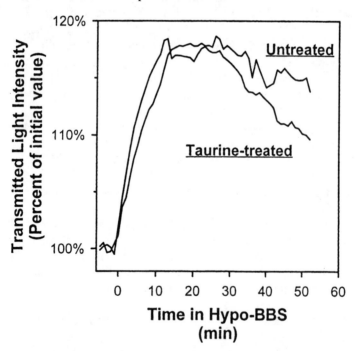

Figure 1. Transmitted light intensity in the CA1 stratum radiatum of taurine-treated and untreated hippocampal slices. At 0 min the perfusion solution was changed from Iso-BBS to Hypo-BBS. Values are expressed as the percent of the average intensity measured in Iso-BBS during the 5 min prior to hypoosmotic exposure.

Table 2. Transmitted light intensity during hypoosmotic exposure

	After 15 min in Hypo-BBS	After 60 min in Hypo-BBS
Taurine-treated Slices	126 ± 4%	118 ± 2%*
Untreated Slices	128 ± 3%	124 ± 4%

Values are expressed as a percentage of the intensity measured in Iso-BBS just prior to the start of Hypo-BBS exposure and are calculated from 12 independent determinations.*Indicates mean values significantly different from that measured after 15 min in Hypo-BBS.

Table 3. CA1 pyramidal cell diameters during hypoosmotic exposure

	End of exposure to Iso-BBS	After 15 min in Hypo-BBS	After 60 min in Hypo-BBS
Taurine-treated Slices	11.1 ± 0.3	11.8 ± 0.8	13.6 ± 0.4*
Untreated Slices	10.4 ± 0.6	14.4 ± 0.4*	13.9 ± 0.4*

Values are given in units of μm and are calculated from measurements on 17-43 different cells in 2-3 independent slices. * Indicates mean values significantly different from that measured at the end of exposure to Iso-BBS.

The mean specific gravity of taurine-treated and untreated slices were 1.0305 ± 0.0006 and 1.0306 ± 0.0007 (N=8), respectively, at the end of exposure to Iso-BBS. Dried hippocampal slices were found to have a specific gravity of 1.35 ± 0.01 (N=9). Using this mean value in the formula of Nelson et al.[12] we calculated slice water contents of 88.7 ± 0.2% and 88.6 ± 0.3% for taurine-treated and untreated slices, respectively. Consistent with the initial changes observed in the light transmission signal, the specific gravity of both taurine-treated and untreated slices decreased to similar values of 1.0260 ± 0.0004 and 1.0263 ± 0.0006, respectively, after 15 min in Hypo-BBS, indicating net increases in slice water content. However, both groups of slices demonstrated no further change in specific gravity during the remainder of the period of exposure to Hypo-BBS.

4. DISCUSSION

Our data indicate that physiological levels of taurine are necessary for cellular volume regulation of hippocampal slices during hypoosmotic swelling. Taurine has been shown to participate in cell volume regulation of cultured astrocytes[15-17] and neurons[2,18]. Because both neurons and glia possess high affinity Na^+- and Cl^--dependent taurine transporters[19-21], the cellular location of taurine accumulated during the incubation periods prior to hypoosmotic exposure is uncertain. We observed that pyramidal cells in taurine-treated slices show delayed swelling during hypoosmotic exposure. Thus, these and other neurons are expected to accumulate sufficient taurine during incubation in Iso-BBS to facilitate volume regulation through taurine efflux when exposed to Hypo-BBS.

The magnitude of the volume regulation observed here is smaller and slower than that reported for isolated cells. The time-course of swelling in these brain slices might be slowed compared with that of cultured cells due to diffusion of the hypotonic solution from the bath into the interior of the slice through the more restricted extracellular space of the brain slices. Rather than swelling in response to a rapid change in extracellular osmolarity, cells within the slice may maintain normal volume for an extended period, similar to the isovolumetric volume regulation previously described in kidney and brain cell preparations exposed to slowly changing osmolarity[22-23].

Our data using the taurine transport inhibitor, NPPB, indicate taurine efflux is critical for volume regulation in the hippocampus. As others have observed, we anticipate some of this taurine will leave the slice and be carried away by the perfusing solution[4,24]. However, the restricted extracellular space and close apposition of neuronal and glial elements in hippocampal slices may facilitate intercellular transfer of taurine from neurons to glia during hypoosmotic exposure similar to the observations of Nagelhus et al.[25] for rat cerebellum. Net transfer of taurine and resulting osmotic water flux between hippocampal neurons and glial cells would be consistent with our observed lack of recovery of slice specific gravity following the initial swelling. Because taurine treatment of the slices promotes recovery of the transmitted light signal during hypoosmotic exposure, our results indicate a differential contribution of neuronal and astroglial volume changes to the net light transmission signal. Others have suggested a greater importance of changes in the volume of neuronal dendrites compared with glial elements to the intrinsic optical signal[26-27]. Analysis of the changes in neuronal and glial cell volumes and corresponding changes in taurine content in this *in vitro* system will further our understanding of the role taurine plays in volume regulation of each cell type.

ACKNOWLEDGEMENTS

Supported by the National Institutes of Health (NS37485).

REFERENCES

1. Kimelberg, H.K., Goderie, S.K., Higman, S., Pang, S., and Waniewski, S., 1990, Swelling induced release of glutamate, aspartate, and taurine from astrocyte cultures. *J. Neurosci.* 10: 1483-1491.

2. Pasantes-Morales, H., Chacón, E., Murray, R.A., and Morán, J., 1994, Properties of osmolyte fluxes activated during regulatory volume decrease in cultured cerebellar granule cells. *J. Neurosci. Res.* 37: 720-727.

3.   Solis, J.M., Herranz, A.S., Herreras, O., Lerma, J., and Martin del Río, R., 1988, Does taurine act as an osmoregulatory substance in the rat brain? *Neurosci Lett* 91:53-58.
4.   Law, R.O., 1994, Taurine efflux and the regulation of cell volume in incubated slices of rat cerebral cortex. *Biochim. Biophys. Acta* 1221: 21-28.
5.   Bedford J.J., and Leader J.P., 1993, Response of tissues of the rat to anisosmolality *in vivo. Am. J. Physiol.* 264:R1164-R1179.
6.   Olson J.E., Banks M., Evers J., and Dimlich R., 1997, Blood-brain barrier water permeability and brain osmolyte content during edema development. *Acad. Emerg. Med.* 4:662-673.
7.   Andrew, R.D., Lobinowich, M.E., and Osehobo, E.P., 1997, Evidence against volume regulation by cortical brain cells during acute osmotic stress. *Exp. Neurol.* 143: 300-312.
8.   Kreisman, N.R., and LaManna, J.C., 1999, Rapid and slow swelling during hypoxia in the CA1 region of rat hippocampal slices. *J. Neurophysiol.* 82: 320-329.
9.   Kreisman, N.R., LaManna, J.C., Liao, S.-C., Yeh, E.R., and Alcala, J.R., 1995, Light transmittance as an index of cell volume in hippocampal slices: optical differences of interfaced and submerged positions. *Brain Res.* 693: 179-186.
10.   Olson, J.E., 1999, Osmolyte contents of cultured astrocytes grown in hypoosmotic medium. *Biochim. Biophys. Acta,* 1453: 175-179.
11.   Lowry O.H., Rosebrough N.J., Farr A.L., and Randall R.J., 1951, Protein measurement with the Folin phenol reagent. *J. Biol.Chem.* 193:265-275.
12.   Nelson S.R., Mantz M.L., and Maxwell J.A., 1971, Use of specific gravity in the measurement of cerebral edema. *J. Appl. Physiol.* 30:268-271.
13.   Palkovitz, M., Eleka, S.I., Lang, T., and Patthy, A., 1986, Taurine levels in discrete brain nuclei of rats. *J. Neurochem.* 47: 1333-1335.
14.   Jackson PS, Strange K. Volume-sensitive anion channels mediate swelling activated inositol and taurine efflux. *Am. J. Physiol.* 265:C1489-C1500, 1993.
15.   Pasantes-Morales, H., and Schousboe, A., 1988, Volume regulation in astrocytes: a role for taurine as an osmoeffector. *J. Neurosci. Res.* 20: 505-509.
16.   Pasantes-Morales, H., Morán, J., and Schousboe, A., 1990, Volume-sensitive release of taurine from cultured astrocytes: properties and mechanism. *Glia* 3: 427-532.
17.   Vitarella, D., DiRisio, D.J., Kimelberg, H.K., and Aschner, M., 1994, Potassium and taurine release are highly correlated with regulatory volume decrease in neonatal primary rat astrocyte cultures. *J. Neurochem.* 63: 1143-1149.
18.   Olson, J.E., and Li G., 2000, Osmotic sensitivity of taurine release from hippocampal glial and neuronal cells. In: *Taurine 4.* Della Corte L., Huxtable R.J., Sgaragli G. and Tipton K.F. (Eds.) Plenum Press, New York, pp. 213-218.
19.   Holopainen, I., Malminen, O., and Kontro, P., 1987, Sodium-dependent high-affinity uptake of taurine in cultured cerebellar granule cells and astrocytes. *J. Neurosci. Res.* 18: 479-483.
20.   Beetsch, J.W., and Olson, J.E., 1996, Hyperosmotic exposure alters total taurine quantity and cellular transport in rat astrocyte cultures. *Biochim. Biophys. Acta* 1290: 141-148.
21.   Pow, D.V., Sullivan, R., Reye, P., and Hermanussen, S., 2002, Localization of taurine transporters, taurine, and ^{3}H taurine accumulation in the rat retina, pituitary and brain. *Glia* 37: 153-168.
22.   Lohr, J.W., and Grantham, J.J., 1986, Isovolumetric regulation of isolated S_2 proximal tubules in anisotonic media. *J. Clin. Invest.* 78: 1165-1172.
23.   Tuz, K., Ordaz, B., Vaca, L., Quesada, O., and Pasantes-Morales, H., 2001, Isovolumetric regulation mechanisms in cultured cerebellar neurons. *J. Neurochem.* 79: 143-151.

24. Oja, S.S., and Saransaari, P., 1992, Taurine release and swelling of cerebral cortex slices from adult and developing mice in media of different ionic composition. *J. Neurosci. Res.* 32: 551-561.
25. Nagelhus, E.A., Lehmann, A., and Ottersen, O.P., 1993, Neuronal-glial exchange of taurine during hypo-osmotic stress: A combined immunocytochemical and biochemical analysis in rat cerebellar cortex. *Neuroscience* 54: 615-631.
26. Andrew, R.D., Jarvis, C.R., and Obeidat, A.S., 1999, Potential sources of intrinsic optical signals imaged in live brain slices. *Methods Enzymol.* 18: 185-196.
27. Johnson, L.J., Hanley, D.F., and Thakor, N.V., 2000, Optical light scatter imaging of cellular and sub-cellular morphology changes in stressed hippocampal slices. *J. Neurosci. Meth.* 98: 21-31.

Regulation of the Volume-Sensitive
Taurine Efflux Pathway in NIH3T3 Mouse Fibroblasts

IAN HENRY LAMBERT,

The August Krogh Institute, Universitetsparken 13, DK-2100, Copenhagen Ø, Denmark,
ihlambert@aki.ku.dk

1.   ACTIVATION OF THE VOLUME-SENSITIVE TAURINE EFFLUX PATHWAY – ROLE OF PLA$_2$ AND 5-LIPOXYGENASE

Osmotically swollen NIH3T3 fibroblasts regulate their cell volume towards their initial value just like most mammalian cells[1]. This back regulation, designated regulatory volume decrease (RVD), is the result of loss of KCl and organic osmolytes[1,2]. It is estimated that the loss of cellular amino acids constitutes 20% of the total loss of osmolytes following exposure to a medium with half osmolarity[2]. The swelling-induced K$^+$ efflux is partly via a K$^+$ selective pathway and partly via a K$^+$, Cl$^-$ cotransporter[1,3], whereas the swelling-induced Cl$^-$ efflux is Ca^{2+} independent and via a non-selective anion pathway, that exhibits moderate outward rectification as well as time dependent inactivation[1,3]. Taurine is an important organic osmolyte in NIH3T3 cells and the swelling-induced taurine release from NIH3T3 cells is Na$^+$-independent and mediated via a leak pathway that accepts a variety of neutral amino acids[2]. The volume-sensitive organic osmolyte channel (VSOAC), that accepts Cl$^-$ as well as a broad range of organic osmolytes, has been proposed as the volume-sensitive taurine efflux pathway in mammalian cells[4]. However, the time courses for the swelling-induced taurine efflux and Cl$^-$ efflux in NIH3T3 cells differ with respect to the time point for maximal activation and time point for inactivation[2]. Furthermore, the swelling-induced Cl$^-$ loss and the swelling-induced taurine loss from NIH3T3 cells diverge with respect to their sensitivity to anion channel blockers[2], kinase inhibitors[3] and expression of constitutive Rho[3]. Thus, taurine and Cl$^-$ are most probably released from NIH3T3 cells via separate pathways in accordance with the findings in Ehrlich ascites tumour cells[5] and HeLa cells[6]. The cellular signal cascade,that is activated by osmotic cell swelling and that leads to activation of selective pathways for inorganic and organic osmolytes is only just being revealed in some cell types. Swelling-induced activation of taurine efflux is demonstrated to require phospholipase A_2 (PLA_2) and 5-lipoxygenase (5-LO) activity[7,8,9,10] and to be modulated by cytoskeletal components[11], Ca^{2+}/calmodulin[5,9,12], protein tyrosine kinases and phosphatases[3,13,14,15], and by GTP-binding proteins[3,5,7].

PLA_2 constitutes a diverse family of enzymes, and the secretory Ca^{2+}-dependent PLA_2 ($sPLA_2$), the ubiquitously expressed cytosolic Ca^{2+}-dependent PLA_2 ($cPLA_{2\alpha}$), and the cellular Ca^{2+}-independent PLA_2 ($iPLA_2$) are often involved with cell signal transduction[17]. From Figs. 1A and 1C is seen that exposing NIH3T3 cells to hypotonic NaCl media (200 mOsm, 150 mOsm) induces a transient increase in the rate constant for taurine release, and that the swelling-induced taurine release is potentiated by the anti-calmodulin drug N-(6-aminohexyl)-5-chloro-1-naphathalene sulphonamide (W7), unaffected by the $cPLA_2$ inhibitor AACOCF$_3$, but reduced in the presence of the $iPLA_2$ inhibitor bromoenol lactone (BEL), the 5-LO inhibitor ETH 615-139, and the anion channel blockers 4,4´-diisothiocyano-2,2´-stilbene acid (DIDS) and arachidonic acid (AA). As binding of calmodulin to $iPLA_2$ results in loss of enzymatic activity[18], and the swelling-induced efflux is blocked by BEL and ETH it has been suggested that the swelling-induced activation of taurine efflux from NIH3T3 cells involves $iPLA_2$ and 5-LO activity[16]. From Figs. 1B and 1C it is seen that NIH3T3 cells also release taurine following exposure to isotonic NaCl medium containing the phospholipase activating protein melittin (0.5 µg/ml, 1.0 µg/ml), and that the melittin-induced taurine efflux is reduced in the presence of W7, AACOCF$_3$, BEL, ETH 615-139, DIDS and AA. Thus, melittin-induced taurine efflux from NIH3T3 cells under isotonic conditions presumably involves activation of various types of PLA_2 ($iPLA_2$, $cPLA_2$), 5-LO and the volume-sensitive taurine efflux pathway. It should be noted that RVD in Ehrlich cells is inhibited by the $cPLA_2$ inhibitor AACOCF$_3$[7] as well as by the $sPLA_2$ inhibitor RO 31-4639[5]. Thus, various types of PLA_2 also participate in the swelling-induced activation of osmolyte releasing pathways in some cell systems.

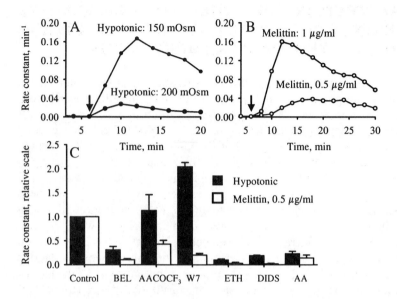

Figure 1. Comparison of swelling- and melittin-induced taurine efflux. Cells, grown to 80% confluence (35 mm diameter polyethylene dishes) in Dulbecco´s Modified Eagle Medium (high glucose) containing heat inactivated fetal bovine serum (10%) and penicillin (100 units/ml), were loaded with [^{14}C]-taurine (80 nCi/ml, 2 hr). The cells were subsequently washed 5 times with isotonic NaCl medium containing in mM: 143 NaCl, 5 KCl, 1 Na$_2$HPO$_4$, 1 CaCl$_2$, 0.1 MgSO$_4$, 5 glucose, and 10 *N*-2-hydroxyethyl piperazine-*N'*-2-ethanesulfonic acid. After the final wash 1 ml of experimental solution was added to the dish, left for 2 min, and transferred to a scintillation vial for estimation of ^{14}C activity. This procedure was repeated throughout the whole experiment, with the solution being replaced at time 6 min (indicated by an arrow) by either hypotonic NaCl medium (200 mOsm, 150 mOsm, NaCl concentration being reduced to 95 mM and 75 mM, respectively, *Panel A*) or isotonic NaCl medium containing melittin (0.5 μg/ml or 1.0 μg/ml, *Panel B*). [^{14}C]-taurine activity remaining in the cells at the end of the efflux experiment was estimated by lyzing the cells with 1 ml NaOH (0.5 M, 1 hr), washing the dishes twice with distilled water and estimating the ^{14}C activity in the NaOH lysate as well as in both water washouts. The rate constant for the initial taurine efflux at a given time point was estimated as the negative slope of a graph, where the natural logarithm to the fraction of ^{14}C activity remaining in the cells at the given time was plotted versus time. *Panel C*: BEL (30 μM), AACOCF$_3$ (40 μM), W7 (50 μM), ETH 615-139 (ETH, 10 μM), DIDS (100 μM) and arachidonic acid (AA, 50 μM) were present in the experimental solution throughout the whole experiment. The maximal rate constant for swelling- and melittin-induced taurine efflux (closed / open bars) in the presence of drugs was estimated and given relative to the control value ± SEM. The number of paired sets of experiments following osmotic cell swelling / melittin addition is 8/5 (BEL), 3/3 (AACOCF$_3$), 3/3 (W7), 9/3 (ETH 615-139), 4/3 (DIDS) and 3/3 (arachidonic acid). Data are reproduced from (16).

2.   INACTIVATION OF THE VOLUME-SENSITIVE TAURINE EFFLUX PATHWAY – ROLE OF ROS AND PROTEIN TYROSINE PHOSPHORYLATION

Reactive oxygen species, i.e., hydrogen peroxide, superoxide anions and hydroxyl radicals are produced as by-products of the general cellular metabolism and by a membrane-associated NAD(P)H oxidase system in most cell types[19]. It has recently been demonstrated that ROS production in NIH3T3 cells is increased following exposure to either hypotonic NaCl medium or isotonic NaCl medium containing melittin[16]. As BEL inhibits the swelling-induced ROS production in NIH3T3 cells[16], and as the PLA$_2$ products arachidonic acid and lysophosphatidyl choline (LPC) generate ROS when added exogenously[20,21, 22], it has been proposed that ROS are produced down-stream to PLA$_2$ activation following osmotic challenge[16]. Oxidative stress has been associated with tyrosine phosphorylation[23,24] and an increase in tyrosine phosphorylation has recently been associated with a prolongation of the open time of the volume-sensitive taurine efflux pathway in NIH3T3 cells[3], and a shift in the osmosensitivity of the taurine efflux in supraoptic astrocytes[15]. Expression of constitutive RhoA (RhoAV14) also increases the activity of the volume-sensitive taurine efflux pathway, but the effect of RhoA is on a mechanism that seems not to be affected by tyrosine kinase inhibition[3].

From Figs. 2A and 2B it is seen that H$_2$O$_2$, just like the tyrosine phosphatase inhibitor vanadate, potentiates the swelling-induced taurine efflux from NIH3T3 cells, and delays the closing/inactivation of the volume-sensitive taurine efflux pathway. The protein tyrosine kinase inhibitor genistein and the antioxidant butulated hydroxytoluene (BHT), on the other hand, inhibit the swelling-induced taurine efflux from NIH3T3 cells (Figs. 2A and 2B). These data indicate that an increase in the ROS production and an increase in the tyrosine phosphorylation of a yet unidentified protein potentiate the swelling-induced taurine efflux in NIH3T3 cells. Addition of H$_2$O$_2$ to cells under isotonic conditions elicits a small and slow release of taurine from NIH3T3 cells, which however is insensitive to DIDS and arachidonic acid[16]. From Fig. 2C it is seen that the potentiating effect of H$_2$O$_2$ on the swelling-induced taurine efflux is unaffected by BEL but reduced in the presence of genistein. Thus, cell swelling and protein tyrosine kinase activity is required in order to see the effect of H$_2$O$_2$.

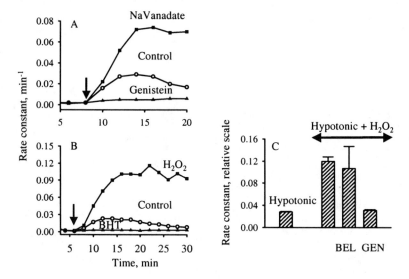

Figure 2. Effect of ROS and tyrosine phosphorylation on the swelling-induced taurine efflux. The cells were prepared and the taurine efflux followed with time in NaCl medium as indicated in the legend to Fig. 1. *Panel A & B*: The extracellular osmolarity was reduced from 300 mOsm to 200 mOsm at the time indicated by the arrow. NaVanadate (50 μM, closed squares, *Panel A*), genistein (100 μM, closed triangles, *Panel A*), H_2O_2 (2 mM, closed squares, *Panel B*) and BHT (0.5 mM, closed triangles, *Panel B*) were present in the experimental media throughout the whole efflux experiment. Control cells (open circles) were exposed to hypotonicity alone. *Panel C*: The maximal rate constant following hypotonic exposure was estimated in control cells (n = 114) and in cells exposed to H_2O_2 (2 mM, n = 19), H_2O_2 plus BEL (30 μM, n = 3), or H_2O_2 plus genistein (GEN, 100 μM, n = 4). Values are given as means ± SEM. Data is reproduced from (16).

Protein tyrosine phosphatases possess an essential cystein residue in the catalytic site, that forms a thiol-phosphate intermediate during the catalytic process. However, catalytic activity of the protein tyrosine phosphatase PTP1B is obstructed by the phosphate analogue vanadate and by H_2O_2[24,25,26]. It is accordingly suggested that the ROS, produced by osmotic cell swelling, inhibit a protein tyrosine phosphatase (PTP1B) and subsequently increase the open probability of the taurine efflux pathway. Protein kinase C (PKC) isoforms have recently been demonstrated to promote assembly and activation of the NAD(P)H oxidase complex[27]. From Fig. 3A it is seen that 10 min pretreatment of NIH3T3 cells with the PKC activator phorbol-myristate-acetate (PMA) potentiates the swelling-induced taurine efflux from NIH3T3 cells, and that the effect of PMA is stored by the cells for more than 20 minutes. Furthermore, the selective inhibitor of the NAD(P)H oxidase, diphenylene iodonium (DI) impairs the effect of PMA (Fig. 3B). It is suggested that the potentiating effect of PMA on the volume-sensitive taurine efflux involves an increased NAD(P)H oxidase activity and a subsequent ROS-mediated inhibition of protein tyrosine phosphatase activity.

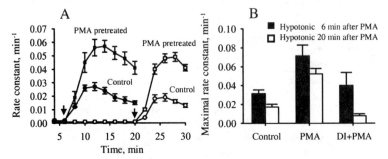

Figure 3. Effect of PMA on the swelling-induced taurine efflux. The cells were prepared and the efflux experiments performed in NaCl medium as indicated in the legend to Fig. 1. *Panel A*: The extracellular osmolarity was reduced from 300 mOsm to 200 mOsm at time 6 min (closed symbols) or time 20 min (open symbols) as indicated by the arrow. PMA pretreatment (squares) was carried out by exposing the ^{14}C-taurine loading cells to 100 nM PMA for 10 min prior to initiation of the efflux experiments. *Panel B*: DI (25 µM) was present from one hour before and throughout the whole efflux experimental period. The maximal rate constant for the swelling-induced taurine efflux was estimated in the absence and in the presence of PMA or DI plus PMA. The data represent 6 (control), 6 (PMA pretreated) and 3 (DI plus PMA pretreated) sets of experiments. The rate constant in PMA treated cells was in all cases significantly ($p < 0.02$) larger then in control cells and cells treated with PMA plus DI.

Figure 4 is a model that illustrates the current hypothesis for the regulation of the volume-sensitive taurine efflux pathway in NIH3T3 cells. PLA_2 and 5-LO activity is required for activation of the volume-sensitive taurine efflux pathway following exposure to hypotonic medium or isotonic medium containing melittin. Activation of PLA_2 generates unsaturated fatty acids, lysophospholipids and ROS. A 5-LO product, not yet identified, is assumed to activate the volume-sensitive osmolyte channel. ROS inhibit a PTP1B and subsequently increase the net tyrosine phosphorylation of a regulatory protein, which either potentiates elements in the swelling-induced signal cascade and/or delays the inactivation/closing of the volume-sensitive taurine efflux pathway. The focal adhesion kinase (FAK) is tyrosine phosphorylated by osmotic cell swelling[28] and following exposure to H_2O_2[24] and could well be the target of PTP1B. PKC is assumed to potentiate ROS production via assembly and activation of the NAD(P)H oxidase complex. Exogenous LPC induces taurine release from various cells, which differs pharmacologically from the swelling-induced taurine efflux and resembles a general permeabilisation of the plasma membrane[9,10,13,14]. LPC is accordingly not considered as an essential second messenger in the swelling-induced taurine efflux.

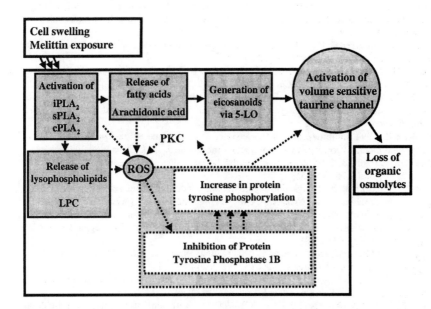

Figure 4. Model for activation and regulation of the volume-sensitive taurine efflux pathway in NIH3T3 cells. See text for details.

ACKNOWLEDGEMENTS

This work was supported by the Danish Research Council and "Fonden af 1870".

REFERENCES

1. Pasantes-Morales, H., Sanchez, O.R., Miranda, D., Moran, J. 1997. Volume regulation in NIH/3T3 cells not expressing P-glycoprotein. I. Regulatory volume decrease. *Am.J. Physiol* 272:C1798-C1803.
2. Moran, J., Miranda, D., Pena-Segura, C., Pasantes-Morales, H.. 1997. Volume regulation in NIH/3T3 cells not expressing P-glycoprotein. II. Chloride and amino acid fluxes. *Am. J. Physiol* 272:C1804-C1809.
3. Pedersen, S.F., Beisner, K.H., Hougaard, C., Willumsen, B.M., Lambert, I.H., Hoffmann, E.K. 2002. Rho family GTP binding proteins are involved in the regulatory volume decrease process in NIH3T3 mouse fibroblasts. *J. Physiol* 541:779-96.
4. Kirk, K. 1997. Swelling-activated organic osmolyte channels. *J. Membr. Biol.* 158:1-16.
5. Lambert, I.H., Hoffmann, E.K. 1993. Regulation of taurine transport in Ehrlich ascites tumor cells. *J. Membr. Biol.* 131:67-79.
6. Stutzin, A., Torres, R., Oporto, M., Pacheco, P., Eguiguren, A..L, Cid, L.P., Sepulveda, F.V. 1999. Separate taurine and chloride efflux pathways activated during regulatory volume decrease. *Am. J. Physiol* 277:C392-C402.
7. Thoroed, S.M., Lauritzen, L., Lambert, I.H., Hansen, H.S., Hoffmann, E.K. 1997. Cell swelling activates phospholipase A2 in Ehrlich ascites tumor cells. *J. Membr. Biol.* 160:47-58.

8. Basavappa, S., Pedersen, S.F., Jorgensen, N.K., Ellory, J.C., Hoffmann, E.K. 1998. Swelling-induced arachidonic acid release via the 85-kDa cPLA$_2$ in human neuroblastoma cells. *J. Neurophysiol.* 79:1441-9

9. Lambert, I.H., Sepulveda, .FV. 2000. Swelling-induced taurine efflux from HeLa cells: cell volume regulation. *Adv. Exp. Med. Biol.* 483:487-95.

10. Lambert, I.H., Nielsen, J.H., Andersen, H.J., Ortenblad, N. 2001. Cellular model for induction of drip loss in meat. *J. Agric. Food Chem.* 49:4876-83.

11. Ding, M., Eliasson, C., Betsholtz, C., Hamberger, A., Pekny, M. 1998. Altered taurine release following hypotonic stress in astrocytes from mice deficient for GFAP and vimentin. *Brain Res. Mol. Brain Res.* 62:77-81.

12. Kirk J, Kirk K. 1994. Inhibition of volume-activated I- and taurine efflux from HeLa cells by P-glycoprotein blockers correlates with calmodulin inhibition. *J. Biol. Chem.* 269:29389-94.

13. Lambert IH, Falktoft B. 2000. Lysophosphatidylcholine induces taurine release from HeLa cells. *J. Membr. Biol.* 176:175-85.

14. Lambert IH, Falktoft B. 2001. Lysophosphatidylcholine-induced taurine release in HeLa cells involves protein kinase activity. *Comp Biochem. Physiol A Mol.Integr. Physiol* 130:577-84.

15. Deleuze C, Duvoid A, Moos FC, Hussy N. 2000. Tyrosine phosphorylation modulates the osmosensitivity of volume- dependent taurine efflux from glial cells in the rat supraoptic nucleus. *J. Physiol* 523 Pt 2:291-9.

16. Lambert IH. 2002. Reactive oxygen species regulate swelling-induced taurine efflux in NIH3T3 mouse fibroblasts. *J. Membrane Biol.* 192: xx-xx.

17. Balsinde J, Balboa MA, Insel PA, Dennis EA. 1999. Regulation and inhibition of phospholipase A2. *Annu. Rev. Pharmacol. Toxicol.* 39:175-89.

18. Jenkins CM, Wolf MJ, Mancuso DJ, Gross RW. 2001. Identification of the calmodulin-binding domain of recombinant calcium- independent phospholipase A$_{2\beta}$. implications for structure and function. *J. Biol. Chem.* 276:7129-35.

19. Thannickal VJ, Fanburg BL. 2000. Reactive oxygen species in cell signaling. *Am. J. Physiol Lung Cell Mol. Physiol* 279:L1005-L1028.

20. Hensley K, Robinson KA, Gabbita SP, Salsman S, Floyd RA. 2000. Reactive oxygen species, cell signaling, and cell injury. *Free Radic. Biol. Med.* 28:1456-62.

21. Kugiyama K, Sugiyama S, Ogata N, Oka H, Doi H, Ota Y, Yasue H. 1999. Burst production of superoxide anion in human endothelial cells by lysophosphatidyl choline. *Atherosclerosis* 143:201-4.

22. Yamakawa T, Tanaka S, Yamakawa Y, Kamei J, Numaguchi K, Motley ED, Inagami T, Eguchi S. 2002. Lysophosphatidylcholine activates extracellular signal-regulated kinases 1/2 through reactive oxygen species in rat vascular smooth muscle cells. *Arterioscler. Thromb. Vasc. Biol.* 22:752-8.

23. Carballo M, Conde M, El Bekay R, Martin-Nieto J, Camacho MJ, Monteseirin J, Conde J, Bedoya FJ, Sobrino F. 1999. Oxidative stress triggers STAT3 tyrosine phosphorylation and nuclear translocation in human lymphocytes. *J. Biol. Chem.* 274:17580-6.

24. Vepa S, Scribner WM, Parinandi NL, English D, Garcia JG, Natarajan V. 1999. Hydrogen peroxide stimulates tyrosine phosphorylation of focal adhesion kinase in vascular endothelial cells. *Am. J. Physiol* 277:L150-L158.

25. Huyer G, Liu S, Kelly J, Moffat J, Payette P, Kennedy B, Tsraprailis G, Gresser MJ, Ramachandran C. 1997. Mechanism of inhibition of protein-tyrosine phosphatases by vanadate and pervanadate. *J. Biol. Chem.* 272:843-51.

26. Meng TC, Fukada T, Tonks NK. 2002. Reversible oxidation and inactivation of protein tyrosine phosphatases in vivo. *Mol. Cell* 9:387-99.

27. Fontayne A, Dang PM, Gougerot-Pocidalo MA, El Benna J. 2002. Phosphorylation of p47phox sites by PKC alpha, beta II, delta, and zeta: effect on binding to p22phox and on NADPH oxidase activation. *Biochemistry* 41:7743-50.

28. Tilly BC, Edixhoven MJ, Tertoolen LG, Morii N, Saitoh Y, Narumiya S, de Jonge HR. 1996. Activation of the osmo-sensitive chloride conductance involves P21rho and is accompanied by a transient reorganization of the F-actin cytoskeleton. *Mol. Biol. Cell* 7:1419-27.

Taurine Counteracts Cell Swelling in Rat Cerebrocortical Slices Exposed to Ammonia *in Vitro* and *in Vivo*

R. O. LAW*, M. ZIELINSKA**, and J. ALBRECHT**

*Department of Cell Physiology and Pharmacology, University of Leicester, Leicester, UK and
**Department of Neurotoxicology, Medical Research Centre, Polish Academy of Sciences, Warsaw, Poland

1. INTRODUCTION

Taurine is believed to fulfil a wide range of physiological functions within the mammalian body[1]. While it is believed that one of these may be neuroprotection, neither its relative significance in this capacity, nor the underlying mechanism, are clear. Data in favour of this concept include taurine's ability to counteract the toxic effect of MPP+ in brain slices[2] and kainite-evoked excitotoxicity in cultured neurons[3] (for exhaustive review see[4]).

The present study is concerned with the protective action of taurine against the effects of ammonia on the CNS. Ammonia is recognized as a neurotoxin implicated in the pathogenesis of hyperammonaemic encephalopathies which may lead to fatal cerebral oedema[5,6], one component of which, both *in vitro* and *in vivo*, involves astrocytic swelling[7,8]. Our experimental model has been the cerebrocortical minislice prepared (i) from normal rats and acutely exposed to exogenous ammonia and (ii) from rats undergoing experimental hepatic encephalopathy (HE), a conditions known to be associated with a significant increase in plasma ammonia[9,10]. We have recently provided evidence suggesting that ammonia-induced cerebral oedema is mechanistically related to ammonia-induced excitotoxicty in brain slices, and that cell swelling can be significantly ameliorated by taurine[11].

---

*Present address: Department of Preclinical Sciences, University of Leicester, Leicester, UK

2.   METHODS

Full details of the induction of pre-coma HE by the injection of thioacetamide (i.p.) over 3 days, the preparation and incubation of cerebrocortical minislices, and the calculation of cell volumes from the equilibrium volume of distribution of [^{14}C]inulin within slices, have been previously published[7,12] and are here presented in outline only.

Adult male Wistar rats (180-230g) were killed by cervical dislocation. Cerebral hemispheres were rapidly excised and placed in ice-cold aerated medium containing (mmol/l) NaCl 126, MgSO$_4$ 1.29, NaH$_2$PO$_4$ 1.29, KCl 5, CaCl$_2$ 0.8, HEPES 15, glucose 10, NaOH 11.7, pH 7.4. Ascorbic acid (4mmol/l) was also included in order prevent cell swelling due to ROS formation[13]. Batches of 4 minislices (300-400 μm thickness weight 4-9mg, were cut freehand from the frontal part of each hemisphere, using a chilled razor blade. After weighing to the nearest 50μg on a torsion balance. Slices were individually incubated in 350μl aerated medium for 60 min at 18-21°C. In the case of slices from Normal (non-HE) rats, the medium was prepared with or without the addition of ammonium acetate ("ammonia", 5mmol/l). In medium containing taurine, this was present at 10mmol/l unless stated otherwise. Other additives were PDC (Tocris Cookson Ltd), bicuculline (Sigma Chemical Co.), GES (a generous gift from Prof. S. Oja) and mannitol.

[^{14}C]Inulin (American Radiolabeled Chemicals) was added to incubation media (final activity approx. 20kBq/ml) for the final 20 min. of incubation. After 60 min. slices were removed, gently blotted using hard filter paper, reweighed, and individually leached for 18 hr in 0.5ml distilled water in order to extract the inulin. The activity of the leaching fluid was compared with that of a 1:500 dilution of the incubation medium, using a Packard 2405 liquid scintillation spectrometer.

The volumes of distribution of inulin, and hence cell volumes, were calculated as previously described[12], and volumes expressed as μl fluid/mg dry weight.

Statistical analysis was performed by one-way ANOVA followed by Dunnet's test for multiple comparisons.

3.   RESULTS

The results of this study are shown in Figs. 1 and 2. Those in Fig. 1 may be summarized as follows –

- 5 mmol/l ammonia causes significant swelling in slices prepared from normal rats.
- Taurine alleviates this swelling in a concentration-dependent manner, the reduction in volume being significant at 10mmol/l (the concentration routinely used in this study) .
- There was no reduction in swelling due to ammonia when 10mmol/l taurine was replaced by equimolar mannitol.
- A comparable swelling profile was observed in slices from HE rats incubated in ammonia-free media (the effect of mannitol was not studied in these animals).

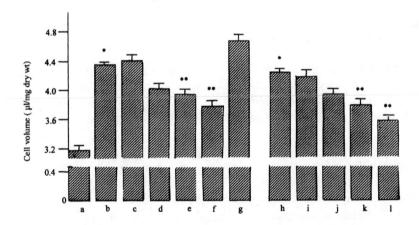

Figure 1. The effect of taurine on cell swelling following acute exposure to ammonia and in HE. a, Control, b, ammonia, c-f, ammonia + taurine at 1, 5, 10 and 20 mmol/l respectively, g, mannitol (10mmol/l), h, HE, i-l, HE + taurine at 1, 5, 10 and 20 mmol/l respectively. Values are mean ± s.e.m. (n=12). *P<0.05 *vs* Control, **P<0.05 *vs* ammonia or HE, as appropriate.

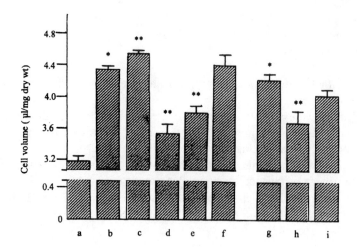

Figure 2. The effect of further incubatory interventions on cell swelling following acute exposure to ammonia or in HE. a, Control, b, ammonia (both from *Figure 1*), c, ammonia + PDC, d, ammonia + PDC + taurine (10mmol/l), e, ammonia + taurine (10mmol/l) + GES (10mmol/l), f, ammonia + taurine (10mmol/l) + bicuculline (100μ mol/l), g, HE (from *Figure 1*), h, HE + taurine (10mmol/l) + GES (10mmol/l), HE + taurine (10mmol/l) + bicuculline (100μ mol/l). Values are mean ± s.e.m. (n=12). *P<0.05 *vs* Control, **P<0.05 *vs* ammonia or HE, as appropriate.

Fig. 2 shows the results of experiments designed to elucidate the molecular mechanism of taurine's neuroprotective effect against ammonia toxicity. Columns a, b and g repeat the swelling effect of exogenous ammonia and HE already shown in Fig. 1. The results of additional incubatory manoeuvres were as follows –

- The swelling effect of ammonia was significantly enhanced by PDC (1mmol/l) – i.e. the effects of ammonia and PDC were additive.
- 10mmol/l taurine significantly attenuated this effect.
- In the presence of taurine, GES (10mmol/l) had no effect on cell volumes either in slices from normal rats incubated in the presence of ammonia, nor in slices from HE animals.
- In slices from normal animals exposed to ammonia, but not in those from HE rats, the alleviation of swelling by taurine was abolished by bicuculline (100 μmol/l).

4.   DISCUSSION

The results of the present study clearly show (i) that cells in cerebrocortical minislices from normal rats swell significantly when acutely exposed to exogenous ammonia, a similar response also being seen in slices from rats undergoing experimental HE, and (ii) taurine alleviates this swelling in a concentration-dependent within a concentration range comparable with that found in cerebral tissue[14]. The fact that in acutely exposed cells equimolar substitution of mannitol for taurine did not reduce cell swelling (Fig. 1) strongly supports a molecule-specific protective role for taurine that is independent of an osmotic effect

In terms of the molecular mechanism, we have previously shown that in the presence of ammonia the swelling effects of the glutamate transport inhibitor PDC, which may be presumed to increase extracellular glutamate levels and its level of interaction with NMDA receptors, are significantly reduced by the NMDA receptor channel blocker MK801[11]. In the present study the fact that taurine also reduced swelling due to ammonia + PDC (Fig. 2) suggests that it, too, may interfere with NMDA receptor function, possibly targeting the NMDA receptor/NO pathway; there is evidence that the latter may be associated with clinical and pathophysiological aspects of ammonia neurotoxicity *in vivo*[15,16].

The observation that taurine transport inhibitor GES was without effect on the attenuation of cell swelling by taurine (Fig. 2), both in the presence of exogenous ammonia and in HE slices, suggests that the neuroprotective effect of taurine does not have a transport-dependent component is suggested by the experiments in which the GABA$_A$ receptor blocker bicuculline was used.

Activation of this receptor by taurine has previously been demonstrated[17,18]. The manner in which this might lead to reduction in cell swelling is not clear. The link between activation of GABA$_A$ receptors and reduction in cell swelling is not clear. It may be recalled that a GABA$_A$ receptor-mediated amelioration of cell damage by taurine was observed in brain slices treated with MPP+, a free radical generated during oxidation of MPTP[2]. However, it is clear from Fig. 2 that the effect of biccuculine was not significant in slices from HE rats.

There are 2 possible explanations of this – (i) the bicuculline-sensitive component of taurine's action, even if present *in vivo* may not have survived the process of slice preparation, or (ii) the effect may already have been masked *in vivo* by one of the other numerous changes in body fluid composition that accompany HE. Nevertheless, it should be noted that the swelling profile of cells from HE rats persists for at least 60 minutes during incubation in ammonia-free media.

5.   CONCLUSIONS

- Cells in cerebrocortical minislices from normal rats undergo significant swelling when acutely exposed to ammonium ions ("ammonia"). Comparable swelling is observed in cells from animals chronically exposed to low concentrations of ammonia in a hepatic encephalopathy (HE) model.

- In both groups swelling is significantly attenuated by taurine in a concentration-dependent manner. This response is not is not observed if taurine is replaced by equimolar mannitol.

- During acute exposure to ammonia swelling is enhanced by the glutamate transport inhibitor PDC.

- The reduction of swelling due to taurine is unaffected by the taurine transport inhibitor GES.

- The effect of taurine in cells acutely exposed to ammonia is abolished by the GABA$_A$ receptor blocker biccuculine, but this response is attenuated in cells from HE animals

- It is concluded that taurine counteracts brain cell swelling induced by toxic concentrations of ammonia by a mechanism unrelated to its osmotic properties.

ACKNOWLEGEMENTS

This study was supported by Wellcome Trust Grant 056593/Z/99/Z (R.O.L.) and a statutory Grant to the Medical Research Centre (M. Z. and J. A.)

REFERENCES

1. Huxtable, R., 1992, Physiological actions of taurine. *Physiol. Rev.*72: 101-163.
2. O'Byrne, M.B., and Tipton, K.F., 2000, Taurine-induced attenuation of MPP+ neurotoxicity in vitro: a possible role for the GABA A subclass of receptors. *J. Neurochem.* 74: 2087-2093.
3. Boldyrev, A. A., Johnson, P., Wei, Y., Tan, Y., and Carpenter, D. O., 1999, Carnosine and taurine protect rat cerebellar granular cells from free radical damage. *Neurosci. Lett.* 263: 169-172.
4. Saransaari, P., and Oja, S., S., 2000, Taurine and neural cell damage. *Amino acids* 19: 509-526.
5. Blei, A.T., Olaffson, S., Therrien, G., and Butterworth, R.F., 2000, Ammonia-induced brain edema and intracranial hypertension in rats after portacaval anastomosis. *Hepatology* 19: 1437-1444.
6. Donovan, J.P., Schafer, D.F., Shaw, B.W., and Sorrell, M.F., 1998, Cerebral oedema and increased intracranial pressure in chronic liver disease. *Lancet* 351:719-721.
7. Zielinska, M., Hilgier, W., Law, R.O., Gorynski, P., and Albrecht, J., 1999, Effects of ammonia in vitro on endogenous taurine efflux and cell volume in rat cerebrocortical minislices: influence of inhibitors of volume-sensitive amino acid transport. *Neuroscience* 91: 631-638.
8. Albrecht, J., and Dolinska, M., 2001, Glutamine as a pathogenic factor in hepatic encephalopathy. *J. Neurosci. Res.* 65: 1-6.
9. Albrecht, J., and Hilgier, W., 1984, Brain carbonic anhydrase activity in rats in experimental hepatogenic encephalopathy. *Neurosci. Lett.* 45: 7-10.
10. Zimmermann, C., Ferenci, P., Pifl, C., Yerdaydin, C., Ebner, J., Lasserman, H., Roth, E., and Hortnagl, H., 1989, Hepatic encephalopathy in thioacetamide-induced acute liver failure in rats: characterization of an improved model and study of amino acid-ergic neurotransmission. *Hepatology* 9: 594-601.
11. Zielinska, M., Law, R.O., and Albrecht, J, 2002, An excitotoxic component of ammonia-induced cell swelling in rat cerebrocortical minislices. *Acta Physiol. Hung.* 89: 197.
12. Law, R.O., 1994, Taurine efflux and the regulation of cell volume in incubated slices of rat cerebral cortex. *Biochim. Biophys. Acta* 1221: 21-28.
13. Brahma, B., Forman, R.E., Stewart, E.E., Nicholson, C., and Rice M.E., 2000. Ascorbate inhibits edema in brain slices. *J. Neurochem.*74: 1263-1270.
14. Palkovits, M., Elekes, I., Lang, T., and Patthy, A., 1986, Taurine levels in discrete brain nuclei of rats. *J. Neurochem.*47: 1333-1335.
15. Marcaida, G., Felipo, V., Hermenegildo, C., Minana, M., and Grisolia, S., 1992, Acute ammonia toxicity is mediated by the NMDA type of glutamate receptors. *FEBS Lett* .296: 67-68.
16. Hermenegildo, C., Monfort, P., and Felipo, V., 2000, Activation of N-methyl-D-aspartate receptors in rat brain following acute ammonia intoxication: characterization by in vivo brain microdialysis. *Hepatology* 31, 709-715.
17. Quinn, M.R., and Harris, C.L., 1995, Taurine allosterically inhibits binding of [35S]-t-butylbicyclophosphorothionate (TBPS) to rat synaptic membranes. *Neuropharmacology* 34: 1607-1613.
18. Del Olmo, N., Bustamante, J., Del Rio, R.M., and Solis, J.M., 2000, Taurine activates GABA A but not GABA B receptors in rat hippocampal CA1 area. *Brain Res* .864: 298-307.

Cloning of Human Intestinal Taurine Transporter and Production of Polyclonal Antibody

HYE SUK AN[*], HEE CHANG HAN[*], TAESUN PARK[**], KUN KOO PARK[***] and HA WON KIM[*]

[*]*Department of Life Science, University of Seoul, Seoul 130-743,* [**]*Department of Food and Nutrition, Yonsei University, Seoul 120-749 and* [***]*PharmacoGenechips Inc., Chunchon, 200-160, Korea*

1.   INTRODUCTION

Taurine (2-ethaneaminosulfonic acid, $^+NH_3CH_2CH_2SO_3^-$) is an endogenous amino acid which functions as a modulator of transmembrane calcium transport, an osmoregulator and, putatively a free radical scavenger[1]. *In utero*, taurine deficiency causes an abnormal development of brain, retina, kidney, and abnormal function of the myocardium, where taurine plays an important role in intracellular calcium homeostasis in the cardiomyocytes[2-4].

Taurine transporter contains 12 transmembrane helices, which are typical of the Na^+- and Cl^--dependent transporter gene family[5-13]. The taurine transporter has been cloned recently from several species and tissues, including Madin-Darby canine kidney (MDCK) cells[5], rat brain[6], mouse brain[7], human thyroid cells[10], human placenta[11], and pig kidney cells (LLC-PK1)[9, 14]. The open reading frame of human thyroid taurine transporter encoded a protein of 619 amino acids with a calculated molecular weight of 69,675 Da[10]. The cloned human placental taurine transporter cDNA consisted of a coding region 1,863 bp long[12]. The cDNA encoded a protein of 620 amino acids with a calculated molecular weight of 69,853. The length of the coding region of the taurine transporter in the human retinal pigment epithelium is 1,863 bp. A protein predicted from this cDNA sequence consists of 620 amino acids and its relative molecular mass is 69,826 Da.

In an attempt to clone Na⁺-dependent taurine transporters from the human intestinal epithelial cells, a partial cDNA clone encoding the human taurine transporter was isolated. The primary structure of the human taurine transporter was determined, and compared with the recently cloned human thyroid, human placental, and human retinal taurine transporters. A specific polyclonal antibody from the rabbit was made against taurine transporter peptides conjugated with keyhole limpet hemocyanin (KLH) carrier proteins. These antibodies may be useful in refinement of clinical immunoassays of taurine transporter or immunohistological study of taurine transporter-synthesizing cells.

2.   MATERIALS AND METHODS

2.1   Materials

HT-29 and RAW264.7 cells were obtained from American Type Culture Collection. RPMI-1640, DMEM, fetal bovine serum (FBS), penicillin-streptomycin, and TRIzol reagent were purchased from Gibco-BRL (Grand Island, NY, USA). [2-^{3}H(N)]-taurine was obtained from NEN (Boston, MA, USA). One step RT-PCR kit, *Eco*RI, *Bam*HI and T4 DNA ligase were purchased from Takara Biomedicals (Japan). PCR primers and sequencing primers were synthesized in TaKaRa-Korea Biomedical Inc. (Korea). pGEX 4T-1 vector, *E. coli* (DH5α) and ECL Western blotting analysis system were purchased from Amercham Pharmacia Biotech (NJ, USA).   Rabbits (New Zealand White, 2.5 kg, male) for polyclonal antibodies were purchased from Jung-Ang Lab Animal, Inc. (Korea). TMB (3, 3', 5, 5' tetramethylbenzidine) substrate reagent set was obtained from Pharmingen (San Diego, USA).

2.2   RNA Isolation

Total RNA was extracted from the cultured HT-29 cells with TRIzol reagent according to the manufacturer's instructions. The cells were lyzed by adding 1 ml of TRIzol reagent to a 3.5 cm diameter dish, and passing the cell lysate several times through a pipette. Chloroform (0.2 ml) was added and the lysates were shaken vigrously for 15 seconds and incubated at 25°C for 3 minutes. The aqueous phase was transferred to a fresh tube. The RNA was precipitated by mixing with isopropyl alcohol. The RNA was briefly dried, dissolved in RNase-free water, and stored at -70°C until use.

2.3   RT-PCR

One step RNA PCR kit was used for the reverse transcription from RNA to cDNA using AMV reverse transcriptase and subsequent amplification in the same tube utilizing AMV-optimized *Taq* DNA polymerase. The PCR primers were designed on the basis of the sequence of human taurine transporter: primer-1 (forward primer): 5'-tcgcggatccataccgtattttattttcctgttt-3', primer-2 (reverse primer): 5'-tctagaattccctgtacgagttatacttgtactt-3', primer-3 (reverse primer): 5'-gagagaattcaagatcaaccaaggatgtgatctg-3'.

2.4   cDNA Cloning and Sequencing

The TAUT-A PCR products were digested with *Eco*RI and *Bam*HI enzymes and purified by phenol/chloroform and ethanol precipitation methods. The TAUT-A cDNA was inserted into *Eco*RI-*Bam*HI digested pGEX 4T-1 vector and the plasmids ligated were used to transform *E. coli* DH5α. After isolation of plasmids from the DH5α transfromants, two clones containing the 732 bp cDNA inserts were selected by an agarose gel electrophoresis. The sequencing primers were designed on the basis of the sequence of pGEX 4T-1 vector: 5' pGEX sequencing primer: 5'-gggctggcaagccacgtttggtg-3', 3' pGEX sequencing primer: 5'-ccgggagctgcatgtgtcagagg-3'. The TAUT-A cDNA was sequenced in both directions of 5' to 3' and 3' to 5' by using the ABI PRISM 377×L automated sequencer (Perkin-Elmer).

2.5   Production of Polyclonal Antibody

A 14-mer conserved synthetic peptide sequence (LFQSFQKELPWAHC, 149 to 162 residues) of predicted human taurine transporter was coupled to keyhole limpet hemocyanin (KLH). A rabbit was immunized subcutaneously with approximately 200 μg of immunogen emulsified in complete Freund's adjuvant on the back of the rabbit. The second booster injection was given with incomplete Freund's adjuvant at 4 weeks. The final booster immunization was administered with an equivalent amount of antigen in PBS 8 weeks later. The rabbit sera were obtained from the ear vein at 0, 2, 6, and 9 weeks to measure the antibody response. Antisera were prepared from the heart of the rabbit at 9 weeks and stored in aliquots at -20°C.

2.6   Western Blot with Polyclonal Antibody

RAW264.7 cells were collected by scraping with PBS buffer and washed once with ice-cold PBS and lysed in triple-detergent lysis buffer. An equal amount of each protein sample (50 μg) was resolved using 10% SDS-PAGE and transferred in transfer buffer onto nitrocellulose membrane by semi-dry transfer (Trans-Blot SD, BioRad Laboratories). After the polyclonal antiserum was applied for one hour, the membrane was washed. Horseradish peroxidase (HRP) conjugated secondary antibody was applied for one hour and the membrane was rewashed. Bound antibody was visualized using enhanced chemiluminescence according to the manufacture's instructions (ECL, Amersham Pharmacia Biotech).

3.   RESULTS

3.1   Cloning of the Human Taurine Transporter Partial cDNA

Plasmids were ligated with TAUT-A cDNA by T4 DNA ligases and transformed into *E. coli* DH5α using a electroporation method. Recombinant plasmids were isolated from bacteria clones, digested with *Eco*RI and *Bam*HI, and analyzed by 1% agarose gel electrophoresis. Two clones of #6 and #8 colonies were identified as positive clones containing pGEX 4T-1 plasmids ligated with TAUT-A cDNA (Fig. 1.)

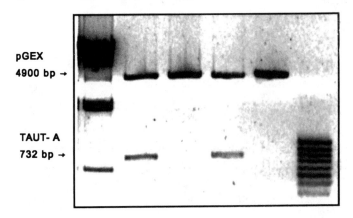

Marker colony# 6 colony# 7 colony# 8 colony# 9 Marker

Figure 1.   Agarose gel electrophoresis of recombinant plasmids after digestion by *Eco*RI and *Bam*HI. Positive clones; colony #6 and colony #8.

3.2   Partial Sequencing of Intestinal Taurine Transporter

The TAUT-A cDNA was cloned, sequenced and registered in NCBI GenBank (accession number is AF346763). The partial sequence of the taurine transporter cDNA obtained from the human intestinal HT-29 cells was almost identical to the previously reported sequences of the human placental and retinal taurine transporters. But 3 nucleotides were different (2 amino acids) from that of the human thyroid. The partial cDNA and amino acid sequences of the human intestinal taurine transporter were compared with those sequences of the human placental, retinal, and thyroid taurine transporters (Table 1).

Table 1. Comparison of human intestinal TAUT-A cDNA sequence with the reported human placental, retinal, and thyroid taurine transporter sequence.

Organ	Site of Amino Acids				
	...	236	...	272	...
Intestine	...	Trp	...	Ala	...
	...	tgg	...	gca	...
Placenta	...	Trp	...	Ala	...
	...	tgg	...	gca	...
Retina	...	Trp	...	Ala	...
	...	tgg	...	gca	...
Thyroid	...	Cys	...	Arg	...
	...	tcg	...	cga	...

3.3 Production of Polyclonal Antibody

For the production of antibodies, synthetic peptides were used as immunogens. The synthetic peptide was LFQSFQKELPWAHC (14 amino acids) and located in secondary extracellular domain. The percent of the polor or charged amino acids consisting of synthetic peptides was 50%. The peptides were corresponded to the position of 149-162 amino acid residues of the taurine transporter. This region was chosen to avoid a steric hindrance of putative glycosylation sites.

The rabbit was bled and serum prepared from whole blood 10 to 14 days after each immunization. After secondary injection, the production of polyclonal antibodies was detected by ELISA analysis. And before bleeding from the heart of the rabbit immunized, antibody response was identified by ELISA (Fig. 2). Using the sera from the heart of the rabbit, polyclonal antibodies bound to the taurine transporter proteins (~70 kDa) were confirmed by an immunoblotting (Fig. 3).

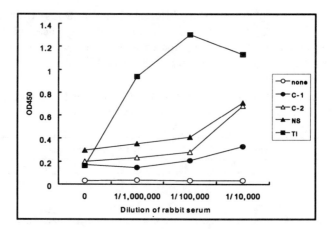

Figure 2. ELISA analysis of polyclonal antibody responses of the rabbit immunized after third boosting. C-1; control only treated with secondary antibodies, C-2; control treated with the serum of a rabbit immunized and secondary antibodies without antigen, NS; normal serum, TI; rabbit serum after third immunization.

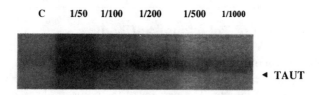

Figure 3. Western blot analysis with polyclonal antibody of taurine transporter protein from mouse macrophage RAW264.7 cells. The serum from a heart of the rabbit immunized was diluted to 1/50, 1/100, 1/200, 1/500 and 1/1000. NS; rabbit normal serum.

4.   DISCUSSION

The partial cDNA encoding the taurine transporter from human intestinal HT-29 cells was produced by RT-PCR, cloned, sequenced and its sequence was registered in the NCBI GenBank. The partial cDNA sequence of the taurine transporter in the human intestinal epithelium appeared to be almost identical to that of human retina and placenta but was different in 3 base pairs (2 amino acids) from the reported sequence of human thyroid. The differences between the partial sequence of the taurine transporter from the human intestinal epithelium and the human thyroid taurine transporter sequence were at the 236th and 272nd amino acids. The 236th and 272nd amino acids were tryptophan (tgg) and alanine (gca) in the PCR product from the human intestinal taurine transporter, and cysteine (tgc) and arginine (cga) in the human thyroid taurine transporter.

Alignment of the human, dog, and rat taurine transporters showed that the taurine transporter was highly conserved among these species. This suggests that taurine plays fundamental roles in mammals[10]. The amino acid sequence of the human thyroid taurine transporter showed 91% identity with that of dog kidney taurine transporter, and 93% identity with that of rat brain taurine transporter[10]. The amino acid sequence of the human retinal taurine exhibited more than 80% homology with the sequences of the canine, rat and murine taurine transporters and was almost identical with that of the taurine transporter from human thyroid cells[11].

From these considerations, the reported thyroid taurine transporter sequences had sequencing errors and the human intestinal taurine transporter sequence seems to be identical to the reported human taurine transporter sequences. The specificity of polyclonal antibody to taurine transporter peptide was confirmed by the immunoblotting assay according to the modified method of Tebele[15]. The taurine transporter protein in RAW264.7 cell lysates was detected and found to be the predicted size, 70 kDa, with the obtained polyclonal antibody. Thus, these polyclonal antibodies can be useful tools in the biological analysis of taurine transporters.

CONCLUSIONS

The partial cDNA encoding the TAUT from HT-29 cells was produced by RT-PCR, cloned and sequenced. The partial cDNA of the human intestinal TAUT was identical to the human retinal and placental TAUT sequence but different (3 bases) from the reported sequence of thyroid TAUT. The rabbit polyclonal antibody recognized 70 kDa TAUT of RAW264.7 cell by the Western blot assay.

ACKNOWLEDGEMENTS

This work was supported by grant No.(R01-1999-000-00128-0) from the Basic Research Program of the Korea Science & Engineering Foundation.

REFERENCES

1. Chesney, R.W., and Budreau, A., 1993, Efflux of taurine from renal brush border membrane vesicles: Is it adaptively regulated?. *Pediatr. Nephrol.* 7: 35-40.
2 Chesney, R.W., Gusowski, N., and Dabbagh, S., 1985, Renal cortex taurine content regulates renal adaptive response to altered dietary intake of sulfur amino acids. *J. Clin. Invest.* 76: 2213-2221.
3. Chesney, R.W., Gusowski, N., and Freidman, A.L., 1983, Renal adaptation to altered amino acid intake occurs at the luminal brush border membrane. *Kidney Int.* 24: 588-594.
4. Chesney, R.W., Helms, R.A., Christensen, M., Budreau, A.M., Han, X., and Sturman, J.A., 1998, An updated view of the value of taurine in infant nutrition. In *Advances in Pediatrics*, edited by Barnes La, pp. 179-193.
5. Uchida, S., Kwon, H.M., Yamauchi, A., Preston, A.S., Marumo, F., and Handler, J.S., 1992, Molecular cloning of the cDNA for an MDCK cell Na^+- and Cl^--dependent taurine transporter that is regulated by hypertonicity. *Proc. Natl. Acad. Sci. USA* 89: 8230-8234.
6. Smith, K.E., Borden, L.A., Wang, C.D., Hartig, P.R., Branchek, T.A., and Weinshank, R.L., 1992, Cloning and expression of a high affinity taurine transporter from rat brain. *Mol. Pharmacol.* 42: 563-569.
7. Liu, Q.R., Lpez-Corcuera, B., Nelson, H., Mandiyan, S., and Nelson, N., 1992, Cloning and expression of a cDNA encoding the transporter of taurine and β-alanine in mouse brain. *Proc. Natl. Acad. Sci. USA* 89: 12145-12149.
8 Vinnakota, S., Qian, X., Egal, H., Sarthy, V., and Sarkar, H.K., 1997, Molecular characterization and in situ localization of a mouse retinal taurine transporter. *J. Neurochem.* 69: 2238-2250.
9. Han, W., Budreau, A.M., and Chesney, R.W., 1998, Molecular cloning and functional expression of an LLC-PK1 cell taurine transporter that is adaptively regulated by taurine. *Adv. Exp. Med. Biol.* 442: 261-268.
10. Jhiang, S.M., Fithian, L., Smanik, P., McGill, J., Tong, Q., and Mazzaferri, E.L., 1993, Cloning of the human taurine transporter and characterization of taurine uptake in thyroid cells. *FEBS Lett.* 318: 139-144.
11. Ramamoorthy, S., Leibach, F.H., Mahesh, V.B., Han, H., Yang-Feng, T., Blakely, R.D., and Ganapathy, V., 1994, Functional characterization and chromosomal localization of a cloned taurine transporter from human placenta. *Biochem. J.* 300: 893-900.
12. Miyamoto, Y., Liou, G.I., and Sprinkle, T.J., 1996, Isolation of a cDNA encoding a taurine transporter in the human retinal pigment epithelium. *Curr. Eye Res.* 15: 345-349.
13. Takeuchi, K., Toyohara, H., and Sakaguchi, M., 2000, A hyperosmotic stress-induced mRNA of carp cell encodes Na^+- and Cl^--dependent high affinity taurine transporter. *Biochim. Biophys. Acta.* 1464: 219-230.
14. Han, X., Budreau, A.M., and Chesney, R.W., 1999, Ser-322 is a critical site for PKC regulation of the MDCK cell taurine transporter (pNCT). *J. Am. Soc. Nephrol.* 10: 1874-1879.
15. Tebele, N., Skilton, R.A., Katende, J., Wells, C.W., Nene, V., McElwain, T., Morzaria, S.P., and Musoke, A.J., 2000, Cloning, characterization, and expression of a 200-kilodalton diagnostic antigen of *Babesia bigemina*. *J. Clin. Microbiol.* 38(6): 2240-2247.

Transactivation of *TauT* by p53 in MCF-7 Cells

The Role of Estrogen Receptors

XIAOBIN HAN, ANDREA BUDREAU PATTERS, and RUSSELL W. CHESNEY
Department of Pediatrics, University of Tennessee Health Science Center, and the Children's Foundation Research Center at Le Bonheur Children's Medical Center, Memphis, TN, USA

1. INTRODUCTION

The p53 tumor suppressor gene functions as a cell cycle checkpoint, blocking cell division in the G1 phase to allow repair of damaged DNA or even triggering apoptosis in cells that have defective genomes [1]. Numerous stimuli trigger increases in the level of p53, including DNA-damaging drugs, ionizing radiation, ultraviolet light, and hypoxia [2-5]. MCF-7, a human breast cancer estrogen receptor-positive (ER+) cell line, expresses a high level of wild-type p53, which is up-regulated by 17-ß-estradiol (E2) [6]. However, p53 function is largely inactivated in MCF-7 cells, caused by misallocation of p53 from nuclei to cytosol [7], which may result in the altered expression of certain p53 target genes. Studies have shown that the taurine transporter gene (*TauT*) is a putative target of p53 [8], and that mutation of *TauT* results in severe and progressive retinal degeneration, a small brain, and shrunken kidneys [9]. In the present study, we show that *TauT* is up-regulated by p53 and E2 in MCF-7 human breast cancer cells in a manner that appears to be mediated by estrogen receptor α (ERα).

2. MATERIALS AND METHODS

2.1 Construction of the Reporter Gene

The promoter region of *TauT* was identified in previous studies [10], and a p53-binding consensus site was found in the *TauT* promoter sequence, located at -663 to -695. In this study, approximately 1.1 kb of the *TauT* promoter region DNA was used as the template for PCR (GenBank accession number AR151716) and the PCR fragment was cloned into the promoterless luciferase vector pGL3-Basic (Promega, Madison, WI) to generate the plasmid p923 for use in transfections and luciferase assays. The conditions used are 30 cycles of 1 min of denaturation at 94°C, 1 min of annealing at 58°C, and 1 min of elongation at 72°C. The sense primer (5'-GGGGTACCTTACTGAAGGTCACACAGC-3') designed for PCR contained a unique site for *KpnI*, and the antisense primer (5'-AAGATC-TTGGCACGGGAG-TTCA-3') contained a unique site for *Bgl II*. PCR products were digested with *KpnI* and *Bgl II* and re-ligated into *KpnI* and *Bgl II* sites of pGL3-Basic to generate plasmids containing segments of the *TauT* promoter sequence extending from the +48 nucleotide corresponding to the transcriptional start site. The constructs were verified by DNA sequencing. The p53-binding site deletion (pGL-563) construct was generated from the p923 plasmid by using sense primer 5'-GGGGTACCGAGTTGGGGAGGGA-3'.The antisense primer used for these constructs was the same as described above.

2.2   Cell Culture

Human breast cancer MCF-7 (ER⁺) and MDA-MB-231 (ER⁻) cells were cultured according to ATCC (American Type Culture Collection) guidelines. Briefly, cells were grown as confluent monolayers in 10 cm diameter tissue culture plates in media specific for each cell line with 10% fetal calf serum at 37°C in the presence of 5% CO_2 in a humidified incubator. The charcoal-treated fetal calf serum was used for the experiments using E2. Cells were plated 18 h before transfection and fed with fresh medium 4 h before transfection.

2.3   Transient Transfection

Plasmid DNA was introduced into cultured mammalian cells using cationic liposomes (Lipofectamine, Life Technologies). Transfection was carried out for 16-18 h, and then cells were washed twice with phosphate-buffered saline and incubated in fresh medium for 24-48 h before harvesting. pGL-control, which contains a luciferase gene driven by the SV40 early region promoter/enhancer, and empty pGL-Basic vectors were used as positive and negative controls, respectively. To standardize the transfection efficiency, 0.1 µg of pRL-CMV vector (pRL Renilla Luciferase control reporter vector, Promega) was cotransfected in all experiments. Cells were harvested 48 h after transfection and lysed in 200 µl of reporter lysis buffer

(Promega). A luciferase assay was performed using a dual-luciferase assay kit (Promega), and activity was measured with an Optocomp 1 luminometer (MGM Instruments, Inc., Hamden, CT). Promoter activity (mean ± SD of four samples in relative light units) of each construct is represented by relative light output normalized to pRL-CMV control. Graphs represent typical results of four separate experiments. The concentration of protein in the cell extracts was determined using the Bradford method (Bio-Rad, Hercules, CA).

2.4   Statistics

All experiments were performed in triplicate. Luciferase assays are expressed in units of relative light output. The data represent the mean ± SEM of 3 or 4 experiments. Statistical comparisons were made using one-way ANOVA and Student's *t* test to determine significant differences in the means.

3.   RESULTS

3.1   Up-regulation of *TauT* Transcription by E2

Two estrogen response element (ERE) half-site sequences are found in the promoter region of *TauT* [10]. One is located upstream of the p53-binding site. To determine if E2 regulates *TauT* expression, transient transfection of the *TauT* promoter (construct p-923, containing both ERE-half sites) was carried out in MCF-7 (ER$^+$) and MDA-MB-231 (ER$^-$) cells. As shown in Figure 1, 17-ß-estradiol increased promoter activity of *TauT* in MCF-7 cells in a dose-dependent manner, but had no effect on the *TauT* promoter function in MDA-MB-231 cells. To precisely define the role of the ERE sequence for E2 regulation of *TauT*, progressive deletion of the *TauT* promoter sequence was performed, and the resulting reporter gene constructs were studied as described above. As shown in Figure 2, the up-regulation of *TauT* promoter activity by E2 (50 nM) was only observed in the reporter gene p-923, suggesting that the ERE sequence located at -949 to -954 is essential for the binding of E2 to the *TauT* promoter.

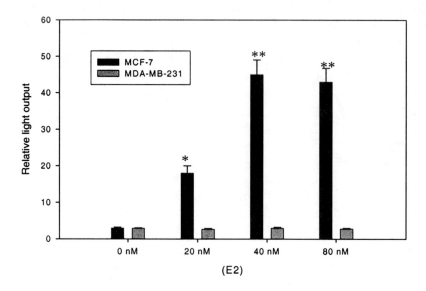

Figure 1. Effect of E2 on *TauT* promoter activity. DNA construct p-923 was transiently transfected into MCF-7 and MDA-MB-231 cells 2 h before E2 was added, and cells were cultured for another 48 h. Luciferase assays were performed as described in the Methods. *P<0.05 vs control, **P<0.01 vs control.

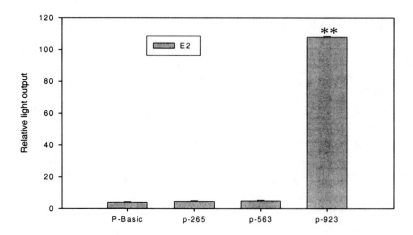

Figure 2. ERE is required for E2 up-regulation of *TauT* promoter. Progressive deletion was performed by PCR using p-923 as a template. The resulting constructs (p-265, and p-563) were transiently transfected into MCF-7 cells treated with E2 for 24 h, and the luciferase assay was performed as described in the Methods. **P<0.01 vs control.

3.2   Transactivation of *TauT* Promoter by p53

Our studies have demonstrated that *TauT* is a putative target gene of p53[8]. The *TauT* gene appears to be differentially regulated by p53 in different types of cells. To further test this observation, reporter gene p-923, which contains a p53-binding sequence at -663 to -695, was transiently transfected into MCF-7 and MDA-MB-231 cells. As shown in Figure 3, p53 enhanced *TauT* promoter activity by about 20-fold in MCF-7 cells, which was synergized by E2, but had no effect on *TauT* promoter function in MDA-MB-231 cells. Deletion of p53 and ERE sites (p-563) abolished such action induced by either E2 or p53 in MCF-7 cells.

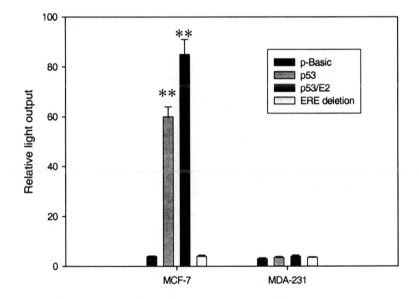

Figure 3. Effect of p53 and E2 on *TauT* promoter activity. Construct p-923 was cotransfected with p53 into MCF-7 and MDA-MB-231 cells treated with or without E2 for 24 h, and then luciferase assays were performed as described in the Methods. **P<0.01 vs control.

3.3   Adaptive Regulation of *TauT* is Impaired by p53

The reporter construct p-923, containing the cis-element for adaptive regulation [11], was transiently transfected into MCF-7 cells cultured in medium containing 0 μM, 50 μM, or 500 μM taurine for 48 h with or without p53 transfection. As shown in Figure 4, the *TauT* promoter activity was adaptively regulated by the external concentration of taurine, i.e., the *TauT* promoter activity was up- or down-regulated when cells were cultured in medium containing 0 μM or 500 μM taurine, as compared to control (50 μM taurine). Transfection of p53 increased *TauT* promoter activity, and abolished adaptive regulation of *TauT* promoter function by taurine.

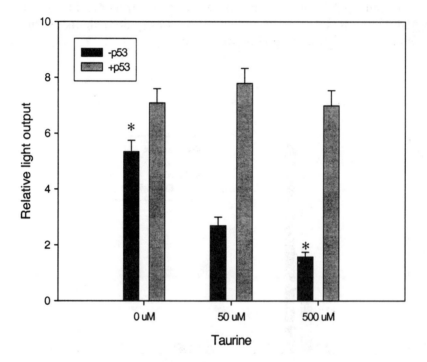

Figure 4. p53 abolished the adaptive regulation of *TauT* by external taurine. MCF-7 cells were cultured in medium containing 0 μM, 50 μM or 100 μM taurine with or without p53 transfection for 24 h. Luciferase assays were performed as described in the Methods. *P<0.05 vs control.

3.4   Role of Estrogen Receptors in Differential Regulation of *TauT* Promoter by p53

Induction of *TauT* promoter activity by p53 was found in MCF-7 cells that express an endogenous ERα, whereas p53 had no effect on *TauT* promoter function in MDA-MB-231 cells (ER⁻). To determine if ERs play a role in differential regulation of *TauT* by p53, reporter gene p-923 was co-transfected with p53, ERα, ERβ2, ERβ2/p53, or ERα/ERβ2 into MDA-MB-

231 cells, respectively. As shown in Figure 5, co-transfection with p53, ERß2, or ERß2/p53 had no effect on *TauT* promoter activity. ERα or ERα/p53 enhanced *TauT* promoter activity, while the effect of ERα on *TauT* promoter activity was blocked by ERß2.

4.   DISCUSSION

Human leukemia cells (HL-60) express a high level of taurine transport activity - a gain of function, compared with the normal mature lymphocytes, which do not express a detectable taurine transporter [12].

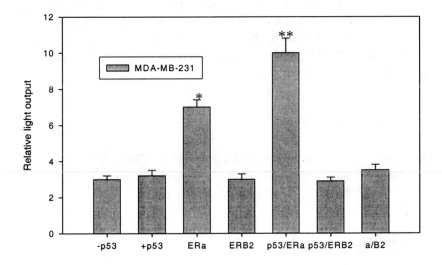

Figure 5. The role of estrogen receptors in the regulation of *TauT* by p53. The p-923 reporter gene was co-transfected with p53, ERα, ERβ2, ERα/p53, ERß2/p53, or ERα/ERß2 intoMDA-MB-231 (ER⁻) cells, respectively, and cells were cultured for another 24 h. Luciferase assays were performed as described in the Methods. *P<0.05 vs control. ** P<0.01 vs control.

Interestingly, HL-60 cells are found to be highly resistant to Fas-mediated apoptosis [13]. Therefore, study of the mechanisms by which the expression of *TauT* is turned on and the consequence of this gain of function may help to define the role of *TauT* in tumorigenesis.

In this study, we found that *TauT* was transactivated by 17-ß-estradiol (E2) and p53 in estrogen receptor-positive MCF-7 human breast cancer cells. This transactivation is apparently mediated by ERα, since transfection of ERα enhanced *TauT* promoter activity in the ER-negative MDA-MB-231 human breast cancer cell line, which could then be blocked by ERß. Recent studies have shown that ERß, especially ERß2, suppress ERα-mediated transcriptional activation [14]. These results indicate that up-regulation of *TauT* by E2 and p53 requires endogenous ERα, which is attenuated by ERß. The mechanism by which the ERs and p53 regulate *TauT* expression remains to be determined.

Studies have shown that *TauT* is adaptively regulated by dietary manipulation in various types of cell lines [15-17]. In the present study we demonstrated that the promoter activity of *TauT* was adaptively regulated by external taurine concentration in MCF-7 cells. p53 enhanced the *TauT* promoter activity, and abolished the adaptive regulation of *TauT* transcription induced by taurine manipulation. The mechanisms and outcomes of this action are unknown. Others studies suggest that the wild-type p53 gene product present in the MCF-7 cell could be inactivated by E2 through nuclear exclusion, which may be involved in the tumorigenesis of estrogen-dependent neoplasm [7].

ACKNOWLEDGMENT

This study was supported by the Le Bonheur Chair of Excellence in Pediatrics and a grant from the Children's Foundation Research Center at Le Bonheur Children's Medical Center, Memphis, TN, USA.

REFERENCES

1.   Hartwell, L.H., and Kastan, M.B., 1994, Cell cycle control and cancer. *Science* 266(5192), 1821-1828.
2.   Pei, X.H., Nakanishi, Y., Takayama, K., Bai, F., and Hara, N., 1999, Benzo[α]pyrene activates the human p53 gene through induction of nuclear factor kappaB activity. *J. Biol. Chem.* 274(49), 35240-35246
3.   Hirao, A., Kong, Y.Y., Matsuoka, S., Wakeham, A., Ruland, J., Yoshida, H., Liu, D., Elledge, S.J., and Mak, T.W., 2000, DNA damage-induced activation of p53 by the checkpoint kinase Chk2. *Science.* 287(5459), 1824-1827.
4.   Nylander, K., Bourdon, J.C., Bray, S.E., Gibbs, N.K., Kay, R., Hart, I., and Hall, P. A., 2000, Transcriptional activation of tyrosinase and TRP-1 by p53 links UV irradiation to the protective tanning response. *J .Pathol.* 190(1), 39-46.
5.   Ashcroft, M., Taya, Y., and Vousden, K.H., 2000, Stress signals utilize multiple pathways to stabilize p53. *Mol .Cell Biol.* 20(9), 3224-3233.

6.   Qin, C., Nguyen, T., Stewart, .J, Samudio, I., Burghardt, R., and Safe, S., 2002, Estrogen Up-Regulation of p53 Gene Expression in MCF-7 Breast Cancer Cells Is Mediated by Calmodulin Kinase IV-Dependent Activation of a Nuclear Factor kappaB/CCAAT-Binding Transcription Factor-1 Complex. *Mol Endocrinol.* Aug;16(8):1793-809.

7.   Molinari, A.M., Bontempo, P., Schiavone, E.M., Tortora, V., Verdicchio, M.A., Napolitano, M., Nola, E., Moncharmont, B., Medici, N., Nigro, V., Armetta, I., Abbondanza, C., and Puca, G.A., 2000, Estradiol induces functional inactivation of p53 by intracellular redistribution. *Cancer Res.* May 15;60(10):2594-7.

8.   Han, X., Budreau, A.M., and Chesney, R.W., 2002, Transcriptional repression of taurine transporter gene (*TauT*) by p53 in renal cells. *J Biol Chem.* (*in press*).

9.   Heller-Stilb, B., van Roeyen, C., Rascher, K., Hartwig, H.G., Huth, A., Seeliger, M.W., Warskulat, U.,and Haussinger, D., 2002, Disruption of the taurine transporter gene (taut) leads to retinal degeneration in mice. *FASEB J.* Feb;16(2):231-223.

10. Han, X., Budreau, A.M., and Chesney, R.W., 2000, Cloning and characterization of the promoter region of the rat taurine transporter (*TauT*) gene. *Adv Exp Med Biol.* 483:97-108.

11. Han, X., Budreau, A.M., and Chesney, R.W., 2000 Identification of promoter elements involved in adaptive regulation of the taurine transporter gene: role of cytosolic Ca2+ signaling. *Adv Exp Med Biol.* 483:535-544.

12. Learn, D.B., Fried, V.A. and, Thomas, E.L., 1990, Taurine and hypotaurine content of human leukocytes. *I Leukoc Biol.* Aug;48(2):174-182.

13. Lang, F., Madlung, J., Uhlemann A.C., Risler, T., and Gulbins, E., 1998, Cellular taurine release trggered by stimulation of the Fas(CD95) receptor in Jurkat lymphocytes. *Pflgers Arch.* Aug;436(3):377-383.

14. Maruyama, K., Endoh, H., Sasaki-Iwaoka, H., Kanou, H., Shimaya, E., Hashimoto, S., Kato, S., and Kawashima, H., 1998, A novel isoform of rat estrogen receptor beta with 18 amino acid insertion in the ligand binding domain as a putative dominant negative regular of estrogen action. *Biochem Biophys Res Commun.* May 8;246(1):142-147.

15. Han, X., Budreau, A.M., and Chesney, R.W., 1997, Adaptive regulation of MDCK cell taurine transporter (pNCT) mRNA: transcription of pNCT gene is regulated by external taurine concentration. *Biochim Biophys Acta.* Apr 10;1351(3):296-304.

16. Jones, D.P., Miller, L.A., and Chesney, R.W., 1995, The relative roles of external taurine concentration and medium osmolality in the regulation of taurine transport in LLC-PK1 and MDCK cells. *Pediatr Res.* Feb;37(2):227-232.

17. Jayanthi, L.D., Ramamoorthy, S., Mahesh, V.B., Leibach, F.H., and Ganapathy, V., 1995, Substrate-specific regulation of the taurine transporter in human placental choriocarcinoma cells (JAR). *Biochim Biophys Acta.* May 4;1235(2):351-360.

Gating of Taurine Transport
Role of the Fourth Segment of the Taurine Transporter

XIAOBIN HAN, ANDREA BUDREAU PATTERS, and RUSSELL W. CHESNEY
Department of Pediatrics, University of Tennessee Health Science Center, and the Children's Foundation Research Center at Le Bonheur Children's Medical Center, Memphis, TN, USA

1.   INTRODUCTION

Taurine (2-aminoethanesulfonic acid) is primarily transported by its carrier, the taurine transporter, the gene for which has been cloned recently from several species and tissues [1-5]. Expression of the taurine transporter gene (*TauT*) is regulated by many factors, including the p53 tumor suppressor gene, Wilms' tumor suppressor gene (WT1), Sp1 transcription factor, and dietary taurine manipulation [6-9]. In contrast, the taurine transporter itself, which contains 12 hydrophobic membrane-spanning domains, is mainly regulated by protein kinase C (PKC) phosphorylation [10]. Our previous studies show that the fourth intracellular segment (S4) of the taurine transporter may be involved in the gating of taurine transport that is modulated by PKC phosphorylation of Ser-322 located in S4 [11]. In the present study, the role of charged amino acid residues located in S4 in the gating of taurine transport was determined using site-directed mutagenesis and the *Xenopus laevis* oocyte expression system.

2.   MATERIALS AND METHODS

2.1   Site-Directed Mutagenesis and *in vitro* RNA Synthesis

An *in vitro* mutagenesis system (QuikChange, Stratagene) was used for site-directed mutagenesis using the MDCK taurine transporter (pNCT) cDNA as a template. Mutations were confirmed by enzymatic nucleotide sequencing (United States Biochemical Corp). Capped RNA transcripts were synthesized *in vitro* with SP6 RNA polymerase following the manufacturer's instructions (mCAP RNA Capping Kit, Stratagene).

2.2   Microinjection of *Xenopus laevis* Oocytes

Ovarian lobes were dissected from anesthetized frogs and oocytes were separated by incubation of ovarian fragments for 40 minutes with 2 mg/ml collagenase type II in calcium-free buffer at room temperature. Defolliculated stage V-VI oocytes were selected and incubated overnight at 18°C in media (50% L-15, Leibovitz, Sigma), 1 mM L-glutamine, 15 mM HEPES, pH 7.6, 100 µg/ml gentamicin sulfate) before injection. Water (30 nl) containing 30 ng of capped transcripts was injected into each oocyte. Injected oocytes were maintained at 18°C in the above media for 3 days prior to taurine uptake experiments.

2.3   Measurement of Taurine Uptake

Oocytes were transferred to a Na^+-containing uptake solution referred to as ND96 (2 mM KCl, 1 mM $MgCl_2$, 96 mM NaCl, 1.8 mM $CaCl_2$, 5 mM HEPES, pH 7.6) to which was added 10 µM unlabelled taurine and 0.5 µCi/ml ^{14}C-taurine (Du Pont, Perkin-Elmer). After incubation for 30 min at room temperature, oocytes were transferred to a 24-well cell culture plate in which they were washed rapidly 5 times with 1 ml of ice-cold Na^+-free uptake solution. Lastly, individual oocytes were transferred to miniscintillation vials, solubilized in 100 µl of 10% SDS, and counted in 2 ml of scintillation cocktail (Aquasol, NEN).

2.4   Oocyte Membrane Isolation and Western Blot Analysis

Oocyte membranes were isolated as described [12]. Briefly, twenty oocytes were lysed in 200 µl of lysis buffer (7.5 mM Na_2HPO_4, pH 7.4, 1 mM EDTA buffer containing 20 µg/ml phenylmethylsulfonyl fluoride, 1 µg/ml pepstatin A, 1µg/ml leupeptin, 1:2000 diisopropylfluorophosphate) by repeatedly vortexing and pipetting the samples. The yolk and cellular debris were pelleted at 750 x *g* for 5 min at 4°C. The membranes were then pelleted from the supernatant at 16,000 x *g* for 30 min at 4°C. The floating yolk and supernatant were removed, and then the membrane pellets were washed once with an equal volume of ice-cold lysis buffer and resuspended in 10 µl lysis buffer. Western blots were performed as described using an affinity-purified antibody [11].

2.5   In Vitro Phosphorylation

Membrane proteins of oocytes were prepared as described above. Protein A (Pierce) and affinity-purified anti-pNCT antibody (1:100 dilution) were used for immunoprecipitation. Immune complexes were resuspended in 50 μl of phosphorylation buffer (4.5 mM $CaCl_2$, 60 μl/ml phosphatidylerine, 6 μl/ml diolein, 10 mM $MgCl_2$, 12.5 mM MOPS, pH 7.2, 12.5 mM β-glycerol phosphate, 5 mM EGTA, 50 mM sodium fluoride, and 250 μM dithiothreitol) and incubated at 37°C in the presence of PKC (10 ng). The reaction was initiated by the addition of 0.5 mM ATP containing 10 μCi of [γ-^{32}P] ATP (6000 Ci/mmol, Du Pont, Perkin-Elmer). After 1 h of incubation, the phosphorylation reaction was terminated by the addition of 5X SDS sample buffer (250 mM Tris-HCl, pH 6.8, 10% SDS, 25% glycerol, 5% β-mercaptoethanol, and 0.02% bromphenol blue). Phosphorylated membrane proteins were analyzed by SDS-polyacrylamide gel electrophoresis. The protein was transferred to nitrocellulose membranes and followed by autoradiography.

2.6 Materials

[^{14}C]-Taurine (92.1 mCi/mmol) was purchased from Du Pont, Perkin-Elmer. Female *Xenopus laevis* frogs were purchased from Xenopus LTD. pNCT was a gift from Dr. S. Uchida of the University of Tokyo.

2.7 Data Analysis

All experiments were performed in triplicate. The data represent the mean ± SEM of taurine uptake by three separate batches of oocytes. Student's t test for paired data was applied to determine significant differences in the means.

3. RESULTS

3.1 Taurine Transport after Mutation of Charged Residues in S4

Our previous studies have shown that a polyclonal antibody directed against this conserved intracellular segment (S4) of the MDCK taurine transporter enhanced taurine uptake by pNCT cRNA-injected oocytes [10]. Mutation of Ser-322 located in S4 abolished PCK modulation of the taurine transport activity, suggesting that S4 is involved in the gating of taurine transport, which is controlled by PKC phosphorylation of Ser-322.

The S4 region also contains several conserved charged amino acid residues. To determine the role of these residues in the gating function, we individually mutated each of the charged residues in pNCT. We then compared the function of the mutants with that of the wild-type transporter by expressing the transporters in *Xenopus laevis* oocytes. As shown in Figure 1, taurine uptake by oocytes expressing mutant K319Q (lysine-319 replaced

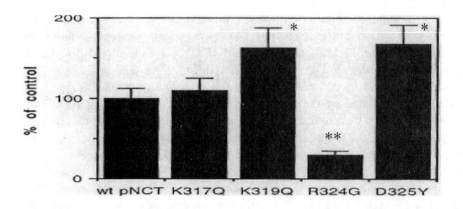

Figure 1. Effect of mutation of charged residues of pNCT on taurine uptake. Oocytes were injected with 30 ng of wt-pNCT or mutant cRNA and assayed for taurine uptake 3 days post-injection. Data for each mutant are given as percentages of the control and are representative of three experiments. Each value represents the mean ± SEM for six oocytes. * $p<0.05$ vs control, ** $p<0.01$ vs control.

by glutamine) and D325Y (aspartic acid-325 replaced by tyrosine) was increased by about 50% compared with the control (wild-type pNCT). In contrast, when arginine-324 was replaced by glycine (mutant R324G), taurine transport was decreased by 70% compared with the control, while replacing lysine with glutamine (mutant D317Q) had no effect on taurine transport activity.

3.2   Kinetic Analysis of Taurine Transport by Mutant pNCT

The mechanism underlying the observed impact on taurine transport by these pNCT mutants was examined by kinetic analysis using Eadie-Hofstee plots. Kinetic studies showed that the V_{max} rather than the K_m of the transporter was changed in mutants K319Q (Figure 2A & 2B), and D325Y (2C & 2D).

However, the decreased taurine transport activity by mutant R324G appeared to   be caused by an increase in the K_m of the mutant transporter (Figure 2E & 2F).

3.3   Effect of Mutation on Membrane Location and Phosphorylation of pNCT

To ascertain if the mutations changed the amount of transporter protein located on the membrane, and/or the phosphorylation status of the transporter, Western blot analysis and phosphorylation assessment of mutant

A

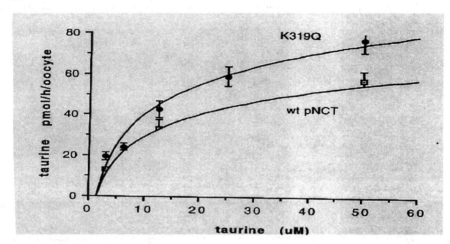

B

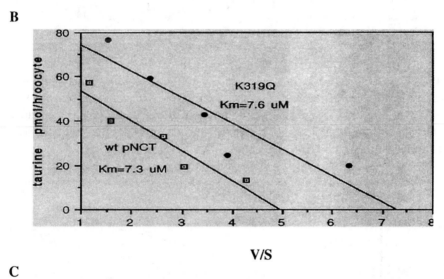

V/S

C

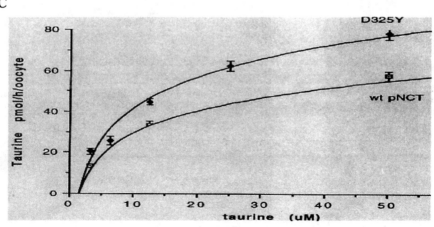

D

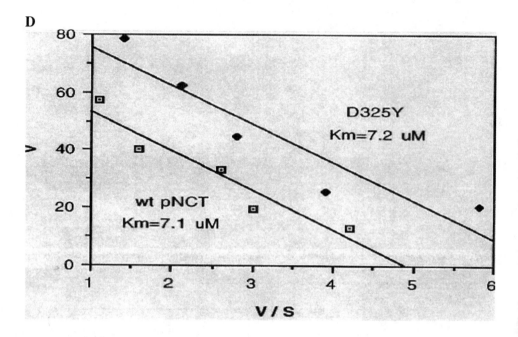

pNCTs were performed as described in our previous studies[11]. As shown in Figure 3A (left), replacement of charged amino acid residues with a neutral amino acid residue did not change the amount of taurine transporter located on the membrane, and had no effect on the phosphorylation status of the transporter expressed in oocytes (Figure 3B, right).

E

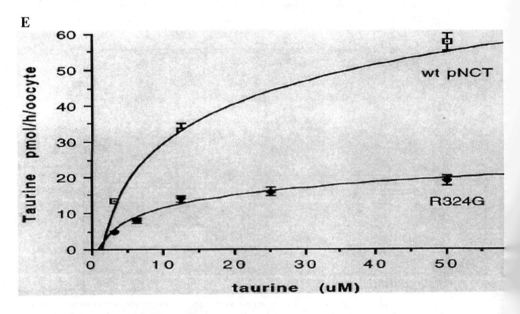

F

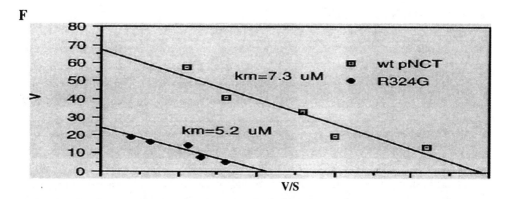

Figure 2. Kinetic analysis of taurine transport by wt-pNCT and mutant pNCT expressed in oocytes. Taurine accumulation was measured for taurine concentrations ranging from 3.125 μM to 50 μM. Oocytes were injected with 30 ng of wt-pNCT or mutant pNCT cRNA and assayed for taurine uptake 3 days post-injection. Each value represents the mean ± SEM for six oocytes and is representative of three experiments.

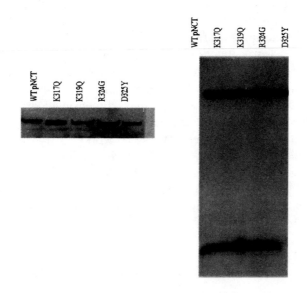

Figure 3. Effect of mutation on membrane location and phosphorylation of the transporters. A.W estern blot analysis of oocyte membrane proteins probed with the antibody against pNCT. B. *In vitro* phosphorylation of oocyte membrane proteins.

4.   DISCUSSION

Taurine is a zwitterionic molecule, and its co-substrates for transport across the plasma membrane are Na^+ and Cl^-. Studies in other transport systems have shown that the highly charged intracellular loops of Shaker K^+ channel and the intracellular loop joining transmembrane segments IV and V in the first membrane-spanning domain of the cystic fibrosis transmembrane conductance regulator (CFTR) play an important role in channel function[13-14]. Two potential overlapping sodium binding motifs are located immediately upstream of S4. These molecular features make S4 a particularly likely candidate for electrostatic or allosteric interactions with other intracellular loops that contribute to the transporter function. As stated previously, we have shown that the fourth intracellular loop of the taurine transporter is directly involved in the gating of taurine transport, which is controlled by PKC phosphorylation of Ser-322 [10-11].

In this study, we demonstrated that charged amino acid residues in S4 play an important role in the gating of taurine transport. Enhanced taurine uptake by K319Q, rather than K317Q, suggests that lysine-319 may play an important role in the formation of the amino acid consensus sequence for taurine transport. Furthermore, the positively charged lysine-319 and negatively charged aspartic acid-325 may attract one another and thus help to maintain the status of the loop, which is important for controlling the gating function of the taurine transporter,

Arginine-324 may function as a binding site for taurine, because kinetic analysis of mutant R324G demonstrated that decreased taurine transport was the result of an increase in the K_m of the transporter. Phosphorylation of Ser-322 by PKC could potentially change the 3-dimensional structure of the gate and block the ionic binding of taurine to Arg-324. The mechanism by which the fourth intracellular segment of pNCT works to transport extracellular taurine into the cell remains to be further investigated.

In summary, the charged amino acid residue Arg-324 appears to play an important role in the gating, or ionic binding of taurine to the transporter, and Ser-322 likely functions as key to that gate controlled by PKC phosphorylation.

ACKNOWLEDGMENT

This study was supported by the Le Bonheur Chair of Excellence in Pediatrics and a grant from the Children's Foundation Research Center at Le Bonheur Children's Medical Center, Memphis, TN, USA.

REFERENCES

1. Liu, Q.R., Lopez-Corcuera, B., Nelson, H., Mandiyan, S. and Nelson, N., 1992, Cloning and expression of a cDNA encoding the transporter of taurine and β-alanine in mouse brain. *Proc Natl Acad Sci USA* 89:12145-12149.
2. Ramamoorthy, S., Leibach, F.H., and Mahesh, V.B., 1994, Functional characterization and chromosomal localization of a cloned taurine transporter from human placenta. *Biochem J* 300:893-900.
3. Smith, K.E., Borden, L.A., Wang, C.D., Hartig, P.R., Branchek, T.A., and Weinshank, R.L., 1992, Cloning and expression a high affinity taurine transporter from rat brain *Mol Pharmacol* 42:563-569.
4. Uchida, S., Kwon, H.M., Yamauchi, A., Preston, A.S., Marumo, F., and Handler, J.S., 1992, Molecular cloning of the cDNA from an MDCK cell Na(+)- and Cl(-)-dependent taurine transporter that is regulated by hypertonicity. *Proc Natl Acad Sci USA* 89:8230-8234.
5. Han, X., Budreau, A.M., and Chesney, R.W., 1998, Molecular cloning and functional expression of an LLC-PK1 cell taurine transporter that is adaptively regulated by taurine. *Adv Exp Med Biol* 442:261-268.
6. Han, X., Budreau, A.M., and Chesney, R.W., 2002, Transcriptional repression of taurine transporter gene (*TauT*) by p53 in renal cells. *J Biol Chem.* (in press).
7. Han, X., Budreau, A.M., and Chesney, R.W., 2000, The taurine transporter gene and its role in renal development. *Amino Acids* 19(3-4):499-507.
9. Han, X., Budreau, A.M., and Chesney, R.W., 2000, Cloning and characterization of the promoter region of the rat taurine transporter (*TauT*) gene. *Adv Exp Med Biol.* 483:97-108.
10. Han, X., Budreau, A.M., and Chesney, R.W.,1997, Adaptive regulation of MDCK cell taurine transporter (pNCT) mRNA: transcription of pNCT gene is regulated by external taurine concentration. *Biochim Biophys Acta* Apr 10;1351(3):296-304.
11. Han, X., Budreau, A.M., and Chesney, R.W., 1996, Role of conserved peptide in taurine transporter inactivation modulated by protein kinase C. *J Am Soc Nephrol* Oct;7(10):2088-2096.
12. Han, X., Budreau, A.M., and Chesney, R.W., 1999, Ser-322 is a critical site for PKC regulation of the MDCK cell taurine transporter (pNCT). *J Am Soc Nephrol.* Sep;10(9):1874-1879.
13. Dascal, N., 1987, The use of *Xenopus* oocytes for the study of ion channels. *CRC Crit Rev Biochem* 22:317-387.
14. Isacoff, E.Y., Jan, Y.N., and Jan, L.Y., 1991, Putative receptor for the cytoplasmic inactivation gate in the shaker K^+ channel. *Nature* 353:86-90.
15. Xie, J., Drumm, M.L., and Ma, J.B.D., 1995, Intracellular loop between transmembrane segments IV and V of cystic fibrosis transmembrane conductance regulator is involved in regulation of chloride channel conductance state. *J Biol Chem* 270:28084-28091.

Finding of TRE (TPA Responsive Element) in the Sequence of Human Taurine Transporter Promoter

KUN-KOO PARK, EUNHYE JUNG, SANG-KEUN CHON, MOOSEOK SEO, HA WON KIM[*] and TAESUN PARK[#]

PharmacoGenechips Inc,. Chunchon 200-160 , [*]Dept. of Life Science, University of Seoul, Seoul 130-743, [#]Dept. of Food and Nutrition, Yonsei University, Seoul 120-749, Korea

ABSTRACT

Activity of the taurine transporter (TAUT) is regulated by signal transduction in response to diverse stimuli including tumor promoters such as phobol ester. Regulation of the transcription rate of TAUT appears to play an important role in exerting biological roles of taurine in mammalian tissues in adverse environments. Although cDNA of human TAUT has been cloned and sequenced in placenta, thyroid cells, and retinal pigment epithelial cells, the promoter region of TAUT has never been reported. In order to clone the upstream region of the human TAUT promoter, we have compared TAUT cDNA sequences with the entire human genome sequence. Polymerase chain reaction (PCR) was performed from genomic DNA prepared from a SK-Hep-1 cell line for the amplification of the TAUT promoter region including the partial exon (150 bp) and the 5' untranslated region (UTR, 380 bp). The PCR product of the promoter region, which was 1800 bp long, was ligated into the pGEM-T vector, and sequenced. The 5' flanking region of the TAUT promoter was analysed for the identification of enhancer and regulation motifs. Surprisingly we found the consensus TPA responsive element (TGAGTCAG) which is responsible for gene regulation by the protein kinase C (PKC)-mediated signal transduction pathway. The well known fact that proto-oncogene AP1 (cFos/cJun heterodimer or cJun/cJun homodimer) binds to TRE implies that TAUT expression might be closely linked to tumor promotion. Since AP1 activity is also tightly regulated in nerve cells, AP1-regulated TAUT transcription might be an important step in nerve cell function. Furthermore, the TFIID binding site, cap signal for transcription initiation, PEA3 motif, heat shock factor binding motif, and many other motifs were found in the TAUT promoter region, and require characterization.

1.   INTRODUCTION

Interactions of cis-acting elements in the promoter region with trans-acting factors are one of the major regulatory mechanisms of gene expression in mammalian cells. AP1 (Jun/Jun homodimer or Jun/Fos heterodimer) is a transcription factor that binds to TRE (TPA responsive element), and is involved in the expression of several genes such as cyclooxygenase-2, IL-8, and PEG-3 in cooperation with another transcription factor, PEA3[1-3].

Regulation of the TAUT gene expression appears to play an important role in exerting biological roles for taurine in mammalian tissues in adverse environments. Activity and gene expression of TAUT are known to be regulated in various cell lines by signal transduction in response to diverse stimuli including hypertonicity, NO, TNFα, and calmodulin[4-6].

cDNA of the human TAUT has been cloned and sequenced in placenta, thyroid cells, and retinal pigment epithelial cells, and is known to have a coding region of 1863 bp long that expresses a 69.9 kDa protein composed of 620 amino acids. Although Han et al.[7,8] cloned the promoter region of rat TAUT which is regulated by p53, the promoter region of the human TAUT has never been reported. Recently, the human genome project revealed the entire sequence of the human genomic DNA. By comparing sequence homology between human TAUT cDNA and the human genomic DNA map, we have successfully cloned the promoter region of human TAUT, and identified promoter elements including TRE which need to be characterized.

2.   MATERIALS AND METHODS

2.1   Reagents

A T-vector ligation kit and a luciferase assay kit were purchased from Promega (Madison, USA). DNA sequencing was performed by using a kit from Applied Biosystem (California, USA). The cloning vector pGL3 promoter, pGL3 basic and all reagents used for electrophoretic mobility shift assays (EMSA) were obtained from Promega (Madison, USA). Radioactive products were purchased from PE Life Sciences (Boston, USA). Tissue culture plasticware was obtained from BD Biosciences (Massachusetts, USA), and all other chemicals were purchased from Sigma-Aldrich (Missouri, USA).

2.2   Cell Culture

Human cervical cancer cells (SiHa) and human hepatocarcinoma cells (SK-Hep-1) were maintained in RPMI 1640 medium with 10% fetal bovine serum and 1% streptomycin/penicillin. Cells were cultured at 37°C in a humidified 95% air - 5% CO_2 atmosphere.

2.3   Cloning and Sequencing of Human TAUT Promoter

Genomic DNA from human SK-Hep-1 cells was prepared as described by the manufacturer (Promega, USA). A genomic DNA fragment of 1800 bp length was amplified by the polymerase chain reaction (PCR) technique using the primers 5' ACAAAGTCGATCTTGCTAGACC3'   and   5'GAGATAGACGACAGG GAAGC 3' referred to as TAU1 and TAU3, respectively. The PCR product was extracted from an agarose gel, and cloned into the respective sites of the pGEM-T vector (Promega, Korea). Sequencing of the cloned DNA (designated pGEM TAU13) was performed on an automatic sequencer (model # ABI377-96, Perkin Elmer) using a kit from Applied Biosystem (California, USA). The amplified DNA fragments were digested with the appropriate enzymes, and analyzed by agarose gels.

2.4   Transient Transfection and Assay of Luciferase Activity

The TAUT-TRE luciferase plasmid was constructed by inserting a tandom array of the AP1 binding site (CGCGC**TGAGTCA** GTC**TGAGTCA**GTCT**GA GTCA**G) into the pGL3 promoter plasmid. SiHa cells (1.0 x 10^6 cells) were cotransfected with a total of 2.0 μg DNA (1.4 μg TAUT-TRE luciferase plasmid, 0.3 μg pCMV β-galactosidase, and 0.3 μg pCMV c-jun plasmid) by the lipofectamine method. Cells were harvested 36 h after transfection. The luciferase activity of each cell lysate was measured using the Luciferase Reporter Plasmid System (Promega, USA). The firefly luciferase activity of each sample was normalized by the value of the β-galactosidase activity. Transfection assays were carried out in triplicate.

2.5   Electrophoretic Mobility Shift Assay (EMSA)

Nuclear proteins were extracted as described by C. Jonat et al.[9]. Routinely, 5 μg of the extract were used for each experiment. A double-stranded oligonucleotide, representing a AP1-TAU element with the sequence 5'

AGCTGCTGACTCAGAGGGATC 3', was labelled and used for the binding assay. The subsequent steps were performed following the recommendation of the distributor (Promega, USA). In order to verify binding of the predicted nuclear protein, competition assays were performed with 100 fold molar excess of the unlabeled double-stranded oligonucleotide which represents an AP1 binding site (5' CGCTTGATGAGTCAGCCGGAA 3').

3.   RESULTS AND DISCUSSION

In order to clone the upstream region of the human TAUT promoter, we had compared TAUT cDNA sequences (NM_003043) with the entire human genome project database (www.ncbi.nlm.nih.gov). From this comparison of sequence homology, we identified the entire genomic sequence and chromosomal location of TAUT. Results indicated that the human TAUT gene is localized in chromosome 3, and is 40 kb-long composed of 13 exons. The TATA box like sequence was identified approimately 380 bp upstream from the ATG (start) codon. Polymerase chain reaction (PCR) was performed from the genomic DNA prepared from SK-Hep-1 cells for the amplification of the TAUT promoter region including the partial exon (150 bp) and 380 bp 5' untranslated region (UTR). Using TAU3 and TAU1 (located at +150 bp in exon 1) primers, a 1800 bp long PCR product of the promoter region was obtained, and ligated into the pGEM-T vector (Fig. 1).

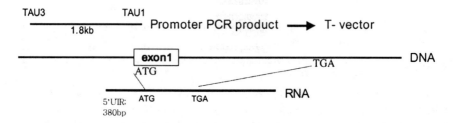

Figure 1. Cloning of the taurine transporter promoter. Genomic structure near the human taurine transporter promoter region is shown. The rectangle represents exon 1. The 1.8 kb region including 380 bp 5'UTR and 5'-flaning TAUT promoter region was amplified by PCR with TAU3 and TAU1 primers.

The 5' flanking region of the human TAUT promoter was sequenced, and analysed for the identification of enhancer and regulation motifs (Fig. 2). Surprisingly, we found the consensus TPA responsive element (TGAGTCAG) which is responsible for gene regulation by the protein kinase C (PKC)-mediated signal transduction pathway. The well known fact that proto-oncogene AP1 (cFos/cJun heterodimer or cJun/cJun homodimer) binds to TRE implies that

TAUT expression might be closely linked to tumor promotion. Since AP1 activity is tightly regulated in nerve cells, AP1-regulated TAUT transcription might be an important step in nerve cell function. Furthermore, PEA3 motif, E12/RFX2 site, TATA box, CAP signal for transcription initiation, and heat shock transcription factor (HSTF) binding motif were found in the TAUT promoter region, and need to be characterized.

TTTCAGTGTTGGCCACCAGTTTGATCATGAAGACAAACAAGTTGTGTCCCAGACTGCTTGGC

TT**GAGTCAG**CCACTTCTGCTGT<u>**CATCCT**</u>TTTTATTCTG<u>**CAGATG**</u>GGGAAAC
  AP1   PEA3   E12/RFX2

TGATGCTCAGAGAGGGGCGAGTAGGCCTAGTAGCTCTGGTTTTCCCAGCTCATCCGTTTTCT

GTGTCCCTCTGTTATGTTGTGGGTCAGCCCTTGAAGGCAGGGGACCCAGCCTTGAGCTGTTTC

CTCACGCTGTGGGCTAGAAATC<u>**AGGATG**</u>CTCTTGGTAACCCTTGTC
                                PEA3

AGCCTGAAACCCCATTCCTCGTTCGTTTCCGTCTATCTGTTTTTTTAACTTACAGT

               ⟶

<u>TATAAA</u>TGGT<u>**CATCCCTG**</u>TGGGCTGGGCACTGTATTAAGGCATTTGCAAAAGAGGAG
TATA Box   CAP signal

CCTCCATTTTTTTAAATACAAATCCAAATTTAGAAAATAGTATCTGTTAAT<u>**TTCTAG**</u>
                                                       HSTF

Figure 2. Nucleotide sequence of 5'-flanking promoter region of human taurine transporter gene. The TATA box as well as AP1 binding site are underlined. The estimated cap-site of the human TAUT promoter is indicated by the arrowhead. The various potential transcription factor binding sites are indicated.

The AP1 binding site found in the human TAUT promoter was 100% conserved. In order to evaluate the functional activity of the TAUT AP1 binding site, TAUT-TRE luciferase plasmid was constructed by inserting the tandom array of AP1 binding site into the pGL3 promoter plasmid. SiHa cells were cotransfected with the TAUT-TRE luciferase plasmid, the c-jun expression vector and the pCMV β-galactosidase plasmid, and assayed for luciferase activity. Activation of the TAUT AP1 binding site by c-Jun proto-oncogene product was evaluated by comparing the luciferase activity of cells transfected with the c-jun expression vector to the value of control cells transfected with the empty pc DNA 3.1 zeo (+) vector. Cells co-trasfected with TAUT-TRE luciferase plasmid along with the c-jun expression vector exhibited a 23 fold increase in luciferase activity compared to the value for cells co-transfected with the empty vector (Fig. 3).

These results indicate that the AP1 binding site in the human TAUT promoter region is functional, and activated by c-Jun. Interestingly, it has been reported that c-Jun is one of the major transcriptional factors responding to changes in cellular osmolarity[10,11]. Taken together with our observations, it is speculated that c-Jun might be involved in the regulation of osmolarity by activating the TAUT promoter, and thereby regulating the intracellular level of taurine, which is known to be an osmolyte.

Reporter Sequence: CGCGC**TGAGTCA**GTC**TGAGTCA**GTC**TGAGTCA**G

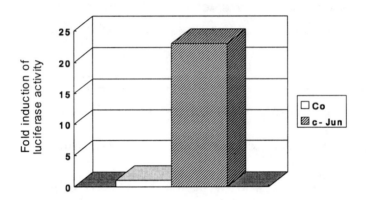

Figure 3. TRE of TAUT promoter is activated by c-Jun proto-oncogene product. TAUT-TRE luciferase plasmid was generated by ligating the oligonucleotide (tandom array of AP1 binding site) in pGL3 promoter plasmid. SiHa cells were co-transfected with TAUT-TRE luciferase plsmid, pCMV c-jun vector (as c-Jun expression vector) and pCMV β-galactosidase, which served as an internal control. The firefly luciferase activity of cells (c-Jun) co-transfected with c-Jun expression vector was normalized by its β-galactosidase activity, and subsequently compared with the value of cells co-transfected with the empty pcDNA3.1 zeo(+) vector (Co).

The electrophoretic mobility shift assay was performed for the evaluation of the binding activity of TAUT-TRE. Nuclear proteins (5 μg) extracted from SiHa cells were incubated for 20 min with the radiolabelled oligonucleotide probe containing the TAUT AP1 binding site, and tested for their electrophoretic mobility. As shown in Fig. 4, AP1 binding activity was observed in nuclear extracts incubated with radiolabelled probes containing TAUT-TRE, while the mobility shift disappeared when nuclear extracts were incubated with 100-fold molar excess of unlabelled nucleotide containing the consensus AP1 element. From these results, the binding activity of nuclear proteins from SiHa cells appears to be specific for TAUT-TRE.

AGCTGC**TGACTCAG**AGGGATC

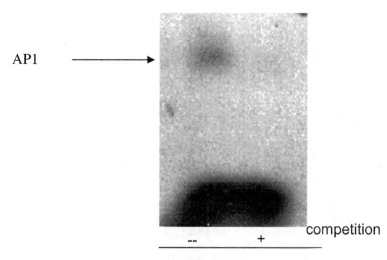

Figure 4. Electrophoretic mobility shift assay using the AP1 binding site sequence of the TAUT promoter. Nuclear proteins were incubated with radiolabeled probe TAUT-TRE. Competition was preformed with a 100-fold molar excess of unlabeled-oligonucleotide containing the consensus AP1 element [+].

ACKNOWLEDGEMENT

This work was supported by a grant No. R01-1999-000-00128-0 from the Basic Research Program of the Korea Science & Engineering Foundation.

REFERENCES

1. Subbaramaiah, K., Norton, L., Gerald, W., and Dannenberg, A.J., 2002, Cyclooxygenase-2 is overexpressed in HER-2/neu-positive breast cancer: evidence for involvement of AP-1 and PEA3. *J Biol Chem* 277(21):18649-57.
2. Iguchi, A., Kitajima, I., Yamakuchi, M., Ueno, S., Aikou, T., Kubo, T., Matsushima K, Mukaida, N., and Maruyama, I., 2000, PEA3 and AP-1 are required for constitutive IL-8 gene expression in hepatoma cells. *Biochem Biophys Res Commun* 279(1):166-71.

3. Su, Z., Shi, Y., and Fisher, P.B., 2000, Cooperation between AP1 and PEA3 sites within the progression elevated gene-3 (PEG-3) promoter regulate basal and differential expression of PEG-3 during progression of the oncogenic phenotype in transformed rat embryo cells. *Oncogene* 19(30):3411-21.
4. Bridges, C.C., Ola, M.S., Prasad, P.D., El-Sherbeny, A., Ganapathy, V., and Smith, S.B., 2001, Regulation of taurine transporter expression by NO in cultured human retinal pigment epithelial cells. *Am J Physiol Cell Physiol* 281(6):C1825-36.
5. Ramamoorthy, S., Del Monte, M.A., Leibach, F.H., and Ganapathy, V., 1994, Molecular identity and calmodulin-mediated regulation of the taurine transporter in a human retinal pigment epithelial cell line. *Curr Eye Res* 13(7):523-9.
6. Satsu, H., Miyamoto, Y., and Shimizu, M., 1999, Hypertonicity stimulates taurine uptake and transporter gene expression in Caco-2 cells. *Biochim Biophys Acta* 1419(1):89-96.
7. Han, X., Budreau, A.M., and Chesney, R.W., 2000, Cloning and characterization of the promoter region of the rat taurine transporter (TauT) gene. *Adv Exp Med Biol* 483:97-108.
8. Han, X., Patters, A.B., and Chesney, R.W., 2002, Transcriptional repression of taurine transporter gene (TauT) by p53 in renal cells. *J Biol Chem* [epub ahead of print]
9. Jonat, C., Rahmsdorf, H.J., Park, K.K., Cato, A.C., Gebel, S., Ponta, H., and Herrlich, P., 1990, Antitumor promotion and antiinflammation: down-modulation of AP-1(Fos/Jun) activity by glucocorticoid hormone. *Cell* 62(6):1189-204.
10. Capasso, J.M., Rivard, C., and Berl, T., 2001,The expression of the gamma subunit of Na-K-ATPase is regulated by osmolality via C-terminal Jun kinase and phosphatidylinositol 3-kinase-dependent mechanisms. *Proc Natl Acad Sci* 98(23):13414-9. U S A.
11. Wojtaszek, P.A., Heasley, L.E., Siriwardana, G., and Berl, T., 1998, Dominant-negative c-Jun NH2-terminal kinase 2 sensitizes renal inner medullary collecting duct cells to hypertonicity-induced lethality independent of organic osmolyte transport. *J Biol Chem* 273(2):800-4.

Protein Kinase C and cAMP Mediated Regulation of Taurine Transport in Human Colon Carcinoma Cell Lines (HT-29 & Caco-2)

SUNGYOUN PARK, HAEMI LEE, KUN-KOO PARK[*], HA WON KIM[#], and TAESUN PARK
Department of Food and Nutrition, Yonsei University, Seoul, 120-749, [*]PharmacoGenechips Inc., Chunchon, 200-160, [#]Department of Life Science, University of Seoul, Seoul 130-743, Korea

1.   INTRODUCTION

Taurine is one of the most abundant free amino acids comprising 38% and 18% of the total free amino acid pool in human duodenal and colonic mucosa, respectively[1]. Recent observations have demonstrated that serum and intestinal taurine concentrations were lower in patients with cancer[2], trauma[3], surgery[4] and critical illnesses[5] compared to healthy controls. These results postulate the possibility that intestinal taurine absorption might be impaired in such stressed conditions. Taurine transport across the intestinal epithelium plays an important role in the regulation of taurine homeostasis in neonates and adults particularly under diverse stressed conditions[6]. The brush border membrane of intestinal mucosal cells express the Na^+- and Cl^--dependent taurine transporter (TAUT) responsible for the active absorption of taurine across the epithelium. TAUT activity in the intestinal epithelium is known to be regulated in response to changes in the availability of dietary sulfur amino acids[6,7].

Taurine transport studies in the intestine, mainly undertaken with brush-border membrane vesicles, were recently expanded by an *in vitro* cell culture system. HT-29 and Caco-2 cell lines derived from human colon carcinomas undergo enterocytic differentiation, and have been shown to be useful experimental models for the study of taurine transport across the intestinal epithelium[8,9]. Recent regulatory studies involving diverse cell lines suggested that TAUT activity is regulated by intracellular and extracellular factors involved in signal transduction, such as protein kinase C[10], intracellular Ca^{2+} concentration [11], calmodulin[12], glucocorticoid[13], and nitric oxide[14]. In the present study, protein kinase C and cAMP mediated regulations of TAUT were compared in the two human colon carcinoma cell lines, HT-29 and Caco-2, and underlying mechanisms were elucidated.

2.   MATERIALS AND METHODS

2.1   Cell Culture

HT-29 and Caco-2 cells were cultured in RPMI-1640 and Dulbeco's modified Eagle's medium, respectively, and supplemented with 10% fetal bovine serum, 100 U/ml penicillin, 100 µg/ml streptomycin, and 25 µg/ml amphotericin B. Cells were maintained at 37 °C in a humidified 95% air-5% CO_2 atmosphere. Subconfluent cultures were treated with phosphate-buffered saline (PBS) containing 0.25% trypsin. For all experiments, cells were plated at an initial density of 2×10^6 cells/well in 6-well plates. Stock solutions of reagent were prepared in deionized water or DMSO. Cultured cells were preincubated in the medium to which was added the desired concentrations of phorbol 12-myristate 13-acetate (PMA), adenosine 3',5'-cyclic monophosphate (cAMP), or actinomycin D for varied time periods.

2.2   Taurine Uptake Study

Taurine upake was measured in confluent monolayer cultures of HT-29 and Caco-2 cell lines 4 days after seeding. Uptake buffer contained (in mM) 25 HEPES-Tris (pH 7.5), 140 NaCl, 5.4 KCl, 1.8 $CaCl_2$, 0.8 $MgSO_4$, 5 glucose, 49 nM non-radioactive taurine and 1 nM [^{3}H]taurine (NEN Life Science Products, MA, USA). Following a 30 min incubation with the uptake buffer, cells were quickly washed three times with PBS, and solubilized in 1ml of 0.1% sodium dodesyl sulfate. Radioactivity was counted by liquid scintillation spectrometry (Beckman LS6500, CA, USA).

For kinetics analyses, taurine uptake was measured over a taurine concentration range of 2~60 µM in the presence of 140 mM NaCl (total uptake) or 140 mM choline chloride (passive diffusion) using a 30 min incubation. Active taurine uptake was calculated by subtracting values for passive diffusion from those for total taurine uptake. Vmax and Km values for TAUT were calculated from the linear regression curve of Lineweaver-Burk plots for active taurine uptake.

2.3   Reverse Transcriptase -Polymerase Chain Reaction

Total RNA was isolated from HT-29 cells treated with PMA (1μM) using the TRIZOL reagent (GIBCO, NY, USA). One μg of total RNA was reverse-transcribed using the Superscript^{TM} II kit (GIBCO, NY, USA) according to the manufacturer's recommendation. A fragment of TAUT cDNA (732 bp) was amplified using the PCR technique under the following condition: 30 cycles (94℃ for 30 s; 59℃ for 30 s; 72℃ for 60 s) in a total volume of 50 μl containing 100 mM KCl, 20 mM Tris-HCl (pH 8.0), 2.5 mM dNTPs, 2.5 mM MgCl$_2$, 1 pmol sense primer (5'-cgc ggatcc ata ccg tat ttt att ttc ctg ttt -3'), 1 pmol antisense primer (5'-tcta gaattc cct gta cga gtt ata ctt gta ctt -3'), and 2.5 units of Taq polymerase (TAKARA, Japan). Primers were also designed to amplify the 600 bp cDNA fragment encoding β-actin (sense: 5'-gtg ggg cgc ccc agg cac cag ggc -3'/ antisense: 5'-ctc ctt aat gtc acg cac gat ttc -3') as an internal control. The PCR product was electrophoresed in a 1% agarose gel, and stained with ethidium bromide.

2.4   Measurement of mRNA Stability

The degradation rate of TAUT mRNA was measured to assess the inherent stability of constitutively expressed TAUT mRNA in HT-29 cells. Growth arrested HT-29 cells were incubated with actinomycin D (0.5 μg/ml), an inhibitor of *de novo* gene transcription, for 1 hr prior to the addition of PMA (10^{-6} M) in the incubation medium. The degradation rate of mRNA was determined by performing the reverse transcriptase-polymerase chain reaction (RT-PCR) technique using total RNA isolated from cells harvested at 4 h following the addition of actinomycin D to the incubation medium.

2.5   Statistical Analysis

Data represent the mean ± SEM of 6 independent measurements. Significance differences between means for treated cells and untreated control cells were tested by the Student's *t*-test at p<0.05.

3.   RESULTS AND DISCUSSION

Repeated experiments demonstrated that the specific activity of taurine uptake was about 15~30 times higher in HT-29 cells than the value observed in Caco-2 cells. Earlier studies by Brandsch et al.[8] comparing the taurine transport activity in T-29 cells with that in Caco-2 cells revealed that the latter cell line expressed TAUT at a much reduced level (about one-fifth compared with HT-29).

Pretreating HT-29 and Caco-2 cells with PMA (1 µM), a potent stimulator of protein kinase C, for 3 hr inhibited taurine uptake by 90% compared to the value obtained for the control cells (p<0.001) (Fig. 1).

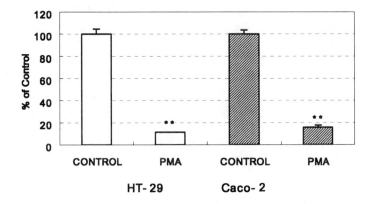

Figure 1. Effect of phorbol 12-myristate 13-acetate (PMA) on taurine uptake in HT-29 and Caco-2 cells. Values are mean ± SEM from two separate experiments done in triplicate. Confluent monolayer cultures were treated for 3 hr with or without 1 µM PMA. Taurine uptake values measured in control HT-29 (5.30 ± 0.60 pmol · mg protein^{-1} · 30 min^{-1}), and Caco-2 cells (0.30 ± 0.05 pmol · mg protein^{-1} · 30 min^{-1}) were considered as 100%. **Significantly different from the value for control cells by Student's t-test at p<0.001.

cAMP (1 µM), an intracellular second messenger, significantly reduced taurine uptake in Caco-2 cells (40% reduction) compared to the value obtained for untreated control cells (p<0.05), but failed to influence taurine uptake in HT-29 cells (Fig. 2). Therefore, TAUT appears to be differentially regulated by cAMP in the two human colon carcinoma cell lines.

We previously observed that pretreating HT-29 and Caco-2 cells with cholera toxin, which is known to increase intracellular cAMP levels, for 24 hr significantly decreased taurine transport activity in both cell lines[15]. However, cholera toxin did not affect taurine transport activity in placenta choriocarcinoma cells (JAR), although it elevated the intracellular cAMP content several fold[16]. *Escherichia coli* heat-stable enterotoxin-induced modulation in taurine transport activity was observed in Caco-2 cells, but not in HT-29 cells[17].

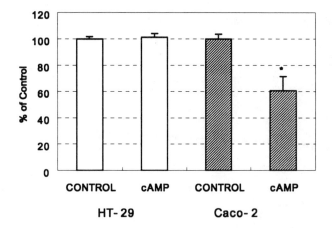

Figure 2. Effect of cAMP on taurine transport in HT-29 and Caco-2 cells. Values are mean ± SEM from two separate experiments done in triplicate. Confluent monolayer cultures were treated for 3 hr with or without cAMP (1 µM). Taurine uptake values measured in control HT-29 (9.0 ± 0.17 pmol · mg protein^{-1} · 30 min^{-1}), and Caco-2 cells (0.30 ± 0.01 pmol · mg protein^{-1} · 30 min^{-1}) were considered as 100%. *Significantly different from the value for control cells by Student's t-test at $p < 0.05$.

Kinetic analyses of taurine uptake indicated that pretreatment of HT-29 cells with PMA for 3 h decreased the maximal velocity (3.33 *vs.* 0.63 nmole·mg protein^{-1}·30 min^{-1}), and increased the Michaelis-Menten constant (16.0 *vs.* 23.0 µM) of TAUT compared to values for control cells (Fig. 3).

In order to determine whether this PMA-induced down-regulation of taurine transport activity was associated with the change in TAUT mRNA levels, RT-PCR was conducted. mRNA levels of β-actin were determined as an internal standard for constitutive expression. Pretreatment of HT-29 cells with PMA (1 µM) for 3 hr resulted in a 50% reduction in the mRNA level of TAUT compared to the value for untreated control cells (Fig. 4). Therefore, PMA-induced changes in the kinetic parameters of TAUT activity seem to be attributed in part to the reduced transcriptional rate of the TAUT gene.

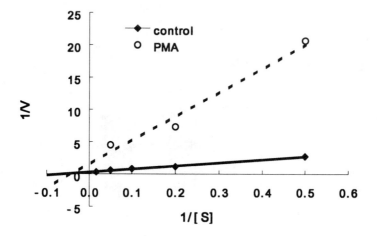

Figure 3. Lineweaver-Burk plot of active taurine uptake in HT-29 cells treated with (open circle) or without (closed rectangle) 1 µM phorbol 12-myristate 13-acetate (PMA) for 3 hr. Values are mean ± SEM from two separate experiments done in triplicate. Taurine uptake was measured in HT-29 cells 4 days after seeding over a taurine concentration range from 2 to 60 µM using a 30 min incubation. Units for V and S are nmole·mg protein^{-1}·30 min^{-1} and µM, respectively.

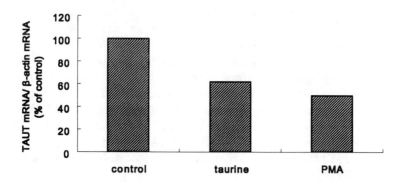

Figure 4. Effect of phorbol 12-myristate 13-acetate (PMA) on the expression of taurine the transporter (TAUT) gene in HT-29 cells. Reverse transcriptase-polymerase chain reaction was performed to measure the TAUT mRNA levels as described in the text. Data were normalized to the densitometric results of β-Actin mRNA levels which were used as an internal control.

We also investigated whether the PMA-induced decrease in TAUT mRNA levels were associated with changes in mRNA stability. Cells treated with actinomycin D (0.5 µg/ml), an inhibitor of *de novo* gene transcription, for 3 hr exhibited reduced levels of TAUT mRNA compared to untreated control cells. However, when cells were treated with PMA (1 µM) for 3 hr following prior treatment with actinomycin D for 1 hr, TAUT mRNA levels were increased compared to cells treated with actinomycin D only. These results indicate that the half-life of TAUT mRNA is relatively short, and PMA enhanced the mRNA stability of TAUT in HT-29 cells.

Therefore, PKC mediated down-regulation of TAUT activity in the human colon carcinoma cell line appears to be exerted at the level of the gene expression and/or a post-translational modification. The latter case might include phosphorylation of the transporter protein, and would partly explain the PMA-induced affinity change of TAUT.

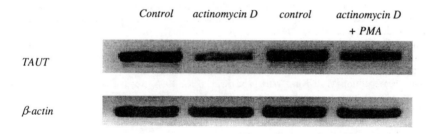

Figure 5. Effect of phorbol 12-myristate 13-acetate (PMA) on mRNA stability of the taurine transporter (TAUT). RT-PCR was conducted to amplify mRNA for TAUT and β-actin as described in the text. β-Actin was used as an internal standard.

ACKNOWLEDGEMENT

This work was supported by a grant No. R01-1999-000-00128-0 from the Basic Research Program of the Korea Science & Engineering Foundation.

REFERENCES

1. Ahlman, B., Leijonmark, C.E., and Wemerman, J., 1993, The content of free amino acids in the human duodenal mucosa. *Clin. Nutr.* 12: 266-271.
2. Gray, C.E., Landel, A.M., and Meguid, M.M., 1994, Taurine supplemented total parenteral nutrition and taurine status of malnourished cancer patients. *Nutrition* 10: 11-15.

3.   Paauw, J.D., and Davis, A.T., 1990, Taurine concentrations in serum of injured patients and age and sex matched healthy controls. *Am. J. Clin. Nutr.* 52: 657-660.

4.   Ahlman, B., Ljunqvist, O., Andersson, K., and Wemerman, J., 1995, Intestinal amino acid content in critically ill patients. *J. Parenteral Enteral Nutr.* 19: 272-278.

5.   Ahlman, B., Ljunqvist, O., Persson, B., and Bindslev, L., 1995, Free amino acid in the human intestinal mucosa: Impact of surgery and critical illness. *Clin. Nutr.* 14: 54-55.

6.   O'Flaherty, L., Stapleton, P.P., Redmond, H.P., and Bouchier-Hayes, D.J., 1997, Intestinal taurine transport: a review. *Eur. J. Clin. Investig.* 27:873-880.

7.   Hideo, S., Hirohito, W., Soichi, A., and Makoto, S., 1997, Characterization and regulation of taurine transport in Caco-2, human intestinal cells. *J. Biochem.* 121: 1082-87.

8.   Brandsch, M., Ramamoorthy, S., Ganapathy, V., and Leibach, F.H., 1995, Regulation of taurine transport in human colon carcinoma cell line (HT-29 and Caco-2) by protein kinase. *Am. J. Physiol.* 264: G939-946.

9.   Tiruppathi, C., Brandsch, M., Miyamoto, Y., Ganapathy, V., and Leibach F.H., 1992, Constitutive expression of the taurine transporter in a human carcinoma cell line. *Am. J. Physiol.* 263: G625-631.

10.   Ganapathy, Y., and Leibach, F.H., 1994, Expression and regulation of the taurine transporter in cultured cell lines of human origin. *Adv. Exp. Med. Biol.* 359: 51-7.

11.   Han, X., Burdreau, A.M., and Chesney, R.W., 2000, Identification of promoter elements involved in adaptive regulation of the taurine transporter gene: role of cytosolic Ca^{2+} signalling. *Adv. Exp. Med. Biol.* 48: 535-44.

12.   Ramamoorthy, S., Del Monte, M.A., Leibach, F.H., and Gemapathy, V., 1994, Molecular identity and calmodulin-mediated regulation of the taurine transporter in a human retinal pigment epithelial cell line. *Curr. Eye Res.* 13(7): 523-9.

13.   Kim, H.W., Shim, M.J., Kim, W.B., and Kim, B.K., 1995, Regulation of tauine transporter activity by glucocorticoid hormone. *J. Biochem. Mol. Biol.* 28(6): 527-532.

14.   Bridges, C.C., Ola, M.S., Prasad, P.D., El-Sherbeny, A., Ganapathy, V., and Smith, S.B., 2001, Regulation of taurine transporter expression by NO in cultured human retinal pigment epithelial cells. *Am. J. Physiol.* 281(6): C1825-36.

15.   Yoon M.Y., Park, S.Y., and Park, T., 2001, Stress-induced changes of taurine transporter activity in the human colon carcinoma cell line (HT-29). *Kor. J. Nutr.* 34: 150-157.

16.   Cool, D.R., Leibach, F.H., Bhalla, V.K., Mahesh, V.B., and Ganapathy, V., 1991, Expression and cyclic AMP-dependent regulation of a high affinity serotonin transporter in the human placental choriocarcinoma cell line (JAR). *J. Biol. Chem.* 266: 15750-15757.

17.   Brandsch, M., Ramammoorthy, S., Marczin, N., Catravas, J.D., Leibach J.W., Ganapathy, V., and Leibach, F.H., 1995, Regulation of taurine transport by *Escherichia coli* heat-stable enterotoxin and guanylin in human intestinal cell lines. *J. Clin. Invest.* 96: 361-369.

Characterization of Transcriptional Activity of Taurine Transporter Using Luciferase Reporter Constructs

KUN-KOO PARK, SANG-KEUN CHON, EUNHYE JUNG, MOOSEOK SEO, HA WON KIM* and TAESUN PARK#

*PharmacoGenechips Inc., Chunchon 200-160 , *Dept. of Life Science, University of Seoul, Seoul 130-743, #Dept. of Food and Nutrition, Yonsei University, Seoul 120-749, Korea*

ABSTRACT

Although taurine transporter (TAUT) activity has been known to be regulated by diverse intracellular and extracellular factors involved in the signal transduction pathway, such as protein kinase C, intracellular Ca concentration, and glucocorticoids, little is known concerning the underlying mechanisms. Evidence suggests that such stimulation-mediated changes in TAUT activity in mammalian cells are partly achieved through the modulation of TAUT transcription activity. In order to better understand the regulation of TAUT transcription activity and subsequently the role of taurine in the signal transduction pathway, we have cloned and sequenced the 5' flanking region of the human TAUT gene, and characterized the TAUT promoter region in human cells. For these reasons, the TAUT luciferase reporter vector was constructed using the 5' flanking region of the TAUT gene (1800 bp). The TAUT luciferase reporter vector was then transfected into SiHa cells, and luciferase activity was measured. The construct containing its own promoter of TAUT (pGL3 b TAU31) showed a 10 fold higher luciferase activity compared to the value found in the empty vector (pGL3 b). This implies a functional transcription of the homologous TAUT promoter. Similar results were obtained with the exon deleted construct [pGL3 b TAU31(-e)]. We also constructed the TAUT luciferase reporter gene (pGL3 pro TAU13) using a heterologous promoter. About 2.5 fold higher luciferase activity was observed in cells transfected with this construct containing the heterologous promoter compared to the value found in the control vector (PGL3 pro).

1.   INTRODUCTION

The taurine transporter (TAUT) plays an important role in the regulation of intracellular taurine concentrations in mammalian tissues. Although TAUT activity is regulated by diverse intracellular and extracellular factors involved in the signal transduction pathway, little is known about the underlying mechanisms. Recent regulatory studies with diverse cell lines have explored the role of protein kinase C^1, intracellular Ca^{2+} concentration[2], hypertonicity[3,4], glucose[5], calmodulin[6], and nitric oxide[7] in the modulation of TAUT function. Evidence suggests that stimulation-mediated changes in TAUT activity in mammalian cells are partly achieved through the modulation of TAUT gene expression[2,3,7]. Since the stability of TAUT mRNA in human cells appears to be relatively short (our unpublished data), regulation at the transcriptional level might be an important control point of TAUT activity.

The promoter region of TAUT gene was first cloned and sequenced from a rat P1 genomic DNA library by Han et al.[8]. Using the reporter constructs containing this rat TAUT promoter region, they reported that TAUT may represent a downstream target gene of p53 that could link the roles of p53 in renal development and apoptosis[9]. In order to better understand the regulation of TAUT gene expression and subsequently the role of taurine in the signal transduction pathway in human cells, we have cloned and sequenced the 5' flanking region of the human TAUT gene for the first time. Transcriptional activity of the TAUT promoter was partially characterized by using constructs of the human TAUT promoter-luciferase reporter.

2.   MATERIALS AND METHODS

2.1   Reagents

The T-vector ligation kit and the luciferase assay kit were purchased from Promega (Madison, USA). DNA sequencing was performed using a kit from Applied Biosystem (California, USA). The cloning vectors pGL3 promoter and pGL3 basic were obtained from Promega (Madison, USA). Tissue culture plasticware was obtained from BD Biosciences (Massachusetts, USA), and all other chemicals were purchased from Sigma-Aldrich (Missouri, USA).

2.2   Cell Culture

Human cervical cancer cells (SiHa) and human hepatocarcinoma cells (SK-Hep-1) were maintained in RPMI 1640 medium with 10% fetal bovine serum and 1% streptomycin/penicillin. Cells were cultured at $37\,^{\circ}C$ in a humidified 95% air - 5% CO_2 atmosphere.

2.3 PCR and Cloning of TAUT Promoter

Genomic DNA from human SK-Hep-1 cells was prepared as described by the manufacturer (Promega, USA). A genomic DNA fragment of 1800 bp length was amplified by the polymerase chain reaction (PCR) technique using the primers 5'ACAAAGTCGATCTTGCTAGACC3' and 5'GAGATAGACGACAGGGA AGC3' referred to as TAU1 and TAU3, respectively. The PCR product was extracted from an agarose gel, and cloned into the respective sites of the pGEM-T vector (Promega, Korea). The resulting recombinant plasmid was designated as a TAU13-T vector.

2.4 Construction of the Luciferase Reporter Plasmid

A fragment of the human TAUT promoter, approximately 1.8 kb, which was derived from the TAU13-T vector was subcloned into the luciferase reporter plasmid pGL3 basic and pGL3 promoter (Promega, Korea), respectively. The resulting human TAU13-luciferase fusion plasmid was designated as pGL3b TAU13 and pGL3pro TAU13, respectively. The construct of pGL3b TAU31 was generated by cutting pGL3b TAU13 with Sac I, and religating. The reporter plasmids with deletion mutation in exon 1, designated as pGL3b TAU13(-e), were generated by cutting pGL3b TAU31 with Pvu II and Bgl II and religating.

2.5 Transient Transfection and Assay of Luciferase Activity

SiHa cells (1.0×10^6 cells) were co-transfected with a total of 1.7 μg DNA (1.4 μg constructed TAUT promoter luciferase plasmid, and 0.3 μg pCMV β-galactosidase) by the lipofectamine method. Cells were harvested 36 h post transfection. Luciferase activity of each cell lysate was measured using the Luciferase Reporter Plasmid System (Promega, USA). Firefly luciferase activity of each sample was normalized by the value of the β-galctosidase activity. All transfection assays were carried out in triplicate.

3. RESULTS AND DISCUSSION

Using human SK-Hep-1 genomic DNA as a template and TAU3 and TAU1 as upstream and downstream primers, respectively, a TAUT promoter region, approximately 1.8 kb long, was amplified (Fig. 1). The PCR product was more efficiently amplified at lower annealing temperatures (from 59°C to 62.6°C) compared to the band obtained at a higher (65.2°C) annealing temperature.

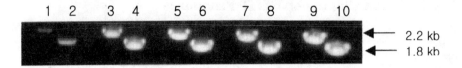

Figure 1. Gel electrophoresis of PCR products of human taurine transpoter promoter. Human SK-Hep-1 genomic DNA was used as a PCR template for cloning of the promoter. PCR was performed at varied annealing temperature (59 °C ~ 65.2 °C) using TAU1 as the sense DNA primer complementary to promoter, and TAU3 as the exon 1-specific antisense primer. The PCR products were electrophoresed on a 1% agarose gel, and stained with ethidium bromide. Lanes 1, 3, 5, 7, 9, 2.2kb size marker; lane 2, annealing temperature of 65.2°C; lane 4, annealing temperature of 62.6°C; lane 6, annealing temperature of 61°C; lane 8, annealing temperature of 59.8°C; lane 10, annealing temperature of 59°C.

In order to characterize the human TAUT promoter, various types of TAUT promoter-luciferase reporter constructs were generated as shown in Fig. 2. The pGL3pro TAU13 plasmid was constructed by inserting the 1.8 kb TAUT promoter fragment derived from TAUT13-T vector by Not I digestion into the pGL3 promoter vector. The pGL3b TAU13 construct was generated by ligating the 1.8 kb TAU13 fragment derived from the TAUT 13-T vector at the SmaI site of the pGL3 basic vector. The luciferase reporter construct containing the TAUT promoter (pGL3b TAU31) was obtained by cutting the pGL3b TAU13 with Sac I, and religating. This construct includes a partial sequence of exon 1 (150 bp) downstream of the TATA box. In order to eliminate this partial exon sequence, 405 bp from the 3' end was removed by cutting pGL3b TAU31 with Pvu II and Bgl II, and religating. As a result, pGL3b TAU31(-e) plasmid containing 1446 bp TAUT promoter region including the TATA box was constructed.

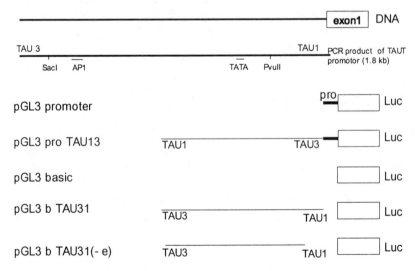

Figure 2. Cloning of the taurine transporter promoter and construction of human taurine transporter (TAUT) promoter-luciferase reporter plasmids. pGL3b TAU31 and pGL3pro TAU13 were generated by subcloning a 1.8 kb DNA fragment of the TAUT promoter in front of the luciferase reporter gene in the pGL3 basic and pGL3 promoter, respectively. The exon-deletion mutant [pGL3b TAU13(-e)] was generated by deleting the Pvu II/Bgl II fragment from the pGL3b TAU31.

Relative luciferase activities of cells transfected with constructed human TAUT promoter reporter plasmids are shown in Fig. 3. SiHa cells were co-transfected with various recombinant human TAUT promoter-reporter plasmids along with the pCMV β-galactosidase vector as an internal control. Luciferase activity of cells transfected with human TAUT promoter-reporter constructs has been normalized by its β-galactosidase activity, and subsequently compared with the value for cells transfected with the empty pGL3 basic or pGL3 promoter vector.

Cells transfected with pGL3pro SV40 showed a 13.5 fold increase in luciferase activity compared to the value for control cells transfected with the empty pGL3 promoter vector. Cells transfected with pGL3b TAU31 or pGL3b TAU31(-e) exhibited 10.5 or 8.1 times higher luciferase activity compared to the value for control cells transfected with the empty pGL3 basic vector. Therefore, the

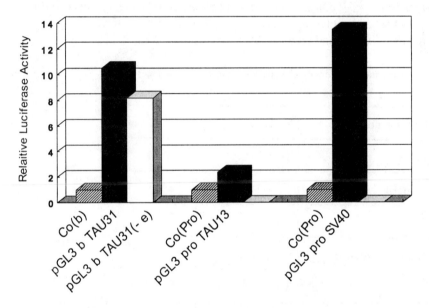

Figure 3. Luciferase activity of human taurine transporter promoter reporter genes. SiHa cells were transfected with different reporter gene constructs along with the pCMV β-galctosidase vector which served as an internal control. The firefly luciferase activity of cells co-transfected with each constructs has been normalized by its β-galactosidase activity, and subsequently compared with the value for cells co-transfected with the empty pGL3 basic or pGL3 promoter vector. **Co(b)**; control cells transfected with pGL3 basic and pCMV β-galctosidase vector/ **Co(pro)**; control cells transfected with pGL3 promoter and pCMV β-galctosidase vector/ **pGL3b TAU31**; cells transfected with pGL3b TAU31 along with pCMV β-galctosidase vector/ **pGL3b TAU31(-e)**; cells transfected with pGL3b TAU31(-e) along with pCMV β-galctosidase vector/ **pGL3pro TAU13**; cells transfected with pGL3pro TAU31 along with pCMV β-galctosidase vector/ **pGL3 pro SV40**; cells transfected with pGL3pro SV40 along with pCMV β-galctosidase vector.

construct containing the own promoter of TAUT (pGL3 b TAU31) showed about 10 fold higher luciferase activity compared to the value found in the empty vector (pGL3 b), and similar results were obtained with the exon deleted construct [pGL3 b TAU31(-e)]. We also constructed the TAUT luciferase reporter gene (pGL3 proTAU13) using the heterologous promotor. The TAUT promoter reporter construct containing the antisense TAUT promoter (pGL3pro TAU13) showed a 2.3 times higher luciferase activity in SiHa cells compared to the value for the empty pGL3 promoter vector. These results imply a functional transcription of the homologous TAUT promoter.

Sequence analysis of the 1.8 kb TAUT promoter region that we had cloned revealed several important motifs and enhancer sequences as shown in Table 1. Consensus sequences of AP1, TATA, ATF-CREB, CAAT, HIV negative enhancer, NF1, and TFIIB-B factor binding sites were found in the human TAUT promoter from 1 to 7 copies. Cooperation of these motifs and enhancers within the human TAUT promoter appears to play a key role in modulating the gene expression of TAUT, and requires further characterization.

Table 1. Motifs and enhancers found in the human taurine transporter promoter.

Motif or enhancer	Sequence	No. of Copy
AP1	TGACTCA	1
ATF – CREB	AGTCA	6
CAAT	CAAT	2
HIV negative enhancer	CAAGG	6
NF1	TTGGC	7
TATA	TATA	6
TFIIB-B factor	TTTATT	2

ACKNOWLEDGEMENT

This work was supported by a grant No. R01-1999-000-00128-0 from the Basic Research Program of the Korea Science & Engineering Foundation.

REFERENCES

1. Ganapathy, V., and Leibach, F.H., 1994, Expression and regulation of the taurine transporter in cultured cell lines of human origin. *Adv Exp Med Biol* 359:51-7.
2. Han, X., Budreau, A.M., and Chesney, R.W., 2000, Identification of promoter elements involved in adaptive regulation of the taurine transporter gene: role of cytosolic Ca^{2+} signaling. *Adv Exp Med Biol* 483:535-44.
3. Satsu, H., Miyamoto, Y., and Shimizu, M., 1999, Hypertonicity stimulates taurine uptake and transporter gene expression in Caco-2 cells. *Biochim Biophys Acta* 1419(1):89-96.
4. Shioda, R., Reinach, P.S., Hisatsune, T., and Miyamoto Y., 2002, Osmosensitive taurine transporter expression and activity in human corneal epithelial cells. *Invest Ophthalmol Vis Sci* 43(9):2916-22.
5. Stevens, M.J., Hosaka, Y., Masterson, J.A., Jones, S.M., Thomas, T.P., and Larkin, D.D., 1999, Downregulation of the human taurine transporter by glucose in cultured retinal pigment epithelial cells. *Am J Physiol* 277(4 Pt 1):E760-71.
6. Ramamoorthy, S., Del Monte, M.A., Leibach, F.H., and Ganapathy, V., 1994, Molecular identity and calmodulin-mediated regulation of the taurine transporter in a human retinal pigment epithelial cell line. *Curr Eye Res* 13(7):523-9.
7. Bridges, C.C., Ola, M.S., Prasad, P.D., El-Sherbeny, A., Ganapathy, V., and Smith, S.B., 2001, Regulation of taurine transporter expression by NO in cultured retinal pigment epithelial cells. *Am J Physiol Cell Physiol* 281(6):C1825-36.
8. Han, X., Budreau, A.M., and Chesney, R.W., 2000, Cloning and characterization of the promoter region of the rat taurine transporter (TauT) gene. *Adv Exp Med Biol* 483:97-108.
9. Han, X., Patters, A.B., and Chesney, R.W., 2002, Transcriptional repression of taurine transporter gene (TauT) by p53 in renal cells. *J Biol Chem* [epub ahead of print].

Isovolumetric Regulation in Mammal Cells: Role of Taurine

B. ORDAZ, R. FRANCO, and K. TUZ

Department of Biophysics, Institute of Cell Physiology, National University of Mexico, Mexico City, Mexico

1. INTRODUCTION

The ability to regulate cell volume is an ancient conserved trait present in essentially all species through evolution. The maintenance of a constant cell volume is a homeostatic imperative in animal cells. Changes in cell water content affecting the concentration of intracellular messenger molecules impair the complex signaling network, crucial for cell functioning and intercellular communication. Although the renal homeostatic mechanisms exert a precise control of extracellular fluid osmolarity, this is challenged in a variety of pathological situations. The intracellular volume constancy is continuously compromised by the generation of local and transient osmotic microgradients, associated with nutrients uptake, secretion, cytoskeleton remodeling and transynaptic ionic gradients[1].

Cell volume disturbances have particularly dramatic consequences in the brain. The limits to expansion imposed by the rigid skull give narrow margins for the buffering of intracranial volume changes. As expansion occurs, the constraining of small vessels generates episodes of anoxia, ischemia, excitotoxicity and neuronal death. In extreme conditions, caudal herniation of the brain parenchyma through the foramen magnum affects brain stem nuclei resulting in death by respiratory and cardiac arrest.

Brain cell swelling occurs either by decreases in external osmolarity as in hyponatremia or by changes in ion redistribution in isosmotic conditions, termed cytotoxic edema. Cells exposed to hyposmotic conditions first swell and then exhibit a volume regulatory response, accomplished by the extrusion of intracellular osmotic solutes, essentially the inorganic osmolytes K^+ and Cl^-, as well as organic osmolytes such as amino acids, polyamines and polyalcohols. This is an active regulatory process (termed regulatory volume decrease, RVD) which occurs in spite of the persistence of the hyposmotic condition. In most studies, RVD is investigated by exposure of cells to abrupt and large decreases in external osmolarity. Although this condition has been useful to characterize and amplify the events occurring during RVD, these drastic changes in external osmolarity only occur seldom, even in pathological situations.

An experimental paradigm developed by Lohr and Grantham[2] in renal cells is that of exposure of cells to small and gradual changes in osmolarity. Under these conditions, cell volume remains constant even after some time of exposure to the osmotic gradient, when the external osmolarity has been considerably reduced. This response was named "isovolumetric regulation" (IVR). We have characterized this phenomenon in various brain cell preparations, including hippocampal slices, cerebellar granule neurons in culture, and C6 glioma cells. That taurine plays an important role in IVR was demonstrated in all cases.

1.1   Mechanism of Isovolumetric Regulation

The volume constancy observed in cells exposed to an osmotic gradient may reflect either the absence of swelling in these conditions or the activation of mechanisms for volume adjustment. The study by Lohr and Grantham supported the last conclusion, since cells exposed to gradual decreases in osmolarity, which present IVR, shrink when returned to an isosmotic medium, indicative of loss of intracellular osmolytes. IVR in renal cells is blocked by ouabain, quinine and barium, indicating the involvement of K^+ as an osmolyte in this process[3]. Also in A6 cells, a 70% loss of intracellular K^+ was found after exposure to an osmotic gradient[4]. No other potential osmolyte was examined in those studies.

Subsequently, Souza[5] estudied IVR in cultured chick embryo cardiomyocytes. In contrast to renal cells, in cardiomyocytes only partial IVR was observed, since cells started swelling when the external osmolarity had dropped by approximately 15%. At the end of the experiment, a 35% decrease in K^+ and a 22% decrease in taurine cell content were observed. IVR seems not to be a general feature of all cells. Trout erythrocytes were unable to regulate volume in the presence of an osmotic change as low as -0.7 mOsml/min. Neither K^+ nor taurine fluxes were activated in these cells which explains the absence of the volume regulatory response[6] .

Hippocampal slices[7] exposed to an osmotic gradient of -2.5 mOsml/min, are also able to maintain a constant volume. In this preparation, no significant change in K+ was found during IVR. In contrast, a marked decrease in the concentrations of amino acids of approximately 60% occurred during IVR. The marked reduction of endogenous taurine and glutamate levels of 80% and 65%, respectively, emphasize their contribution to IVR (Fig. 1). Radiolabelled amino acids were used to examine the time course of taurine, glutamate and GABA release in response to the osmotic gradient. [³H] Taurine release is observed almost immediately after the reduction in osmolarity, with a threshold of -5 mOsm, while that of glutamate (traced by ³H-aspartate) and [³H] GABA were -20 mOsm. Also the efflux rate of taurine is notably higher, with a maximal increase of 7-fold over the isosmotic efflux while those of glutamate and GABA were only 2- and 3-fold higher, respectively. Interestingly, a similar early release of taurine has been observed *in vivo* in rat brain in response to hyponatremia[8]. Altogether, these results support the contribution of amino acids to IVR,. The apparent lack of contribution of K⁺ to the regulatory process in hippocampal slices is somewhat surprising since this ion is actively participating in IVR in renal cells, A6 cells and cardiomyocytes. It is possible that in an integrated preparation such as the hippocampal slice, an active buffering mechanism of extracellular K⁺ masks the osmosensitive release occurring gradually during IVR. Due to the key role of K⁺ in the control of neuronal excitability, its extracellular levels have to be strictly controlled.

Full IVR occurs in cultured cerebellar granule neurons exposed to gradual changes in external osmolarity of -1.8 mOsml/min.[9]. No cell volume increase is observed even when the external osmolarity has dropped by 50%. The contribution of K⁺, Cl⁻ to IVR and of some amino acids has been examined in this preparation. As in hippocampal slices, taurine exhibited the earliest efflux threshold, of -2% mosm. Other amino acids are also released in response to the osmotic gradient, with thresholds of -19 and -30 mOsm for glutamate and glycine, respectively. Cl⁻ (followed as ¹²⁵I) and K⁺ (as ⁸⁶Rb) efflux showed a release threshold of -25 and -29 mOsm, respectively. These results suggest a contribution by amino acids in the early phases of IVR, while ions have a contribution only late in the process.

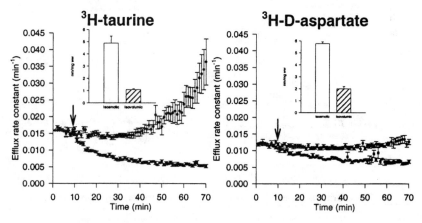

Figure 1. Efflux of ³H-taurine and ³H-D-aspartate from hippocampal slices exposed (arrow) to medium of decreased osmolarity (•) at a rate of 2.5 mOsm min-1. Controls (■) were superfused with isosmotic medium. Data are expressed as efflux rate constants (min⁻¹) and are means ± SEM (n = 8-10). The insets show the decrease of endogenous levels ± SEM after expose to isosmotic medium (empty bars) or to the osmotic gradient conditions (dashed bars).

To evaluate the active contribution of these different osmolytes, suppression of their fluxes by several experimental maneuvers was attempted. The rational for these experiments is that a reduction in osmolyte extrusion may result in an increase in cell volume, corresponding to the fraction impaired by the blockade of the osmolyte involved. K^+ efflux is blocked by Ba^{+2} and in the presence of this agent, cell swelling occurs at external osmolarity reduced by 24%, coincident with the ^{86}Rb release threshold (Fig. 2B). Similarly cell treatment with solutions in which all Cl^- was replaced by gluconate, lead to cell swelling when the external osmolarity is reduced by -24% , progressively increasing to reach a maximum of 29% when the external osmolarity is -50%. Niflumic acid is a blocker of both, taurine and Cl^- efflux during IVR. In the presence of this compound, swelling is observed from the first minutes of exposure to the osmotic gradient, and this increase continues to attain a maximal of 50%. This observation indicates that the early phase of volume regulation occurs entirely by a niflumic-acid sensitive mechanism (Fig. 2A). Comparing cell swelling in a medium without Cl^-, which affects only the Cl^- contribution but not that of organic osmolytes, and swelling in niflumic-acid containing solutions, gives an estimate of the contribution of organic osmolytes. The differences between the two curves suggest an early, continuous and substantial contribution of these types of osmolytes throughout the IVR process. The particular features of taurine efflux, being more sensitive and more robust than that of other amino acids, underline its predominant role in volume regulation under IVR conditions.

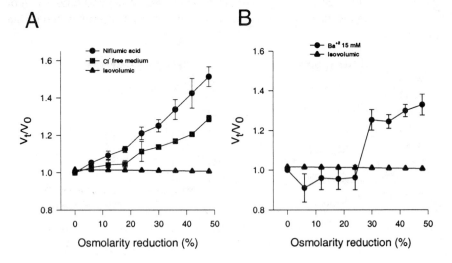

Figure 2. Impairment of IVR by niflumic acid, Ba^{2+}, and Cl-free medium.. **A**. cell volume measured in cells exposed to the osmotic gradient (▲), in the presence of 600 μM niflumic acid (•), or Cl^--free medium (Cl- replaced by gluconates) (■). In **B**, cell volume in the presence of 15 mM Ba^{+2} (•). Data are means ± SEM of 3-4 separate experiments

Preliminary studies in C6 cells further support this notion, although mainly by negative results. In these cells, the osmotic gradient effectively elicits Cl⁻ and K⁺ currents, which, however, are not sufficient to maintain a constant cell volume. Interestingly, the osmotic gradient is unable to evoke taurine efflux. Not even sudden changes in osmolarity of -30% evoke any detectable efflux of taurine. Only very large decreases in osmolarity mobilize taurine.

Altogether, these observations stress the importance of taurine as a volume regulatory element in conditions which, as mentioned before, may better fit the physiological and pathological situations.

ACKNOWLEDGEMENTS

This work was supported by grants No. 3488-6M from CONACYT and IN-204900 from DGAPA-UNAM.

REFERENCES

1. Lang, F., Busch, G. L., Ritter, M., Völki, H., Waldegger, S., Gulbins, E., and Häussinger, D. 1998, Functional significance of cell volume regulatory mechanisms. *Physiol. Rev.*78, 247-306.
2. Lohr, J. W., and Grantham, J. J. 1986, Isovolumetric regulation of isolated S₂ proximal tubules in anisotonic media. *J. Clin. Invest.* 78, 1165-1172. 5
3. Lohr J. W. 1990, Isovolumetric regulation of renal proximal tubules in hypotonic medium. *Renal Physiol. Biochem.* 13, 233-240.
4. Van Driessche, W., De Smet, P., Li, J., Allen, S., Zizi, M., and Mountian, I. 1997, Isovolumetric regulation in a distal nephron cell line (A6). *Am. J. Physiol.* 272, C1890-C1898.
5. Souza, M..M., Boyle, R.T., and Lieberman, M. 2000, Different physiological mechanisms control isovolumetric regulation and regulatory volume decrease in chick embryocardiomyocytes. *Cell Biol. Int.* 24, 713-721.
6. Godart, H., Ellory J. C., and Motais, R. 1999, Regulatory volume response of erythrocytesexposed to a gradual and slow decrease in medium osmolarity. *Pflugers Arch. Eur. J. Physiol.* 437, 776-779.
7. Franco R., Quesada, O., and Pasantes-Morales, H. 2000, Efflux of osmolyte amino acids during isovolumic regulation in hippocampal slices. *J Neurosci. Res.* 61, 701-11.
8. Solís, J.M., Herranz, A.S., Herraz, O., Lerma, J., and Del Rio, R.M. 1988, Does taurine acts as an osmoregulatory substance in the rat brain?. *Neurosci. Lett.* 91:53-58
9. Tuz, K., Ordaz, B., Vaca, L., Quesada, O., and Pasantes-Morales, H. 2001, Isovolumetric regulation mechanims in cultured cerebellar granule neurons. *J. Neurochem.* 79: 1-10.

Osmosensitive Taurine Release
Does Taurine Share the Same Efflux Pathway With Chloride and Other Amino Acid Osmolytes?

RODRIGO FRANCO

Biophysics, Institute of Cell Physiology, National University of Mexico, Mexico City, Mexico

1.   INTRODUCTION

Swelling subsequent to hyposmotic conditions activates a process of volume regulation present in most cell types. This volume adjustment is accomplished by osmolyte translocation towards the extracellular space to reach a new osmotic equilibrium. Molecules involved in this homeostatic mechanism have been broadly classified into two categories: organic and inorganic osmolytes. Inorganic osmolytes comprise mainly the intracellular ions K^+ and Cl^-. Cell swelling-induced activation of separate K^+ and Cl^- channels has been described in most preparations.   Organic osmolytes are grouped in three categories: amino acids, polyalcohols and methylamines. These osmolytes, particularly taurine, are present in high intracellular concentrations and may also play a role as cytoprotectants[1]. Amino acids are part of the organic osmolyte pool contributing to RVD in most cells[1,2]. Among them, taurine has been studied in detail mainly due to its metabolic inertness, and it is often considered as representative of all osmolyte amino acids.

1.1   Volume Sensitive Organic Osmolyte and Anion Pathway

Volume regulatory loss of organic osmolytes has been characterized in a wide range of cell types. Efflux of these osmolytes seems to be mediated by passive concentration gradient-driven pathways, which do not exhibit saturation in their efflux profile and are not susceptible to trans-stimulation[3]. It is now generally accepted that swelling-activated organic osmolyte release is achieved by diffusion through membrane pores, rather than by carrier transport. Transport pathways for organic osmolytes are in general Na^+-independent and non-stereoselective. A particular feature of organic osmolyte release is its sensitivity to general anion channel blockers[4-6]. This has raised the question of whether the volume-sensitive Cl^- channel may be the permeation pathway for these osmolytes. Electrophysiological evidence has shown that amino acids and some other organic osmolytes permeate through the volume-sensitive anion/Cl^- channel (VSAC), which has broad permeability, and the necessary size pore (8-9 Å)[2,7] to allow translocation of amino acids as large as glutamate, as well as other structurally unrelated osmolytes. This channel was named volume-sensitive organic osmolyte/anion channel (VSOAC) by Strange and coworkers[8]. Experimental evidence for the existence of this common pathway is not conclusive so far, and as the characterization of the volume-sensitive Cl^- channel and the organic osmolyte fluxes progresses, evidence against the notion of a common Cl^-/osmolyte pathway becomes less consistent. In this review, we address basic questions that remain unanswered: 1) may the volume-sensitive Cl^- channel be a common pathway for both Cl^- and taurine?; and 2) are taurine and other amino acid osmolytes translocated through the same efflux pathway? We present here results pertaining to these two possibilities and discuss the current state of the field.

2. EVIDENCE AGAINST A COMMON PATHWAY FOR SWELLING-ACTIVATED TAURINE AND Cl^- RELEASE.

Taurine is found at concentrations of up to tens of millimolar in vertebrate cells under physiological conditions. It has a pK2 of 8.82 and is therefore present in the cytosol predominantly as an electroneutral zwitterion[9]. Swelling-activated taurine release (SATR) has been observed in almost all preparations studied to date. SATR occurs via a non-saturable Na^+-independent transport pathway, inhibited by both anion exchanger blockers (DIDS, SITS, niflumic acid and pyridoxal phosphate), and chloride channel blockers (NPPB, dideoxyforskolin and tamoxifen), and modulated by tyrosine kinases and PI3K activity[for rev. see 10 and 11], as has been also reported for VSAC[12-14].

There is recent evidence against a common pathway for both osmolytes. In Ehrlich ascites cells SATR and VSAC fluxes are pharmacological distinct[15], the former being inhibited by DIDS and niflumic acid, and stimulated by arachidonic acid and LTD4, while the later is inhibited by arachidonic acid, tamoxifen and insensitive to DIDS and niflumic acid[16]. Studies in rat mammary gland and skate erythrocytes demonstrate SATR release without an activation of VSAC[17-20], and the opposite was observed in human biliary cell line[21]. In skate erythrocytes, the osmosensitive taurine and Cl^- fluxes appear mediated by different pathways, taurine through a channel and Cl⁻ by cotransporter pathways[20]. Different pathways for SATR and VSAC were also suggested in HeLa cells[22], where SATR was more sensitive to DIDS than I^{125} efflux. In this study SATR and I^{125} efflux elicited differences in their kinetic activation and inactivation profiles. Differences between VSAC and SATR in their sensitivity to Cl⁻ channel blockers have been found in NIH3T3 and CHO cells[23,24]. In NIH3T3 cells, the small G-protein Rho A modulates the Cl⁻ channel conductance while SATR remains almost unaffected[25]. SATR, but not swelling-activated Cl⁻ conductance, occurs in Xenopus laevis oocytes [26].

3. THE SWELLING-ACTIVATED TAURINE AND AMINO ACID RELEASE

Taurine is convenient for studies aimed to characterize osmolyte fluxes, because it is abundant in cells and tissues and is essentially metabolically inert, and has been often considered as representative of amino acids and other organic osmolytes. Besides taurine, other amino acids translocate in response to hyposmotic swelling. In the same way, glycine, β-alanine, GABA, leucine, glutamine, aspartate, and glutamate permeate through the swelling-activated Cl⁻/anion pathway with Paa/Pcl ranging from 0.25 to 0.78 in cell types including MDCK cells, inner medullary collecting duct IMCD cells and glial cells[27-31]. Amino acid release as for the case of taurine is inhibited by Cl⁻ channel blockers (SITS, DIDS, L644-711, niflumic acid, NPPB, dideoxyforskolin and tamoxifen, furosemide, 9-AC and dipyridamole) in different preparations including primary cortical and cerebellar astrocytes, cerebellar granule neurons and cultured cortical neurons, neuroblastoma CHP-100, cortex, hippocampal slices, chick retina, NIH3T3 cell and endothelial cells[23,32-31]. Hyposmotic-induced N-acetylaspartate release has been studied in rat striatum preparations, hippocampal slices *and in vivo* studies; similar to SATR, N-acetylaspartate release was also inhibited by Cl⁻ channel blockers[40,43-46]. These results may suggest a common efflux pathway for SATR and other amino acids and would, in principle, allow the extrapolation of SATR results to all other osmolyte amino acids.

Recent results from our and other laboratories indicate that the osmosensitive taurine efflux properties clearly differ from those of amino acids such as GABA and glutamate which may act as neurotransmitters. The efflux rate of taurine, glutamate and D-aspartate were similar in cultured astrocytes regarding the time course and the effect of Cl-channel inhibitors, but are differentially modulated by tyrosine kinase blockers[34-37,47]. More important differences between SATR and other amino acids were found in hippocampal[39,48] and cortex slices (Figure 1) including: 1) The osmosensitive amino acid neurotransmitter release shows a kinetic release profile notably different from SATR. Taurine efflux exhibits delayed activation and inactivation while that of glutamate and GABA fully activate immediately after the stimulus and also rapidly inactivate, 2) GABA and glutamate efflux are insensitive to Cl⁻ channel blockers, which typically inhibit SATR, 3) SATR is modulated by signaling cascades involving tyrosine phosphorylation events, including an important role for the tyrosine kinase target PI3K; while GABA and glutamate fluxes are not responsive to tyrosine phosphorylation state nor to the influence of PI3K activity, and 4) swelling-activated neurotransmitter release, but not SATR, is influenced by the cytoskeleton depolymerization and manipulation of PKC activity.

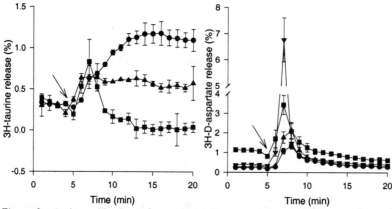

Figure 1. Amino acid release from cortex slices exposed to 30% hyposmotic medium. Slices preloaded with ³H-taurine or ³H-D-aspartate were superfused 5 min with isosmotic medium. Thereafter (arrow), the medium was replaced by 30% hyposmotic solution. One-min fractions were collected during 20 min. ³H-Taurine release was inhibited by tyrosine kinase blockers (50μM AG18 (■)) and PI3K inhibitors (wortmannin 100nM (▲)). In contrast, glutamate release (followed as ³H-D-aspartate) was insensitive to these agents but potentiated by PKC activation (100nM PMA (▼)). (●) Controls with vehicle. Data represent the radioactivity released per min expressed as percentage of the total incorporated and are means ± SE (n=4-6).

In the same way, in rat cerebral cortex dialysates and bullfrog sympathetic ganglia different effects were obtained between SATR and SAAAC in the presence of several Cl⁻ channel blockers, SATR being the most widely sensitive and that of GABA, D-aspartate and glutamate rather inhibited in the presence of these agents[49,50]. Moreover, *in vivo* studies show that glutamate and aspartate are preferentially released during acute hyponatremia while taurine release is sensitive to both chronic and acute hyponatremia[51]. In cultured neurons taurine efflux is more sensitive to osmolarity perturbations than glutamate and GABA [52]. Hyposmotic-induced amino acid neurotransmitter release in excitable cells has been suggested to involve an exocytotic mechanism elicited in response to cell depolarization by either Cl⁻ release or by activation of non-selective cation channels; or by calcium-independent vesicle fusion mechanisms[reviewed in 53].

4. CONCLUDING REMARKS

The molecular identification and characterization of the swelling-activated efflux pathway for organic osmolytes release has been attempted by several research groups in the field. During their efforts there was initial evidence that suggested a common efflux pathway named VSOAC for both swelling activated-organic osmolyte release and VSAC. This led to the common extrapolation of experimental results from one osmolyte release (either taurine or chloride) to others (like amines, amino acids or polyalcohols). In this review, we summarize increasing evidence that points to different efflux pathways for SATR with respect to the VSAC. We also show that for the organic osmolyte group of amino acids, care must be taken in the further extrapolation of SATR to other swelling-activated release of amino acids at least for the case of neurotransmitter amino acids in excitable cells. In this case, it is clear that in brain preparations, different pathways mediate the release of amino acid neurotransmitter release (like GABA and glutamate) with respect to that of SATR. In brain preparations, the different regulatory mechanisms and efflux pathways involved in the swelling-activated amino acid neurotransmitter release with respect to that of taurine may be of physiological relevance because high extracellular concentrations of these neurotransmitters may lead to exocytotic insults, during conditions of cell swelling. It must be taken into account that several efflux pathways may be activated during cell swelling conditions for different osmolytes or even for the same one[54-56]. All this data challenge the hypothesis of VSOAC as the common efflux pathway for the release of organic osmolytes and anion conductance elicited by cell swelling conditions, and even for a common pathway for taurine and other amino acid osmolytes in excitable cells.

ACKNOWLEDGEMENTS

This work was supported by grants No. 3488-6M from CONACYT and IN-204900 from DGAPA-UNAM

REFERENCES

1. Lang, F., Busch, G.L., Ritter, M., Volkl, H., Waldegger, S., Gulbins, E., and Haussinger, D., 1998, Functional significance of cell volume regulatory mechanisms. *Physiol. Rev.* 78:247-306.
2. Pasantes-Morales, H., 1996, Volume regulation in brain cells: cellular and molecular mechanisms. *Metab. Brain Dis.* 11:187-204.
3. Kirk, K., and Strange, K., 1998, Functional properties and physiological roles of organic solute channels. *Annu. Rev. Physiol.* 60:719-739.
4. Kirk, K., Ellory, J.C., and Young, J.D., 1992, Transport of organic substrates via a volume-activated channel. *J. Biol. Chem.* 267:23475-23478.

2. Pasantes-Morales, H., 1996, Volume regulation in brain cells: cellular and molecular mechanisms. *Metab. Brain Dis.* 11:187-204.
3. Kirk, K., and Strange, K., 1998, Functional properties and physiological roles of organic solute channels. *Annu. Rev. Physiol.* 60:719-739.
4. Kirk, K., Ellory, J.C., and Young, J.D., 1992, Transport of organic substrates via a volume-activated channel. *J. Biol. Chem.* 267:23475-23478.
5. Junankar, P.R., and Kirk, K., 2000, Organic osmolyte channels: a comparative view. *Cell. Physiol. Biochem.* 10:355-360.
6. Kirk, K., 1997, Swelling-activated organic osmolyte channels. *J. Membr. Biol.* 158:1-16.
7. Strange, K., and Jackson, P.S., 1995, Swelling-activated organic osmolyte efflux: a new role for anion channels. *Kidney Int.* 48:994-1003.
8. Strange, K., Emma, F., and Jackson, P.S., 1996, Cellular and molecular physiology of volume-sensitive anion channels. *Am. J. Physiol.* 270:C711-C730.
9. Huxtable, RJ., 1992, Physiological actions of taurine. *Physiol. Rev.* 72:101-163.
10. Pasantes-Morales, H., Franco, R., Torres-Marquez, M.E., Hernandez-Fonseca, K., and Ortega, A., 2000, Amino acid osmolytes in regulatory volume decrease and isovolumetric regulation in brain cells: contribution and mechanisms. *Cell. Physiol. Biochem.* 10:361-370.
11. Pasantes-Morales, H., and Franco, R., 2002, Influence of protein tyrosine kinases on cell volume change-induced taurine release. *The Cerebellum.* 1:103-109.
12. Lepple-Wienhues, A., Szabo, I., Wieland, U., Heil, L., Gulbins, E., and Lang, F., 2000, Tyrosine kinases open lymphocyte chloride channels. *Cell. Physiol. Biochem.* 10:307-312.
13. Feranchak, A.P., Roman, R.M., Doctor, R.B., Salter, K.D., Toker, A., and Fitz, J.G., 1999, The lipid products of phosphoinositide 3-kinase contribute to regulation of cholangiocyte ATP and chloride transport. *J. Biol. Chem.* 274:30979-30986.
14. Voets, T., Manolopoulos, V., Eggermont, J., Ellory, C., Droogmans, G., and Nilius, B., 1998, Regulation of a swelling-activated chloride current in bovine endothelium by protein tyrosine phosphorylation and G proteins. *J. Physiol.* 506:341-352.
15. Lambert, I.H., and Hoffmann, E.K., 1994, Cell swelling activates separate taurine and chloride channels in Ehrlich mouse ascites tumor cells. *J. Membr. Biol.* 142:289-298.
16. Hoffmann, E.K., 2000, Intracellular signalling involved in volume regulatory decrease. *Cell. Physiol. Biochem.* 10:273-288.
17. Shennan, D.B., and Thomson, J., 2000, Further evidence for the existence of a volume-activated taurine efflux pathway in rat mammary tissue independent from volume-sensitive Cl-channels. *Acta Physiol. Scand.* 168:295-299.
18. Shennan, D.B., Cliff, M.J., and Hawkins, P., 1996, Volume-sensitive taurine efflux from mammary tissue is not obliged to utilize volume-activated anion channels. *Biosci. Rep.* 16:459-465.
19. Shennan, D.B., McNeillie, S.A., and Curran, D.E., 1994, The effect of a hyposmotic shock on amino acid efflux from lactating rat mammary tissue: stimulation of taurine and glycine efflux via a pathway distinct from anion exchange and volume-activated anion channels. *Exp. Physiol.* 79:797-808.
20. Davis-Amaral, E.M., Musch, M.W., and Goldstein, L., 1996, Chloride and taurine effluxes occur by different pathways in skate erythrocytes. *Am. J. Physiol.* 271:R1544-R1549.
21. Roman, R.M., Wang, Y., and Fitz, J.G., 1996, Regulation of cell volume in a human biliary cell line: activation of K+ and Cl- currents. *Am. J. Physiol.* 1996 271:G239-G248.
22. Stutzin, A., Eguiguren, A.L., Cid, L.P., and Sepulveda, F.V., 1997, Modulation by extracellular Cl- of volume-activated organic osmolyte and halide permeabilities in HeLa cells. *Am. J. Physiol.* 273:C999-C1007.
23.. Moran, J., Miranda, D., Pena-Segura, C., and Pasantes-Morales, H., 1997, Volume regulation in NIH/3T3 cells not expressing P-glycoprotein. II. Chloride and amino acid fluxes. *Am. J. Physiol.* 272:C1804-C1809.
24. Sanchez-Olea, R., Fuller, C., Benos, D., and Pasantes-Morales, H., 1995, Volume-associated osmolyte fluxes in cell lines with or without the anion exchanger. *Am. J. Physiol.* 269:C1280-C1286.

25. Pedersen, S.F., Beisner, K.H., Hougaard, C., Willumsen, B.M., Lambert, I.H., and Hoffmann, E.K., 2002, Rho family GTP binding proteins are involved in the regulatory volume decrease process in NIH3T3 mouse fibroblasts. *J. Physiol.* 541:779-796.

26. Stegen, C., Matskevich, I., Wagner, C.A., Paulmichl, M., Lang, F., and Broer, S., 2000, Swelling-induced taurine release without chloride channel activity in Xenopus laevis oocytes expressing anion channels and transporters. *Biochim. Biophys. Acta.* 1467:91-100.

27. Banderali, U., and Roy, G., 1992, Anion channels for amino acids in MDCK cells. *Am. J. Physiol.* 263:C1200-C1207.

28. Roy, G., and Banderali, U., 1994, Channels for ions and amino acids in kidney cultured cells (MDCK) during volume regulation. *J. Exp. Zool.* 268:121-126.

29. Boese, S.H., Wehner, F., and Kinne, R.K., 1996, Taurine permeation through swelling-activated anion conductance in rat IMCD cells in primary culture. *Am. J. Physiol.* 271:F498-F507.

30. Roy, G., 1995, Amino acid current through anion channels in cultured human glial cells. *J. Membr. Biol.* 147:35-44.

31. Olson, J.E., and Li, G.Z., 1997, Increased potassium, chloride, and taurine conductances in astrocytes during hypoosmotic swelling. *Glia.* 20:254-61.

32. Pasantes-Morales, H., Chacon, E., Murray, R.A., and Moran, J., 1994, Properties of osmolyte fluxes activated during regulatory volume decrease in cultured cerebellar granule neurons. *J. Neurosci. Res.* 37:720-727.

33. Sanchez-Olea, R., Morales, M., Garcia, O., and Pasantes-Morales, H., 1996, Cl channel blockers inhibit the volume-activated efflux of Cl and taurine in cultured neurons. *Am. J. Physiol.* 270:C1703-C1708.

34. Kimelberg, H.K., Goderie, S.K., Higman, S., Pang, S., and Waniewski, R.A., 1990. Swelling-induced release of glutamate, aspartate, and taurine from astrocyte cultures. *J. Neurosci.* 10:1583-1591.

35. Sanchez-Olea, R., Pena, C., Moran, J., and Pasantes-Morales, H., 1993, Inhibition of volume regulation and efflux of osmoregulatory amino acids by blockers of Cl- transport in cultured astrocytes. *Neurosci. Lett.* 156:141-144.

36. Pasantes-Morales, H., Murray, R.A., Sanchez-Olea, R., and Moran, J., 1994, Regulatory volume decrease in cultured astrocytes. II. Permeability pathway to amino acids and polyols. *Am. J. Physiol.* 266:C172-C178.

37. Rutledge, E.M., Aschner, M., and Kimelberg, HK., 1998, Pharmacological characterization of swelling-induced D-[3H]aspartate release from primary astrocyte cultures. *Am. J. Physiol.* 274:C1511-C1520.

38. Basavappa, S., Huang, C.C., Mangel, A.W., Lebedev, D.V., Knauf, P.A., and Ellory, J.C., 1996, Swelling-activated amino acid efflux in the human neuroblastoma cell line CHP-100. *J. Neurophysiol.* 76:764-769.

39. Franco, R., Quesada, O., and Pasantes-Morales, H., 2000, Efflux of osmolyte amino acids during isovolumic regulation in hippocampal slices. *J. Neurosci. Res.* 61:701-711.

40. Bothwell, J.H., Rae, C., Dixon, R.M., Styles, P., and Bhakoo, K.K., 2001, Hypo-osmotic swelling-activated release of organic osmolytes in brain slices: implications for brain oedema in vivo. *J. Neurochem.* 77:1632-1640.

41. Pasantes-Morales, H., Ochoa de la Paz, L.D., Sepulveda, J., and Quesada, O., 1999, Amino acids as osmolytes in the retina. *Neurochem. Res.* 24:1339-1346.

42. Manolopoulos, V.G., Voets, T., Declercq, P.E., Droogmans, G., and Nilius, B., 1997, Swelling-activated efflux of taurine and other organic osmolytes in endothelial cells. *Am. J. Physiol.* 273:C214-C222.

43. Sterns, R.H., Baer, J., Ebersol, S., Thomas, D., Lohr J.W., and Kamm, D.E., 1993, Organic osmolytes in acute hyponatremia. *Am. J. Physiol.* 264:F833-F836.

44. Taylor, D.L., Davies, S.E., Obrenovitch, T.P., Doheny, M.H., Patsalos, P.N., Clark, J.B., and Symon, L., 1995, Investigation into the role of N-acetylaspartate in cerebral osmoregulation. *J. Neurochem.* 65:275-281.

45. Davies, S.E., Gotoh, M., Richards, D.A., and Obrenovitch, T.P., 1998, Hypoosmolarity induces an increase of extracellular N-acetylaspartate concentration in the rat striatum. *Neurochem. Res.* 23:1021-1025.

46. Baslow, M.H., 2002 Evidence supporting a role for N-acetyl-L-aspartate as a molecular water pump in myelinated neurons in the central nervous system. An analytical review. *Neurochem. Int.* 40:295-300.

47. Mongin, A.A., Reddi, J.M., Charniga, C., and Kimelberg, H.K., 1999, [3H]taurine and D-[3H]aspartate release from astrocyte cultures are differently regulated by tyrosine kinases. *Am. J. Physiol.* 276:C1226-C1230.

48. Franco, R., Torres-Marquez, M.E., and Pasantes-Morales, H., 2001, Evidence for two mechanisms of amino acid osmolyte release from hippocampal slices. *Pflugers Arch.* 442:791-800.

49. Sakai, S., and Tosaka, T., 1999, Analysis of hyposmolarity-induced taurine efflux pathways in the bullfrog sympathetic ganglia. *Neurochem. Int.* 34:203-212.

50. Estevez, A.Y., O'Regan, M.H., Song, D., and Phillis, J.W., 1999, Effects of anion channel blockers on hyposmotically induced amino acid release from the in vivo rat cerebral cortex. *Neurochem. Res.* 24:447-452.

51. Verbalis, J.G., and Gullans, S.R., 1991, Hyponatremia causes large sustained reductions in brain content of multiple organic osmolytes in rats. *Brain Res.* 567:274-282.

52. Pasantes-Morales, H., Alavez, S., Sanchez-Olea, R., and Moran, J., 1993, Contribution of organic and inorganic osmolytes to volume regulation in rat brain cells in culture. *Neurochem. Res.* 18:445-452.

53. Pasantes-Morales, H., Franco, R., Ochoa, L., and Ordaz, B., 2002, Osmosensitive release of neurotransmitter amino acids: relevance and mechanisms. *Neurochem. Res.* 27:59-65.

54. Zhang, J.J., and Jacob, T.J., 1997, Three different Cl- channels in the bovine ciliary epithelium activated by hypotonic stress. *J. Physiol.* 499:379-389.

55. Staines, H.M., Godfrey, E.M., Lapaix, F., Egee S., Thomas, S., and Ellory, J.C., 2002, Two functionally distinct organic osmolyte pathways in Plasmodium gallinaceum-infected chicken red blood cells. *Biochim. Biophys. Acta.* 1561:98-108.

56. Banderali, U., and Ehrenfeld, J., 1996, Heterogeneity of volume-sensitive chloride channels in basolateral membranes of A6 epithelial cells in culture. *J. Membr. Biol.* 154:23-33.

Electrophysiological Properties of the Mouse Na⁺/Cl⁻-Dependent Taurine Transporter (mTauT-1): Steady-State Kinetics

Stoichiometry of Taurine Transport

HEMANTA K. SARKAR*, ERNEST M. WRIGHT[#], KATHRYN J. BOORER[#] and DONALD D. F. LOO[#]

Department of Chemistry & Biochemistry, University of Massachusetts at Dartmouth, North Dartmouth, MA 02747 and [#]Department of Physiology, The David Geffen School of Medicine at UCLA, Los Angeles, CA 90095

1.   INTRODUCTION

Taurine, one of the most abundant free amino acids found in many cells, is found in millimolar concentrations in the brain, heart, eye and liver[1,2]. Among the many postulated roles of taurine (such as a neurotransmitter, a cytoprotector and an essential nutrient for the fetal and neonatal brain development), the best understood role is as an osmolyte in the kidney and brain[1,2]. Na⁺ and Cl⁻ -dependent transport of taurine have been described in a variety of tissues including kidney, heart, retina, brain and choroid plexus[2]. Over the last several years, the cDNA encoding the Na⁺/Cl⁻-dependent taurine transporter has been cloned from rat brain[3], mouse brain[4], dog MDCK epithelial cells[5], human thyroid[6], human placental JAR cells[7], mouse retina[8] and bovine aortic endothelial cells[9]. The proteins encoded by these clones show high degree of sequence homology at the amino acid level[9]. The Na⁺/Cl⁻-dependent taurine transporter belongs to the Na⁺/Cl⁻/GABA cotransporter family of proteins[10]. The hallmark of this family of proteins is that they all predicted to contain twelve transmembrane helices crossing the membrane in a zigzag fashion, and both the N- and C-terminal ends are predicted to face the cytoplasm milieu.

We previously cloned a taurine transporter encoding cDNA from the mouse retina (mTauT-1) and determined the biochemical and pharmacological properties of the transporter expressed in *Xenopus* oocyte[8]. In this study, we have used the two-electrode voltage-clamp technique to characterize the electrophysiological properties of the mTauT-1 clone expressed in *Xenopus* oocytes. Results presented here suggest that mTauT-1 transports taurine, Na^+ and Cl^- across the cell membrane in a $2Na^+:1Cl^-$:1taurine stoichiometry.

2.   MATERIALS AND METHODS

2.1   Oocytes

Xenopus laevis oocytes (stage VI) were defolliculated and injected with 50 nl (1 ng/nl) of *in vitro* transcribed synthetic RNA (cRNA) prepared from the mTauT-1 cDNA as described[8,11]. Injected oocytes were maintained in Barth's solution [concentration in mM: 88 NaCl, 1 KCl, 0.33 $Ca(NO_3)_2$, 0.41 $CaCl_2$, 0.82 $MgSO_4$, 2.4 $NaHCO_3$, 10 HEPES/Tris, pH 7.4, 100 µg/ml gentamycin sulfate] at 18 °C for 2-10 days before they were used for electrophysiological experiments.

2.2   Electrophysiological Methods

Electrophysiological experiments were performed with the membrane under voltage clamp using the two-electrode voltage clamp method as previously described[11-13]. During experiments, oocytes were bathed in the NaCl buffer [(in mM): 100 NaCl, 2 KCl, 1 $CaCl_2$, 1 $MgCl_2$ and 10 HEPES (pH 7.4)]. To obtain current-voltage relations, 100 ms voltage steps were applied using pCLAMP software (Axon Instruments, Foster City, CA) from the holding potential of - 50 mV to a series of test voltages from -150 mV to +50 mV in 20 mV increments. Currents were low-pass filtered at 500 Hz (Bessel filter model LP902, Frequency Devices, Haverhill, Mass), digitized at 100 µs/point, and the average current of three sweeps were recorded.

The effects of external substrate concentration ($[taurine]_o$, $[Na^+]_o$, and $[Cl^-]_o$) on the steady-state kinetics were determined by non-linear curve fitting (SigmaPlot, SPSS Inc., Chicago, IL) of the substrate-induced currents (I) at each membrane potential to the Hill equation:

$$I = I_{max}^S \cdot ([S]_o)^n / ((K_{0.5}^S)^n + ([S]_o)^n) \qquad (1)$$

where $[S]_o$ is $[taurine]_o$, $[Na^+]_o$, or $[Cl^-]_o$, I_{max}^S is the maximal substrate-induced current, n is the Hill coefficient and $K_{0.5}^S$ is the half-maximal substrate concentration.

2.3 Total Inward Charge/Taurine Flux Ratio

The ratio of the total inward charge to taurine flux was determined by simultaneous measurement of taurine-induced inward current and influx of [^{3}H]taurine under voltage clamp condition as described[13].

3. RESULTS

3.1 Pre-Steady-State and Steady-State Currents

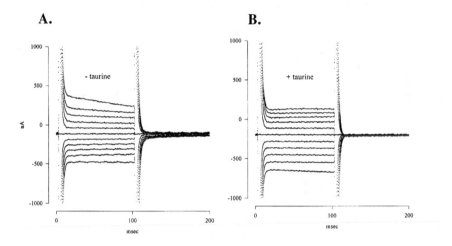

Figure 1. Current records of the taurine-induced current. Current records in a mTauT-1 cDNA expressing oocyte bathed in 100 mM NaCl buffer: (A) in the absence of taurine (pre-steady-state currents) and (B) in the presence of 500 μM of taurine (steady-state currents). Oocyte membrane potential was held at -50 mV.

Applied voltage steps evoked two types of currents in oocytes expressing the mTauT-1: pre-steady-state currents and steady-state inward currents. The pre-steady-state currents were observed in the absence of taurine in mTauT-1 expressing oocytes. Figure 1 shows typical current traces from an cRNA-injected oocyte bathed in NaCl buffer as the membrane voltage was stepped from the holding potential (V_m) of -50 mV to various testing voltages. In the absence of taurine, the current relaxation consisted of the initial capacitance spike followed by a slower decay to the steady state (Figure 1A). The pre-steady-state transient was not observed in control H₂O-injected oocytes (results not shown) and was most pronounced at the large depolarizing potentials (i.e., +50 mV), suggesting that they were associated with the taurine transporter. Addition of taurine (500 μM) to the bath solution induced an inward steady-state current and abolished the pre-steady-state current (Figure 1B). However, no taurine-induced inward current was observed in Na⁺-free buffer (results not shown). Although kinetic analysis of the pre-steady-state currents provided useful information on the taurine transporter, in this communication we will present only the results obtained from the recordings of the steady-state currents.

3.2   Current-Voltage (I-V) Relation

The current-voltage (I-V) relation of the taurine-induced inward current was obtained from the difference of the steady-state current at each test potential in the presence and absence of taurine. Figure 2 shows a series of I-V relation plots of taurine-induced currents in buffers containing different anions.

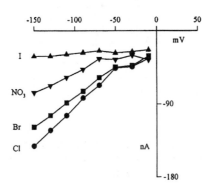

Figure 2. Current-voltage (I-V) relation plots generated using 500 μM of taurine and bath solutions containing 100 mM Na⁺ and 100 mM of different anions.

As shown in Figure 2, the induced current increased with hyperpolarizing membrane potentials (except in I⁻ buffer) and did not saturate at the most negative voltage tested (-150 mV). In the depolarizing direction, the taurine-induced current tended to asymptote towards zero. The current level was highest in Cl⁻ containing buffer at most test potentials. At -150 mV, the selectivity ratio of Cl⁻:Br⁻:NO₃⁻:I⁻ was 1:0.8:0.5:0.2. No taurine-induced current was however observed when Cl⁻ was replaced by gluconate or methanesulfonate (results not shown). Our results suggest that anions of small ionic radii than Cl⁻ can substitute for Cl⁻. Our results also suggest that under physiologic condition Cl⁻ is an essential anion for the transport of taurine via the taurine transporter.

3.3 Activation of Taurine-Induced Current by Taurine, Na^+ and Cl^-

To determine the stoichiometry of taurine transport, we measured the activation of taurine-induced current by $[taurine]_o$, $[Na^+]_o$ and $[Cl^-]_o$. Figure 3A shows the plot of normalized current as a function of different external taurine concentration $([taurine]_o)$ at -50 mV. The normalized current increased as the $[taurine]_o$ concentration was increased from 1 µM - 5 mM. The smooth curve was drawn by fitting the experimental data points to the Hill equation with n = 1, I_{max} = 80 nA, and $K_{0.5}$ (taurine) = 10 µM. The $K_{0.5}$ (taurine) and the Hill coefficient n for taurine, virtually remained unchanged over a range of test membrane potentials (results not shown). Figure 3B and 3C show plots of normalized currents at different external Na^+ ($[Na^+]_o$) and Cl^- ($[Cl^-]_o$) concentrations, respectively. The Na^+ activation of taurine-induced current showed a sigmoidal dependence on $[Na^+]_o$, and the data points were best fitted (Figure 3B, smooth curve) to Hill equation with n = 1.9, I_{max} = 98 nA, and $K_{0.5}$ (Na^+) = 77 mM. In contrast, the Cl^--activation of taurine-induced current showed a hyperbolic dependence on $[Cl^-]_o$ measured at two different membrane holding potentials, -50mV and -110 mV (Figure 3C). The experimental data points obtained at -110 mV holding potential were best fitted to the Hill equation with n = 1.1, I_{max} = 52 nA, and $K_{0.5}$ (Cl^-) = 50 mM (Figure 3C, top smooth curve). Similarly, the experimental data points obtained at -50 mV holding potential were best fitted (Figure 3C, bottom smooth curve) to the Hill equation with n = 1.0, I_{max} = 42 nA, and $K_{0.5}$ (Cl^-) = 107 mM.

3.4 Total Inward Charge/Taurine Flux Ratio

To determine the ion/substrate stoichiometry of mTauT-1, we next determined the ratio of the total inward positive charge to taurine flux by simultaneously measuring taurine uptake and taurine-induced current under voltage clamp condition. Figure 3D shows the result from a typical experiment where the current induced by [³H]taurine (15 µM) in NaCl buffer was continuously recorded at holding membrane potential of -100 mV. After 10 min, the taurine was washed out and the current was allowed to return to the baseline. The oocyte was subsequently washed and solubilized, and the amount of [³H]taurine transported into the oocyte during the current recording period was determined by scintillation counting. The

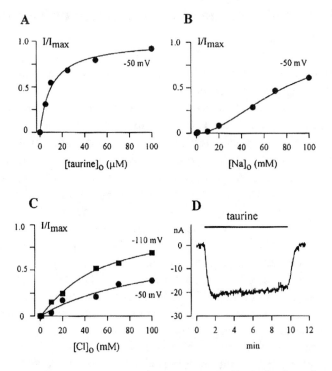

Figure 3. Stoichiometry of mTauT-1. (A), (B) and (C): Activation of taurine current by taurine, Na^+ and Cl^-. Normalized taurine current is plotted as a function of (A) $[taurine]_o$, (B) $[Na^+]_o$ and (C) $[Cl^-]_o$. In (A), the induced currents were measured at different $[taurine]_o$ (1 μM to 5 mM) and keeping the $[NaCl]_o$ constant at 100 mM (V_m = -50 mV). In (B), current was measured (V_m = -50 mV) at different $[Na^+]_o$ (varied from 0 - 100 mM with choline replacement) while the $[Cl^-]_o$ and $[taurine]_o$ were maintained at 100 mM and 500 μM, respectively. In (C), currents were measured (V_m = -50 and -100 mV, respectively) at different $[Cl^-]_o$ (varied from 0 - 100 mM with MES replacement) while maintaining the $[Na^+]_o$ and $[taurine]_o$ at 100 mM and 500 μM, respectively. The smooth curves in (A), (B) and (C) are best fits to Hill equation. (D): Ratio of total inward charge to taurine flux. A time course of the taurine-induced inward current in a mTauT-1 expressing oocyte perfused with 100 mM NaCl buffer containing 15 μM $[^3H]$taurine (V_m = -100 mV).

total inward positive charge moved by the transporter during the same period was calculated by time integral of the taurine-evoked current. The ratio of total inward positive charge to the total $[^3H]$taurine flux was found to be 1.22 ± 0.07 (12 oocytes; $\pm$ SE).

3.5   Taurine-Mediated Uptake of Chloride

To determine whether Cl^- is transported by mTauT-1, we measured taurine-mediated uptake of $^{36}Cl^-$. As shown in Table 1, mTauT-1 expressing oocytes accumulated significantly higher level of $^{36}Cl^-$ than the control oocytes.

Table 1. Taurine-mediated uptake of $^{36}Cl^-$ into oocytes. Experiments were performed using non-clamped oocytes (N = 9) bathed in (mM): 100 Na^+, 40 Cl^-, 60 MES, 100 µM taurine, HEPES/pH 7.4.

H₂O-injected (Control)	mTauT-1 cRNA-injected
(nM/h/oocyte)	(nM/h/oocyte)
1.0 ± 0.1	6.0 ± 0.6

4.   DISCUSSION

Previous kinetic analysis of mTauT-1 expressed in *Xenopus* oocytes using [³H]taurine as a transport substrate suggested that $2Na^+$ and $1Cl^-$ are required to transport one taurine molecule across the cell membrane[8]. These results also predicted that taurine transport is electrogenic. Observation of taurine-induced inwardly directed current in mTauT-1 expressing oocytes confirms this idea (Figure 1). That taurine induces inward current in mTauT-1 expressing oocytes, also suggests net movement of positive charge (i.e., Na^+) with taurine.

Analyses of $[taurine]_o$, $[Na^+]_o$ and $[Cl^-]_o$ activation of the taurine-induced currents according to Hill equation revealed a ratio of Hill coefficients of 1taurine, $2Na^+$ and $1Cl^-$, which was consistent with the previously determined ratio from biochemical analysis[8]. However, the substrate ratios derived only from the determination of Hill coefficients may not truly represent the transport stoichiometry of the transporter since they cannot distinguish between the substrate-mediated activation of the transporter and the actual transport of the substrate(s).

To obtain a better idea of the ion/substrate stoichiometry of mTauT-1, we then simultaneously measured the total inward charge and total [³H]taurine flux (Figure 3D) and found that their ratio is ~ 1.0. Since the Hill coefficient for Na^+ is 2, these results imply that Cl^- is transported with Na^+ and taurine during the transport cycle. That Cl^- is indeed transported by mTauT-1, is demonstrated by the taurine-mediated $^{36}Cl^-$ influx studies, where taurine was shown to induce significantly higher level of $^{36}Cl^-$ influx into the mTauT-1 expressing oocytes than the control oocytes (Table 1). Taken together, these results suggest a transport stoichiometry of $2Na^+:1Cl^-:1$taurine for mTauT-1.

ACKNOWLEDGEMENTS

This study was funded in part by AHA Grant-in-Aid No. 95012109 (to HKS) and in part by NIH DK-19567 (to EMW).

REFERENCES

1.   Chesney, R.W., 1985, *Adv. Pediatr.* 32:1-42.
2   Huxtable, R.J., 1992, *Physiol. Rev.* 72:101-163.
3.   Smith, K.E., Borden, L.A., Wang, C.-H., Hartig, P.R., Branchek, T.A., and Weinshank, R.L., 1992, *Mol. Pharmacol.* 42:563-569.
4.   Liu, Q.R., Lopez-Corcuera, B., Nelson, H., Mandiyan, S., and Nelson, N., 1992, *Proc. Natl. Acad. Sci. USA* 89:8230-8234.
5.   Uchida, S., Kwon, H.M., Yamaguchi, A., Preston, A.S., Marumo, F., and Handler, J.S., 1992, *Proc. Natl. Acad. Sci. USA* 89:8230-8234.
6.   Jhiang, S.M., Fithian, L., Smanik, P., McGill, J., Tong, Q., and Mazzaferri, E.L., 1993, *FEBS Lett.* 318:139-144.
7.   Ramamoorthy, S. Leibach, F.H., Mahesh, M.J., Ham, Y., Yang-Feng, T., Blakely, R.D., and Ganapathy, V., 1994, *Biochem. J.* 30:893-900.
8.   Vinnakota, S., Qian, X., Egal, H., Sarthy, V., and Sarkar, H.K., 1997, *J. Neurochem.* 69:2238-2250.
9.   Qian, X., Vinnakota, S., Edwards, C., and Sarkar, H.K., 2000, *Biochim. Biophys. Acta* 1509:324-334.
10.   Nelson, N., 1998, *J. Neurochem.* 71:1785-1803.
11.   Loo, D.D.F, Hazama, A., Supplisson, S., Turk, E., and Wright E.M., 1993, *Proc. Natl. Acad. Sci. USA* 90:5767-5771.
12.   Loo, D.D.F., Hirsch, J.R., Sarkar, H.K., and Wright, E.M., 1996, *FEBS Lett.* 392:250-254.
13.   Loo, D.D.F., Eskandari, S., Boorer, K.J., Sarkar, H.K., and Wright, E.M., 2000, *J. Biol. Chem.* 275:37414-37422.

Taurine Uptake and Release by the Pancreatic β-Cells

Taurine Transport in β-Cells

SHYAMALA VINNAKOTA[#] and HEMANTA K. SARKAR[#*]

[#]*Department of Molecular Physiology & Biophysics, Baylor College of Medicine, Houston, TX 77054;* [*]*Current Address: Department of Chemistry & Biochemistry, University of Massachusetts at Dartmouth, North Dartmouth, MA 02747*

1. INTRODUCTION

The β-amino acid taurine is abundantly present in its free form in various mammalian cells and tissues[1], including the islet of Langerhans[2]. Although the exact physiological role of taurine in the pancreas is not understood, a number of studies suggest a possible role of taurine in glucose metabolism. Taurine has been shown to attenuate streptozotocin-induced diabetes in animals[3], to increase the stimulatory effects of conventional secretagogues in cultured fetal β-cells[4], and to protect against β-cells injury by NO and IL-1 beta[5]. Most interestingly, hyposmotic challenge that induces taurine release in many cells[1] has been shown to increase secretion of insulin[6,7].

Despite these important observations, information on taurine transport in β-cells is virtually lacking. Given the relationship between the cell volume and insulin secretion, it is important to characterize and understand the mechanisms of taurine release in the β-cells. In this communication, we have characterized the endogenous taurine uptake and taurine release pathways in one of the most widely used insulin-secreting cultured β-cells, namely the hamster pancreatic insulinoma HIT-T15 cells.

2.   MATERIALS AND METHODS

2.1   Cell Culture

HIT-T15 cells (American Type Culture Collection, Rockville, MD) were plated in 6-well culture dishes at a density of ~ 1.5 - 2.0 x 10^6 cells/well in F-12K medium (87.5% HAM's medium, 2.5% FBS, 10% dialyzed horse serum) supplemented with 1% penicillin/streptomycin and grown for 5-7 days at 37 $^\circ$C in the incubator (95% air/5 % CO_2). Medium was replaced every third day. Taurine uptake and release studies were performed using cell passages between 66-74.

2.2   Chemicals

1,2-[^{3}H]Taurine (1.18 TBq/mmol) was purchased from Amersham Life Sciences (Arlington Heights, IL). Taurine, EGTA, β-alanine, nipecotic acid, anthracene-9-carboxylic acid (9-AC), 4,4'-diisothiocyanatostilbene-2,2'-disulfonic acid (DIDS), 5-nitro-2(3-phenylpropylamino)benzoic acid (NPPB) were purchased from Sigma Chemicals (St. Louis, MO). Ionomycin, 8-bromoadenosine-3',5'-cyclic monophosphate (8-Br-cAMP) and A23187 were purchased from Calbiochem (San Diego, CA). FK-12 medium was purchased from Life Technologies (Gaithersburg, MD). All chemicals used in buffers were ACS pure grade and were purchased from one of the general chemical suppliers such as Sigma Chemicals (MO), Fisher Scientific Co. (Pittsburgh, PA), and Life Technologies (MD).

2.3   Buffer Compositions

Taurine uptake was performed in the NaCl buffer [(in mM): 137 NaCl, 1 $CaCl_2$, 2 KCl, 5 $MgCl_2$, 5 HEPES/pH 7.5]. The choline chloride (ChCl) and LiCl buffers were prepared by replacing the Na^+ in the NaCl buffer with the equimolar amount of choline and lithium, respectively. The compositions of buffers containing different anions were same as the NaCl buffer and were prepared using the respective sodium salt and gluconate salts of Ca^{2+}, Mg^{2+} and K^+.

The main base release buffer was NaCl-1 buffer [(in mM): 147 NaCl, 5 HEPES/pH 7.5]. The other base release buffers were sodium acetate-1 (NaOAc-1) [(in mM): 147 NaOAc, 5 HEPES/pH 7.5], KCl-1 [(in mM): 120 KCl, 27 NaCl, 5 HEPES/pH 7.5], and the potassium acetate-1 (KOAc-1) [(in mM): 120 KOAc, 27 NaOAc, 5 HEPES/pH 7.5]. Release buffers of various other compositions were prepared by adding chemicals and drugs directly to one of the base release buffers. All buffers were isosmotic with the growth medium (~ 310 mOsm).

2.4   Taurine Uptake and Release Assays

Taurine uptake studies were performed essentially as described earlier using 5 μM of [³H]-taurine as the uptake substrate[8]. Following uptake (60-min), cells were washed with ice-cold ChCl buffer, solubilized in 1% SDS and the amount of radiolabeled taurine in the cells was determined by liquid scintillation counting.

For release assay, cells were first loaded with 5 μM of [³H]-taurine in NaCl buffer for 3-4 h at 37° C. Subsequently, the cells were washed quickly in NaCl-1 buffer and the release process was elicited by incubating the cells in 1 ml of a given isosmotic release buffer. After 30-min incubation at room temperature, the release buffer was removed, and the amount of [³H]-taurine in the buffer was determined by liquid scintillation counting. The cells were washed and then solubilized in 1 ml of 1% SDS. The amount of residual [³H]-taurine remained in the cell was determined by liquid scintillation counting. Fraction of taurine release (F) was calculated from $F = R/(R + RC)$, where R is the amount of taurine released in the buffer and RC is the amount of taurine remained in the cell.

3.   RESULTS

3.1   Taurine Uptake in HIT Cells

[³H]Taurine uptake was measured to characterize the HIT-T15 cell endogenous taurine transporter. Figure 1 shows the Na^+- and Cl^--dependence of [³H]-taurine uptake. As shown, taurine uptake in HIT cells is highest in NaCl buffer, and the uptake is virtually ablated when the Na^+ is replaced with either Li^+ or choline (Figure 1A). The uptake pattern of anion-dependence (Figure 1B) is, however, different from that was observed for the mouse taurine transporter (mTauT-1)[9] and the bovine aortic endothelial cell taurine transporter (bTauT-1)[8] cDNA clones expressed in *Xenopus laevis* oocytes. Thus, the uptake level is highest in buffer containing Cl^-. In HIT cells, significant level of taurine uptake activity is also observed in buffers containing other anions (bromide, iodide, acetate, gluconate, formate and sulfate), except in buffers containing fluoride (F^-) or citrate. Nevertheless, Cl^- is likely the preferred anion under physiological condition.

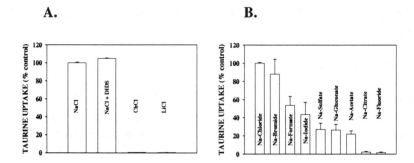

Figure 1. Cation and anion dependence of taurine uptake. Sixty minute taurine uptake assays were performed in different cation (A) and anion (B) containing buffers as described[8]. Results are average of 2 or more (± SE) independent experiments.

3.2   Inhibition of Taurine Uptake

Taurine uptake in HIT cells is inhibited significantly (~ 70%) by known taurine transport inhibitors β-alanine (100 μM) and GES (200 μM), but not by the GABA transport inhibitor nipecotic acid (10 μM) (results not shown). Taurine uptake remained unchanged from the control level in the presence of 1 mM DIDS, a Cl⁻ transport blocker (Figure 1A). Raising the osmolarity of the growth medium to ~ 510 mOsm by adding 100 mM NaCl resulted in an increase in taurine uptake, which was time dependent and reached 70 - 80% higher value than that of the control level by 24 h (results not shown).

3.3   Kinetics of Taurine Uptake

The taurine uptake was measured at different external substrate concentrations by varying the concentration of one of the substrates ([taurine]$_o$ was varied from 1-30 μM; [Na⁺]$_o$ and [Cl⁻]$_o$ were varied from 0-137 mM, respectively) while keeping the concentrations of the other two substrates constant as described[9]. The results were then analyzed by curve fit analysis according to Hill equation[8,9]. The [taurine]$_o$ dependent taurine uptake results (not shown) fitted best to the Hill equation with a $K_{0.5}$ (taurine) ~ 9.31 ± 1.5 μM   and Hill coefficient n = 1.0 for taurine (average of 3 independent experiments ± SE). The [Na⁺]$_o$ dependence of taurine uptake data points fitted best to the Hill equation with a $K_{0.5}$ (Na⁺) = 67.4 mM, n = 2.1 and V_{max} = 1180 pmol/10⁶ cells (average of 2 independent experiments). The [Cl⁻]$_o$ dependence of taurine uptake data showed different patterns depending on the anion used to replace Cl⁻ (Figure 2). When the [Cl⁻]$_o$ was varied using acetate, the uptake data fitted best to Hill equation (smooth curve, Figure 2A) with $K_{0.5}$ (Cl⁻) = 18.5 mM, n = 1.4 and V_{max} = 617.5 pmol/10^6 cells. In contrast, when F⁻ was used to vary [Cl⁻]$_o$, the data points fitted best to the Hill equation (smooth curve, Figure 2B) with $K_{0.5}$ (Cl⁻) = 97.0 mM, n = 4.5 and V_{max} = 782.0 pmol/10^6 cells).

A.   **B.**

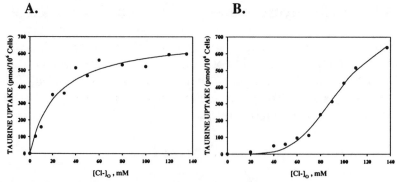

Figure 2. Dependence of taurine uptake on [Cl⁻]$_o$. External concentrations of Cl⁻ ([Cl⁻]$_o$) was varied from 0-137 mM with (A) acetate and (B) F⁻ as the replacement ion. Smooth curves are best fits of the experimental data points to Hill equation (see text).

3.4   Taurine Release

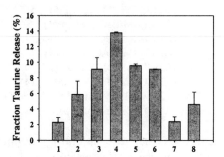

Figure 3. Taurine release. Cells were loaded with 5 μM [³H]taurine in NaCl buffer for 3 - 4 h, and a 30-min release was initiated in release buffers of different compositions: 1, NaCl; 2, NaOAc; 3, NaCl-1; 4, NaOAc-1; 5, ChCl-1; 6, KOAc-1; 7, NaCl-1/ CaCl$_2$ (1 mM); and 8, KOAc-1/Ca(OAc)$_2$ (1 mM). Results are average of 2 or more (± SE) independent experiments.

HIT cells were loaded with [³H]taurine in NaCl buffer (isosmotic with the growth medium) for 3- 4 h, and release was initiated in isosmotic buffers of various compositions (Figure 3). When the loading buffer was used as the release buffer (control), ~ 2.3% of the total taurine was released in 30 min (basal release). Replacement of Cl⁻ by acetate (NaOAc buffer) resulted in about 2-fold increase in taurine release; where as, removal of divalent cations (Ca²⁺ and Mg²⁺) from the NaCl buffer (NaCl-1 buffer) resulted in an ~ 4-fold increase in taurine release over the basal level. Replacement of Na⁺ in NaCl-1 buffer by choline (ChCl) or replacement of both Na⁺ and Cl⁻ by K⁺ and acetate (OAc⁻) did not cause any further increase in release. Addition of 1 mM Ca^{2+} back to the NaCl-1 buffer completely inhibited the increase in release; where as, addition of 1 mM Ca^{2+} to the KOAc buffer partially inhibited the increase in release.

3.5 Effects of Glucose and 8-Br-cAMP on Taurine Release

Addition of 20 mM glucose to the NaCl-1/$CaCl_2$ or KCl-1/$CaCl_2$ (depolarizing condition) buffer did not result in any increase in basal taurine release (results not shown). Similarly, pretreatment of HIT cells with 8-Br-cAMP (100 µM), a stimulator of vesicular secretion in pancreatic β-cells[10], also did not affect the basal taurine release.

3.6 Effect of Ionomycin on Taurine Release

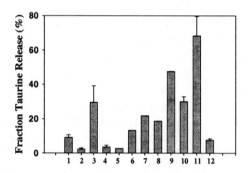

Figure 4. Ionomycin-induced taurine release. A 30-min release assay was performed in the absence and presence of 10 µM ionomycin (IM). Release buffers: 1, NaCl-1; 2, NaCl-1/$CaCl_2$; 3, NaCl-1/$CaCl_2$/IM; 4, NaCl-1/$CaCl_2$/A23187 (10 µM); 5, NaCl-1/$CaCl_2$/IM/$LaCl_3$ (1 mM); 6, NaCl-1/$CaCl_2$/IM/$CdCl_2$ (200 µM); 7, NaCl-1/$CaCl_2$/IM/9-AC (1 mM); 8, NaCl-1/$CaCl_2$/IM/ DIDS (100 µM); 9, NaCl-1/$CaCl_2$/IM/NPPB (100 µM); 10, NaOAc-1/Ca(OAc)₂/IM; 11, KOAc/Ca(OAc)₂/IM; and 12, NaCl-1/$MgCl_2$ (5 mM)/IM/EGTA (1 mM). Results are expressed as average of 2 or more (± SE) independent experiments.

Maximum taurine release was achieved when the calcium ionophore ionomycin (IM) was added to the NaCl-1 release buffer containing 1 mM Ca^{2+} (Figure 4). In contrast, the Ca^{2+} ionophore A23187 was unable to mimic the effect of ionomycin. The IM-induced increase in release was mostly inhibited when Ca^{2+} was replaced with Mg^{2+} and any external Ca^{2+} was chelated with EGTA. The IM-induced increase in release was completely blunted by 1 mM La^{3+} and partially inhibited by chloride channel blockers Cd^{2+} (200 µM), DIDS (100 µM) and 9-AC (1 mM), but not by NPPB (100 µM).

4. DISCUSSION

Taurine uptake results described above suggest that HIT cells possess a Na$^+$- and Cl$^-$-dependent high-affinity taurine transporter. Biochemical properties of this transporter differ from those of the mouse retinal taurine transporter[9] in the anionic dependence pattern of taurine uptake. In contrast to the mouse retinal taurine transporter, the taurine uptake in HIT cells is maintained up to a significant level in several different buffers containing anions other than Cl$^-$. An interesting observation is the behavior of taurine uptake as a function of different [Cl$^-$]$_o$, which differed markedly depending on whether acetate or F$^-$ was used as a replacement anion to vary the [Cl$^-$]$_o$. A reasonable explanation for this difference is that while the acetate is acting as a replacement ion for Cl$^-$, the F$^-$ may be strongly interfering with the Cl$^-$ binding to the transporter. Exposure to hyperosmolarity also increased taurine uptake in HIT cells. This result, in combination with the observation that hyposmotic challenge induces taurine release from β-cells[11,12], suggests an osmolyte role for taurine in insulin secreting pancreatic HIT cells.

Taurine release by hyposmotic challenge is believed to occur via a volume-sensitive anion channel[11-13]. Since we have examined taurine release from HIT cells in isosmotic buffers, the observed increase in taurine release is not likely via this volume-sensitive anion channel when the Cl$^-$ in the release buffer was replaced by acetate or the divalent cations were removed from the release buffer. We propose that the increase in taurine release under these experimental conditions is likely due to an enhanced reversal of the HIT cell endogenous taurine transporter. Failure of glucose (a stimulator for insulin secretion[14]) and 8-Br-cAMP (an activator of vesicular secretion in β-cells[10]) to stimulate taurine release in HIT cells rules out the possibility of any taurine release via the insulin secretary pathway.

Maximum level of taurine release was induced by ionomycin (IM) in the presence of Ca^{2+}. Ionomycin has been shown to increase expression of taurine transporter and an IM-sensitive promoter element has recently been identified[15]. Ionomycin has also been shown to evoke taurine release and elicit volume-sensitive Cl$^-$-channel activation in 9HTE$_o$- human tracheal cells[16]. The complete inhibition of IM-induced release by La^{3+} may be due to the block of Ca^{2+} entry and inhibition of the Cl$^-$ channel. Since 9-AC and DIDS, at concentrations known to blunt chloride channel activity, were unable to completely inhibit IM-induced increase in taurine release in HIT cells, the 9-AC and DIDS insensitive portion of the release may result from the IM-induced increase in taurine transporter expression. Thus, we propose that the mechanism for the IM-induced taurine release in the pancreatic β-cells may be due to a combined activation of the endogenous taurine transporter and the taurine permeable anion channel.

ACKNOWLEDGEMENTS

This work was in part supported by a generous gift from Dr. Vernon Knight (Baylor College of Medicine, Houston, TX) and in part by USDA/ARS Grant No. 58-6255-6001 (to HKS). Authors thank Cynthia Edwards for technical help with cell culture.

REFERENCES

1. Huxtable, R.J., 1992, *Physiol. Rev.* 72:101-163.
2. Briel, G., Gylfe, E., Hellman, B., and Neuhoff, V., 1972, *Acta Physiol. Scand.* 84:247-253.
3. Tokunaga, H., Yoneda, Y., and Kuriyama, K., 1983, *Eur. J. Pharmacol.* 18:237-243.
4. Cherif, H., Reusens, B., Dahri, S., Remacle, C., and Hoet, J.J., 1996, *J. Endocrinol.* 151:501-506.
5. Merezak, S., Hardikar, A.A., Yajnik, C.S., Remacle, C., and Reusens, B., 2001, *J. Endocrinol.* 171:299-308.
6. Blackard, W.G., Kikuchi, M., Rabinovitch, A., and Renold, A.E., 1975, *Am. J. Physiol.* 228:706-713.
7. Miley, H.E., Sheader, E.A., Brown, P.D., and Best, L., 1997, *J. Physiol.* 504:191-198.
8. Qian, X., Vinnakota, S., Edwards, C., and Sarkar, H.K., 2000, *Biochim. Biophys. Acta* 1509:324-334.
9. Vinnakota, S., Qian, X., Egal, H., Sarthy, V., and Sarkar, H.K., 1997, *J. Neurochem.* 69:2238-2250.
10. Takahashi, N., Kadowaki, T., Yazaki, Y., Ellis-Davies, G.C., Miyashita, Y., and Kasai, H., 1999, *Proc. Natl. Acad. Sci. USA* 96: 760-765.
11. Best, L., Sheader, E.A., and Brown, P.D., 1996, *Pflugers Arch.* 431:363-370.
12. Grant, A.C.G., Thomson, J., Zammit, V.A., and Shennan, D.B., 2000, *Mol. Cell. Endocrinol.* 162:203-210.
13. Drews, G., Zempel, G., Krippeit-Drews, P., Britsch, S., Busch G.L., Kaba, N.K., and Lang, F., 1998, *Biochim. Biophys. Acta* 1370:8-16.
14. Asfari, M., Janic, D., Meda, P., Li, G., Halban, P.A., and Wollheim, C.B., 1992, *Endocrinology* 130:167-178.
15. Han, X., Budreau, A.M., and Chesney, R.W., 2000, *Adv. Exp. Med. Biol.* 483:535-544.
16. Galietta, L.J., Falzoni, S., Di Virgilio, F., Romeo, G., and Zegarra-Moran, O., 1997, *Am. J. Physiol.* 273:C57-C66.

Regulation of Intestinal Taurine Transporter by Cytokines

MAKOTO SHIMIZU, TETSUNOSUKE MOCHIZUKI, and HIDEO SATSU
Department of Applied Biological Chemistry, The University of Tokyo, Bunkyo-ku, Tokyo 113-8657, Japan

1.   INTRODUCTION

Taurine is one of the most abundant free amino acids in animal cells/tissues and is thought to have such functions as antioxidation, anti-inflammation, osmoregulation and nerve regulation. Taurine transporter (TAUT) plays an important role in regulating the intracellular taurine concentration. The intestinal TAUT is particularly important because the intestinal epithelium is the tissue required for the absorption of dietary taurine in the body. We have studied regulatory properties of intestinal TAUT by using human intestinal epithelial Caco-2 cells as a model. Various aspects of TAUT regulation in Caco-2 cells have been observed.

1.1   Regulation of TAUT by Extracellular Taurine[1,2]

TAUT was down-regulated by culturing the cells in a taurine-rich medium and was up-regulated in a taurine-free (serum-free) medium as shown in Fig. 1a. This adaptive regulation was associated with changes in both Vmax and Km values of taurine transport. A change in the mRNA level of TAUT in this regulation was also observed. These results suggest that the intracellular taurine concentration is maintained at a certain level and TAUT is responsible for this regulation.

1.2   Regulation of TAUT under Hyperosmotic Conditions[2,3]

Caco-2 cells were cultured in the media containing 0-100 mM raffinose (corresponding to 325-442 mOsm/kg of H_2O) for 24 h. The activity of TAUT was increased with increasing the raffinose concentration (Fig. 1b).   This up-regulation was accompanied by the increase in the intracellular taurine content, suggesting that the cells under hypertonic conditions require higher concentrations of intracellular taurine, thereby protecting the cells.

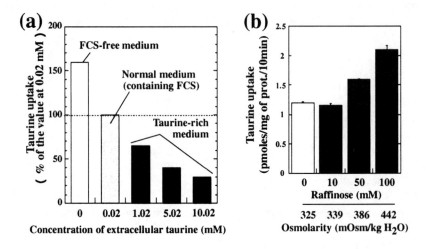

Figure 1. Effects of the extracellular taurine concentration (a) and osmotic pressure (b) on the taurine uptake activity of Caco-2 cells.

1.3   Regulation of TAUT by Food Factors[4,5]

TAUT in Caco-2 cells was also affected by food substances. We have reported a search for food factors that could enhance or suppress the TAUT activity. A factor that inhibited taurine uptake by Caco-2 cells was detected in an alcohol extract of sesame seeds; the active substance was identified as lysophosphatidylcholine (LPC) [4]. The inhibitory activity was dependent on the bound fatty acid, and surface-active LPC with fatty acids of C14 or longer was highly inhibitory. The phosphorylcholine residue was also likely to have played an important role[5]. These results suggest that the interaction of LPC with TAUT in the intestinal cell membrane is the cause of the reduced taurine uptake. LPC is easily produced from phosphatidylcholine by hydrolysis with phospholipase A2 secreted into the intestinal tract. Therefore, the lipid (phospholipid)-rich diet may interfere with the intestinal TAUT.

2.　　REGULATION OF TAUT BY CYTOKINES[6]

A highly developed immune system is known to work in the intestinal tract, where T lymphocytes, macrophages, fibroblasts and intestinal epithelial cells secret various cytokines in response to extracellular stimulation. Under such pathogenic conditions as inflammatory bowel diseases, the cytokine profile may dramatically change. It is therefore possible that intestinal epithelial cells are frequently exposed to unusually high concentrations of cytokines. However, the regulatory properties of intestinal TAUT by cytokines have not previously been reported. The regulation of taurine uptake by cytokines was therefore investigated.

Caco-2 cells after 14 days of culture were incubated with various cytokines for 24 h , and uptake experiments using [³H]taurine were then performed. Among the cytokines tested, TNF-α and IL-1β increased the taurine uptake activity as shown in Figure 2. Taurine uptake increased in a time-dependent manner from 0 to 24 h and reached a plateau after 24 h. The activity reached nearly 200% of the control value when the cells were exposed to a 10 ng/ml or higher concentration of TNF-α. Northern blot analysis indicated that the expression level of TAUT mRNA markedly increased by the TNF-α treatment in a time-dependent manner.

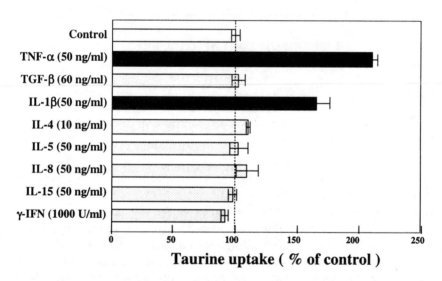

Figure 2. Effects of cytokines on the taurine uptake activity of Caco-2 cells.

TAUT is expressed not only in the intestines but also in other tissues[7]. We therefore investigated whether taurine uptake by human hepatic (HepG2), renal (HEK293), and macrophage-like (THP-1) cells was up-regulated by TNF-α. However, little or no change was observed in taurine uptake by these cells (Fig.

3), suggesting that the TNF- α-induced regulation was specific to the intestinal epithelial cells. However, Chang et al.[8], have recently reported that the taurine uptake by rat astrocytes was also increased by TNF-α. This finding may suggest that taurine plays a similar or common physiological role in these two types of cells.

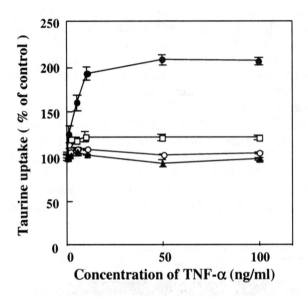

Figure 3. Effect of TNF-α treatment on the taurine uptake in Caco-2, HepG2, HEK293, and THP-1 cells. Three cell lines (Caco-2 (●), HepG2 (○), HEK293 (▲), THP-1 (□)) were precultured for 24 h in a medium containing various concentrations of TNF-α, respectively. Uptake experiments were then performed.

The detailed mechanisms for the TNF-α-induced up-regulation of TAUT are not known. Since TNF-α is known to activate nuclear factor κB, activation of the NF-κB pathway may be involved in this regulation. Our preliminary experiments have demonstrated that the TNF-α-induced increase in taurine uptake is diminished by treating the cells with specific inhibitors of the NF-κB pathway.

The physiological significance of this phenomenon is most interesting. Taurine is a well-known antioxidant and also an anti-inflammatory agent. It is reported that taurine blunted lipopolysaccharide-induced increases in TNF-α by Kupffer cells[9]. The intestinal epithelium is known to be an organ exposed to various cytokines secreted from immune cells beneath the epithelium. The term "controlled inflammation" is often used to express the physiological condition of the intestinal epithelium exposed to various proinflammatory cytokines. If taurine suppresses the excess production of inflammatory cytokines such as TNF-α, the up-regulation of TAUT in the intestinal epithelial cells by TNF-α would be a reasonable response.

3. CONCLUSION

We found for the first time that human intestinal TAUT was up-regulated by TNF-α and that the expression of TAUT mRNA was also up-regulated. Studying the regulatory mechanism and its physiological significance may lead to finding new functions for taurine.

ACKNOWLEDGEMENTS

The work was supported by a grant from Taisho Pharmaceutical Co. Ltd.

REFERENCES

1. Satsu, H., Watanabe, H., Asai, S., and Shimizu, M., 1997, Characterization and regulation of taurine transport in human intestinal cell, Caco-2. *J. Biochem.* 121: 1082-1087.
2. Shimizu, M., and Satsu, H., 2000, Physiological significance of taurine and taurine transporter in intestinal epithelial cells. *Amino Acids*, 19: 605-614.
3. Satsu, H., Miyamoto, Y., and Shimizu, M., 1999, Hypertonicity stimulated taurine uptake and transporter gene expression in Caco-2 cells. *Biochim. Biophys.Acta*, 1419: 89-96.
4. Ishizuka, K., Kanayama, A., Satsu, H., Miyamoto, Y., Furihata, K., and Shimizu, M., 2000, A taurine transport inhibitory substance in sesame seeds. *Biosci. Biotechnol. Biochem.*, 64: 166-1172.
5. Ishizuka, K., Miyamoto, Y., Satsu, H., Sato, R., and Shimizu, M., 2002, Characteristics of lysophosphatidylcholine in its inhibition of taurine uptake by human intestinal Caco-2 cells. *Biosci. Biotechnol. Biochem.*, 66: 730-736.
6. Mochizuki, T., Satsu, H., and Shimizu, M., 2002, Tumor necrosis factor a stimulates taurine uptake and transporter gene expression in human intestinal Caco-2 cells. *FEBS Lett.* 517: 92-96.
7. Satsu, H., and Shimizu, M., 2002, Characterization and regulation of taurine transport in human intestinal cell, Caco-2. *Amino acids.* 121: 1082-1087.
8. Chang, R.C., Stadlin, A., and Tsang, D., 2001, Effects of tumor necrosis factor alpha on taurine uptake in cultured rat astrocytes. *Neurochem. Intern.*, 38: 249-254
9. Seabra, V., Stachlewitz, R.F., and Thurman, R.G., 1998, Taurine blunts LPS-induced increases in intracellular calcium and TNF-alpha production by Kupffer cells. *J. Leukoc. Biol.*, 64: 615-621.

Part 4:

Taurine Synthesis, Localization, Determination and Nutrition

Determination of Taurine and Hypotaurine in Animal Tissues by Reversed-Phase High-Performance Liquid Chromatography after Derivatization with Dabsyl Chloride

HIRONORI NAKAMURA* and TOSHIHIKO UBUKA*

*Department of Clinical Nutrition, Kawasaki University of Medical Welfare, Kurashiki, Okayama, Japan

1. INTRODUCTION

Taurine and sulfate are major end products of L-cysteine metabolism in mammals[1], and they play various roles as tissue constituents and physiologically important compounds. Taurine is found in high concentrations in skeletal and heart muscle and brain[2]. It is utilized for the synthesis of taurine conjugates of bile acids. It is believed that most taurine is produced via the cysteinesulfinate pathway[3] and cysteamine pathway[4]. In both pathways, hypotaurine is an obligatory precursor leading to taurine. Hypotaurine content in animal tissues is far less than the taurine content. However, it has been reported that hypotaurine is found in rat brain[5], accumulates in the liver of cysteine-administered rats[3] and in the regenerating rat liver[6], increases in blood of rats injected with L-cysteine[7], and is excreted in the urine of rats fed L-cystine[8]. When hypotaurine was injected into rats, urinary taurine excretion increased in a dose-dependent manner[9].

It has been proposed that hypotaurine may act as an antioxidant[10, 11], but the mechanism of the oxidation of hypotaurine to taurine in the tissue is not fully understood[1,12]. Thus, studies on the physiological role(s) of hypotaurine independent of its role as a precursor of taurine are needed. In order to investigate the mode of oxidation and the physiological function(s) of hypotaurine, it is necessary to develop a method for the simultaneous microdetermination of taurine and hypotaurine in animal tissues.

Separation and determination of hypotaurine have been performed by means of paper chromatography[3], ion-exchange chromatography[13], amino acid analysis[6,7,14], radioactivity measurement[12], high-performance liquid chromatography (HPLC)[15,16] and gas chromatography[17]. In contrast to the stability of taurine, hypotaurine can be oxidized to taurine spontaneously[3,10] or via ultraviolet irradiation[18]. The instability of hypotaurine is a problem in determining its exact content.

Previously we reported a method for the determination of hypotaurine and taurine in rat urine using reversed-phase HPLC (RP-HPLC), in which hypotaurine oxidation during the analysis was minimized[19]. In this paper, we report a further development of simultaneous microanalysis of taurine and hypotaurine in animal tissues using RP-HPLC.

2.   METHODS

2.1   Preparation of Tissue Extracts

A rat tissue (up to 150 mg) placed in an Eppendorf tube was homogenized with 3 volumes of 10 mM solution of asparagine (an internal standard) in water and 6 volumes of acetone (final concentration, 60%) at 0 °C using a Polytron homogenizer (for approximately 30 sec at controller dial 2). The homogenate was placed in a freezer at –60 °C overnight. After thawing, coagulated proteins were eliminated by centrifugation at 10,000 xg for 15 min at 0 – 4 °C. The resulting supernatant was used for derivatization of amino acids.

2.2   Derivatization of Amino Acids

A 100 μl portion of the supernatant obtained above was mixed, in a 10 ml-round bottomed flask, with 400 μl of 0.1 M sodium bicarbonate buffer (pH 9.0) and 1,000 μl of 0.2 mM dabsyl chloride (4-dimethylamino-azobenzene-4'-sulfonyl chloride) solution in acetone. The mixture was incubated at 40 °C for 30 min and then it was evaporated to dryness with a flash evaporator at 40 °C. The resulting residue was dissolved in 300μl of 70% ethanol and filtered through a 0.2 μm filter. The filtrate was used for chromatography.

2.3   RP-HPLC

An RP-HPLC system consisted of a Hitachi LaChrom (Hitachi Co., Tokyo) and TSKgel ODS-80Ts (4.6mm diameter x 150 mm length) with a guard column TSKguardgel ODS-80Ts (3.2 mm diameter x 15 mm length) (Tosoh Co., Tokyo). The Hitachi LaChrom system was composed of a gradient pump L-7100, a UV-visible detector L-7420 and a data processor D-7500. The column temperature was maintained at 20 °C using a column oven 505 of FROM Co. (Tokyo). Ten μl of a sample were injected, and chromatography was performed using a mixture of 50 mM sodium acetate buffer (pH 4.00, solvent A) and acetonitrile (solvent B). The concentration of solvent B was increased as follows: 0-3 min, 28%; 3-23 min, linear gradient from 28-31%; 23-47 min, linear gradient from 31-32% (flow rate: 0.7 ml / min). Detection of dabsyl-amino acids was performed at 430 nm.

3.   RESULTS

Figs. 1a, and b show the effect of dabsyl chloride concentration on the formation of dabsyl-taurine and -hypotaurine when the dabsylation reaction was performed at 40 °C for 30 min. As shown in Fig. 1a, the formation of dabsyl–taurine increased with the increase in dabsyl chloride concentration. However, when hypotaurine was treated with dabsyl chloride, dabsyl-taurine was also formed at the same time as the formation of dabsyl-hypotaurine. This result indicates that hypotaurine was oxidized during the dabsylation reaction. Therefore, the reaction must be performed under conditions in which the oxidation of hypotaurine is minimal, namely, at a dabsyl chloride concentration of 0.13 mM.

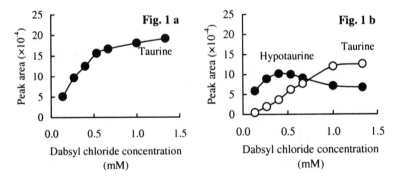

Figure 1. Effect of dabsyl chloride concentration on the dabsylation reaction of taurine (Fig. 1a) and hypotaurine (Fig. 1b). Dabsyl- taurine was formed during the reaction with hypotaurine (Fig. 1b).

Figs. 2a and b show the effect of temperature on the dabsylation reaction of taurine and hypotaurine when the reaction was performed for 30 min with 0.13 mM dabsyl chloride. The formation of dabsyl-taurine decreased with an increase in temperature. In the case of hypotaurine, the sum of dabsyl-hypotaurine and -taurine also decreased as the temperature increased. These results seem to indicate that with the increase in temperature the hydrolysis of dabsyl chloride reduced its concentration during the dabsylation reaction. When enough dabsyl chloride was used, such a decrease was not observed, and the oxidation of hypotaurine to taurine was increased. Therefore, a reaction temperature of 40 °C was chosen in the present study.

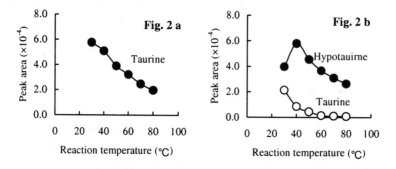

Figure 2. Effect of temperature on the dabsylation reaction of taurine (Fig. 2a) and hypotaurine (Fig. 2b). Dabsyl-taurine was formed during the reaction with hypotaurine (Fig.2b).

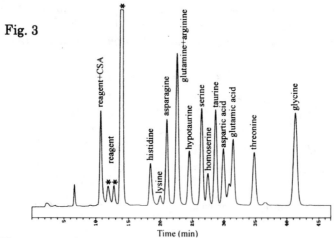

Figure 3. Chromatogram of dabsylated standard amino acids. *: Reagent peak.

Fig. 3 is a typical example of a chromatogram of standard dabsyl amino acids showing a good separation of the peaks. Dabsyl-hypotaurine, -taurine, and –asparagine peaks were clearly separated.

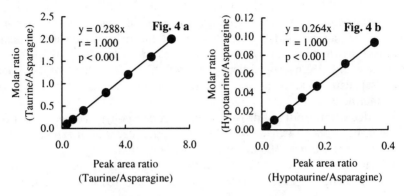

Figure 4. Standard curves of taurine (Fig.4a) and hypotaurine (Fig. 4b).

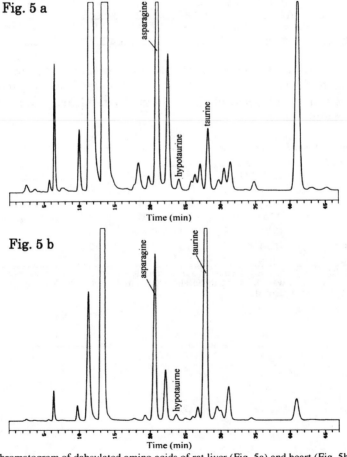

Figure 5. Chromatogram of dabsylated amino acids of rat liver (Fig. 5a) and heart (Fig. 5b).

Figs. 4a and b show the standard curves for taurine and hypotaurine, in which the molar ratio of taurine and hypotaurine relative to asparagine was plotted against the peak area ratio of taurine and hypotaurine to asparagine.

Figs. 5a and b are examples of chromatography of dabsylated amino acids of rat liver and heart, showing the good separation of dabsylated taurine, hypotaurine and the other amino acids. Identification of the dabsyl-hypotaurine peak was performed by treating dabsylated samples with hydrogen peroxide[19]. Upon treatment with 0.2% hydrogen peroxide at 40 °C for 2 h, the dabsyl-hypotaurine peak disappeared and there was a corresponding increase in the size of the dabsyl-taurine peak.

The detection limits of dabsyl-taurine and dabsyl-hypotaurine were 4.0 pmol at a signal-to-noise ratio of 3. Based on the present method, 1 nmol of hypotaurine or taurine / 100 mg of tissue was detected.

As shown in Table 1, the recovery of a known amount of taurine and hypotaurine using the present method was 103.3% and 101.8%, respectively,.

Table 1. Recovery of taurine and hypotaurine examined using liver

	Amount (μ mol / g)		
Compound	Added	Recovered	Recovery (%)
Taurine	29.84 ± 0.59	30.45 ± 0.64	103.3 ± 1.9
Hypotaurine	1.52 ± 0.03	1.54 ± 0.05	101.8 ± 5.4

Table 2 summaries the content of taurine and hypotaurine in some rat tissues. High amounts of taurine were contained in heart and skeletal muscle, but hypotaurine content was relatively low. In the liver, relatively high amounts of hypotaurine were detected.

Table 2. Content of taurine and hypotaurine in tissues of Wistar strain male rats (8 weeks of age) [a]

Tissue	Taurine	Hypotaurine	
	(μ mol / g)	(μ mol / g)	(% of taurine)
Liver	1.58 ± 0.50	0.18 ± 0.13	11.4
Heart	27.52 ± 3.03	0.11 ± 0.06	0.4
Kidney	10.71 ± 1.18	0.36 ± 0.11	3.4
Whole brain	6.34 ± 0.80	n.d.[b]	(0.0)
Skeletal muscle[c]	27.99 ± 2.84	d	(1.8)

a: mean ± SD, n = 5; b: not detected; c: soleus muscle; d: an unknown peak overlapped with hypotaurine peak which was estimated to be about 0.5 μ mol / g by the decrease of the peak by oxidation with hydrogen peroxide.

4. DISCUSSION AND CONCLUSION

Since taurine is a stable amino acid, it can be easily determined by an automatic amino acid analyzer and various methods using HPLC. On the other hand, as mentioned above, hypotaurine is easily oxidized to taurine either spontaneously[3, 10] or via ultraviolet irradiation[18]. The instability and low tissue contents of hypotaurine have hindered the development of methods for hypotaurine determination.

In the present method, deproteinization was performed with 60% acetone at minus 60 °C in Eppendorf tubes. Using this method, the sample volume could be reduced because the derivatization reaction in 60% acetone could be carried out successively. The temperature of derivatization with dabsyl chloride was performed at 40 °C and the concentration of dabsyl chloride was limited at 0.13 mM. These conditions minimized the oxidation of hypotaurine during the derivatization procedure. Since dabsyl-hypotaurine is as stable as the other dabsyl-amino acids, many samples can be prepared at the same time, and stored at room temperature for at least several weeks until HPLC analysis.

In conclusion, the present method enables microdetermination of hypotaurine and taurine in animal tissues. It should prove useful for further studies on the mechanism of hypotaurine oxidation and the physiological roles of hypotaurine and taurine.

REFERENCES

1. Griffith, O.W., 1987, Mammalian sulfur amino acid metabolism: Overview. *Methods Enzymol.* 143: 366-376.
2. Jacobsen, J.G., and Smith, L.H., Jr., 1968, Biochemistry and physiology of taurine and taurine derivatives. *Physiol. Rev.* 48: 424-511.
3. Awapara, J., 1953, 2-Aminoethanesulfonic acid: An intermediate in the oxidation of cysteine in vivo. *J. Biol. Chem.*, 203: 183-188
4. Cavallini, D., Scandurra, R., Dupre, S., Santoro, L., and Barra, D., 1976, A new pathway of taurine biosynthesis. *Physiol. Chem. Phys.* 8: 157-160.
5. Bergeret, B., and Chatagner, F., 1954, Sur la presence d'acide cysteinesulfinique dans le cerveau du rat normal. *Biochim. Biophys. Acta* 14: 297.
6. Sturman J.A., 1980, Formation and accumulation of hypotaurine in rat liver regenerating after partial hepatectomy. *Life Sci.*, 26: 267-272.
7. Yuasa, S., Akagi, R., and Ubuka, T., 1990, Determination of hypotaurine and taurine in blood plasma of rats after the administration of L-cysteine. *Acta Med. Okayama* 44: 47-50.

8. Cavallini, D., Mondovi, B., and De Marco, C., 1955, The isolation of pure hypotaurine from the urine of rats fed cysteine. *J. Biol. Chem.*, 216: 577-582.

9. Fujiwara, M., Ubuka, T., Abe, T., Yukihiro, K., and Tomozawa, M., 1995, Increased excretion of taurine, hypotaurine and sulfate after hypotaurine loading and capacity of hypotaurine metabolism in rats. *Physiol. Chem. Phys. & Med. NMR* 27: 131-137.

10. Fellman J.H., and Roth E.S., 1985, The biological oxidation of hypotaurine to taurine: hypotaurine as an antioxidant. In *Taurine: Biological Actions and Clinical Perspectives* (S.S. Oja, L. Ahtee, P. Kontro and M.K. Paasonen, eds.), Allan R. Liss, New York, pp. 71-82.

11. Green, T.R., Fellman J.H., Eicher, A.L., and Pratt, K.L., 1991, Antioxidant role and subcellular location of hypotaurine and taurine in human neutrophils. *Biochim. Biophys. Acta* 1073: 91-97.

12. Oja, S.S., and Kontro, P., 1981, Oxidation of hypotaurine in vitro by mouse liver and brain tissues. *Biochim. Biophys. Acta* 677: 350-357.

13. Cavallini, D., De Marco, C., and Mondovi, B., 1958, Experiments with D-cysteine in the rat. *J. Biol. Chem.*, 230: 25-30.

14. Perry, T.L., and Hansen, S., 1973, Quantitation of free amino compounds of rat brain: Identification of hypotaurine. *J. Neurochem.*, 21: 1009-1011.

15. Hirschberger, L.L., De La Rosa, J., and Stipanuk, M.H., 1985, Determination of cysteinesulfinate, hypotaurine and taurine in physiological samples by reversed-phase high-performance liquid chromatography. *J. Chromatogr.*, 343: 303-313.

16. Masuoka, N., Ubuka, T., Yao, K., Kinuta, M., and Wakimoto,M., 1993, Determination of 3, 5-dinitrobenzoylated taurine and hypotaurine by high-performance liquid chromatography. *Amino Acids*, 5: 133-134.

17. Kataoka, H., Yamamoto, H., Sumida, Y., Hashimoto, T., and Makita, M., 1986, Gas chromatographic determination of hypotaurine. *J. Chromatogr.*, 382: 242-246.

18. Ricchi, G., Dupre, S., Federici, G., Spoto, G., Matarese, R.M., and Cavallini, D., 1978, Oxidation of hypotaurine to taurine by ultraviolet irradiation. *Physiol. Chem. Phys.*, 10: 435-441.

Immunohistochemical Localization of Taurine in the Rat Stomach

NING MA, XIAOHUI DING, TATSUO MIWA and REIJI SEMBA
Department of Anatomy, Mie University School of Medicine, Tsu 514-0001, Mie, Japan

1.   INTRODUCTION

Taurine is one of the most abundant free amino acids in animal tissues. It is not incorporated into proteins but found mainly in the free form. Taurine has long been regarded as a mere end product of sulphur amino acid metabolism with no physiological role except that of bile acid conjugation. Yet, recent studies have implicated taurine in many physiological events, including osmoregulation, anti-oxidation, detoxification, membrane stabilization and neuromodulation[1-4]. However, its localization and function in the stomach has not been previously analysed in detail. The aim of this study is to reveal the distribution of taurine in the rat stomach, and to define taurine-containing cells in this organ by immunohistochemical techniques.

2.   MATERIALS AND METHODS

2.1   Antibody

Taurine specific antibody was prepared in our laboratory. In brief, taurine was coupled with rabbit serum albumin (RSA) via glutaraldehyde (GAL) and taurine-GAL-RSA conjugate, was emulsified with an equal volume of complete Freund's adjuvant and repeatedly injected intracutaneously into multiple sites on the back of a rabbit. Taurine antibody was purified by affinity chromatography with taurine immobilized on cellulofine GCL-2000m, and specificity of this antibody was examined by a dot GCL-immunobinding assay (Fig.1). The antibody demonstrated

Taurine 5: Beginning the 21st Century
Edited by Lombardini *et al.*, Kluwer Academic/Plenum Publishers, New York 2003

a significant reactivity only with taurine. To confirm immunoreactivity of the antibody in histological sections, rat cerebellar sections were studied with this antibody.As in previous studies,Purkinje cells and their dendrites exhibited intense immunostaining (Fig. 2)

2.2   Fixation and Sectioning of Tissue

Ten adult rats (five males and five females) were deeply anaesthetized with an IP injection of sodium pentobarbital and were perfused transcardially with fixatives. The composition of the fixatives used were 4% paraformaldehyde, 0.5% GAL for the double labelling studies and 4% paraformaldehyde, 1% GAL for the other studies. After perfusion, stomachs were removed and placed in the fixative for 4 hours. Then they were then rinsed several times with 50 mM Tris-HCl buffer, pH 7.6, dehydrated with a graded alcohol series and acetone, and embedded in paraffin. Five-μm thick sections were mounted on albumin-coated slides. Thirty-μm thick vibratome sections were also prepared.

Dot immunobinding assay

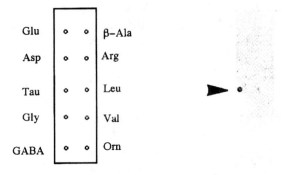

Figure 1. Dot immunobinding assay of taurine-specific antibody. Specificity of the antibody was studied by a dot immunobinding assay on a nitrocellulose membrane. Rabbit serum albumin conjugates of various amino acids, including glutamate (Glu), aspartate (Asp), taurine (Tau), glycine (Gly), GABA, β-alanine (β-Ala), arginine (Arg), leucine (Leu), valine (Val) and ornithine (Orn), were applied to a nitrocellulose membrane and incubated with an affinity-purified taurine antibody at a dilution of 0.1 μg/ml. Antigen-antibody reactions were visualized by the peroxidase-anti-peroxidase method. Purified taurine antibody gave a strong immunostaining on the spot of taurine-GAL-RSA (arrowhead).

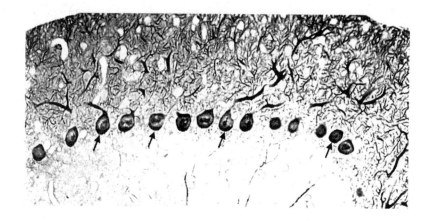

Figure 2. Immunohistochemical detection of taurine in the rat cerebellum. Purkinje cells (arrows) and their dendrites showed strong immunoreactivity.

2.3   Immunocytochemical Processing

Sections were deparaffinized in xylene and left in 0.3% H_2O_2 in methanol for 30 min. They were rehydrated through a graded alcohol series and washed twice for 5 min each in NaCl-TB (50mM Tris-HCl buffer containing 500 mM NaCl pH 7.6). Sections were incubated in normal goat serum that had been diluted 1:50 in NaCl-TB for 30 min at room temperature. The sections were then incubated with purified taurine antibody (0.1 μg/ml NaCl-TB) for 24 hours, and for 3 hours with antibodies against rabbit IgG that had been raised in goat (diluted 1:200 in NaCl-TB). After a 60 min incubation with the peroxidase-anti-peroxidase complex, they were washed in NaCl-TB. Finally, the sections were incubated with a solution that contained 3'-diaminobenzidine tetrahydrochloride (DAB) as chromogen.

2.4   Immunofluorescent Double Labelling

Double immunofluorescent labelling studies were performed to characterize the taurine-positive cells. Rabbit anti-taurine antibody and mouse anti-H^+, K^+-ATPase antibody were applied as primary antibody. The sections were then incubated with goat anti-rabbit IgG-Alexa 488 and goat anti-mouse IgG-Alexa 594. After washing, the sections were examined using a Nikon fluorescence microscope.

2.5   Electron Microscopic Immunocytochemistry

Thirty-μm-thick vibratome sections were treated with Lugol's iodine solution for 10 min, rinsed with 2.5% sodium thiosulfate, then incubated with 10% normal goat serum for 10 min. Sections were incubated with taurine antibody overnight at room temperature, washed with PBS, and left in 1-nm gold-labelled goat anti-rabbit IgG for 3 hours. The sections were then developed with the silver enhancement kit for 15 min and treated with 0.5% osmium tetroxide for 30 min. Areas of interest were selected, dehydrated, and embedded in Epon812 resin. Seventy-nm-thick ultra-thin sections were cut, counter-stained in uranyl acetate and lead acetate, and examined with an electron microscope H-800 (Hitachi).

3.   RESULTS

In the fundic glands, taurine immunoreactive cells were observed along the length of the gland. They appeared as large, pyramidal or spherical cells with a centrally oriented nucleus showing the typical morphology of parietal cells (Fig 3). Double immunofluoresecence labelling studies showed that those cells were also positive for H^+, K^+-ATPase, a marker for parietal cells (Fig 4). In the lamina propria, taurine was detected in vascular endothelial cells (Fig 3, 4). Some small cells scattered in the fundic glands were also immunolabeled (Fig 3). These cells were identified by electron microscopy as intraepithelial macrophages (Fig 5). In the myenteric plexuses, some of nerve cell bodies and their fibres were also labelled (Fig 6). No immunoreactive cells were found when non-immune serum was used instead of the taurine antibody.

4.   DISCUSSION

The distribution of taurine in the stomach was determined by means of immunohistochemical techniques. In our laboratory, the preparation of anti-taurine antibody, which has been previously utilised for immunohistochemical studies[5] and dot immunobinding assays, does not readily cross-react with other related amino acids.

The reason for the high concentration of taurine in parietal cells is not fully understood. There is evidence suggesting that taurine plays an important role in the stabilization of membranes[1, 2] and protects against excessive cell swelling and damage resulting from osmotic-linked shifts in water and ions.

It has been reported that taurine protects gastric mucosa from damage caused by

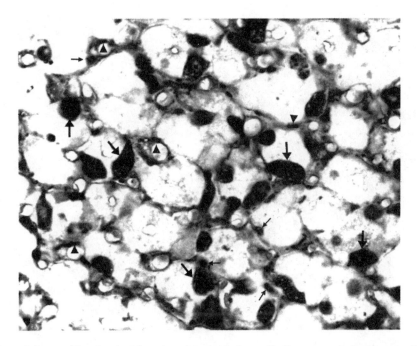

Figure 3. Immunohistochemical detection of taurine in the rat fundic mucosa. Intensely stained large cells (big arrows) were found in the fundic glands. Taurine was also detected in vascular endothelial cells (arrowheads). Some small cells (small arrows) scattered in the oxyntic glands were intensely immunolabelled.

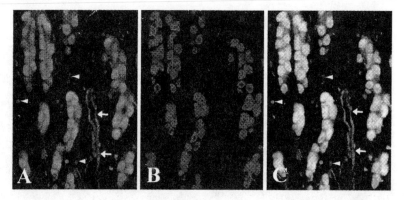

Figure 4. Double immunofluorescence labelling with taurine antibody (A) and H^+, K^+-ATPase antibody (B). Double-exposed micrograph (C) shows that the distribution pattern of taurine imunoreactive cells is similar to that of H^+, K^+-TPase-positive cells, the exception being the vascular endothelial cells (arrows) and other interstitial cells (arrowheads).

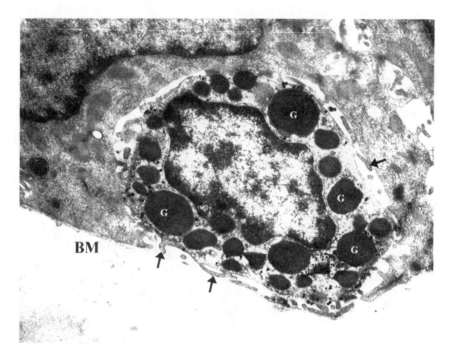

Figure 5. Observation of taurine immunoreactive small cells in the oxyntic gland by electron microscopy. Taurine immunoreactive cells containing many cytoplasmic processes (arrows) and granules (G) were identified as intraepithelial macrophage. BM: basement membrane.

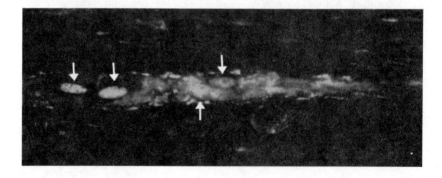

Figure 6. Some of the taurine immunoreactive cells were observed in the myenteric ganglia (arrows).

monochloramine[6]. Therefore, taurine stored in parietal cells may protect cells from self-destruction during events that generate oxidants.

There is an exceptionally high concentration of taurine in neutrophils[7], a cell can enzymatically produce oxidants during phagocytosis[8]. It has been reported that taurine is rich in macrophages in the male reproductive system[9,10]. The presence of taurine in gastric intraepithelial macrophages is consistent with these reports. The high concentration of taurine in the macrophage indicated that it may function as an antioxidant in the macrophage.

Taurine has been proposed to be a neurotransmitter or neuromodulator in the CNS[11]. The present finding that taurine was present in some neurons of the myenteric plexuses raises the possibility that taurine may function as a modulator of gastric motility.

5. CONCLUSION

Recent studies have shown that taurine may play a role in osmoregulation, membrane stabilization, detoxification, antioxidation and as an inhibitory neurotransmitter or neuromodulator. In the stomach, taurine has been shown to protect gastric mucosa against certain lesions. The aim of this study was to characterize the distribution of taurine in the stomach, and to define taurine-containing cells in this organ using immunohistochemical methods. Using affinity purified antibodies against taurine, it was shown that taurine is localized in parietal cells of the fundic glands. Some of nerve cell bodies and fibres in the myenteric plexuses were also labelled. Moreover, many stromal cells including fibroblasts, vascular endothelial cells and macrophages were also stained. Results of the present study imply that taurine may play an important role in gastric secretion and motility.

ACKNOWLEDGEMENTS

The excellent technical assistance of S. Ogawa is gratefully acknowledged.

REFERENCES

1. Gaull GE, Rassin DK., 1979, Taurine and brain development: human and animal correlate. In: Meisami E, Braziers MAB, eds. *Developmental neurobiology.* Raven Press, New York, pp.461-477.
2. Huxtable RJ., 1992, Physiological actions of taurine. *Physiol Rev.* 72:101-163.

3. Barbeau A, Inour N, Tsukada Y, Butterworth RF., 1975, The neuropharmacology of taurine. *Life Sci.* 17:669-677.
4. Wright CE, Tallan HH, Lin YY., 1986, Taurine: biological update. *Annu. Rev Biochem.* 55:427-453.
5. Ma N, Aoki E and Semba R., 1994, An immunohistochemical study of aspartate, glutamate, and taurine in rat kidney. *J Histochem Cytochem.* 42:621-626.
6. Kodama M, Tsukada H, OoyaM, Onomura M, Saito T, Fukuda K, Nakamura H, Taniguchi T, Tominaga M, Hosokawa M, Fujita J, Seino Y., 2000, Gastric mucosal damage caused by monochloramine in the rat and protective effect of taurine: endoscopic observation through gastric fistula. *Endoscopy.* 32:294-299.
7. Fukuda K, Hirai Y, Yoshida H, Nakajima T, Usui T., 1982, Free amino acid content of lymphocytes and granulocytes compared. *Clin Chem.* 28:1758-1761.
8. Clark RA, Klebanoff SJ., 1977, Myeloperoxidase—H_2O_2—halide system: cytotoxic effect on human blood leukocytes. *Blood.* 50:65-70.
    Lobo MV, Alonso FJ, Latorre A, del Rio RM., 2001, Immunohistochemical localization of taurine in the rat ovary, oviduct, and uterus. *J Histochem Cytochem.* 49:1133-1142.
10. Lobo MV, Alonso FJ, del Rio RM., 2000, Immunohistochemical localization of taurine in the male reproductive organs of the rat. *J Histochem Cytochem.* 48:313-320.
11. Hanretta AT, Lombardini JB., 1987, Is taurine a hypothalamic neurotransmitter?: A model of the differential uptake and compartmentalization of taurine by neuronal and glial cell particles from the rat hypothalamus. *Brain Res.* 434:167-201.

Modulation of Taurine on CYP3A4 Induction by Rifampicin in a HepG2 Cell Line

KYOKO TAKAHASHI, HIDEYASU MATSUDA, KAYOKO KINOSHITA, NORIKAZU MATSUNAGA, AKIHIKO SUMITA, TAKAHISA MATSUDA, KOICHI TAKAHASHI*, and JUNICHI AZUMA

*Clinical Evaluation of Medicine and Therapeutics, Graduate School of Pharmaceutical Sciences, Osaka University, Osaka JAPAN, and *Department of Pharmaceutics, School of Pharmaceutical Sciences, Mukogawa Women's University, Hyogo, JAPAN*

1.   INTRODUCTION

There has recently been an increased demand for nutritional supplements and health drinks containing taurine. The Japanese diet also provides high levels of taurine. In addition, 60% of the over-the-counter (OTC) drugs classified as vitamin-mixture medication contain taurine (Figure 1). Because taurine intake is so high, there has been concern that it might interact with other drugs. A common form of drug-drug interaction involves the induction or inhibition of specific drug-metabolizing enzymes such as cytochrome P450 (CYP)[1]. However, little is known about the influence of taurine on the metabolism of clinically important drugs. The antibiotics rifampicin (RFP) is a strong CYP3A4 inducer in humans[2]. In the present study we examined whether taurine could affect the levels of CYP3A4 mRNA in the presence and absence of RFP.

2.   MATERIALS AND METHODS

2.1   Cell Culture and Drug Treatment

HepG2 cells were cultured in defined medium as described previously[4]. Experiments initiated after 100-110 passages in culture. Cells were treated with reagents for 3 to 72 hr. After treatment, the cells were harvested and total RNA was obtained using the TRIzol RNA extraction kit (Life Technologies). The concentration of DMSO in the medium was 0.2% in both the control and the experimental groups.

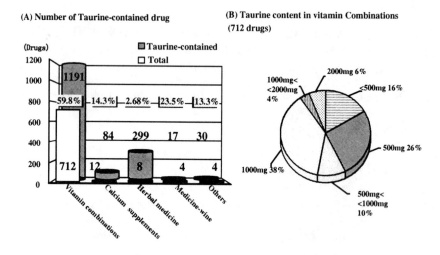

Figure 1. OTC-drugs Containing Taurine. These drugs were summarized from "Drugs in Japan: OTC-drugs 12ed. 2000-02" edited by Japan Pharmaceutical Information Center[3].

2.2   Primer Design and Generation of DNA Competitor

Forward and reverse primers were designed to amplify a 247 bp fragment from a specific region of the CYP3A4 gene and a 190 bp fragment of a competitor[4]. A pair of primers was also designed to amplify a 275 bp fragment from a specific region of the β-actin gene and a 340 bp fragment of a competitor[4]. The DNA competitors were generated using reagents supplied with a commercial kit (Competitive DNA Construction Kit, TaKaRa, Kyoto, Japan).

2.3 Reverse Transcription-Competitive PCR

cDNA synthesis was carried out as described previously[4]. Competitive PCR was performed using a series of 10-fold dilutions of a DNA competitor (CYP3A4 ranging from 10^1 to 10^3 copies/μl; β-actin ranging from 10^4 to 10^6 copies/μl) with a constant amount of the first-strand cDNA. The housekeeping gene β-actin was used for normalizing the expression of CYP mRNA to enable a cross-comparison among the different samples.

2.4 Intracellular Rifampicin Concentration with or without Taurine Pretreatment

The intracellular RFP content of HepG2 was measured by the procedure of Suzuki R et al[5] using an HPLC system consisting of an AS-400 Intelligent Auto Sampler, L-5020 Column Oven, L-6020 Intelligent Pump (Hitachi) and SI-1 NANOSPACE UV detector (Shiseido). The intracellular taurine content of cells was measured by the procedure of Jones & Gilligan[6] using a HPLC system (JASCO880-PU) equipped with a Hitachi F1000 fluorescence detector and Hitachi D-2500 integrator.

2.5 Analysis of PCR Products

Analysis of PCR products was performed according to the previously described protocol[4]. The copy number of each CYP isoform was expressed as copies/10^5 copies of β-actin.

2.6 Statistical Analysis

Statistical evaluation of data was performed by one-way analysis of variance followed by the Bonferroni/Dunn multiple range test. Significant differences from untreated control cells are indicated as follows : $*p < 0.05$ and $**p < 0.01$.

3. RESULTS AND DISCUSSION

3.1 Effects of 20mM Taurine on CYP3A4 mRNA Induction by RFP

HepG2 cells treated for 72 hr with 50μM RFP showed a 5-fold increase in

the amount of CYP3A4 mRNA in comparison with the untreated control cells
(Fig.2). Taurine treatment increased the induction of CYP3A4 mRNA by RFP in
a concentration-dependent manner (Fig.2); taurine (20 mM) doubled the levels of
CYP3A4 mRNA in the presence of RFP.   No effect was observed in cells treated
solely with taurine over a concentration range of 0.2-20 mM (not data shown).
Although RFP was capable of increasing CYP3A4 mRNA content of HepG2 cells
72 hr after the initial treatment in medium containing or lacking taurine, the
induction of CYP3A4 occurred more rapidly in the cells co-treated with RFP and
taurine (Fig.2). β-alanine, an analogue of taurine, did not affect CYP3A4 mRNA
expression (not data shown).

The impact of taurine on the induction of CYP3A4 mRNA by RFP was dose-
and time-dependent. Although previous studies have examined the effect of
taurine on mRNA induction of other proteins[7-9], this is the first report that taurine
affects the induction of CYPs mRNA.

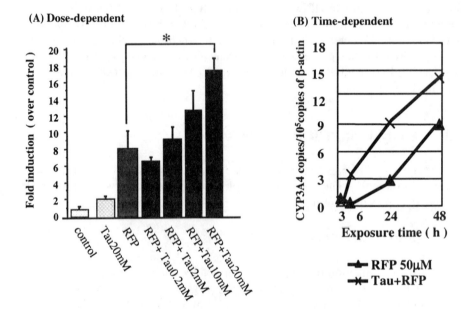

Figure 2. Dose- and Time-response of CYP3A4 mRNA Enhancement by Rifampicin and Taurine.
(A)Total RNA was extracted from HepG2 cells cultured for 72 hr with or without 50 μM RFP.
These cells were treated with 0.2, 2, 10 or 20 mM taurine before adding RFP. Values are means ±
SE of 8 different experiments.   Significance was assessed statistically by one-way analysis of
variance followed by Bonferroni/Dunn's PLSD multiple range test. *Significant difference between
groups (p< 0.01). (B) Total RNA was extracted from cells cultured for 3, 6, 24, or 48 hr with or
without 50 μM RFP. Each point is the mean of 2 different experiments.

3.2 The Influence of Taurine on the Cellular Responses induced by RFP in HepG2 cells

RFP is known to increase CYP3A4 mRNA content in a concentration-dependent manner[10,11]. We examined whether taurine would affect the accumulation of RFP by the HepG2 cell following 72 hr of incubation with 50 μM RFP. However, no significant difference in the concentration of RFP in RFP-treated (123 ± 45 mg/g protein) and RFP/taurine co-treated (156 ± 68 mg/g protein) cells was observed. This result strongly supports the notion that taurine might influence CYP3A4 mRNA induction rather than elevating the intracellular concentration of RFP in the HepG2 cell.

Subsequently, we measured the intracellular taurine content of cells exposed to medium containing both 20 mM taurine and either 0 or 50 μM RFP. RFP was found to significantly elevate (1.7-fold) the intracellular content of taurine (Fig.3).

We have previously reported that there was a significant correlation between enzyme activity and the mRNA content of CYP3A4 in human liver[12]. Although the HepG2 cell line derived from human hepatoblastoma expresses several drug-metabolizing enzymes, their activities are low[13]. One of these drug metabolizing enzymes in HepG2 cells is CYP 3A. As seen in Figure 3, RFP was found to induce CYP3A4, an effect that was enhanced in cells treated with taurine.

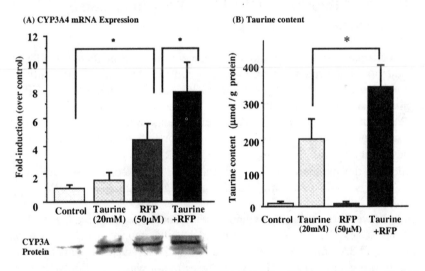

Figure 3. The Influence of Taurine on RFP-mediated Cellular Changes in HepG2 cells. (A) CYP3A protein expressed in HepG2 cells by Western blot analysis. The profile of CYP3A4mRNA expression is shown Fig.2. These cells were treated with 50 μM RFP, 20 mM taurine and the combination of RFP and taurine for 72 hr. (B)Effect of RFP on taurine content in HepG2 cells. Incubation time: 72 hr. *p<0.01.

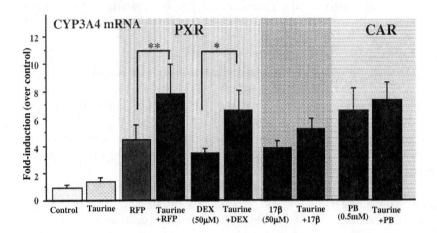

Figure 4. Effect of Taurine on CYP3A4 mRNA Expression in the Presence of Several Inducers. Cells were co-treted with taurine (20 mM) and several inducers for a period of 72 hr. HepG2 cells were pretreated for 24 hr with taurine before inducer treatment. Values are means ± SE of 4-7 different experiments. *p<0.05, **p<0.01.

3.3   Effects of Taurine on CYP3A4 mRNA Induction by Phenobarbital

The mechanism underlying the enhancement of RFP-mediated CYP3A4 mRNA induction by taurine remains an area of considerable interest. Recently, it was reported that the pregnane X receptor (PXR) was a key regulator of CYP3A transcription in a wide variety of species[14-16]. PXR and retinoid X receptor (RXR) heterodimer binds to DNA response elements and stimulates CYP3A4 gene transcription. PXR is a promiscuous receptor that mediates signals from xenobiotics and natural hormones, including RFP, pregnenolone 16 α - carbonitrile and dexamethasone[14-16]. On the other hand, constitutive activated receptor (CAR) has been implicated in the induction of CYP3A by phenobarbital[17,18].

Therefore, we were intrigued by the idea that taurine might activate a specific transcription factor. Therefore, we examined whether taurine would affect the induction of CYP3A4 mRNA by an inducer of CYP3A4 such as phenobarbital, 17β-estradiol or dexamethasone (Fig.4). 17β-estradiol was of particular interest because it induces CYP3A4 by activating both CAR and PXR. HepG2 cells were treated with the CYP3A4 inducers in the absence and presence of 20mM taurine. All of the inducers examined elevated the CYP 3A4 mRNA content of HepG2 cells. Interestingly, the amount of CYP3A4 mRNA in phenobarbital-treated cell was increased 7-fold relative to the control cell. However, taurine did not affect the induction of CYP3A4 mRNA by phenobarbital (Fig.4). Our results clearly demonstrate that taurine enhances RFP-mediated induction of CYP3A4 mRNA, but has no effect on the induction of CYP3A4 by phenobarbital. Thus, it is plausible that taurine influences PXR-modulated transcription activity, but not that of CAR.

These studies do not reveal the mode of taurine's effect on RFP-mediated induction of CYP3A4. RFP is known to inhibit the hepatocellular bile salt export pump, subsequently increasing the concentration of bile acids in the hepatocyte. As a result, RFP causes cholestasis in rat liver[19]. Since bile acids readily conjugate with taurine, their composition might be influenced by taurine[20,21].

4. CONCLUSION

In conclusion, we have shown that taurine enhances the induction of CYP3A4 by RFP. Therefore, attention should be paid to drug interactions that occur when taurine containing medicinal agents, food and drink are administered together with RFP or drugs that are metabolized mainly by CYP3A4. It is also important to emphasize that taurine may improve hepatic cholestasis caused by RFP and facilitate the detoxification of xenobiotics. Nonetheless, the *in vivo* significance of this effect should be examined.

REFERENCES

1. Chiba, H., 1995, Drug interaction via cytochrome P450. FARUMASHIA, 31, 922-996 .
2. Venkatesan, K., 1992, Pharmacokinetic drug interactions with rifampicin *Clin. Pharmacokinet.* 22, 47-65.
3. DRUGS IN JAPAN OTC-DRUGS 12ed. 2000-02, 2000, Japan Pharmaceutical Information Center, ed., JIHO, Tokyo, pp449-571.
4. Sumida, A., Fukuen, S., Yamamoto, I., Matsuda, H., Naohara, M. and Azuma, J. 2000, Quantitative analysis of constitutive and inducible CYPs mRNA expression in the HepG2 cell line using reverse transcription-competitive PCR. *Biochem. Biophys. Res. Commun.* 267,756-760.
5. Ohwada, S., Kishi, F., Hayashi, N., and Takada, M., 1994, The enhancement of rectal absorption of rifampicin by sodium *p*-aminosalicylic acid in a human. *Yakugaku zasshi*, 114, 894-900.
6. Jones, B.N. and Gilligan, J.P., 1983, *o*-Phthaldialdehyde precolumn derivatization and reversed-phase high-performance liquid chromatography of polypeptide hydrolysates and physiological fluids. *J. Chromatogr.* 266, 471-482.
7. Labudova, O., Yeghiazarjan, C., Hoger, H., Lubec, G., 1999, Taurine modulates expression of transporters in rat brain and heart. *Amino Acids.* 17,301-313.

8. Bitoun, M., Tappaz, M., 2000, Taurine down-regulates basal and osmolarity-induced gene expression of its transporter, but not the gene expression of its biosynthetic enzymes, in astrocyte primary cultures. *J. Neurochem.* 75,919-924.
9. Moenkemann, H., Labudova, O., Yeghiazarian, K., Rink, H., Hoeger, H., Lubec, G., 1999, Evidence that taurine modulates osmoregulation by modification of osmolarity sensor protein ENVZ--expression. *Amino Acids.* 17,347-355.
10. Kolars, J.C., Schmiedlin-Ren, P., Schuetz, J.D., Fang, C. and Watkins, P.B., 1992, Identification of rifampin-inducible P450IIIA4 (CYP3A4) in human small bowel enterocytes. *J. Clin. Invest.* 90,1871-1878.
11. Muntane-Relat, J., Ourlin, J.C., Domergue, J. and Maurel, P., 1995, Differential effects of cytokines on the inducible expression of CYP1A1, CYP1A2, and CYP3A4 in human hepatocytes in primary culture. *Hepatology* 22,1143-1153.
12. Sumida, A., Kinoshita, K., Fukuda, T., Matsuda, H., Yamamoto, I., Inaba, T. and Azuma, J. 1999, Relationship between mRNA levels quantified by reverse transcription-competitive PCR and metabolic activity of CYP3A4 and CYP2E1 in human liver. *Biochem. Biophys. Res. Commun.* 262,499-503.
13. Schuetz, E.G., Schuetz, J.D., Strom, S.C., Thompson, M.T., Fisher, R.A., Molowa, D.T., Li, D. and Guzelian, P.S., 1993, Regulation of human liver cytochromes P-450 in family 3A in primary and continuous culture of human hepatocytes. *Hepatology.* 18, 1254-.
14. Kliewer, S.A., Moore, J.T., Wade, L., Staudinger, J.L., Watson, M.A., Jones, S.A., McKee, D.D., Oliver, B.B., Willson, T.M., Zetterstrom, R.H., Perlmann, T. and Lehmann, J.M.,1998, An orphan nuclear receptor activated by pregnanes defines a novel steroid signaling pathway. *Cell* 92,73-82.
15. Lehmann, J.M., McKee, D.D., Watson, M.A., Willson, T.M., Moore, J.T.and Kliewer, S.A., 1998, The human orphan nuclear receptor PXR is activated by compounds that regulateCYP3A4 gene expression and cause drug interactions. *J. Clin. Invest.* 102,1016-1023.
16. Pascussi, J.M., Drocourt, L., Fabre, J.M., Maurel, P. and Vilarem, M.J., 2000, Dexamethasone induces pregnane X receptor and retinoid X receptor-alpha expression in human hepatocytes: synergistic increase of CYP3A4 induction by pregnane X receptor activators. *Mol. Pharmacol.* 58,361-372.
17. Sueyoshi, T. and Negishi, M., 2001, Phenobarbital response elements of cytochrome P450 genes and nuclear receptors. *Annu. Rev. Pharmacol. Toxicol.* 41,123-143.
18. Waxman, D.J.,1999, P450 gene induction by structurally diverse xenochemicals: central role of nuclear receptors CAR, PXR, and PPAR. *Arch. Biochem. Biophys.* 369,11-23.
19. Stieger, B., Fattinger, K., Madon, J., Kullak-Ublick, G.A. and Meier, P.J., 2000, Drug- and estrogen-induced cholestasis through inhibition of the hepatocellular bile salt export pump (Bsep) of rat liver. *Gastroenterology.* 118,422-430.
20. Hardison, W.G.and Proffitt, J.H., 1977, Influence of hepatic taurine concentration on bile acid conjugation with taurine. *Am. J. Physiol.* 232,E75-E79.
21. Hardison, W.G., 1978, Hepatic taurine concentration and dietary taurine as regulators of bile acid conjugation with taurine. *Gastroenterology* 75,71-75.

Effect of Acute Ethanol Administration on *S*-Amino Acid Metabolism: Increased Utilization of Cysteine for Synthesis of Taurine Rather Than Glutathione

YOUNG S. JUNG, HYE E. KWAK, KWON H. CHOI, and
YOUNG C. KIM*
Seoul National University, College of Pharmacy, San 56-1 Shinrim-Dong, Kwanak-Ku, Seoul, KOREA

Alterations in the hepatic metabolism of *S*-amino acids were examined in male rats injected with a single dose of ethanol (3 g/kg, ip). The hepatic concentrations of methionine and *S*-adenosylhomocysteine (SAH) were increased, but *S*-adenosylmethionine (SAM), cysteine, and glutathione (GSH) decreased rapidly following ethanol administration. The activities of methionine adenosyltransferase (MAT), cystathionine β-synthase (CβS) and cystathionine γ-lyase (CγL) were all inhibited. γ-Glutamylcysteine synthetase (GCS) activity was increased from t = 8 hr, but hepatic glutathione (GSH) level did not return to control for 48 hr. Both hepatic hypotaurine and taurine levels were increased immediately, which were reduced to below control from t = 18 hr. Changes in the serum concentration of taurine were consistent with results observed in the liver. Cysteine dioxygenase (CDO) activity was increased rapidly, but declined from t = 24 hr. The results indicate that an acute dose of ethanol induces significant alterations in the metabolism of *S*-amino acids in the liver. Ethanol depresses the cysteine availability for GSH synthesis not only by inhibiting the transsulfuration reactions but also by enhancing its irreversible catabolism to taurine via hypotaurine. The physiological significance of this finding is discussed.

1. INTRODUCTION

S-Amino acid metabolism occurs primarily by the transsulfuration pathway. The first step is formation of *S*-adenosylmethionine (SAM) that is then catalyzed by methionine adenosyltransferase (MAT). SAM serves as a methyl donor for biological methylation reactions and the co-product of

---

*Corresponding author

Taurine 5: Beginning the 21st Century
Edited by Lombardini *et al.*, Kluwer Academic/Plenum Publishers, New York 2003

transmethylation, *S*-adenosylhomocysteine (SAH), is hydrolyzed to yield homocysteine. Transsulfuration of homocysteine to cysteine via cystathionine is mediated by cystathionine β-synthase (CβS) and cystathionine γ-lyase (CγL), consecutively. Cysteine is irreversibly metabolized in liver to yield either taurine, inorganic sulfate, or glutathione (GSH). Cysteine dioxygenase (CDO) catalyzes the oxidation of this amino acid to cysteine sulfinate that is mainly converted to taurine via hypotaurine by the activity of cysteine sulfinate decarboxylase (CDC). Synthesis of GSH is mediated by γ-glutamylcysteine synthetase (GCS) and GSH synthetase.

It has been long known that alcohol consumption is associated with impaired *S*-amino acid metabolism. Patients with cirrhosis of different causes, including alcohol, often have hypermethioninemia and delayed plasma clearance of methionine. Reduction of GSH levels in the liver by a single or repeated treatment of ethanol has been well documented both in experimental animals and in humans. Various mechanisms have been proposed to explain the ethanol-induced inhibition of hepatic GSH. Since ethanol administration is associated with lipid peroxidative damage, it has been suggested that increased lipoperoxidation or binding to acetaldehyde would be responsible for the depletion of hepatic GSH[1,2]. It has been also suggested that ethanol reduces the GSH levels by increasing their efflux from the liver and/or by inhibiting the biosynthesis[3,4]. Recently we examined the time-course of changes in hepatic GSH following acute ethanol administration, and observed that the hepatic GSH efflux quantitatively played the most important role in the reduction of GSH[5].

In the present study we examined the effects of acute ethanol administration on the metabolism of *S*-amino acids by determining the metabolic products and the enzyme activities involved in the transsulfuration reactions. It was of special interest to evaluate the dynamics of cysteine, the essential substrate for the synthesis of GSH.

2.   METHODS

2.1   Animals and Treatments

Adult male Sprague-Dawley rats (Animal Breeding Center, Seoul National University), weighing 250–300 g, were used. Rats were acclimated in temperature (22 ± 2 °C) and humidity (55 ± 5 %) controlled rooms for 1 week prior to use. Laboratory rat chow and tap water were allowed *ad libitum*. Ethanol was diluted with physiological saline (40 % v/v).

2.2   Measurements of Metabolites and Enzyme Activities

Liver was homogenized in a 4-fold volume of 1 M perchloric acid. Denatured protein was removed by centrifugation. Total GSH concentration was determined using a HPLC method[6]. Cysteine levels were estimated by the acid-ninhydrin method[7]. The method of She et al.[8] was employed to determine SAM and SAH concentrations.

Liver was homogenized in a 5-fold volume of methanol. Serum was diluted with a 3 to 5-fold volume of methanol. Free amino acids, hypotaurine and taurine were determined using a HPLC system with a fluorescence detector after derivatization with *O*-phthalaldehyde/2-mercaptoethanol. Free amino acids were separated by using the method of Rajendra[9]. The method of Ide[10] was used to separate hypotaurine and taurine.

Liver was homogenized in a 3-fold volume of buffer consisting of 0.154 M KCl/50 mM Tris-HCl and 1 mM EDTA (pH 7.4). The homogenate was centrifuged at 10,000 g for 20 min. The supernatant fraction was further centrifuged at 104,000 g for 60 min. Enzyme activities were determined using the cytosol fraction.

MAT activity was estimated by quantifying the production of SAM and SAH. Reaction mixtures consisted of 80 mM Tris-HCl/50 mM KCl (pH 7.4), 5 mM ATP, 40 mM $MgCl_2$, 0.1 mM methionine and 1 mg protein of enzyme solution in a final volume of 1 ml. The incubation was carried at 37 °C for 30 min.

CβS activity was determined by the method of Kashiwamata and Greenberg[11]. The method of Matsuo and Greenberg[12] was used to determine CγL activity. GCS activity was determined using the method of Sekura and Meister[13]. Formation of γ-glutamylcysteine was quantified using a HPLC method[14] after *O*-phthalaldehyde derivatization. The method of Bagley et al.[15] was used to measure CDO activity. CDC activity was estimated by measuring the formation of hypotaurine. Protein (100 μg) was added to a test tube, made up to 0.1 ml of 0.1 M potassium phosphate buffer (pH 7.4) containing 5 mM dithiothretol and 0.2 mM pyridoxal 5-phosphate. The enzyme reaction, triggered by adding 50 μl 20 mM cysteine sulfinate, continued for 30 min.

3.   RESULTS AND DISCUSSION

This study examined the alterations in *S*-amino acid metabolism induced by acute ethanol administration in rats. The methionine concentration was increased from t = 4 hr (Table 1). However, the levels of SAM were decreased significantly, and did not return to control values for at least 48 hr. The hepatic cysteine levels were also reduced from t = 6 hr. Corresponding changes in the activities of enzymes involved in the transsulfuration reactions were observed (Table 2). The activity of MAT was reduced from t = 4 to 8 hr. Ethanol also decreased the activities of the two enzymes, CβS and CγL, critical for the synthesis of cysteine.

In rats given ethanol repeatedly, induction of the hepatic MAT and CβS activities was observed[16]. Recently increases in the mRNA level of both liver

Table 1. Effect of ethanol on concentrations of sulfur-containing amino acids and the metabolites

Hr	Group	Methionine	SAM	SAH	Cysteine	GSH	Hypotaurine	Taurine	Taurine
		nmole/g liver				μmole/g liver			μmole/ml serum
2	Control	64±2	108±4	25±2	106±4	6.70±0.16	0.20±0.02	2.45±0.43	0.36±0.02
	Ethanol	58±2	96±4	24±1	97±3	6.20±0.17	0.20±0.05	3.30±0.52	0.40±0.03
4	Control	57±6	107±5	24±2	103±2	6.30±0.17	0.16±0.01	3.40±0.33	0.38±0.02
	Ethanol	83±5*	110±5	27±3	104±4	4.70±0.13***	0.46±0.04**	5.06±0.25**	0.85±0.13*
6	Control	59±7	111±4	25±1	117±6	6.18±0.38	0.17±0.01	2.04±0.32	0.33±0.02
	Ethanol	79±3*	96±2**	26±2	85±3**	4.04±0.25**	0.46±0.06**	6.92±0.76**	0.92±0.05***
8	Control	66±3	101±5	22±1	121±8	5.87±0.21	0.19±0.02	2.57±0.61	0.40±0.04
	Ethanol	98±12*	85±7*	45±3***	100±2*	2.96±0.34***	0.90±0.16**	5.55±0.41**	1.32±0.09***
18	Control	N/D	111±2	26±2	108±5	6.88±0.21	0.17±0.02	2.43±0.67	N/D
	Ethanol		96±4**	33±3**	82±3**	5.41±0.21**	0.11±0.01**	2.04±0.58	
24	Control	62±4	108±5	25±2	106±5	6.68±0.19	0.20±0.02	3.25±0.77	0.44±0.05
	Ethanol	56±7	92±3*	37±1**	93±7	5.65±0.22*	0.12±0.02*	2.92±0.63	0.41±0.04
48	Control	64±5	99±4	23±1	108±3	7.19±0.18	0.22±0.02	2.21±0.42	0.40±0.03
	Ethanol	72±6	76±4**	23±2	86±6**	6.94±0.76	0.10±0.02**	1.09±0.13*	0.25±0.01**
120	Control	N/D	98±4	23±1	113±4	6.64±0.32	0.18±0.02	2.58±0.54	0.29±0.03
	Ethanol		99±2	25±1	123±4	7.10±0.12	0.19±0.01	1.01±0.21*	0.25±0.01*
168	Control	N/D	107±3	24±1	110±3	7.17±0.14	0.20±0.01	2.28±0.47	N/D
	Ethanol		105±6	26±1	121±1*	7.56±0.28	0.21±0.05	2.69±0.34	

Each value represents the mean ± SEM for more than four rats.

*, **, ***Significantly different from the control at $P < 0.05$, 0.01, 0.001, respectively (Student's t-test).

specific and non-liver specific MAT were demonstrated in rats fed with ethanol[17]. It was suggested that the increase in methionine catabolism reflects the need of cells to replenish GSH to prevent the cell components from ethanol-induced free radical damage[18]. But changes in the *S*-amino acid metabolism in early stages of alcoholic liver injury remain mostly unknown. In the present study both the MAT activity and the SAM generation in liver were depressed rapidly. The activities of CβS and CγL were also decreased. These results suggest that an acute dose of ethanol leads to inhibition of cysteine synthesis, which could affect the generation of final products in the transsulfuration pathway.

In liver the cysteine concentration is regulated by a balance between the rates of its synthesis, hepatic uptake from blood, and metabolism to GSH, inorganic sulfate and taurine[19]. The cysteine availability was suggested to be a major determinant for partitioning of cysteine sulfur to GSH, taurine or inorganic sulfates[20]. Low cysteine availability would favor its utilization for the synthesis of GSH; high cysteine availability enhances its catabolism to inorganic sulfate and taurine. Acute ethanol treatment decreased the hepatic cysteine concentration in this study. However, the consequence is contrary to the suggestion made earlier in that the decrease in cysteine is accompanied by a massive elevation in hypotaurine and taurine concentrations.

Table 2. Effect of ethanol on activities of MAT, CβS, CγL, GCS, CDO and CDC in liver

Hr	MAT	CβS	CγL	GCS	CDO	CDC
2	91±9	92 ± 9	77 ± 8**	92 ± 9	96.4 ± 10	115 ± 12
4	75± 0*	110 ± 9	78 ± 4*	103 ± 10	135 ± 6*	95 ± 10
6	75± 0**	110 ± 4	88 ± 2*	100 ± 4	162 ± 18*	128 ± 8*
8	73 ± 9*	84 ± 11*	93 ± 12	147 ± 18*	116 ± 15	95 ± 12
24	100± 0	102 ± 3	83 ± 4	142 ± 5**	63.0 ± 7**	99 ± 7
48	110 ± 0	99 ± 2	89 ± 6	186 ± 5***	39.1 ± 9**	95 ± 9
120	100 ± 8	111 ± 13	109 ± 4	106 ± 4	66.7 ± 8**	109 ± 8
168	100 ± 0	111 ± 8	110 ± 13	103 ± 9	117 ± 9	80 ± 13

Enzyme activities were shown as percents (%) of the control levels determined simultaneously. Each value represents the mean ± SE for more than four rats.
*,**,*** Significantly different from the control at $P < 0.05, 0.01, 0.001$, respectively (Student's *t*-test).

The hepatic GSH levels were reduced significantly following ethanol treatment. On the other hand, the hepatic hypotaurine/taurine levels were increased rapidly. Elevation of hypotaurine was reversed at t = 18 hr. The taurine level was decreased below control from t = 48 hr.

Synthesis of GSH in liver is limited mostly by two factors, the availability of cysteine and the activity of GCS[21]. GCS, which catalyzes a rate limiting step in the synthesis of GSH, is regulated by the feedback inhibition of GSH and also by the availability of its substrate, cysteine. The GCS activity was elevated from t = 8 hr, which could account for the gradual recovery in hepatic GSH. On the other hand, cysteine was decreased rapidly suggesting that the reduction of cysteine availability plays a significant role in the ethanol-induced decrease of the GSH levels in the liver.

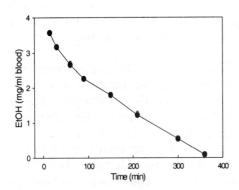

Fig 1. Disappearance of ethanol from blood

Of particular interest is the remarkable elevation of hypotaurine/taurine syntheses. Ethanol was cleared from blood linearly at 0.50 mg ± 0.02 mg/ml/hr, and consequently, disappeared completely in 8 hr (Fig. 1). Therefore, the increase in hypotaurine/taurine concentrations was significant while ethanol was being actively metabolized. Serum taurine levels were also elevated indicating that the increase in hepatic taurine could not be explained by impediment of the transport of this substance across the membranes. Considering the well-known role of GSH in the detoxification of reactive oxygen species and reactive metabolites, the ethanol-induced increase in the hypotaurine/taurine syntheses at the expense of the reduced generation of this tripeptide is paradoxical.

It has been known that taurine exhibits considerable cytoprotective effects, which are often attributed to its antioxidant potential[22]. However, taurine has minimal direct scavenging activities against oxygen-derived radicals[23]. Instead, hypotaurine at a concentration of several hundred μM has been shown to be an excellent scavenger of reactive oxygen species including hydroxyl radical, hypochlorous acid, peroxynitrite and singlet oxygen[24].

Ethanol is first biotransformed to acetaldehyde by several enzyme systems. Acetaldehyde itself or oxygen radicals generated during breakdown of ethanol have been suggested to be responsible for the initiation of lipid peroxidation[25]. Accordingly the ethanol-induced lipid peroxidative damage would increase the demand for antioxidant potential, which has been proposed as the mechanism of GSH depletion by some authors[1,2]. However, the results in this study show that cysteine, the essential substrate for GSH synthesis, is preferentially utilized for synthesis of taurine rather than GSH in liver challenged with ethanol.

In conclusion the present study shows that a single dose of ethanol causes significant alterations of *S*-amino acid metabolism, some of which persist for a much longer period beyond disappearance of ethanol from blood. Inhibition of the transsulfuration reactions by ethanol leads to reduction of the SAM and cysteine concentrations. The decrease in cysteine availability for GSH synthesis is further aggravated by the increased utilization of this *S*-amino acid for the production of taurine, which appears to play a significant role in the depletion of hepatic GSH. The significance of the elevation in hypotaurine/taurine syntheses remains to be investigated. Whether this is a mere consequence of ethanol effect or due to a potential role of cysteine catabolism to taurine in alcoholic liver injury is being studied in this laboratory.

ACKNOWLEDGEMENTS

This work was supported in part by a grant (HMP-00-CD-02-0004) from the Ministry of Health and Welfare, Korea.

REFERENCES

1. Kera, Y., Komura, S., Ohbora, Y., Kiriyama, T., and Inoue, K., 1985, Ethanol induced changes in lipid peroxidation and nonprotein sulfhydryl content. Different sensitivities in rat liver and kidney. *Res. Commun. Chem. Pathol. Pharmacol.* **47**:203-209.
2. Videla, L.A., Fernandez, V., Ugarte, G., and Valenzuela, A., 1980, Effect of acute ethanol intoxication on the content of reduced glutathione of the liver in relation to its lipoperoxidative capacity in the rat. *FEBS Lett.* **111**:6-10.
3. Fernandez-Checa, J.C., Ookhtens, M., and Kaplowitz, N., 1987, Effect of chronic ethanol feeding on rat hepatocytic glutathione. Compartmentation, efflux, and response to incubation with ethanol. *J. Clin. Invest.* **80**:57-62.
4. Speisky, H., MacDonald, A., Giles, G., Orrego, H., and Israel, Y., 1985, Increased loss and decreased synthesis of hepatic glutathione after acute ethanol administration. Turnover studies. *Biochem. J.* **225**:565-572.
5. Choi, D.W., Kim, S.Y., Kim, S.K., and Kim, Y.C., 2000, Factors involved in hepatic glutathione depletion induced by acute ethanol administration. *J. Toxicol. Environ. Health A* **60**:459-469.
6. Neuschwander-Tetri, B.A., and Roll, F.J., 1989, Glutathione measurement by high-performance liquid chromatography separation and fluorometric detection of the glutathione-orthophthalaldehyde adduct. *Anal. Biochem.* **179**:236-241.
7. Gaitonde, M.K., 1967, A spectrophotometric method for the direct determination of cysteine in the presence of other naturally occurring amino acids. *Biochem. J.* **104**:627-633.
8. She, Q.B., Nagao, I., Hayakawa, T., and Tsuge, H., 1994, A simple HPLC method for the determination of S-adenosylmethionine and S-adenosylhomocysteine in rat tissues: the effect of vitamin B6 deficiency on these concentrations in rat liver. *Biochem. Biophys. Res. Commun.* **205**:1748-1754.
9. Rajendra, W., 1987, High performance liquid chromatographic determination of amino acids in biological samples by precolumn derivatization with O-phthaldehyde. *J. Liq. Chromat.* **10**:941-955.
10. Ide, T., 1997, Simple high-performance liquid chromatographic method for assaying cysteinesulfinic acid decarboxylase activity in rat tissue. *J. Chromatogr. B Biomed. Sci. Appl.* **694**:325-332.
11. Kashiwamata, S., and Greenberg, D.M., 1970, Studies on cystathionine synthase of rat liver. Properties of the highly purified enzyme. *Biochim. Biophys. Acta.* **212**:488-500.

12. Matsuo, Y., and Greenberg, D.M., 1957, A crystalline enzyme that cleaves homoserine and cystathionine. I. isolation procedure and some physicochemical properties. *J. Biol. Chem.* **230**:545-560.

13. Sekura, R., and Meister, A., 1977, γ-Glutamylcysteine synthetase. Further purification, "half of the sites" reactivity, subunits, and specificity. *J. Biol. Chem.* **252**:2599-2605.

14. Yan, C.C., and Huxtable, R.J., 1995, Fluorimetric determination of monobromo-bimane and O-phthalaldehyde adducts of γ-glutamylcysteine and glutathione: application to assay of γ-glutamylcysteinyl synthetase activity and glutathione concentration in liver. *J. Chromatogr. B Biomed. Appl.* **672**:217-224.

15. Bagley, P.J., Hirschberger, L.L., and Stipanuk, M.H., 1995, Evaluation and modification of an assay procedure for cysteine dioxygenase activity: High-performance liquid chromatography method for measurement of cysteine sulfinate and demonstration of physiological relevance of cysteine dioxygenase activity in cysteine catabolism. *Anal. Biochem.* **227**:40-48.

16. Finkelstein, J.D., Cello, J.P., and Kyle, W.E., 1974, Ethanol-induced changes in methionine metabolism in rat liver. *Biochem. Biophys. Res. Commun.* **61**:525-531.

17. Lu, S.C., Huang, Z.Z., Yang, J.M., and Tsukamoto, H., 1999, Effect of ethanol and high-fat feeding on hepatic γ-glutamylcysteine synthetase subunit expression in the rat. *Hepatology* **30**:209-214.

18. Trimble, K.C., Molloy, A.M., Scott, J.M., and Weir, D.G., 1993, The effect of ethanol on one-carbon metabolism: increased methionine catabolism and lipotrope methyl-group wastage. *Hepatology* **18**:984-989.

19. Garcia, R.A., and Stipanuk, M.H., 1992, The splanchnic organs, liver and kidney have unique roles in the metabolism of sulfur amino acids and their metabolites in rats. *J. Nutr.* **122**:1693-1701.

20. Kwon, Y.H., and Stipanuk, M.H., 2001, Cysteine regulates expression of cysteine dioxygenase and γ-glutamylcysteine synthetase in cultured rat hepatocytes. *Am J. Physiol. Endocrinol. Metab.* **280**:E804-815.

21. Huang, C.S., Chang, L.S., Anderson, M.E., and Meister, A., 1993, Catalytic and regulatory properties of the heavy subunit of rat kidney γ-glutamylcysteine synthetase. *J. Biol. Chem.* **268**:19675-19680.

22. Huxtable, R.J., 1992, Physiological actions of taurine. *Physiol. Rev.* **72**:101-163.

23. Mehta, T.R., and Dawson, R. Jr., 2001, Taurine is a weak scavenger of peroxynitrite and does not attenuate sodium nitroprusside toxicity to cells in culture. *Amino Acids* **20**:419-433.

24. Shi, X., Flynn, D.C., Porter, D.W., Leonard, S.S., Vallyathan, V., and Castranova, V., 1997, Efficacy of taurine based compounds as hydroxyl radical scavengers in silica induced peroxidation. *Ann. Clin. Lab. Sci.* **27**:365-374.

25. Castillo, T., Koop, D.R., Kamimura, S., Triadafilopoulos, G., and Tsukamoto, H., 1992, Role of cytochrome P-450 2E1 in ethanol-, carbon tetrachloride- and iron-dependent microsomal lipid peroxidation. *Hepatology* **16**:992-996.

Effects of Dietary Taurine Supplementation on Hepatic Morphological Changes of Rats in Diethylnitrosamine-Induced Hepatocarcinogenesis

KYUNG JA CHANG, CHAI HYEOCK YU*, and MIWON SON**

*Department of Food and Nutrition, Inha University, Incheon, Korea, *Department of Biology, Inha University, Incheon, Korea,**Research Laboratories of Dong-A Pharmaceutical Co., Ltd., Kyunggi-do, Korea*

1. INTRODUCTION

Taurine (2-aminoethane sulfonic acid) is a sulfur-containing β-amino acid that is the most abundant free amino acid in many mammalian tissues. It has many physiological functions including conjugation with bile acids, modulation of calcium levels, maintenance of osmolarity, antioxidation, membrane protection, and many others[1,3]. It has been reported that both endogenous and exogenous taurine were effective in cancer prevention through antioxidation in the pro-oxidation condition induced in carcinogenesis[4,5] which may result in damage to the cellular antioxidant defense enzymes as well as the membranes[6]. It has also been reported that taurine showed chemoprevention against colon and hepatic cancer[7]. The protection of hepatocytes by taurine showed a correlation with the inhibition of lipid peroxidation[8] and ornithine decarboxylase activity[9]. However, there were insignificant results in the chemopreventative effects of taurine[10]. Therefore, the present study was conducted in order to examine the effects of dietary taurine supplementation at various levels and taurine depletion by 5% dietary β-alanine on hepatic morphological changes during diethyl-nitrosoamine(DEN)-induced carcinogenesis.

2.   MATERIALS AND METHODS

2.1   Animals and Diet

Weanling male Sprague-Dawley rats weighing 40-60g were supplied from Animal Care Facility of Seoul National University (Seoul, Korea), and they were kept in polycarbonate cages in a room with controlled temperature ($23\pm2°C$), humidity ($55\pm5\%$), and light-dark-cycle (0700-1900). According to the experimental diets and treatments, they were divided into 6 groups (Figure 1), and fed one of the experimental diets for 11 weeks. All groups, except the control group, were given an intraperitoneal injection of DEN (200mg/kg body weight; Sigma, U.S.A) at 6-weeks old, and 2/3 partial hepatectomy was performed after 3 weeks. The animals were sacrificed by decapitation at the 11th week, and their livers were used for morphological observations.

The liver of each rat was removed and tissue samples were fixed in Bouin solution for paraffin sections, and prefixed in 2.5% glutaraldehyde and postfixed in 1% osmium tetroxide. Following dehydration, the samples were treated with Epon 812. Thin sections were stained with uranyl acetate and lead citrate and examined with a Jeol-100CX electron microscope.

Taurine or β-alanine was added to the basal diet (Table 1). The levels of taurine supplemented were 1, 2, and 3% (w/w), and that of β-alanine was 5%. The basal diet was composed of 20% casein protein (Sigma, U.S.A), 54.7% corn starch, 15% corn oil, 5% α-cellulose, 3.5% AIN-76 Mineral mixture (ICN, U.S.A.), 1% AIN-76 fortified Vitamin mixture (ICN, U.S.A), and 0.3% DL-methionine, of which calorie density was 433.8kcal per 100g diet. DEN-C group (DEN control group) were fed the basal diet; DEN-T1, DEN-T2, and DEN-T3 groups were fed the 1, 2, and 3% taurine diet, respectively; DEN-B group 5% β-alanine diet.

3.   RESULTS AND DISCUSSION

The hepatocytes of the carcinogen-treated rats not exposed to taurine contained normal-shaped nuclei, but the nuclear membrane and endoplasmic reticulum (ER) were almost destroyed, and the mitochondria (Mi) were partially destroyed (Figure 2). By contrast, the hepatocytes from the 1% dietary taurine-supplemented groups contained some irregular contour nuclei, normal ER and Mi. However, the hepatocytes from the 2% dietary taurine-supplemented groups contained some irregular contour nuclei, partially-destroyed ER and normal Mi.

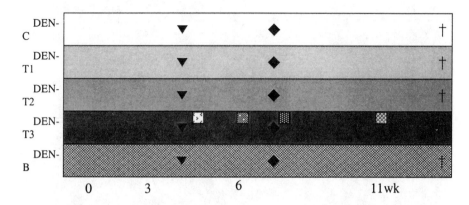

Figure 1. Experimental protocol. Hepatocarcinogenesis was induced by i.p. injection of DEN (▼), and 2/3 partial hepatectomy (◆). All animals were fed with experimental diet (basal diet, □; 1, 2, or 3% taurine supplemented diet, ▨, ▩,▦;5% β-alanine diet, ▨) for 11 weeks, and sacrificed (+).

Table 1. Experimental diet composition

Component	(g/100g diet)
Corn starch	54.7
Casein	20.0
Corn oil	15.0
α-Cellulose	5.0
Mineral Mixture	4.0
Vitamin Mixture	1.0
DL-Methionine	0.3

Taurine was added to 100g diet at the levels of 1, 2, or 3g. β-Alanine was added with 5g.

The hepatocytes from the 3% dietary taurine-supplemented groups contained some irregular contour nuclei, swollen ER and normal Mi. In the hepatocytes from 5% β-alanine taurine-depleted groups, ER was more swollen and glycogen granules were darker compared to other groups.

Taurine is one of the organosulfur compounds with antioxidant activity, and its biological functions are closely related to the chemopreventive potential. These results suggest that the chemopreventive potential of taurine still remains and more than a 1% dietary level of taurine appears to be excessive. In the case of hepatic cancer prevention, taurine supplementation with a pharmaceutical dose,

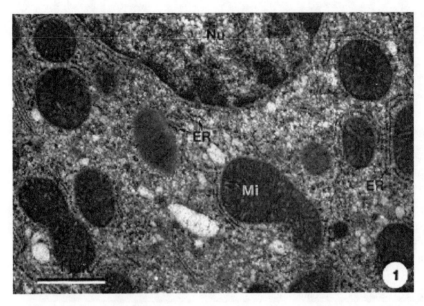

Figure 2-1. Effect of taurine on hepatocyte morphology of carcinogen-treated rat, DEN- 0% taurine group (DEN-C)×19,000.

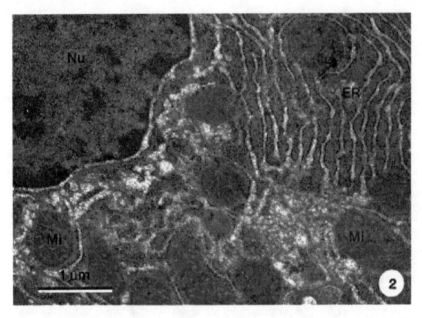

Figure 2-2. Effect of taurine on hepatocyte morphology of carcinogen-treated rat, DEN- 1% taurine (DEN-T1) ×21,000.

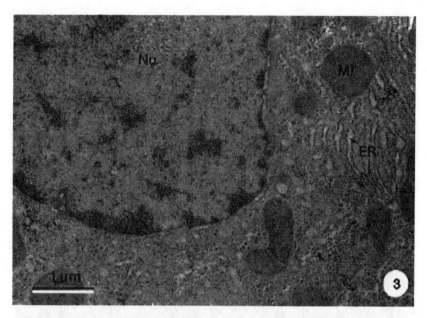

Figure 2-3. Effect of taurine on hepatocyte morphology of carcinogen-treated rat, DEN- 2% taurine (DEN-T2) ×17,000.

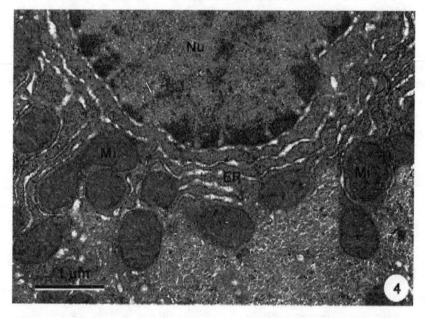

Figure 2-4. Effect of taurine on hepatocyte morphology of carcinogen-treated rat, DEN- 3% taurine (DEN-T3) ×19,000.

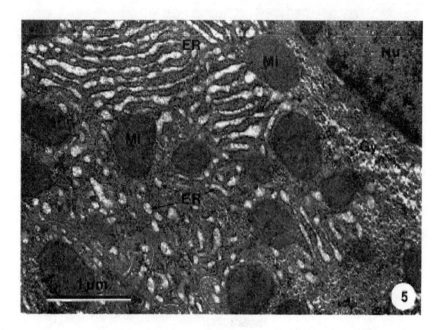

Figure 2-5. Effect of taurine on hepatocyte morphology of carcinogen-treated rat, DEN- % β-alanine (DEN-B) ×24,000.

as much as 2 or 3% of total dietary intake, might induce proliferation of cancer cells initiated by DEN. Nevertheless, the potency of the chemopreventive property owing to taurine at less than 1%, remains, which is suggested from the fact that a 1% taurine diet worked as an activator of GST and increased GSH levels. Further studies should be initiated to investigate the effect of taurine with lower than 1% levels on the prevention of hepatic cancer, and the role of taurine in the proliferation and growth of cancer cells.

ACKNOWEDGEMENTS

This work was supported by a Korea Research Foundation Grant, 1998-1999. We thank Dong-A Pharmaceutical Co. for donation of taurine.

REFERENCES

1.   Huxtable, R.J., 1992, Physiological actions of taurine. *Physiol Rev* 72: 101-163.
2.   Redmond, H.P., Stepleton, P.P., Neary, P., Bouchier-Hayes, D., 1998, Immunonutrition: The role of taurine. *Nutrition* 14: 599-604.

3. Monte, M.J., El-Mir, M.Y., Sainz, G.R., Bravo, P., Marin, J.J., 1997, Bile acid secretion during synchronized rat liver regeneration. *Biochim. Biophys. Acta*. 1362, 56-66.
4. Raschke, P., Massoudy, P., Becker, B.F., 1995, Taurine protects the heart from neutrophil-induced reperfusion injury. Free Radic. *Biol. & Med*. 19, 461-471.
5. Cunningham, C., Tipton, K. F., Dixon, H. B. F., 1998, Conversion of taurine into N-chlorotaurine and sulphoacetaldehyde in response to oxidative stress. *Biochem, J*. 330, 939-945.
6. Koh, Y. D., Yoon, S. J., Park, J.-W., 1999, Inactivation of copper, zinc superoxide dismutase by the lipid peroxidation produce malondialdehyde and 4-hydroynonenal *J Biochem. Mol. Biol*. 32, 440-444.
7. Reddy, B. S., Rao, C. V., Rivenson, A., Kelloff, G., 1993, Chemoprevention of colon carcinogenesis by organosulfur compouds. *Cancer Res*. 53, 3493-3498.
8. You, J. S. and Chang , K. J. 1998, Taurine protects the liver against lipid peroxidation and membrane disintegration during rat hepatocarcinogenesis. *Ady. Exp. Med. Biol*. 442, 105-112.
9. Okamoto, K., Sugie, S., Ohnishi, M., Makita, H., Kawamori, T., Watanabe, T., Tanaka, T., Mori, H., 1996, Chemopreventive effects of taurine on diethylnitrosamine and Phenobarbital-induced hepatocarcinogenesis in male F344 rats. *Jpn. J. Cancer. Res*. 87, 30-36.
10. Lubec, B., Hoeger, H., Kremser, K., Amann, G., Koller, D. Y., and Gialamas, J., 1996, Decreased tumor incidence and increased survival by one year oral low dose arginine supplementation in the mouse. *Life Sciences* 58, 2317-2325.

Effect of Taurine on Cholesterol Degradation and Bile Acid Pool in Rats Fed a High-Cholesterol Diet

WEN CHEN[1], NAOMICHI NISHIMURA[2], HIROAKI ODA[3] and HIDEHIKO YOKOGOSHI[1]

[1]School of Food and Nutritional Sciences, The University of Shizuoka, Shizuoka 422-8526
[2]Department of Human Life and Development, Nayoro City College, Nayoro 096-8641 and
[3]Department of Applied Biological Sciences, Nagoya University, Nagoya 464-8601, Japan

1. INTRODUCTION

Taurine, one of the most abundant free amino acids in many animals tissues, is well known to have several beneficial physiological actions, including antioxidation, detoxification, osmoregulation, cell membrane stabilization, and neuromodulation, etc[1-5]. In lipid metabolism, the function of taurine is considered primarily to be the conjugation with bile acids in the liver[6]. Bile acid is the degrading metabolite of cholesterol, and is secreted into the small intestine to promote lipid absorption. Therefore, 7α-hydroxylase (CYP7A1), the rate-limiting enzyme in the conversion of cholesterol to bile acid, becomes a major point in many studies on the effects of taurine on lipid metabolism.

Up to the present, numerous studies have been performed to determine the effect of taurine on cholesterol metabolism by using animal models with hypercholesterolemia[7,8]. We previously also reported the action of taurine on hypercholesterolemia. Taurine demonstrates inverse effects on different types of hypercholesterolemia[9,10]. Although the effects of taurine on liver are widely reported, the effects of taurine on the intestine to affect bile acid distribution are still unclear.

In the present study, we investigated the actions of taurine on the activity of CYP7A1, the constituent changes of apolipoprotein, and bile acid distribution in the intestine.

Taurine 5: Beginning the 21st Century
Edited by Lombardini *et al.*, Kluwer Academic/Plenum Publishers, New York 2003

2.   MATERIALS AND METHODS

2.1   Experimental Design

Experiment 1. Young male Wistar rats, weighing approximately 120g (SLC Inc., Hamamatsu, Japan), were housed in individual cages at 24℃ with a 12-hour cycle of light (0700-1900) and dark (1900-0700). In order to accustom the rats to our experimental conditions, they were fed with 20% casein diet (control diet) for 3 days before being divided into groups. Then they were fed with the control (C) diet, 5% taurine-supplemented (T) diet, high cholesterol (HC) diet or HC diet supplemented with 5% taurine (HCT) for 2 weeks. The compositions of the experimental diets are shown in Table 1.

Table 1. Compositions of diets in experiment 1 (g/kg).

Ingredients	C	T	HC	HCT
Casein	200	200	200	200
Corn Starch	425.7	392.4	417.3	384
Sucrose	212.8	196.1	208.7	192
Corn Oil	50	50	50	50
AIN-76 vitamin mixture	10	10	10	10
AIN- 93G mineral mixture	50	50	50	50
Choline chloride	1.5	1.5	1.5	1.5
Cellulose	50	50	50	50
Cholesterol	0	0	10	10
Sodium cholate	0	0	2.5	2.5
Taurine	0	50	0	50

Experiment 2. The same strain rats were used as in experiment 1 (Wistar rats). The body weight at the start of the experiment was approximately 100g. The experimental environment was the same as in experiment 1. Rats were fed with the high-fat basal control (F) diet, 1% taurine-supplemented (FT) diet, high cholesterol (FC) diet or FC diet supplemented with 1% taurine (FCT) for one week. The compositions of the diets are shown in Table 2.

2.2   Biochemical Analyses

Liver lipids were extracted by the method of Folch et al[11]. The concentrations of total cholesterol, HDL-cholesterol, triglycerides and phospholipids in serum, liver or feces were determined by commercial kits (Wako Pure Chemical, Osaka, Japan). Bile acids in fecal or intestinal digests were measured by the method of Sheltway and Losowsky[12]. The activity of CYP7A1 was determined by the method of Ogishima and Okuda[13]. CYP7A1 mRNA was determined as we described previously[14]. Lipoprotein fractions were obtained by gradient ultracentrifugation. Apolipoproteins were separated by SDS-PAGE.

Table 2. Compositions of diets in experiment 2 (g/kg).

Ingredients	F	FT	FC	FCT
Casein	200	200	200	200
Corn Starch	368.9	362.3	362.3	355.7
Sucrose	184.6	181.2	181.2	177.8
Corn Oil	50	50	50	50
AIN-93 vitamin mixture	10	10	10	10
AIN- 93G mineral mixture	35	35	35	35
Choline chloride	1.5	1.5	1.5	1.5
Cellulose	50	50	50	50
Cholesterol	0	0	10	10
Lard	100	100	100	100
Taurine	0	10	0	10

2.3 Statistics

The means and SEM of 4-8 rats per group are reported. Statistical analyses were performed by two-way ANOVA followed by Duncan's multiple range test. P-values of less than 0.05 are considered to be of statistical significance.

3. RESULTS

3.1 Effect of Taurine on Lipid Levels in Serum and Liver and the Degradation of Cholesterol (Experiment 1).

Body weight, body weight gain and food intake did not differ among the four groups. The liver of rats fed with the HC diet was significantly greater than that of the C group, and it was markedly suppressed by taurine supplementation (data are not shown). Dietary cholesterol significantly increased serum total cholesterol concentration and taurine notably reduced them (Table 3). Liver lipid levels (cholesterol, triglycerides and phospholipids) were significantly elevated by the high cholesterol diet and all levels of these 3 lipids were significantly decreased by taurine supplementation. The excretion of fecal cholesterol and bile acid were markedly increased in rats fed with HC diet compared to the control group, and the excretion of bile acid was further increased by taurine supplementation. The high cholesterol diet raised CYP7A1 mRNA levels and taurine further increased the levels. The changes of apolipoproteins in HDL and VLDL fractions are shown in Fig. 1 and Fig. 2, respectively. ApoE in the HDL fraction was notably decreased by the HC diet, and it was slightly improved by taurine. ApoA-I and ApoA-IV were observed in the VLDL fraction in hypercholesterolemic rats; the levels of ApoA-I and ApoA-IV were significantly reduced by taurine.

3.2   Effect of Taurine on Lipid Levels in Serum and Liver, the Degradation of Cholesterol, and the Distribution of Bile Acids in the Intestine (Experiment 2).

The results of experiment 2 are shown in Table 4. There were no differences in body weight and food intake among the four groups. The cholesterol levels of serum and liver were significantly increased by the high cholesterol diet, and they were notably reduced by taurine supplementation. The excretion of fecal cholesterol and bile acid were markedly increased by the high cholesterol diet and the excretion levels were further increased by taurine supplementation. Bile acid distribution in the intestine was determined. The results show that bile acid levels in the cecum and large intestine, but not in the small intestine, were significantly increased by a high cholesterol diet and further raised by taurine supplementation.

Table 3. Effect of taurine on lipid concentrations in serum and liver, the degradation of cholesterol, and the levels of CYP7A1 mRNA in experiment 1.

	C	T	HC	HCT	Two-way ANOVA Chol	Tau	Interaction
Serum lipids mmol/L							
Total cholesterol	3.06±0.12	3.07±0.36	3.97±0.42*	2.31±0.19#	NS	0.01	0.01
HDL-cholesterol	1.39±0.06	1.27±0.21	0.71±0.09	0.81±0.07	0.01	NS	NS
Triglyceride	2.29±0.15	1.69±0.14*	1.56±0.14	1.16±0.18	0.01	0.01	NS
Phospholipids	2.19±0.14	2.04±0.20	1.90±0.20	1.46±0.12	0.05	NS	NS
Liver lipids, umol/g liver							
Cholesterol	11.8±0.9	11.1±0.4	30.1±2.1**	22.2±1.7#	0.01	0.01	0.05
Triglyceride	15.9±1.8	12.3±1.2	47.6±6.6**	21.2±1.9##	0.01	0.01	0.01
Phospholipids	45.9±1.4	46.6±1.1	131±11**	63.9±2.7##	0.01	0.01	0.01
Fecal steriod, umol/3d							
Cholesterol	40.7±3.8	53.1±7.3	273±14**	286±16	0.01	NS	0.01
Bile acids	27.6±1.1	38.2±2.4	156±15**	248±16##	0.01	0.01	0.01
mRNA (arbitrary unit)							
CYP7A1	1.00±0.28	1.20±0.27	1.70±0.17	3.32±0.47##	0.01	0.01	0.05

1. Values are means ± sem, n = 5−8.
2. *,** Values are significantly different from that of the control group (P<0.01, P<0.001 respectively).
3. #,## Values are significantly different from that of the HC group (P<0.01, P<0.001 respectively).

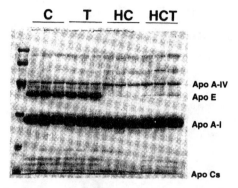

Figure 1. The changes of apolipoproteins in HDL fraction of each group in experiment 1.

Table 4. Effect of taurine on lipid concentrations in serum and liver, the degradation of cholesterol, and CYP7A1 activity in experiment 2.

	F	FT	FC	FCT	Two-way ANOVA Chol	Tau	Interaction
Serum lipids mmol/L							
Total cholesterol	2.00±0.06	1.89±0.07	2.42±0.14*	1.78±0.12#	NS	0.01	0.01
HDL-cholesterol	0.83±0.06	0.95±0.07	0.86±0.05	0.92±0.05	NS	NS	NS
Triglyceride	1.41±0.15	1.19±0.13	1.53±0.14	1.09±0.08	NS	0.05	NS
Phospholipids	1.92±0.07	1.86±0.03	1.84±0.08	1.76±0.07	NS	NS	NS
Liver lipids, umol/g liver							
Cholesterol	4.76±0.2	3.85±0.1	37.5±1.4**	24.4±1.6#	0.01	0.01	0.01
Triglyceride	13.3±0.7	11.8±0.5	37.8±1.2**	30.6±1.8#	0.01	0.01	0.01
Phospholipids	40.4±0.6	36.5±0.6	41.3±1.1	36.6±1.0#	NS	0.01	0.01
Fecal steriod, umol/d							
Cholesterol	1.42±0.24	1.24±0.2	206±13.0**	240±8.8##	0.01	0.05	0.01
Bile acids	16.5±1.5	19.6±0.8	48.4±2.4**	70.8±2.6##	0.01	0.01	0.01
CYP7A1 activity (pmol/min/mg protein)							
	51.8±4.7	74.1±4.8	66.2±10.8	89.0±11.9	NS	0.05	NS
Bile acid, umol/g intestine digest							
Small intestine	52.0±5.9	49.7±2.6	68.5±5.0	65.9±16.7	NS	NS	NS
Cecum	7.53±0.69	6.86±0.52	15.0±1.41**	20.0±1.55##	0.01	NS	0.05
Large intestine	16.0±1.20	17.9±1.36	42.3±3.17**	58.8±1.63##	0.01	0.01	0.01

1. Values are means ± sem, n= 5-6.
2. *,** Values are significantly different from that of the F group (P<0.01, P<0.001 respectively).
3. #,## Values are significantly different from that of the FC group (P<0.01, P<0.001 respectively).

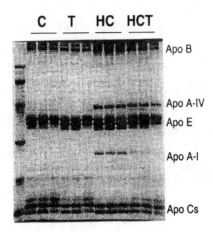

Figure 2. The changes of apolipoproteins in the VLDL fraction of each group in experiment 1.

4.   DISCUSSION

It is well known that a dietary taurine supplement significantly lowers the cholesterol levels in hypercholesterolemic rats induced by a high cholesterol diet. We have previously reported that taurine demonstrated a reverse action in different types of hypercholesterolemia, *i.e.*, a cholesterol-lowering effect and a cholesterol-amplifying effect in hypercholesterolemic rats induced by a high cholesterol diet and polychlorinated biphenyl (PCB), respectively[9,10]. The cholesterol-lowering effect of taurine is mainly considered as (1) the increase in the conversion of cholesterol to bile acids due to the enhancement of CYP7A1 activity, which is the rate-limiting enzyme for bile acid synthesis, and (2) the increase in bile acid conjugation by the enhancement of taurine levels in the liver and with the subsequent increase in biliary bile acid secretion[14].

In the present study, taurine significantly reduced serum and liver cholesterol levels, and increased the activity of mRNA levels of CYP7A1 and the excretion of bile acid in the feces. These results agree with the above mechanism of a hypocholesterolemic effect for taurine. The results of apolipoprotein analysis also demonstrate that taurine modifies the apolipoprotein composition in the VLDL and HDL fractions, especially taurine decreases ApoA-I and ApoA-IV in the VLDL fraction in hypercholesterolemic rats. The modifying mechanism is not clear, however, it is probably related to the activity of receptors in the cholesterol metabolism pathway. Undoubtedly, taurine acts on the liver to regulate cholesterol metabolism. Since intestinal reabsorption of bile acid, which is mediated by a 48-kd sodium-dependent bile acid co-transporter[15], plays a crucial role in lipid metabolism, it is also considered that taurine may affect the reabsorption of bile acids. The decrease of the bile acid pool in enterohepatic circulation is also thought to induce CYP7A1 activity. In our experimental conditions of no exogenous bile acids added to the high cholesterol diets, the results show that the bile acid content in the cecum and feces was significantly increased by taurine in the hypercholesterolemic rats, but not in the small intestine. CYP7A1 activity is also increased by taurine. These results suggest that taurine suppresses the reabsorption of bile acids in the small intestine and enhances the excretion of bile acid in the feces. However, the mechanism of the suppressing effect of taurine on bile acid reabsorption is not yet clear .

The overall results of our experiments suggest that taurine not only has effects on the liver to improve cholesterol degradation, but also affects the reabsorption of bile acids from the small intestine. Further studies are necessary to gain further insight into the action of taurine on lipid metabolism and the reabsorption of bile acids.

REFERENCES

1. Wright, C.E., Tallan, H.H., Lin, Y.Y., 1986, Taurine: biological update. *Annu. Rev Biochem.* 55:427-453.
2. Huxtable,R.J., 1992, Physiological actions of taurine. *Physiol. Rev.* 72:101-163.
3. Kuriyama, K., 1980, Taurine as a neuromodulator. *Federation Proceeding.* 39:2680-2684.
4. Thurston, J.H., Hauhart, R.E., Dirgo, J.A., 1980, Taurine: A role in osmotic regulation of mammalian brain and possible clinical significances. *Life Sci* 26:1561-1568.
5. Pasantes Morales,.H., Wright, C.E., Gaull, G.E., 1985, Taurine protection of lymphoblastoid cells from iron-ascorbate-induced damage. *Biochem. Pharmacol.* 34:2205-2207.
6. Danielsson, H., 1963, Present states of research on catabolism and excretion of cholesterol. *Adv. Lipid Res.* 1:335-385.
7. Sugiyama, K., Ohishi, A., Ohnuma, Y., Muramatsu, K., 1989, Comparison between the plasma cholesterol-lowering effects of glycine and taurine in rats fed on high cholesterol diet. *Agric Biol Chem.* 53: 1647-1652.
8. Yan, C.C., Bravo, E., Cantafora, A., 1993, Effect of taurine levels on liver lipid metabolism: an in vivo study in the rat. *Proc Soc Exp Biol Med.* 202: 88-96.
9. Nanami, K., Oda, H., Yokogoshi, H., 1996, Antihypercholesterolemic action of taurine on streptozotocin-diabetic rats or on rats fed a high cholesterol diet. *Taurine 2, edited by Huxtable et al., Plenum Press, New York,* pp.561-568.
10. Mochizuki, H., Oda, H., Yokogoshi, H., 1998, Amplified effect of taurine on PCB-induced hypercholesterolemia in rats. *Taurine 3: 285-290, Plenum Press, New York* .
11. Folch, J., Lees, M., Sloane-Stanley, G.H., 1957, A simple method for the isolation and purification of total lipids from animal tissues, *J. Biol. Chem.,* 226:497-509.
12. Sheltawy, M.J., Losowsky, M.S., 1975, Determination of fecal bile acid by an enzymatic method., *Clin. Chim. Acta.* 64:127-132.
13. Ogishima, T., Okuda, K., 1986, An improved method for cholesterol 7α-hydroxylase activity, *Analytical Biochemistry,* 158:228-232.
14 Sugiyama, K., Ohishi, A., Ohnuma, Y., Muramatsu, K., 1989, Comparison between the plasma cholesterol-lowering effects of glycine and taurine in fats fed on high cholesterol diets, *Agri. Biol. Chem.,* 53:1647-1652.
15. Shneider, B L., 2001, Intestinal bile acid transport: biology, physiology, and pathophysiology, *J Pediatr Gastroenterol Nutr.,* 32(4):407-41.

Effect of α-Tocopherol and Taurine Supplementation on Oxidized LDL Levels of Middle Aged Korean Women During Aerobic Exercise

C. S. AHN and E. S. KIM [*]

Department of Food and Nutrition, Ansan College, Ansan 425-701 and [*]Department of Food Science and Nutrition, Dankook University, Seoul 140-714, Korea

1. INTRODUCTION

The risk factors of atherosclerosis have been reported as elevated levels of total cholesterol and low density lipoprotein-cholesterol (LDL-C) and low levels of high density lipoprotein-cholesterol (HDL-C) in the serum. Oxidized low density lipoprotein (oxLDL) is reported to be an important independent predictor of atherosclerosis[1].

It has also been demonstrated that high levels of oxLDL 'Atherosclerosis of the aorta' exist inside the blood vessels of patients, and autoantibodies against oxLDL exist in their serum. These data thus indicate that oxLDL is closely connected to the generative mechanism of atherosclerosis[2]. Many studies have been conducted on the functions and operations of antioxidants that influence LDL oxidation, and it has been reported that antioxidant intake is related to coronary artery disease[3]. These present studies measured the levels of blood lipids, a factor of coronary artery disease, after supplying antioxidants to patients and non-patients. Recently, the relationship between serum oxLDL and antioxidants has been studied[4].

Physical exercise is conducted to prevent and treat geriatric diseases and promote health, but too much exercise increases the rate of muscle-oxidation, which increases the levels of free radicals and lipid peroxides that are harmful to the human body by injurying somatic cells[5].

Taurine 5: Beginning the 21st Century
Edited by Lombardini *et al.*, Kluwer Academic/Plenum Publishers, New York 2003

The state of peroxidation in the human body encourages the aging process as well as causes cancer or cardiovascular diseases. In order to prevent these processes and cure these diseases, antioxidants such as vitamin E, beta-carotene, and vitamin C are recommended. Antioxidants decrease free radicals in the human body and restrain the generation of peroxides. In particular, vitamin E is known to restrain the aging process, cancer, ischemic cardiopathy, arthritis, cataracts, *etc.* and protect the body from excess exercise and air-pollution. Thus nutrients are frequently studied as antioxidant substances[6].

Taurine, 2-amino ethanesulfonic acid, has many physiological functions. It helps emulsify and absorb fat and its bile-acid excretory process increases the cholesterol-excretion rate. Taurine was also reported to effectively protect against oxidizing damage in tissues[7]. Therefore, it is important to supply taurine and vitamin E to the human body either together or separately and to examine their antioxidant effects on the serum lipid peroxide concentration changes. Particularly, it is an important task to predict the possibility of an atherosclerotic attack by comparing concentrations of serum oxLDL.

This study aims to determine the concentrations of serum oxLDL (Anti-oxLDL ab) and thiobarbituric acid reactive substances (TBARS), etiological factors of atherosclerosis, after supplying vitamin E and taurine to subjects involved in moderate aerobic exercise. The time course of this study will be 4 weeks and comparisons of before and after supplying antioxidants will be determined in order to prevent and treat cardiovascular diseases.

2.   SUBJECTS AND METHODS

The subjects of this study are women prior to the period of menopause, aged 42.0 ± 4.7 (range 33 to 54 years); the sedentary group (S) consists of 13 persons, and the 4 aerobic groups (A) consist of 44 persons. Aerobic groups have performed aerobic exercise for over 6 months at an exercising intensity of VO_2 max. 60%.

The control aerobic group (Aerobics, control: AC) consists of 12 subjects who took a placebo (corn oil and lactose) instead of the antioxidant nutrients. In the group who took the antioxidant nutrients, 10 subjects took vitamin E (Aerobics + vitamin E : AE), 400IU per day; 11 subjects took taurine (Aerobics + taurine : AT), 3g per day; 11 subjects took both vitamin E and taurine together (Aerobics + vitamin E + taurine : AET), vitamin E 400IU and taurine 3g a day. All agents were administered for 4 weeks.

In a dietary-intake survey using the 24 hour-dietary recall method, the nutrients taken by the subjects were calculated by a computer program (CAN Pro, The Korean Nutrition Society, Korea).

In order to determine the intake amounts of vitamin E and taurine, a food intake frequency questionnaire survey was used in calculating the intake amount per day.

In measuring the concentration of serum lipids, total cholesterol was analyzed by the cholestezyme-V enzyme method, and HDL-cholesterol was analyzed by the phosphotungstic acid-$MgCl_2$ method (Commercial kit. Shinyaung Co., Korea). The concentration of the serum TBARS was measured by the Yagi method[8]. The serum Anti-oxLDL antibody was measured by the ImmuLisa method of ELISA (enzyme-linked immunosorbent assay)[9]. The α-tocopherol levels in serum were measured by the Isocratic HPLC (high performance liquid chromatography) method[10]. The blood plasma taurine levels were measured by the Dabsyl-Cl (4-dimethylamino azobenzene-4-sulfonyl-chloride) pre-column derivatization method and RP (reversed phase) HPLC[11].

Data were expressed as mean ± SD and statistical analyses were performed with the Statistical Analysis System (SAS), version 6.12. Concentration differences between serum α-tocopherol and plasma taurine were examined by paired t-test before and after taking antioxidant nutrients. The data from each group were compared by the Duncan's multiple range test. The analysis of factors influencing the levels of serum TBARS and oxLDL was conducted using the stepwise multiple regression analysis. The ANOVA test was used to compare the means. The p-value of α=0.05 was considered as statistically significant.

3. RESULTS AND DISCUSSION

3.1 Status of Dietary-Intake

Status of dietary-intake of subjects: the amount of the average energy-intake by the 24 hour-dietary recall method is 1659.4 ± 202.0 kcal/day, which is 82.8% of the recommended dietary allowance (RDA) for 30~49 year old Korean women. The amount of average protein-intake is 72.7 ± 13.7 g/day, which is 133.2% of the RDA, 55 g/day.

Status of the cholesterol intake of subjects: the amount of the average cholesterol-intake was 236.4 ± 98.3 mg/day, which is lower than the recommended dietary allowance for Americans, 300 mg. Subjects took in dietary cholesterol mainly from egg yolk.

Status of the vitamin E, β-carotene and taurine intake of subjects: In subjects' taking antioxidant nutrients following the semiquantitative food frequency questionnaire survey, the average amount of their vitamin E-intake was 13.3 ±

7.3 mg/day, which is higher than the recommended dietary allowance for 30~40 year old Korean women. The amount of the average taurine-intake was 177.5± 61.7 mg/day, which was higher than the intake of Korean college women and that of vegetarians, while it was not significantly different than the Japanese (Table 1).

The source of blood taurine is dietary intake and also by biosynthesis *in vivo*. However, the concentration of blood taurine reflects dietary taurine. We considered that the state of the taurine intake varies by residential area (urban and rural), the intake frequency, and the type of animal food. The subjects were people who live in the city with an economic level of mid-to-upper class. According to the dietary survey of the subjects, they have eaten various kinds of food. The intake of a high percentage of animal protein especially results in the high intake rate of taurine as well.

Table 1. Dietary vitamin-E, β-carotene, and taurine intakes by food intake frequency questionnaire survey of the aerobic subjects before antioxidant supplementation (per day).

Groups Antioxidants	Aerobics				P-value
	AC (n=12)	AE (n=10)	AT (n=11)	AET (n=11)	
Vitamin-E (mg)	12.1 ± 8.2	12.3 ± 5.9	12.7 ± 6.6	14.0 ± 7.4	0.924
β -carotene (mg)	2107.3 ± 1255.4	2024.2 ± 1038.0	2687.7 ± 1310.4	2876.0 ± 1351.4	0.308
Taurine (mg)	175.7 ± 53.8	155.5 ± 37.5	169.0 ± 58.0	188.0 ± 41.9	0.498

Values are Mean ± SD.
P-values from the Analysis of Variance (ANOVA) test among groups.

3.2   Comparison of Blood Lipid Peroxide Concentration Before and After Supplying Antioxidant Nutrients

Concentration changes of serum α-tocopherol and plasma taurine: the concentration changes of serum α-tocopherol before and after supplying antioxidant nutrients indicate that the AE and AET groups who took vitamin E, showed a very significant difference ($p < 0.01$ by paired t-test), which reflects concentration changes (Fig. 1). The change of the plasma concentrations of taurine indicates that the AT (15.4%) and AET (18.2%) groups who took taurine, showed a significant difference ($p < 0.05$ by paired t-test)(Fig. 2).

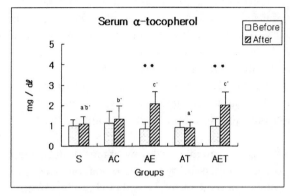

Fig.1. Serum α-tocopherol levels before and after antioxidants supplementation.
S : sedentary
AC : aerobics, control
AE : aerobics + vitamin E
AT : aerobics + taurine
AET : aerobics + vitamin E and taurine

**: Significantly different between before and after antioxidant supplementation at P<0.01 by paired t-test. Values with different letters are significantly different at α=0.05 level by Duncan's multiple range test among groups.

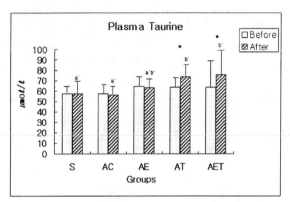

Fig.2. Plasma taurine levels before and after antioxidants supplementation.
S : sedentary
AC : aerobics, control
AE : aerobics + vitamin E
AT : aerobics + taurine
AET : aerobics + vitamin E and taurine

*: Significantly different between before and after antioxidant supplementation at P<0.05 by paired t-test. Values with different letters are significantly different at α=0.05 level by Duncan's multiple range test among groups.

Changes in the levels of serum TBARS and Anti-oxLDL antibody: Serum TBARS levels changed before and after taking antioxidant nutrients. The AE and AET groups decreased significantly (p<0.01); those of the AT group did not decrease (Fig. 3). However, if subjects took taurine for a long time, the figures would decrease significantly. Vitamin E had a powerful effect as an antioxidant. On the other hand, taurine played the role of an antioxidant, but its effect was not higher than that of vitamin E.

The AE and AET groups showed a significant decrease (p<0.05) in their serum Anti-oxLDL ab titers. The values of the AT group did not decrease (Fig. 4). The average serum Anti-oxLDL ab titers of all subjects were lowered from 6.8 ± 4.0 Eu/ml to 6.4 ± 3.6 Eu/ml after taking antioxidant nutrients. All of the subjects had a normal health such that the status of atherosclerosis risk factor was negative (<25 Eu/ml); these data were lower than that of Americans. The concentration of serum oxLDL depends on eating habits, age, smoking, blood antioxidant levels and exercise. In the case of atherosclerosis and essential hypertension, it was reported that there was a high level of antibodies against oxLDL in the serum[12]. Furthermore, oxLDL works as an antigen, generating cell-mediated immunity reactions and autoantibodies, which are consumed by macrophages. These facts contribute to cholesterol accumulation in the artery endothelium.

The continuous aerobic exercise at middle and low intensity that these subjects perform is thought to increase susceptibility to LDL oxidation. Therefore, moderate exercise as well as complex antioxidant-intake is thought to be effective in preventing and treating atherosclerosis.

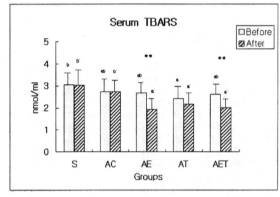

Fig.3. Serum TBARS (Thiobarbituric acid reactive substances) concentration before and after antioxidants supplementation.
S : sedentary
AC : aerobics, control
AE : aerobics + vitamin E
AT : aerobics + taurine
AET : aerobics + vitamin E and taurine

**: Significantly different between before and after antioxidant supplementation at P<0.01 by paired t-test. Values with different letters are significantly different at α=0.05 level by Duncan's multiple range test among groups.

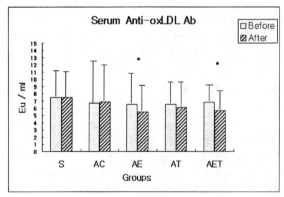

Fig.4. Serum anti-oxLDL ab levels before and after antioxidants supplementation.
S : sedentary
AC : aerobics, control
AE : aerobics + vitamin E
AT : aerobics + taurine
AET : aerobics + vitamin E and taurine
Eu/ml : Enzyme unit/ml

*: Significantly different between before and after antioxidant supplementation at P<0.05 by paired t-test. Values with different letters are significantly different at the α=0.05 level by Duncan's multiple range test among groups.

3.3 Analysis on Factors Influencing Serum Anti-oxLDL Antibody and TBARS

The factors influencing serum oxLDL (Anti-oxLDL ab) statistically had a negative correlation (p=0.005) with HDL-C before the subjects' started taking antioxidant nutrients. After taking antioxidant nutrients, the effects also occurred in a negative correlation with serum α-tocopherol which was caused by supplying vitamin E.

Factors statisticaly influencing serum TBARS before supplying antioxidant nutrients: total cholesterol (TC) was affected in a positive correlation, HDL-C was affected in a negative correlation. In particular, TC (p=0.0016) proved to be an important factor that influences serum TBARS and serum α-tocopherol, and plasma taurine, a factor, that influences serum TBARS had a negative effect. In particular, α-tocopherol was definitively a negative factor, p=0.0003 (Table 2).

Table 2. Stepwise regression for lipidperoxide and blood atherogenic risk factors.

Factors affected by serum TBARS

	Factor)	Estimate	Partial R^2	P-value
	Hb	0.1016	0.0450	0.0867
before (R^2=0.2629)	HDL-C	-0.0086	0.0332	0.1318
	TC	0.0086	0.1342	0.0016
	Intercep	1.1234		0.1408
	TC	0.0077	0.1050	0.0043
after (R^2=0.3745)	α-tocopherol	-0.4567	0.1149	0.0003
	Taurine	-0.0856	0.1546	0.0185
	Intercep	2.5438		0.0001

Factors affected by serum oxLDL

	Factor	Estimate	Partial R^2	P-value
	Hb	0.8858	0.0635	0.0459
before (R^2=0.1794)	HDL-C	-0.1079	0.1159	0.0050
	Intercep	1.3215		0.8229
	PLT	-0.0276	0.1192	0.0049
after (R^2=0.1774)	α-tocopherol	-1.3959	0.0581	0.0559
	Intercep	15.3948		0.0001

PLT : blood platelet,   Hb : hemoglobin,   TC : total cholesterol,
before and after : before and after antioxidant supplementation

4. CONCLUSION

After administering vitamin E (400I U/day) and taurine (3 g/day) for 4 weeks to middle-aged women who were doing aerobic exercise at moderate intensity, we studied the effects on atherosclerosis by measuring serum oxLDL's changes. In conclusion, vitamin E influenced both serum oxLDL and TBARS noticeably, but taurine decreased the level of serum oxLDL and TBARS; taurine did not have a significant influence on oxLDL. Therefore, if the amount and time period of taurine-intake are adjusted, better antioxidant effectiveness could be obtained.

ACKNOWLEDGEMENTS

We would like to thank Dr. Son Mi-Won for the determination of plasma taurine and Dong-A pharmaceutical Co. Ltd., Korea, for the supply of taurine tablets and vitamin E capsules,

REFERENCES

1. Esterbauer, H., Waeg, G., and Puh, I. H., 1993, Lipid Peroxidation and Its Role in Atherosclerosis. *British Med. Bulletin* 49 : 566-576.
2. Arja,T. E., Outi, N., Seppo, L., Matti, I. J., and Uusitupa, S.Y., 2000, Autoantibodies Against oxidized low density lipoprotein and cardiolipin in patients with coronary heart disease. *Arterioscler. Thromb. Vasc. Biol.* 20 : 204-209.
3. Stampfer, M. J., and Rimm, E. B., 1995, Epidemiologic evidence for vitamin E in prevention of cardiovascular disease. *Am. J. Clin. Nutr.* 62 : 1365S-1369S.
4. Hirano, R., Kondo, K., Iwamoto, T., Igarash, O., and Itakura, H., 1997, Effects of antioxidants on the oxidative susceptibility of low-density lipoprotein. *J. Nutr. Sci. Vitaminol.* 43(4) : 435-444.
5. Sjokin, B., Westing, Y. H., and Apple, F. S., 1990, Biochemical mechanisms for oxygen free radical formation during exercise. *Sports Med.* 10(4) : 236-254.
6. Panganamala, R.V., and Cornwell, D. G., 1982, The effect of vitamin E on arachidonic acid metabolism. *Ann. NY. Acad. Sci.* 393 : 376-393.
7. Koyama, I., Nakamura, T., Ogasawara, M., Namoto, M., and Yoshida, T., 1992, The protective effect of taurine on the biomembrane against damage produced by the oxygen radical. *Adv. Exp. Med. Biol.* 315 : 355-359.
8. Yagi, K.,1984, Lipid peroxides in Biology and Medicine, Academic Press, New York, pp. 328-331.
9. ImmuLisa, 2001, Anti-oxLDL antibody ELISA. IMMCO Diagnostics, Inc., Catalog No 1158, New York.
10. Arnaud, J., Fortis, I., Blachier, S., Kia, D., and Favier,A., 1991, Simultaneous determination of retinol, α -tocopherol and ß-carotene in serum by isocratic high-performance liquid chromatography. *J. Chromatography* 572 : 103-116.
11. Stochi, V., Palma, F., Piccoli, G., Biagiarelli, B., Cuchiarimi, L., and Magnani, M., 1994, HPLC analysis of taurine in human plasma sample using the Dabsyl-Cl reagent with sensitivity at picomole level. *J. Liquid chromatography* 17(2) : 347-357.
12. Craig, W. Y., Poulin, S. E., Nelson, C. P., and Ritchie R. F., 1994, An ELISA method for the detection and quantitation of IgG antibody against oxidized low density lipoprotein. *Clin.Chem.* 40 : 882-888

Dietary Taurine Intake and Serum Taurine Levels of Women on Jeju Island

E.S. KIM, J.S. KIM, M.H. YIM, Y. JEONG, Y.S. KO*, T. WATANABE**, H. NAKATSUKA***, S. NAKATSUKA***, N. MATSUDA-INOGUCHI***, S. SHIMBO**** and M. IKEDA*****

*Dept. Food Science and Nutrition, Dankook University, Seoul 140-714, Korea. *Dept. Food and Nutrition, Jeju National University, Jeju 690-756, Korea. **Miyagi University of Education, Sendai 980-0845, Japan. ***Miyagi University, Taiwa-cho 981-3298, Japan. ****Kyoto Women's University, Kyoto 604-8472, Japan, *****Kyoto Industrial Health Association, Kyoto 604-8472, Japan.*

ABSTRACT

The purpose of this study was to investigate the dietary taurine intake and serum taurine levels of women on Jeju Island in Korea. Sixty six married women aged 43.5 ± 7.1 volunteered for this study: 34 from the city area and 32 from two fishing-farming areas. Diet samples were collected from the participants; the samples included three meals (breakfast, lunch and supper), including snacks, drinks and whatever else the participants had eaten for 24 hours. Taurine levels in the diet and serum were determined as the dabsyl derivative by HPLC with a Rf-detector. The intake of taurine ranged from 8.4 to 767.6 mg/day and its mean value was 163.9 ± 150.2 mg/day (mean ± SD). There was a significant difference between the two groups: 114.9 ± 78.7 for the women from the city area and 215.9 ± 187.9 mg/day for the women from the fishing-farming areas (p<0.001). The taurine intake of the total diet, including all snacks and drinks, was 2300 ± 584 g/day for the city area and 2342 ± 528 g/day for the fishing-farming areas. The daily protein intake was 58.8 ± 16.4 g for the women of the city area and 65.5 ± 17.1 g for the women of the fishing-farming areas. There was a significant correlation between the intake of fish/shellfish and taurine (p=0.001) while there was no correlation between the intake of protein and taurine (p=0.057). The taurine levels in serum ranged from 68.6 to 261.6 µmol/L and the mean value was 169.7 ± 41.5 µmol/L. There was no significant difference between the women from the city area and the women from the fishing-farming areas in serum taurine levels. The correlations of serum taurine levels with serum retinol levels (p=0.016) and α-tocopherol (p=0.014) levels were significant. These results suggest that taurine intake is dependent on the fish/shellfish intake and that taurine may play an important role in the retention of antioxidative nutrients.

1.   INTRODUCTION

The role of taurine in human nutrition and physiology have been intensively reviewed[1-4]. The intake of taurine in infants from breast milk was reported by Kim et al.[5-7], and in adolescents and adults by Park et al.[8] using a food recording method with a database on taurine established in Korea. Also the intake of taurine was reported by Kibayashi et al.[9] in Japan and by Zhao et al.[10] in China using a 24 hour recall survey with a database on taurine. Inspite of the important roles of taurine in humans, there has not been many studies on taurine intake. Meals in Jeju Island are rich in seafoods and vegetables, and poor in animal meats (pork being preferred over beef) as compared with other areas in Korea[11]. The purpose of this study was to investigate the dietary taurine intake and serum taurine levels of women on Jeju Island in Korea and the relationships between taurine intake and the serum levels of taurine, retinol and α-tocopherol.

2.   MATERIALS AND METHODS

Dupliate food samples over 24 h[12] were collected on Jeju Island, Jeju-do, Korea, in September, 2000. Sixty six adult women aged 43.5 ± 7.1 (aged 29 to 54years) volunteered for this study: 34 from the city area and 32 from two fishing-farming areas. Diet samples eaten during a 24 hour period were collected from the participants; the samples included three meals (breakfast, lunch and supper), snacks, soft and hard drinks (although most of the women did not take alcoholic beverages) and even fresh water if consumed. A portion of each homogenate was placed in an acid-washed clean 100-mL plastic bottle, and stored at -70°C until anayzed. The serum was prepared by allowing a 5ml fasting blood sample to clot in a serum separator tube for 30 min and then centrifugation at 11,000 x g for 10 min. The collected diets were blended, centrifuged and, then deproteinized. Levels of taurine in the diet and serum were determined as a dabsyl derivative by HPLC with a Rf-detector[13]. Serum retinol and α-tocopherol were determined by HPLC[14]. Proximate composition was determined by official methods in Korea[15]. Daily intake of each nutrient (g/day) was calculated as the amount of the nutrient per weight of a homogenate (g/g) multiplied by total weight of the homogenate (g/day). Daily energy intake (kcal/day) was calculated from the amount of the nutrients and the weight to energy conversion factors of 4, 9 and 4 kcal/g for protein, fat and carbohydrate, respectively.

2.1 Statistical Analysis

Significant differences between diet intakes and major nutrients, and serum taurine contents in the two areas were tested by ANOVA and the t-test. The correlations between taurine intake and fish/shelfish intake, between taurine intake and meat intake, between serum taurine levels and serum retinol levels, and between serum taurine levels and serum α-tocopherol levels were assessed by regression analysis. The significance level was set at $p < 0.05$.

3. RESULTS AND DISCUSSION

3.1 Intake of Energy and Three Major Nutrients

Intakes of energy, protein, fat and carbohydrate are shown in Table 1 with basic demographic data of the women who volunteered for this study. The intake of total diet including all snacks and drinks was 2300 ± 554 g/day for the city area and 2342 ± 528 g/day for the fishing-farming areas. Daily protein intake was 58.4 ± 16.4 g for the city area and 65.5 ± 17.1 g for the fishing-farming areas. The intake in total diet, energy, protein and carbohydrate were lower in the city area than in the fishing-farming areas. These results might be due to differences in theire daily labor. The intake of energy and protein were 80.6% and 112.5% of Korean RDA. The ratios (%) of energy intake from protein, fat and carbohydrate were 15.4 : 15.0 : 69.6.

Table 1. Demography of the women who volunteered for the study and their daily intake of energy and major nutrients

Parameter		City area	Fishing-farming area
Ages(year)		43.5±6.3 (34-54)	43.4±8.0 (29-54)
Body weight(kg)		56.0±7.2	56.5±7.3
BMI		23.1±2.9	24.0±3.3
Total diet intake(g)	Wet basis[1]	2299.8±584.2	2341.7±527.5
	Dry basis	376.8 ± 85.4	391.8 ± 98.4
Energy intake (kcal)		1585.8±374.6	1639.9±416.8
Protein intake(g)		58.4±16.4	65.5±17.1
Fat intake(g)		27.0±14.6	26.9±17.1
Carbohydrate intake(g)		277.4±66.7	283.9±88.2

mean ± SD, ( ): range, [1]:included water and drinks

3.2   Taurine Intake and Serum Taurine Levels

Table 2 shows the intake of taurine and the levels of taurine in serum. The intake of taurine in the two areas ranged from 8.4 to 767.6 mg/day and its mean value was 163.9 ± 150.2 mg/day. There was a significant difference between the two groups: 114.9 ± 78.7 for the city area and 215.9 ± 187.9 mg/day for the fishing-farming areas (p<0.001). The taurine levels in serum ranged from 68.6 to 261.6 μmol/L and its mean value was 169.7 ± 41.5 μmol/L. There was no significant difference in serum taurine levels between the city area and the fishing-farming areas.

Park et al[8] reported that the daily taurine intake was 216 mg for adult men and 181mg for adult women in the Seoul area by a food recording method with a database on taurine. In Japan, Kibayashi et al[9] reported that the daily taurine intake was 226mg for men and 163mg for women living in Toyama city where seafood intake is the third highest in Japan. Mean taurine intake (164 mg) of women on Jeju Island, Korea, was similar to that of women in Toyama, Japan (163 mg). In China, Zhao et al[10] reported that the daily taurine intake of average Chinese men in the 4 different test areas ranged from 34 to 80 mg. They calculated the daily taurine intake based on the amount of the food intake obtained from the dietary survey database on taurine. Park et al[8] reported the plasma taurine levels: 145±7.2 μmol/L for adult men and 126±99 μmol/L for adult women.

Table 2. The intake of taurine and levels of taurine in the serum.

Parameter	Total	City area	Fishing-farming area
Taurine intake (mg/day)	163.9 ± 150.2 (8.4 – 767.6)	114.9±78.7	215.9±187.9*
Serum taurine (μmol/L)	169.7 ± 41.5 (68.6 - 261.6)	172.7±43.2	166.5±40.0

Mean±S, ( ): range
* vs city area (p < 0.001)

3.3   The Relationships Between the Intake of Taurine and Fish/Shelfish, Between the Intake of Taurine and Meat, Between the Intake of Taurine and Protein, and Between the Intake of Taurine and Ash

There were significant correlations between the the fish/shelfish intake and the taurine intake (p=0.001) (Fig 1), and between taurine intake and ash intake (p=0.001) (Fig 3) while there were no significant correlations between meat intake and taurine intake (Fig 2), and between protein intake and taurine intake (p=0.057) (Fig 4). These results are consistent with those reported by Kibayashi et al[9]. However, Park et al[8] reported that there were significant correlations between taurine intake and protein intake, and between taurine intake and meat intake.

The reason why there was no significant correlation between protein intake and taurine intake might be that vegetable protein intake is more than 50% of the total protein intake.

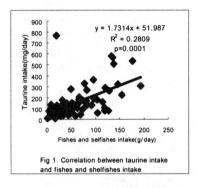

Fig 1. Correlation between taurine intake and fishes and shelfishes intake

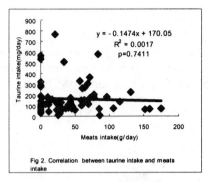

Fig 2. Correlation between taurine intake and meats intake

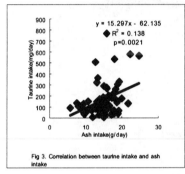

Fig 3. Correlation between taurine intake and ash intake

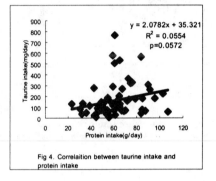

Fig 4. Correlaition between taurine intake and protein intake

Figure 1. Correlation between taurine intake and fish/shelfish intake.
Figure 2. Correlation between taurine intake and meat intake.
Figure 3. Correlation between taurine intake and protein inake.
Figure 4. Correlation between taurine intake and ash intake.

There was no significant correlation between the taurine intake and serum taurine levels. This result suggests that the taurine levels in serum were affected by exogenous taurine as well as endogenous taurine. It was reported that taurine, although not sufficient, can be biosynthesized from sulf-containing amino acids such as methionine and cysteine in the human body[16,17]. Also these results may mean that renal control mechanisms for serum taurine concentrations are very important and complex. It is known that the homeostatic control for taurine is established by absorption through the small intestine, bile acid conjugation and excretion, and renal reabsorption[18] and that taurine homeostasis is especially controlled by the taurine transporter the in renal proximal tubule[19, 20].

3.4   The Relationships Between Levels of Taurine and Retinol, and Between Levels of Taurine and α-Tocopherol in Serum

The serum taurine levels were significantly correlated with serum retinol (p=0.016) (Fig 5) and serum α-tocopherol (p=0.014) levels (Fig 6). These results suggest that taurine may play an important role in the retention of antioxidative nutrients. More research on retention of serum retinol and serum α-tocopherol are needed to clarify the mechanism.

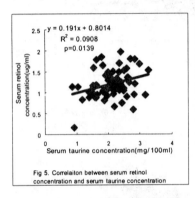

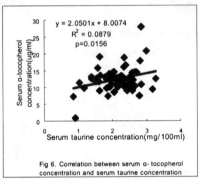

Fig 5. Correlaiton between serum retinol concentration and serum taurine concentration

Fig 6. Correlation between serum α-tocopherol concentration and serum taurine concentration

Figure 5. Correlation between serum taurine and serum retinol
Figure 6. Correlation between serum taurine and serum α-tocopherol

ACKNOWLEGEMENTS

This work was supported by grants from The Ministry of Education, Science, Sports and Culture, the Government of Japan and Dong-A Pharmaceutical Company Ltd. Korea.

REFERENCES

1. Hayes, K.C., Trautwein, E,A., 1994, Taurine. In *Modern Nutrition in health and disease* 8th ed.,(Shils ME, Olson JA, Shike M. ed.), Lea and Febiger, Philadelphia, pp. 477-485.
2. Huxtable, R.J., 1992, Physiological actions of taurine. *Physiological Reviews* 72 : 101-163.
3. Wright, C.E., Tallan, H.H,, Lin, Y.Y., Gaull, G.E., 1986, Taurine: Biogical uptodate *Ann. Rev .Biochem*, 55 : 427-453.
4. Sturman, J.A., Hayes, K.C., 1980, The biology of taurine in nutrition and development. *Adv. Nutr. Res.* 3 : 231-299.
5. Kim, E.S., Kim, J.S., Cho, K.H., 1998, Taurine level in human milk and estimated intake of taurine during the early period of lactation. *Korean J. Nutrition* 31(3) : 363-368.

6. Kim, E.S., Kim, J.S., Cho, K.H., Tamari, Y., 1998, Quantitation of taurine and selenium levels in human milk and estimated intake of taurine by breast-fed infants during the early period lactation. *Adv. Exp. Med. Biol.* 442 : 477-486.
7. Kim, E.S., Lee, J.S., Choi, K.S., Cho, K.H., Seol. M.Y., Park. M.A., Lee, K.H., 1993. Longitudinal study on taurine intake of breast-fed infants from Korean non-vegetarian and lacto-ovo-vegetarian. *Korean J. Nutrition* 26(8) : 967-973.
8. Park, T.S., Kang, H., Sung, M., 2001, Taurine intake, plasma levels and urinary excretions of taurine adolescents and adults residing in Seoul area. *Korean J Nutrition* 34(4) : 440-448.
9. Kibayashi, E., Yokogoshi, H., Miura, K., Yoshita, K., Nakagawa, H., Naruse, Y., Sokejima, S., Kagamimori, S., 2000, Daily dietary taurine intake in Japan. *Adv. Exp. Med. Biol.* 483 : 137-142.
10. Zhao, X.H., Jia, J.B., Lin, Y., 1998, Taurine content in Chinese food and daily taurine intake of Chinese men. *Adv. Exp. Med. Biol.* 442: 501-505.
11. Jeju-do Folklore and Natural History Museum, ed., 1995, *Foods in Jeju.* Daeyoung Press, Jeju, p. 19.(in Korean)
12. Acheson KJ, Cambell IT, Edholm OG, Miller DG, Stock MJ. The measurement of food and energy intake in man-an evaluation of some techniques. *Am J Clin Nutr* 1980:33, 1147
13. Stochi, V., Palma, F., Piccoli, G., Biagiarelli, B., Cucciarimi, L., and Gagnani, M., 1994, HPLC analysis of taurine in human plasma sample using the DABS-Cl reagent with sensitivity at picomole level, *J..Liq..Chromatogr.* 17(2):347-357
14. Bieri, M.A., Tolliver, T.J, Catignani, G.L., 1979, Simultaneous determination of α-tocopherol and retinol in plasma and red cells by high performance liquid chromatography. *Am. J. Clin. Nutr.* 32:2143-2149.
15. Korean Food and Drug Administration (KFDA), 2000, *Korean Food Code*(Official methods in Korea), Moonyoung-sa press, Seoul.
16. Gaull G.E., Rassin D.K., Raiha N.C.R., Heinonen K., 1977, Milk protein quantity and quality in low-birthweight infants III. Effect on sulfur amino acids in plasma and urine. *J Pediatr* 90 : 348-355.
17. Sturman J.A., Hays K.C., 1980, The biology of taurine in nutrition and development. *Adv. Nutr. Res.,* 3: 231-299.
18. Park, T., Rogers, Q.R., Morris, J.G., 1989, Chesney RW. Effect of dietary taurine on renal taurine transport by proximal tubule brush border membrane vesicles in the kitten. *J. Nutr.* 119:1452-1460.
19. Rentschler, L.A., Hirschberger, L.L., Stipanuk, M.H., 1986, Response of the kitten to dietary taurine depletion: Effects on renal reabsorption, bile acid conjugation and activities of enzymes involved in taurine synthesis. *Comp. Biochem. Physiol.* 84B: 319-325.
20. Chesney, R.W., Lippincott, S., Gusowski, N., Padilla, M., Zelikovic, I., 1986 Studies on renal adaptation to altered dietary amino acid intake: Tissue taurine responses in nursing and adult rats. *J. Nutr.* 116: 1965-1976.

Effect of the Obesity Index on Plasma Taurine Levels in Korean Female Adolescents

MI YOUNG LEE, SUN HEE CHEONG, KYUNG JA CHANG, MI JA CHOI[*], and SOON KI KIM[#]

*Department of Food and Nutrition, Inha University, Incheon, Korea, *Department of Food and Nutrition, Keimyung University, Taegu, Korea, #Department of Pediatrics, Inha University Hospital, Incheon, Korea*

1.   INTRODUCTION

Obesity is currently an increasing disease that affects many countries in the world including the United States where this condition is responsible for 300,000 deaths annually[1]. Adolescent obesity has rapidly increased in recent years in Korea due to westernized lifestyle[2].

The most documented and the first established physiological function of taurine is that it plays an important role in promoting micelle formation and fat absorption as one of the conjugated bile acids in the liver[3]. There are reports that plasma cholesterol or triglyceride concentration is significantly lowered by dietary taurine supplementation in hypercholesterolemic rats[4]. It has also been shown that oral taurine supplementation attenuated the increase in total cholesterol and LDL-cholesterol in healthy men on high fat cholesterol diets[5]. Obinata et al. reported that taurine was effective in treating fatty liver of children with simple obesity[6].

The purpose of this study was to estimate the plasma taurine concentrations in Korean female adolescents with respect to obesity.

2.   SUBJECTS AND METHODS

2.1   Subjects

The subjects were 51 female adolescent students residing in Incheon. The subjects were divided into 3 subgroups by BMI based on Korean Society of Obesity (1999): as being underweight (BMI<18.5kg/m^2), normal (BMI 18.5-23kg/m^2) and obese (BMI>25kg/m^2).

2.2   Anthropometric Assessment

Anthropometric parameters such as height, weight, triceps skin-fold thickness (TSF), mid-upper arm circumference (MAC), subscapular skin-fold thickness (SST), waist circumference, and hip circumference were measured. Total body fat was measured using a bioelectrical impedence analyzer (Tanita, TBF-611, Japan). The sitting blood pressure (BP) was measured in a quiet room 5 min after the cuff had been placed on the right forearm.

2.3   Biochemical Analysis

Fasting blood samples (7ml) were drawn from the subjects before the analysis. The following biochemical parameters were measured by blood autoanalyzer (Hitachi 747, Japan): hemoglobin, hematocrit, mean corpuscular volume (MCV), mean corpuscular hemoglobin (MCH), mean corpuscular hemoglobin concentration (MCHC), total iron binding capacity (TIBC), transferring saturation (TS), serum ferritin (SF), total cholesterol, triglyceride, HDL- cholesterol and LDL- cholesterol.

2.4   Amino Acid Analysis

Plasma samples were deproteinized by adding 10% sulphosalicylic acid and was then centrifuged at 12,000 rpm for min at 4C. Plasma free amino acid concentrations were determined by an automated amino acid analyser based on ion exchange chromatography (Biochrom 20, England).

2.5   Statistical Analysis

The statistical analysis was conducted using the SPSS 10.0 program. Three subgroups were compared using one-way analysis variance (ANOVA) for continuous variables. Duncan's multiple range test was used for multiple comparisons. The Pearson correlation coefficient was calculated to evaluate the relationship between plasma taurine and blood analysis data.

3. RESULTS AND DISCUSSION

Distribution of subjects by age and BMI classification is shown in Table 1. The subjects were aged from 12-18 years old. As for prevalence of obesity, 21.5% of subjects were underweight, 47.1% were normal and 31.4% were obese.

Table.1 Distribution of the subjects.

Age	Underweight	Normal	Obesity	Total
12	3	1		4
13	2	10	2	14
14			4	4
15	1	2	4	7
16		4	3	7
17	1	7	3	11
18	4			4
Total	**11(21.5)**	**24(47.1)**	**16(31.4)**	**51(100.0)**

Anthropometric characteristics of the subjects by BMI classification are shown in Table 2. The average height and weight of subgroups were 156cm, 40.2kg, 159cm, 52.3kg and 157.6cm, 68.5kg, respectively. As BMI was high, obesity index, body fat, triceps skin fold thickness, subscapular skin fold thickness, mid-upper arm circumference and W/H ratio were significantly increased. Diastolic BP was not different among the 3 subgroups but systolic BP was significantly different; it was the highest in the obesity subgroup compared to other subgroups.

Plasma taurine and amino acid concentrations by BMI classification are shown in Table 3. According to the BMI classification, the level of taurine concentration was 221.22±19.07μmol/L in the underweight subgroup, 200.64±30.92μmol/L in the normal subgroup and 194.26±44.08μmol/L in the obesity subgroup. There was a significant difference in plasma taurine concentration among the subgroups; it was the lowest in the obesity subgroup compared to the other subgroups. Plasma taurine concentration of subjects was higher compared to levels (135±5.9μmol/L) of adolescents and adults residing in the Seoul area[7]. Among other amino acids, the concentrations of histidine, ornithine, tyrosine and ethanolamine were significantly lower in the underweight subgroup compared to other subgroups.

Table 2. Anthropometric characteristics of the subjects.

	Underweight	Normal	Obesity
Height	156.05±6.02	158.97±5.26	157.58±6.11
Weight	40.15±5.37[a]	52.32±4.40[b]	68.51±8.79[c]
BMI	16.43±1.36[a]	20.66±0.68[b]	27.48±2.04[c]
Obesity index(%)	-20.89±6.66[a]	-0.67±2.40[b]	33.57±8.13[c]
Body fat	19.09±6.14[a]	27.38±2.50[b]	43.50±5.80[c]
Triceps skin fold thickness	17.18±5.27[a]	25.38±4.47[b]	32.38±8.05[c]
Subscapular skin fold thickness	12.18±4.33[a]	18.31±5.00[b]	27.81±7.70[c]
Mid-upper arm circumference	21.10±2.45[a]	23.75±1.45[b]	27.86±2.10[c]
Diastolic BP	75.36±8.55	71.64±8.22	73.23±9.80
Systolic BP	116.10±10.86[a]	121.27±10.29[ab]	128.08±9.77[b]
W/H	0.77±0.006[a]	0.73±0.003[b]	0.82±0.005[c]

Values are mean ± SD. Mean with same letter in each group was not significantly different by Duncan's Multiple Range test.

Table 3. Plasma taurine and other amino acid concentration.

	Underweight	Normal	Obesity
EAA		μmol/L	
Histidine	78.49±13.60[a]	100.00±26.66[b]	127.81±32.93[c]
Isoleucine	61.56±14.85	81.49±49.51	63.69±53.08
Leucine	126.10±28.11	106.22±32.40	103.29±93.02
Lysine	140.12±28.55	127.02±32.70	153.58±48.43
Methionine	24.29±12.04[a]	12.79±7.64[ab]	17.70±11.04[b]
Phenylalanine	83.96±18.74	75.18±30.63	78.10±31.45
Valine	198.22±37.04	161.91±84.72	192.40±61.89
NEAA			
Taurine	**221.22±19.07[a]**	**200.64±30.92[ab]**	**194.26±44.08[b]**
Alanine	515.77±41.80	364.91±229.86	396.05±216.85
Asparagine	69.84±39.24	74.96±44.79	77.95±33.59
Cysteine	20.93±25.85	32.37±20.49	17.45±10.80
α-Aminoadipic acid	35.27±4.13[a]	79.64±119.21[ab]	58.64±95.84[b]
α-Aminobutric acid	21.31±6.33	80.16±101.24	86.46±122.02
Cystathione	22.92±15.0	27.61±33.58	27.47±27.89
Glutamate	56.81±41.87	78.85±65.80	85.96±71.63
Glycine	296.90±44.36	399.43±202.64	340.73±133.97
Ornithine	91.46±38.41[a]	125.73±68.45[ab]	156.41±62.8[b]
Proline	231.1±35.20[a]	254.29±81.35[ab]	291.84±70.02[b]
Serine	90.13±11.07	117.77±27.27	124.19±75.98
Tyrosine	65.69±12.68[a]	79.73±28.40[ab]	95.63±33.66[b]
Phosphoserine	22.73±2.85	22.72±7.2	21.71±7.84
Ethanolamine	44.97±2.62[a]	94.16±79.11[ab]	100.35±78.57[b]

Values are mean ± SD. Mean with same letter in each group was not significantly different by Duncan's Multiple Range test.

Correlations between taurine and blood analysis data are shown in Table 4. There was a significant negative correlation between plasma taurine and mean corpuscular hemoglobin concentrations (MCHC) which indicates the average weight of hemoglobin as compared to the cell size. There was a significant positive correlation between plasma taurine and urea. But there was no correlation between plasma taurine and triglyceride concentrations. For the plasma taurine/glycine ratio, there was no correlation with MCHC, urea and triglyceride. But there was a significant positive correlation between the plasma taurine/glycine ratio and serum LDL cholesterol.

Table 4. Correlation coefficients between taurine and blood analysis data.

	Plasma taurine	Plasma Taurine/Glycine ratio
Serum cholesterol		
Total cholesterol	0.073	0.208
HDL cholesterol	0.169	-0.236
LDL cholesterol	-0.024	0.352*
Triglyceride	0.012	0.003
Mean corpuscular hemoglobin concentration (MCHC)	-0.325*	0.185
Urea	0.501**	-0.190
Plasma free amino acid		
Histidine	-0.125	-0.572**
Isoleucine	0.178	-0.734**
Leucine	0.512**	0.208
Lysine	0.349*	0.015
Methionine	0.159	0.518**
Phenylalanine	0.391**	0.359*
Valine	0.406**	0.167
Alanine	0.273	0.655**
Asparagine	-0.034	0.514**
α-aminoadipic acid	-0.120	-0.828**
α-aminobutyric acid	-0.063	-0.643**
Cystathione	0.061	-0.730**
Glutamate	-0.148	0.413**
Ornithine	0.065	-0.573**
Proline	0.120	-0.549**
Tyrosine	0.084	-0.466**

* p < 0.05   ** p < 0.01   *** p < 0.001

As for the plasma free amino acids, plasma taurine has a significant positive correlation with leucine, lysine, phenylalanine and valine. In addition, the plasma taurine/glycine ratio has a significant positive correlation with methionine, alanine and asparagine and a negative correlation with histidine, isoleucine, α-aminoadipic acid, α-aminobutyric acid, cystathione, ornithine, proline and tyrosine. However, plasma taurine levels were not correlated with anthropometric measurements and other blood analysis data.

4.   CONCLUSION

From the results of this study, it was found that plasma taurine concentrations were the lowest in the obesity subgroup. Therefore, further study on larger populations may be required to determine the relationship between plasma taurine concentrations and obesity.

REFERENCES

1.   Allison DB, Fontaine KR, Monson JE, Stevens J, Vanltallie TB, 1999, Annual deaths attributable to obesity in the United States, *J. Am. Med. Assoc.*; 282: 1530-1538.
2.   Park HS, Yim K, Cho SI, 2002, Familial aggregation of obesity-related phenotypes and nutrient intake in Korean adolescent, *The Korean Journal of Obesity*, 11(1): 97-98.
3.   Danielsson, H., 1963, Present states of research on catabolism and excretion of cholesterol, *Adv. Lipid Res.*, 1: 335-385.
4.   Sugiyama, k., Ohishi A., Ohnuma T., Muramatsu K., 1989, Comparison between the plasma cholesterol-lowering effects of glycine and taurine in rats fed on high cholesterol diets, *Agric Biol Chem.*, 53: 1647-1652.
5.   Mizushimas, Nara Y, Sawamura M, Yamori Y, 1996, Effects of oral taurine supplementation on lipids and sympathetic nerve tone. *Adv Exp Med & Bio* 403: 615-622.
6.   Obinata K, Maruyama T, Hayashi M, Watanabe T, Nittono H, 1991, Effect of taurine on the fatty liver of children with simple obesity, *Metabolism* 40(4): 385-390.
7.   Park T, Kang HW, Park J, Cho S, 2001, Dietary intakes, plasma levels and urinary excretions of taurine in adolescents and adults residing in Seoul area, *The Korean Nutrition Society* 34(4): 440-448.

Regional Differences in the Dietary Taurine Intake in Korean College Students

JEONGHEE LEE, HEAEUN YOU, HYUNI SUNG, JINOH KWAK, KYUNG JA CHANG, SEOYOUN KIM*, HYUNJUNG KIM*, YEOJIN LEE*, and TAESUN PARK*

*Department of Food and Nutrition, Inha University, Incheon, Korea, *Department of Food and Nutrition, Yonsei University, Seoul, Korea*

1.   INTRODUCTION

Taurine (2-aminoethanesulfonic acid) is a free β-amino acid that is normally present in high concentrations in many tissues of man and other animal species. However, taurine is not present in most plant tissues with certain exceptions[1-3].

It is difficult to calculate the dietary intake of taurine, because most of the taurine content in commonly used foods has been analyzed in meat and fish[4,5]. Taurine content has been estimated from 11 food items of meat and fish[6]. The intake of taurine by infants has been estimated from the amount of infant milk consumed[7].

In Korea, a taurine content database of commonly used foods has been developed. A taurine content database for 221 food items of fish and shellfish and 19 food items of seaweeds was developed by the National Fisheries Research and Development Institute[8]. Also a taurine content database for 16 food groups excluding fish and shellfish and seaweeds, commonly used 280 food items, was developed by Park et al[4,5,9]. Therefore, the purpose of this study was to estimate the dietary intake level of taurine in Korean college students.

2.   SUBJECTS AND METHODS

2.1   Subjects

The subjects were 802 college students residing in different provinces [Seoul (n=249), Kyunggi (n=143), Incheon (n=98), Chungnam (n=126), Chungbuk (n=94), Kyungnam (n=64), Pusan (n=28)], and attending a nutrition education class *via* internet. Incheon, Kyungnam and Pusan are provinces near the sea. A cross-sectional study was carried out using a self-administered questionnaire from November 5, 2001 to December 5, 2001.

2.2   Dietary Assessment

A three day recall method was used for dietary assessment (2 weekdays and 1 weekend). Dietary intake levels of taurine and other nutrients were estimated using the computer-aided nutrition (CAN) program inputed with a taurine content database of commonly used Korean foods developed by Park et al[4,5,9].

2.3   Statistical Analysis

Statistical analysis was conducted using the SPSS Program. Data were expressed as means and standard deviations. Significance of difference was determined by analysis of variance (ANOVA) with Duncan's multiple range test. Pearson's correlation coefficient was used to estimate the relation between the variables.

3.   RESULTS AND DISCUSSION

3.1   Dietary Intake Levels of Taurine

Average dietary intake levels of taurine are shown in Figure 1. Average dietary intake levels of taurine in male and female subjects were 278.5±141.4 mg/day and 256.0±153.5 mg/day, respectively. The average taurine intake level of male subjects residing in Kyungnam province (seaside area) (391.8±229.9 mg/day) was significantly higher compared to those for the subjects residing in other provinces; 296.6±180.8 mg/day in Pusan, 284.6±117.9 mg/day in Seoul, 278.9±118.1 mg/day in Incheon, 266.7±140.6 mg/day in Kyunggi, 249.0±98.1 mg/day in Chungnam, 242.9±112.8 mg/day in Chungbuk. Zhao et al.[10] reported that dietary taurine intake of Chinese man ranged from 34-80 mg/day. Kibayshi et al.[11] reported that the dietary taurine intake of Japanese was 225.5 mg/day in males and 162.2 mg/day in females. Another study conducted in Korea by Park et al.[9] reported that the dietary taurine intake was 219 ± 16.9 mg/day for 16-19 years old and 177 ± 18.1 mg/day for adults above 20 years old. Kim et al.[12] reported that the dietary taurine intake was 163.9 ± 150.2 mg/day (range from 8.4 to 767.6 mg/day) in Jeju island.

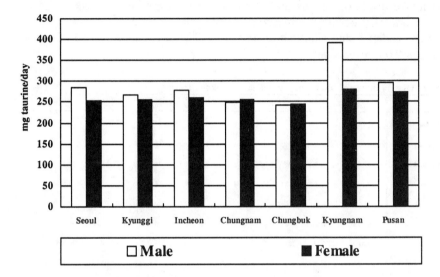

Figure 1. Taurine intake of the subject in different province

The average taurine intake contributed from each food group in the subjects are as follows; 159.7 ± 136.2 mg/day from fish and shellfish, 43.3 ± 31.7 mg/day from vegetables, and 33.3 ± 30.8 mg/day from meat and meat products. Dietary intake of fish and shellfish and dietary taurine intake from food groups of fish and shellfish in the subjects residing in Kyungnam province (seaside area) were significantly higher compared to those for the subjects residing in other provinces. Taurine was not detected in radish root and Korean cabbage[5], but Kimchi or Kkakduki contained taurine ranging from 5.80-32.0 mg/100g wet wt because various kinds of salted fish were included. It is considered that the contribution of taurine intake from vegetables is higher due to the intake of Kimchi in Korea.

3.2   Correlation Between Taurine Intakes and Food Groups of the Subjects

Correlations between dietary taurine intake and food groups of the subjects are shown in Table 2. The dietary taurine intake level was positively correlated with intake levels of cereals and grain products, sugars and sweets, pulse and pulse products, vegetables, seasoning, vegetable oil and fat, meat and meat products, and fish and shellfish at $p<0.01$, and with the intake level of various fruit at $p<0.05$, respectively.

Table 1. Dietary intake of food group and dietary intake of taurine

	Seoul (n=249)	Kyunggi (n=143)	Incheon (n=98)	Chungnam (n=126)	Chungbuk (n=94)	Kyungnam (n=64)	Pusan (n=28)	Total (n=802)
Vegetable food								
Cereals & grain products	[1]293.4([2]10.4)	302.6(9.9)	317.9(11.7)	297.8(11.6)	281.4(9.0)	279.3(10.9)	295.2(9.1)	296.2(10.5)
Potato & starches	28.5(0.6)	22.7(0.8)	26.8(0.4)	31.2(1.1)	24.0(0.3)	24.0(0.3)	20.1(1.3)	26.5(0.7)
Sugar & sweets	7.6(0.5)	7.0(1.5)	8.0(0.2)	8.3(0.4)	6.9(0.0)	7.4(0.2)	7.6(0.6)	7.5(0.3)
Pulse & Pulse products	34.7(0.1)	36.8(0.1)	40.4(0.1)	27.9(0.1)	34.4(0.1)	36.7(0.3)	27.3(0.1)	34.6(0.1)
Nuts & seed	1.8(0.3)	1.4(0.3)	2.0(0.3)	2.2(0.5)	0.3(0.0)	0.9(0.2)	0.5(0.5)	1.6(0.3)
Vegetables	250.7(38.7)	272.5(45.0)	271.9(43.2)	260.1(47.9)	278.0(44.0)	294.9(47.8)	302.7(43.0)	267.2(43.3)
Fungi & mushrooms	1.7(0.4)	1.5(0.4)	1.7(0.5)	1.4(0.5)	2.7(0.2)	2.4(0.4)	1.1(0.3)	1.8(0.4)
Fruits	216.2(1.9)	200.5(0.6)	177.1(0.5)	203.6(0.9)	206.7(0.5)	224.5(1.3)	172.2(0.4)	204.7(1.1)
Seaweeds	2.2(7.0)	2.5(6.0)	1.8(6.4)	3.6(5.9)	2.7(6.7)	3.1(8.0)	2.7(8.0)	2.6(6.7)
Beverages	176.2(0.6)	179.6(0.3)	167.8(1.1)	142.2(1.0)	141.5(0.2)	163.5(0.4)	153.8(0.1)	164.5(0.6)
Seasoning	23.2(0.6)	24.5(0.2)	24.0(0.5)	23.5(1.2)	25.1(0.3)	25.0(0.2)	23.8(0.2)	24.0(0.5)
Vegetable oil & fat	7.4(0.0)	7.6(0.0)	7.8(0.0)	8.4(0.0)	8.3(0.0)	8.9(0.0)	8.5(0.0)	7.9(0.0)
Others	4.2(0.0)	4.8(0.0)	1.2(0.1)	5.4(0.0)	5.9(0.0)	9.5(0.0)	5.0(0.1)	4.8(0.0)
Total(vegetable origin)	1048.0(61.2)	1063.9(63.8)	1048.5(65.1)	1015.6(71.2)	1018.1(61.4)	1080.2(70.0)	1020.5(63.7)	1043.9(64.5)
Animal food								
Meats & meats products	77.2(27.4[a])	97.9(35.7[ab])	91.5(41.4[b])	90.3(34.2[ab])	85.5(35.0[ab])	91.4(34.2[ab])	78.8(32.5[ab])	86.8(33.3[ab])
Eggs	33.9(1.6)	37.0(2.0)	36.7(2.3)	36.1(2.6)	37.2(1.2)	39.4(2.3)	40.6(1.6)	36.2(1.9)
Fishes & shellfishes	47.3(162.3[a])	41.8(155.1[a])	44.6(157.7[a])	40.2(141.3[a])	42.7(142.4[a])	3.5(214.1[b])	52.1(179.8[ab])	45.0(159.7)
Milk & milk products	127.8(4.7)	117.0(4.5)	110.9(6.0)	113.5(4.5)	105.2(3.7)	118.4(5.4)	90.3(4.2)	116.9(4.7)
Animal oil & fat	1.8(0.0)	1.9(0.0)	2.3(0.0)	2.2(0.0)	1.7(0.0)	2.5(0.0)	1.9(0.0)	2.0(0.0)
Total(animal origin)	287.9(96.0)	295.6(197.4)	286.1(207.4)	282.2(182.7)	272.3(182.4)	305.3(256.1)	263.6(218.0)	286.9(199.6)
Total	1335.9(257.2)	1359.5(261.2)	1334.6(272.4)	1297.8(253.9)	1290.5(243.7)	1385.5(326.0)	1284.2(281.8)	1330.7(264.1)

1) Average food group intakes (g).
2) Average taurine intake contributed from each food group (mg).
3) Means with different letters are significantly different at $p<0.05$ by Duncan's multiple range test.

Table 2. .Correlation between dietary taurine intakes and food groups

Food group	Correlation Coefficients	Food group	Correlation Coefficients
Cereals & grain products	0.127**[1)]	Beverages	0.011
Potato & starches	0.036	Seasoning	0.199**
Sugar & sweets	0.132**	Vegetable oil & fat	0.110**
Pulse & Pulse products	0.104**	Vegetable others	−0.015
Nuts & seeds	0.020	Meats & meats products	0.197**
Vegetables	0.220**	Eggs	0.053
Fungi & mushrooms	0.024	Fishes & shellfishes	0.420**
Fruits	0.073*	Milk & milk products	0.018
Seaweeds	−0.022	Animal oil & fat	0.007

1) *, ** Significantly correlated by Pearson correlation at $p < 0.05$ and $p < 0.01$, respectively.

4. CONCLUSION

These results are the first report providing nation-wide information on dietary taurine intake in Korean college students. Therefore, these data will be useful for future studies in estimating dietary intake levels of taurine in various populations.

REFERENCES

1. Huxtable, R.J., 1992, Physiological actions of taurine. *Physiol Rev* 72: 101-163.
2. Schweiigen, R.G., 1967, Low-molecular-weight compounds in *Macrocystis pyrifera*, a marine algae. *Arc Biocem* 118: 383.
3. Kataoka, H., Ohnishi, N., 1986, Occurrence of taurine in plants. *Agric Biol Chem* 50: 1887.
4. Park, T., Park, J.E., 1999, Taurine contents in beverages, milk products, sugars and condiments consumed by Korean. *J Kor soc Food Sci & Nutr* 28(1): 9-15.
5. Park, T., Park, J.E., Chang, J.S., Son, M.W., Shon, K.H., 1998, Taurine contents in Korean foods of plant origin. *J Kor soc Food Sci & Nutr* 27(5): 801-807.
6. Roe, D.A., 1966, Taurine intolerance in Psoriasis. *J Invest Dermatol* 46: 420-430.
7. Kim, E.S., Cho, K.H., Park, M.A., Lee, K.H., Mon, J., Lee, Y.N., Ro, .K., 1996,.Taurine intake of Korean breast-fed infants during lactation. *Adv Exp Med & Biol* 403: 571-77
8. National Rural Living Science Institute, 1996, R.D.A: *Food Composition Table*. Fifth revision.

9.   Park, T., 2000, Studies on novel activities of taurine and the development of taurine content database of foods. Final Reports of Korean Health Research and Development Project, pp.99-106, Ministry of Health and Welfare, Republic of Korea.
10. Zhao, X.H., Jia, J.B, 1998, Taurine content in Chinese food and daily intake of Chinese men. *Adv Exp Med Biol* 442: 501-505.
11. Kibayashi, E., Yokogoshi, H., Mizue, H., Miura, K., Yoshita, K., Nakagawa. H,, Naruse. Y., Sokejima, S., Kagamimori, S., 2000, Daily dietary taurine intake in Japan. *Adv Exp Med Biol* 483: 137-142.
12. Kim, E.S., Kim, J.S., Yim, M.Y., Roh, H.J., Jeong, Y., Ko, Y.S., 2001, International symposium on food, nutrition and health for 21st century, *The Korean Society of Food Science and Nutrition*, p. 82.

Taurine Concentration in Human Blood Peripheral Lymphocytes

Major Depression and Treatment with the Antidepressant Mirtazapine

LUCIMEY LIMA*, FRANCISCO OBREGÓN*, MARY URBINA*, ISABEL CARREIRA[#], EDITH BACCICHET[#] and SOLISBELLA PEÑA[#]

Laboratorio de Neuroquímica, Centro de Biofísica y Bioquímica, Instituto Venezolano de Investigaciones Científicas, Apdo. 21827, Caracas 1020-A and [#]Centro de Salud Mental del Este, El Peñón, Caracas, Venezuela. llima@cbb.ivic.ve

Summary

Major depression is a serious disease with various systemic effects, including dysfunction of the immune response. Taurine has been known to be related to certain modifications of the immune system. The aim of this study was to determine the taurine concentration in lymphocytes of patients with major depression and to evaluate the influence of the antidepressant treatment with mirtazapine for six weeks on the levels of taurine. γ-Aminobutyric acid, aspartate, glutamate and glutamine were also determined. Taurine, aspartate and glutamine levels were increased in the lymphocytes of depressed patients before mirtazapine treatment compared to the control group, and were normalized after treatment. γ-Aminobutyric acid and glutamate did not differ between patients and controls. There was a significant and positive correlation between the severity of the disorder, measured by the Hamilton Rating Scale, and the concentration of taurine in the lymphocytes of depressed patients before treatment. This correlation was not observed after treatment and neither was there a correlation observed for the other amino acids. The present observations could be an indication of the relevance of taurine as a protective agent in the lymphocytes of patients with severe depression, and could be the result of modifications of taurine transport or efflux processes.

1.   INTRODUCTION

Taurine has been known to inhibit the proliferation of human cultured lymphocytes stimulated by the mitogen phytohemaglutinin[1]. Moreover, in mice, aging results in a decrease in the proliferation of T and B lymphocytes produced by phorbol myristate acetate and ionomycin. However, taurine increases the reduced proliferation of T cells[2]. In the rat, aging produces suppression or impairment in the lymphocytic response to the mitogen concanavalin A, but not to lypopolysacharide, increased B cells, decreased T cells, or increased levels of serotonin, aspartic acid, glutamic acid and taurine mainly in the lymph nodes[3]. Relevant research performed in cats revealed that taurine deficiency results in leukopenia, increased mononuclear leukocytes, decreased phagocytosis, increased serum immunoglobulins and histological changes in lymph nodes and spleen[4]. In addition, it has been proposed that taurine plays an important role in infant nutrition, related to the intestinal mucosal immune system[5]. Taurine and arginine stimulate the immune system at the lymphocyte level, probably by blocking the formation of lipid peroxidation products[6]. Previous studies support a role for taurine in the proper function of the immune system.

There are a number of reports indicating modifications of taurine concentrations in circulating cells in specific conditions. For instance, in humans after lengthy parenteral nutrition there is a reduction of the taurine concentrations in plasma, platelelets, lymphocytes, granulocytes and erythrocytes[7,8]. In addition, cats submitted to a diet depleted in taurine present low levels of taurine in plasma, platelets, granulocytes and erythrocytes, without linear correlation between plasma and cellular levels[9]. Finally, taurine concentrations in human platelets and leukocytes have been reported to be around 10 to 20 mM, 500 times higher than plasma levels. This difference may be related to protection against antioxidants[10].

Major depression is a serious disorder characterized by a depressed mood and loss of interest or pleasure which lasts for at least two weeks[11]. The prevalence is reported to be 4.4 – 20%[12]. Considerable evidence supports that major depression is a disease which causes a decrease in the quality of life and productivity, and an increase in the development of medical illnesses, including activation of the hypothalamic-pituitary-adrenal axis and abnormal secretion of cytokines[13].

In peripheral lymphocytes of the blood of psychiatric patients, there are modifications of adrenergic receptors[14] and of the serotonin transporter[15,16], as well as in the levels of plasma and lymphocyte serotonin after antidepressant treatment[15,17]. For instance, serotonin transporters are decreased in patients with major depression and the administration of antidepressants, such as fluoxetine[15] or mirtazapine[17] produces an increase in the number of transporters, as indicated by the binding of [³H]paroxetine.

Due to the lack of information concerning taurine, lymphocytes and major depression, the aim of the present work was to determine the concentration of taurine and other amino acids in patients with major depression before and after the treatment with the antidepressant mirtazapine.

2. MATERIALS AND METHODS

2.1 Subjects

Twenty seven patients (7 men) with ages ranging from 18 to 60 years (41.30 ± 3.80) were diagnosed according to DSM-IV criteria for major depression[11], but without other major psychiatric disorders or somatic illness. They were free of drugs for at least two weeks prior to taking blood samples and initiating the antidepressant treatment. The severity of the depressive symptoms was determined by the Hamilton Rating Scale for Depression (HAM-D, NIH)[18]. Patients with a score of 18 or more were included in the study. The antidepressant treatment consisted of 30 mg/day of mirtazapine, an antagonist of α_2-adrenergic receptors which increases serotonergic and adrenergic transmission[19,20], for six weeks. Controls were composed of 25 and 20 subjects alternating with patients before and after the treatment, 24-59 years (17 men), age 34.80 ± 2.34.

2.2 Preparation of Blood Peripheral Lymphocytes

Blood samples were taken by venipuncture between 7 and 9 a.m. (50 ml + 0.6 ml of heparine, 1000 units/ml), before initiation of the antidepressant treatment and six weeks after receiving 30 mg/day of mirtazapine. Preparation of lymphocytes was done as previously described[16]. The lymphocytes were resuspended in a buffer solution for HPLC, homogenized, acidified with 20% sulfosalycilic acid (50 μl per 300 μl of cell preparation), and centrifuged. The supernatant was stored at -80^0C.

2.3 Determination of Amino Acids

Taurine, γ-aminobutiric acid, aspartic acid, glutamic acid and glutamine were determined by HPLC with a fluorescent detector as previously described[21].

300   *Lucimey Lima et al.*

2.4   Statistical Analysis

Results are expressed as mean ± SEM. Analysis of variance was performed and comparison between groups was performed by the Student's *t* test. Linear correlation was carried out for the amino acid concentrations and severity of the depression symptoms as determined by the score on the Hamilton Rating Scale. Significance if P < 0.05.

3.   RESULTS

Taurine concentration was increased in the lymphocytes of patients with major depression without any antidepressant treatment for two weeks as compared to the corresponding control group. Six weeks after receiving the antidepressant mirtazapine there was a decrease in the score on the Hamilton Scale for depression as well as a clinical improvement in the depression symptoms, and there also was a decrease in the concentration of taurine in the lymphocytes as compared to the patients before the treatment. However, there was no statistical significant difference with the corresponding control groups. The levels of taurine in the lymphocytes were not statistically different between the two groups of controls (Fig. 1).

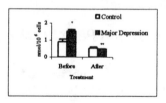

Figure 1. Taurine concentrations in blood peripheral lymphocytes of control and major depression patients before and after the antidepressant treatment with mirtazapine for six weeks. $F_{(3,95)}$ = 24.43, P < 0.001. * P < 0.05 respecting corresponding Control, ** P < 0.05 with respect to Major Depression after treatment.

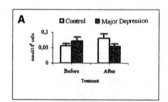
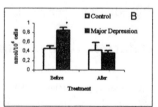

Figure 2. A) γ-Aminobutyric acid, and B) aspartate acid concentrations in blood peripheral lymphocytes of control and major depression patients before and after the antidepressant treatment with mirtazapine for six weeks. γ-Aminobutyric acid, $F_{(3,88)}$ = 1.61, aspartic acid,. $F_{(3,96)}$ = 4.99, P < 0.01. * P < 0.05 with respect to Major Depression after treatment.

The levels of GABA did not present any significant difference between the four groups, controls and patients before and after treatment (Fig. 2A). Aspartate levels were increased before treatment and not different from controls after treatment (Fig. 2B). Glutamine concentrations were higher in the lymphocytes of patients with major depression, but were lower after the treatment compared to patients before the treatment (Fig. 3A). Glutamate concentrations were not statistically different between patients and controls, but were higher in the two groups before treatment than after the administration of the antidepressant (Fig. 3B).

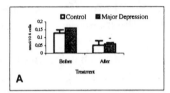 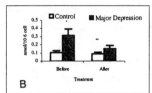

Figure 3. A) Glutamate, and B) glutamine concentrations in blood peripheral lymphocytes of control and major depression patients before and after the antidepressant treatment with mirtazapine for sis weeks. Glutamate, $F_{(3,96)}$ = 8.78, P < 0.001, glutamine, $F_{(3,65)}$ = 5.24, P < 0.01. * P < 0.05 respecting corresponding Control, ** P < 0.05 with respect to Major Depression after treatment.

There was a significant and positive correlation between the concentration of taurine in th lymphocytes of patients with major depression and the score on the Hamilton Scale for Depression. However, a correlation was not observed after treatment with the antidepressant mirtazapine (Table I). There was no significant correlation between the severity of the disorder and the levels of GABA, aspartate, glutamate or glutamine.

Table I. Correlation coefficients of amino acid concentrations *versus* severity of depression symptoms measured by the Hamilton Rating Scale for depression.

	Before treatment	After treatment
Taurine	0.4076*	0.2117
GABA	0.0647	0.0710
Aspartate	0.1056	-0.0540
Glutamate	0.1104	0.0847
Glutamine	0.0580	0.0900

* P < 0.05

4.   DISCUSSION

Studies relating plasma amino acids and major depression have been mainly performed on tryptophan or tyrosine levels, since they are related to the disorder as precursors of serotonin and catecholamine synthesis. Total plasma tryptophan and the ratio of tryptophan/large neutral amino acids (T/LNA) are decreased in depressed patients[22] and are associated with a therapeutic response to specific antidepressants[23,24,25]. The T/LNA ratio seems to be useful in distinguishing major depression from schizophrenia, but the tyrosine/LNA ratio shows very limited usefulness[23]. In addition, a significant higher tyrosine/LNA ratio and a lower T/LNA ratio could represent an index of a beneficial response to fluvoxamine, an antidepressant inhibiting serotonin uptake[26].

Supersensitivity of platelet glutamate receptors has been observed in major depression[27] and glutamate, taurine and lysine plasma levels, and aspartate, serine and lysine platelet levels are higher[26]. In addition, lower serum levels of aspartate, asparagine, serine, threonine and taurine are related to a no response to antidepressants[28].

Aspartate and taurine concentrations were found to be higher in the lymphocytes of patientes with major depression as compared to controls. Also the treatment with the antidepressant mirtazapine reduced the concentrations of both amino acids in lymphocytes. The significance of these findings are unknown at the present, but it could be secondary to transport modifications, to efflux changes, or to a metabolic dysfunction in depression. GABA and glutamine levels were unchanged in the lymphocytes of depressed patients with respect to the control group. Low plasma GABA levels have been proposed as a marker for mood disorders[29].

Control levels were not significantly different in the groups before and after the antidepressant treatment, except for glutamate, which was lower in controls after the treatment. This observation could be due to variations in the year, and for this reason controls should be alternated with patients in relatively long term studies.

Beside the lack of correlation between the amino acid concentrations in lymphocytes and the severity of the depression[30] taurine levels were significantly and positively correlated with the Hamilton Rating Scale score before the treatment, which might be interpreted as a protective effect of taurine in the lymphocytes of depressed patients. Although differences in aspartate and glutamine levels between patients and controls were found, no correlation with severity of the disorder was detected for the rest of the amino acids determined in the present group of patients.

The present results encourage continued research on taurine and lymphocytes in patients with major depression to hopefully achieve a better understanding of this disorder and potentially to determine the influence of taurine on therapeutic approaches.

ACKNOWLEDGEMENTS

This work was supported by Grant G-2000-1387 from FONACIT. We appreciate the secretarial assistance of Mrs. Isabel Otaegui.

REFERENCES

1. Pasantes-Morales, H. and León-Cázeres, J. M., 1981, Effect of taurine on mitogen response of human lymphocytes. *Experiencia*, 37:993-994.
2. Nishio, S., Negoro, S., Hosokawa, T., Hara, H., Tanaka, T., Deguchi, Y., Ling, J., Awata, N., Azuma, J. and Aoike, A., 1990, The effect of taurine on age-related immune decline in mice: the effect of taurine on T cell and B cell proliferative response under costimulation with ionomycin and phorbol myristate acetate. *Merch. Ageing Dev.* 52:125-139.
3. Bonacho, M. G., Cardinali, D. P., Castrillon P., Cutrera, R. A. And Esquifino, A. I., 2001, Aging-induced changes in 24-h rhythms of mitogenic responses, lymphocyte subset populations and neurotransmitter and amino acid content in rat submaxillary lymph nodes during Freund's adjuvant arthritis. *Exp. Gerontol.* 36:267-282.
4. Schuller-Levis, G., Mehta, P. D., Rudelli, R. and Sturman, J., 1990, Immunologic consequences of taurine deficiency in cats. *J. Leukoc. Biol.* 47:321-331.
5. Chesney, R. W., Helms, R. A., Christensen, M., Budeau, A. M., Han, X. and Sturman, J. A., 1998, An updated view of the value of taurine in infant nutrition. *Adv. Pediatr.* 45:179-200.
6. Lubec, B., Hoeger, H., Kremser, K., Amann, G., Koller, D. Y. and Gialamas, J., 1996, Decreased tumor incidence and increased survival by one year oral low dose arginine supplementation in the mouse. *Life Sci.* 58:2317-2325.
7. Vinton, N. E., Laidlaw, S. A., Ament, M. E. and Kople, J. D., 1986, Taurine concentrations in plasma and blood cells of patients undergoing long-term parenteral nutrition. *Am. J. Clin. Nutr.* 44:398-404.
8. Vinton, N. E., Laidlaw, S. A., Ament, M. E. and Kople, J. D., 1987, Taurine concentration in plasma, blood cells, and urine of children undergoing long-term total parenteral nutrition. *Pediatr. Res.* 21:399-403.
9. Laidlaw, S. A., Sturman, J. A. and Kopple, J. D., 1987, Effect of dietary taurine on plasma and blood cell taurine concentrations in cats. *J. Nutr.* 117:1945-1949.
10. Learn, D. B., Fried, V. A. and Thomas, E. L., 1990, Taurine and hypotaurine content of human leukocytes. *J. Leukoc. Biol.* 48:174-182.
11. American Psychiatry Association. DSM-IV: Diagnostic and Statistical Manual for Mental Disorders, 3rd ed. Washington, DC: APA, 1994.
12. Bakish, D. New standard of depression treatment: remission and full recovery. *J. Clin. Psychiatry* 62 Suppl 26:5-9.
13. Leonard, B. E. The immune system, depression and the action of antidepressant. *Prog. Neuropsychopharmacol. Biol. Psychiatry* 25:767-780.
14. Mazzola-Pomietto, P. And Azorin, J.-M., 1994, Relation between lymphocyte β-adrenergic responsivity and severity of depressive disorders. *Biol. Psychiatry* 35:920-925.
15. Hernández, E., Lastra, S., Urbina, M., Carreira, I. and Lima, L., 2002, Serotonin transporter and concentration in blood peripheral lymphocytes of patients with generalized anxiety disorder. *Int. J. Immunopharmacol.* (in press).
16. Urbina, M., Pineda, S., Piñango, L., Carreira, I. and Lima, L., 1999 [³H]Paroxetine binding to human peripheral lymphocyte membranes of patients with major depression before and after treatment with fluoxetine. *Int. J. Immunopharmacol.* 21:631-646.

17. Baccichet, E. and Peña, S., 2001, El transportador de serotonina en linfocitos de sangre periférica de pacientes con trastorno depresivo mayor antes y después del tratamiento con mirtazapina. Trabajo Especial de Investigación, Universidad Central de Venezuela e Instituto Venezolano de Investigaciones Científicas.

18. Handbook of Psychiatric Measurements (HPM), American Psychiatric Association, 1st. edn., 2000.

19. Gorman J. M., 1999, Mirtazapine: clinical overview. *J. Clin. Psychiatry* 60:9-13.

20. Kent, J. M., 2000, SnaRIs, NaSSAs, and NaRIs: new agents for the treatment of depression. *Lancet* 355:911-918.

21. Lima, L., Obregón, F. and Matus, P., 1998, Taurine, glutamate and GABA modulate the outgrowth from goldfish retinal explants and its concentrations are affected by the crush of the optic nerve. *Amino Acids* 15:195-209.

22. Cowen, P. J., Parry-Billings, M. and Newsholme, E. A., 1989, Decreased plasma tryptophan levels in major depression. *J. Affect. Disord.* 16:27-31.

23. Lucca, A., Lucini, V., Catalano, M. And Smeraldi, E., 1995, Neutral amino acid availability in two major psychiatric disorders. *Prog. Neuropsychopharmacol. Biol. Psychiatry* 19:615-626.

24. Lucini, V., Lucca, A., Catalano, M. And Smeraldi, E., 1996, Predictive value of tryptophan/large neutral amino acids ratio to antidepressant response. *J. Affect. Disord.* 36:129-133.

25. Moller, S. E., 1993, Plasma amino acid profiles in relation to clinical response to moclobemide in patients with major depression. *J. Affect. Disord.* 27:225-231.

26. Mauri, M. C., Ferrara, A., Boscati, L., Bravin, S., Zamberlan, F., Alecci M. And Invernizzi, G., 1998, Plasma and platelet amino acid concentrations in patients affected by major depression and under fluvoxamine treatment. *Neuropsychobiology* 37:124-129.

27. Berk, M., Plein, H. and Ferreira, D., 2001, Platelet glutamate receptor supersensitivity in major depressive disorder. *Clin. Neuropharmacol.* 24:129-132.

28. Maes, M., Verkerk, R., Vandoolaeghe, E., Lin, A. and Scharpe, S., 1998, Serum levels of excitatory amino acids, serine, glycine, histidine, threonine, taurine, alanine and arginine in treatment-resistant depression: modulation by treatment with antidepressants and prediction of clinical responsivity. *Acta Psychiatr. Scand.* 97:302-308.

29. Goddard, A. W., Narayan, M., Woods, S. W., Germine, M., Kramer, G. L., Davis, L. L. and Petty, F., 1996, Plasma levels of gamma-aminobutyric acid and panic disorder. *Psychiatric Res.* 63:223-225.

30. Altamura, C., Maes, M., Dai, J. And Meltzer, H. Y., 1995, Plasma concentration of excitatory amino acids, serine, glycine, taurine and histidine in major depression. *Eur. Neuropsychopharmacol.* 5 Suppl:71-75.

Part 5:

Taurine: Growth, Antioxidants and Inflammation

Why Is Taurine Cytoprotective?

STEPHEN SCHAFFER[1], JUNICHI AZUMA[2], KYOKO TAKAHASHI[2], and MAHMOOD MOZAFFARI[3]
[1]Department of Pharmacology, University of South Alabama, College of Medicine, Mobile, AL, USA
[2]Department of Clinical Evaluation of Medicines and Therapeutics, Osaka University, Osaka, Japan
[3]Department of Oral Biology and Maxillofacial Pathology, Medical College of Georgia School of Dentistry, Augusta, GA, USA

1. INTRODUCTION

The concept that taurine exhibits cytoprotective activity was introduced in 1981. Although several studies at the time had supported the notion that taurine was capable of modulating Ca^{2+} movement, most of those studies focused on the transporter affected by taurine. It was only after taurine was found to prevent Ca^{2+}-induced cellular necrosis that its cytoprotective activity was recognized[1]. Since that date, numerous reports have documented that high levels of extracellular taurine render cells resistant to an array of damaging stimuli, including ischemia-reperfusion, hyperglycemia, reactive oxygen species, heat shock, toxic xenobiotics, cellular excitotoxicity and osmotic derangements. The ability of taurine to counteract these toxic stimuli has been attributed to four actions of taurine: (1) osmoregulation, (2) anti-oxidant/membrane stabilizing activity, (3) conjugation and (4) regulation of $[Ca^{2+}]_i$. Indeed, osmotic stress, oxidative stress and Ca^{2+} overload are major causes of cellular necrosis and apoptosis. Nonetheless, cell death is a complex phenomenon, involving interactions between these and several other factors. For example, oxidative stress can damage membranes leading to changes in $[Ca^{2+}]_i$. Conversely, high $[Ca^{2+}]_i$ can cause enhanced free radical generation by the mitochondria. It is also known that osmotic stress affects the activity of ion transporters that regulate $[Ca^{2+}]_i$. Despite these and other interactions, it is clear that taurine affects each of the four factors individually. Thus, the aim of this review is to discuss the importance of each factor in the cytoprotective actions of taurine.

Taurine 5: Beginning the 21st Century
Edited by Lombardini *et al.*, Kluwer Academic/Plenum Publishers, New York 2003

2.   OSMOREGULATION AS A CYTOPROTECTIVE MECHANISM

Taurine is a very important organic osmolyte in mammalian cells. Generally, when cells are subjected to a hyposmotic insult, a rapid efflux of organic and ionic osmolytes ensues, partially normalizing the osmotic balance[2]. This process, known as the regulatory volume decrease, serves as a safety valve to prevent damage caused by excessive cell swelling. Because the efflux of taurine from the hyposmotically stressed cell is very rapid, taurine loss is an important contributor to the regulatory volume decrease.

Although the regulatory volume decrease partially normalizes the osmotic balance across the cell membrane, cell swelling can still represent a major clinical problem. Among the complications associated with cell swelling are the compression of capillaries, the prevention of reperfusion (no-reflow phenomenon), the alteration in a cell's electrophysiology, the elevation in cell stiffness and the disruption of the cell membrane[3]. These effects are minimized by either a decrease in the intracellular osmotic load or an increase in the osmolality of the interstitium. Indeed, intravenous administration of mannitol has been found to reduce cellular edema, minimize swelling-induced disruption of the cell membrane and eliminate the no-reflow phenomenon in the ischemic myocardium. Similarly, a reduction in the intracellular osmotic load through taurine depletion reduces cerebral edema during acute hyponatremia[4] and protects the heart against ischemia[5]. Thus, the acute effects of taurine loss can significantly benefit the osmotically stressed cell.

Taurine loss, however, can also have an adverse effect. The volume decrease that accompanies apoptosis appears to be an exaggerated activation of the regulatory volume decrease[6]. According to Lang et al.[7] stimulation of Fas-mediated apoptosis in Jurkat T-lymphocytes is accompanied by taurine release. This is an important step in the apoptotic cascade because preloading cells with taurine significantly inhibits Fas-induced DNA fragmentation and apoptotic cell shrinkage. Although taurine loading does not prevent some of the early apoptotic events, such as caspase activation, it can prevent the apoptotic cascade from proceeding beyond the cell shrinkage step. These findings are consistent with earlier studies suggesting that the loss of organic osmolytes might participate in the dismantling of the cell during apoptosis[8].

Taurine also contributes to volume regulation following a hyperosmotic insult. In response to an abrupt increase in extracellular osmolality, the cell rapidly accumulates Na^+, K^+ and Cl^-. If the hyperosmotic insult is severe enough, osmotically sensitive cells may experience very severe damage and even die. Ozasa and Gould[9] found that only 30% of chimpanzee spermatozoa survive a one hour hyperosmotic insult produced by doubling the extracellular osmolality with NaCl. Yet, inclusion of 2 mM taurine in the incubation medium increased the number of living cells to 55%.

Several pathological conditions, including chronic hypernatremia, are associated with a hyperosmotic imbalance. The treatment of chronic hypernatremia requires special care because rapid rehydration can lead to an osmotic imbalance associated with brain damage and the risk of seizures, coma and even death. As one would predict, taurine transport plays a central role in chronic hypernatremia; taurine uptake is enhanced and taurine efflux is retarded[10]. Trachtman et al.[11] found that taurine deficient cats are very susceptible to hypernatremic dehydration, experiencing an increased mortality rate, significant brain cell shrinkage and an enhanced rate of seizure activity. In addition, a recent study showed that chronic taurine deficiency modulates renal function and total body fluid homeostasis through the actions of arginine vasopressin on the kidney[71].

Diabetes mellitus is another clinical condition with a hyperosmotic component. High extracellular glucose causes an osmotic imbalance by increasing both the extracellular osmolality and the accumulation of glucose end-products, such as sorbitol. The rise in intracellular sorbitol is associated with a corresponding decrease in the levels of other osmolytes, such as taurine[12,13]. This loss of taurine appears to contribute to specific complications of diabetes, as evidenced by the amelioration of hyperglycemia-induced nephropathy[14,15] and neuropathy[16] and vascular injury[17] following chronic administration of taurine. Although these beneficial effects of taurine may be related in part to an improvement in the osmotic balance, hyperglycemia also leads to oxidant stress and the accumulation of advanced glycosylation end-products, effects that are also altered by taurine (see below).

In summary, there is little doubt that acute changes in taurine content affect a cell's osmotic balance in a way that can either be beneficial or detrimental. More studies are warranted to define conditions in which taurine treatment might be beneficial.

3. ANTI-OXIDANT AND MEMBRANE STABILIZING ACTIVITY OF TAURINE

It is generally accepted that taurine treatment protects cells against oxidative injury. Most studies that have focused on the anti-oxidant activity of taurine, have used taurine as an antidote against specific xenobiotics. However, the first study recognizing the anti-oxidant activity of taurine examined light-induced damage to photoreceptor outer segments[18]. Pasantes-Morales and Cruz[18] showed that taurine treatment dramatically reduced the number of disrupted rod outer segments while preventing light-induced lipid peroxidation. Surprisingly, a year earlier taurine had been found to prevent Fe^{2+}-induced rod outer segment damage, but not by reducing the degree of lipid peroxidation[19].

Oxidative damage to the pulmonary system is also responsive to taurine therapy. The lungs are very susceptible to free radical injury due to their exposure to high oxygen tension and toxic gases, such as ozone and the nitrogen oxides. According to the work of Banks et al.[20] nearly 40% of alveolar macrophages exposed to 0.45 ppm of ozone for 30 min in the absence of taurine were found to lose viability. However, pretreatment of the cells with taurine reduced the extent of ozone-mediated cell death to about 15%, an effect associated with a dramatic reduction in specific measures of oxidative stress, such as lipid peroxidation and reduced glutathione loss. In a related study, Gordon et al.[21] found that oral taurine administration protected hamster bronchioles from acute NO_2-induced injury, a condition also associated with severe oxidative stress. Taurine has also been found to protect the lungs against a number of free radical generating xenobiotics. The antineoplastic agent, bleomycin, forms an intracellular bleomycin-Fe^{2+} complex that generates oxygen free radicals and produces a pneumonitis and fibrotic lesions[22,23]. Addition of taurine and niacin to the drinking water reduced the inflammatory response in animals treated with bleomycin[22]. The protective effect was reflected in the degree of lipid peroxidation, nitric oxide production, collagen accumulation and the appearance of acid phosphatase in the bronchoalveolar lavage fluid. Although the authors attributed the cytoprotection to a reduction in oxidative stress, taurine treatment also prevented the increase in total lung calcium content, suggesting that multiple mechanisms might contribute to the beneficial effects of taurine[22,23]. Similar experiments using monocrotaline[24] and amiodarone[25] as toxic agents have also revealed the anti-oxidant activity of taurine.

Xenobiotic toxicity is a major problem in the liver, the site of xenobiotic metabolism, as well as taurine biosynthesis. Although the hepatic cytochrome P450 system detoxifies most xenobiotics, some xenobiotics are bioactivated by the cytochrome P450 system. A classical example of the latter is carbon tetrachloride, which is converted by the cytochrome P450 system into a trichloromethyl radical. Like most reactive free radicals, the trichloromethyl radical triggers lipid peroxidation and reacts with a number of proteins to inactivate them. Although the damage from the trichloromethyl radical is restricted to the microsomes, a crucial consequence of the toxicity is the inactivation of microsomal Ca^{2+} accumulation. Consequently, $[Ca^{2+}]_i$ increases and some key Ca^{2+} dependent enzymes are activated[26]. The hepatocyte eventually dies from disintegration of the cell membrane by the Ca^{2+} dependent enzymes. Waterfield et al.[27] have found that rat hepatocytes incubated with medium containing various concentrations of taurine are resistant to carbon tetrachloride toxicity, with the higher concentrations of taurine (10-20 mM) exhibiting more protection than intermediate concentrations (5 mM). A similar cytoprotective dose-response curve was observed for hydralazine and naphthoquinone toxicity. Timbrell et al.[28] proposed that the mechanism underlying the cytoprotection appears to involve the modulation of calcium movement, osmoregulation or membrane stabilization. However, Wu et al.[29] implicated polyamines in the observed protection against hydrazine and carbon tetrachloride toxicity, although it is unclear how taurine can affect polyamine biosynthesis. Other toxicity studies appear to agree with the Timbrell hypothesis, as taurine was found to minimize oxidative damage and hepatotoxicity in rats exposed to cadmium[30], thioacetamide[31] and oxidized fish oil[32].

While the liver regulates taurine content through biosynthesis, the kidney regulates whole body taurine through excretion. Like the liver, the kidney benefits from the cytoprotective activity of taurine. Trachtman et al.[33] found that chronic puromycin aminonucleoside administration induces a proteinuric renal disease that resembles focal segmental glomerulosclerosis. The drug-induced nephropathy is characterized by the infiltration of mononuclear cells and neutrophils into the renal parenchyma. The inflammatory response is important because neutrophils produce HOCl, a highly reactive oxygen species that can be scavenged by taurine. Therefore, it is not surprising that taurine treatment reduces the degree of segmental glomerulosclerosis and improves creatinine clearance[33]. However, taurine protection is selective, having little effect on the excretion of protein and albumin. Based on these findings, Trachtman et al.[33] proposed that the cytoprotective activity of taurine must be related in part to the scavenging of HOCl. However, the effects of taurine may also include the suppression of neutrophil activity by N-chlorotaurine (see below). In this regard, Trachtman et al.[33] found that a number of other reactive oxygen species and free radicals are produced during the inflammatory response. Based on electron spin resonance measurements, taurine treatment reduces the levels of these other free radical species in the kidney. Whether the anti-oxidant effect of taurine is related to a reduction in free radical generation by the neutrophil or a secondary effect of taurine on free radical generation remains to be determined.

Recent studies suggest that oxidative stress is responsible for many of the complications of diabetes. In support of this notion, Trachtman et al.[14] found that the malondialdehyde content of kidneys from type 1 diabetic rats is significantly elevated. When they treated the diabetic animals with either insulin or taurine there was a decline in renal malondialdehyde content and an improvement in renal function, as evidenced by reductions in proteinuria, glomerular scarring and tubulointerstitial injury. Taurine treatment was also found to prevent lipid peroxidation and collagen production in mesangial cells exposed to high glucose[34]. In those studies, epithelials cells (MDCK and LLC-PK$_1$) were used as controls because they do not respond to high glucose; consequently, they were also unaffected by taurine. Yet, LLC-PK$_1$ cells are damaged by a pro-oxidant, such as iron and L-dopa; taurine protects against these agents[35]. Although these findings are consistent with a role of taurine as an anti-oxidant, it is important to note that in the diabetic rat, it was found that the free radical scavenger, vitamin E, exacerbates the diabetic nephropathy[14]. Therefore, the effects of taurine appear to be more complex than merely a reduction in oxidative stress.   One possibility is the neutralization of advanced glycation products by taurine (see below).

The heart contains low levels of the antioxidant defenses, therefore, it is also susceptible to oxidative damage.   Two agents that produce oxidative stress and respond favorably to taurine thereapy are doxorubicin and isoproterenol.   Toxic concentrations of isoproterenol have been shown to elevate malondialdehyde content, reduce the levels of glutathione and decrease the activity of glutathione peroxidase, taurine treatment partially prevents these changes[36].   Similarly, taurine treatment prevents the elevation in malondialdehyde content in the doxorubicin treated mouse, but does not prevent the decline in glutathione peroxidase activity[37].   While these data support a role for taurine as an anti-oxidant, taurine also attenuates the excessive accumulation of Ca^{2+} in hearts exposed to the two toxins[36,37].

There has been a great deal of debate in the literature regarding the anti-oxidant activity of taurine.   Aruoma et al.[38] clearly showed that taurine does not directly react with superoxide, H_2O_2 or hydroxyl radical. Yet, taurine has been shown to protect hepatocytes against H_2O_2-induced damage [39].   If taurine cannot <u>directly</u> scavenge the classical reactive oxygen species, it is logical to assume that the actions of taurine within the cell are <u>indirect.</u> There are several mechanisms that could explain the anti-oxidant activity of taurine.   First, taurine could elevate the levels of the anti-oxidant defenses. Unfortunately, most studies that have examined the effect of taurine on the anti-oxidant defenses were initially designed to evaluate its effectiveness in reversing the actions of a toxic xenobiotic.   In one such study, Giri et al.[40] found that bleomycin increased the activity of superoxide dismutase in hamster lung but the increase was prevented in hamsters fed a diet containing elevated levels of taurine and niacin.   Nonetheless, there are a handful of studies that have directly tested the effects of taurine on the anti-oxidant defenses.   Vohra and Hui[41] found that intragastric administration of taurine elevated superoxide dismutase and glutathione peroxidase activity in certain brain regions.   They suggested that the increase in glutathione peroxidase activity might account for taurine's effectiveness in preventing carbon tetrachloride toxicity.   Similarly, Balkan et al.[31] found that taurine improved the status of the anti-oxidant defenses in the liver of thioacetamide-treated rats by elevating the levels of vitamin E and increasing the activity of glutathione peroxidase.   In another study, Nonaka et al.[42] found that pretreatment of vascular smooth muscle cells with 10 mM taurine prevented homocysteine-mediated reductions in the expression of superoxide dismutase. Mochizuki et al.[43], who focused on the organic anti-oxidants, found that taurine altered the metabolism of ascorbic acid while Trachtman et al.[14] observed a beneficial effect of taurine treatment on serum free iron concentration. While these studies raise the possibility that taurine might increase the levels of the endogenous anti-oxidants, a recent study by Pitari et al.[44] questions this notion. Pitari et al.[44] found that taurine treatment reduced superoxide dismutase and catalase activity in muscle of rats maintained under normobaric hyperoxic conditions. Even under normoxic conditions, a reduction in catalase activity was found after taurine treatment. Clearly, further studies are warranted to clarify the effect of taurine treatment on the anti-oxidant defense system.

Although taurine is incapable of scavenging the classical reactive oxygen species, but readily reacts with HOCl to form N-chlorotaurine[45]. This reaction is catalyzed by myeloperoxidase, an enzyme found in the neutrophil. Neutrophils are also rich in both taurine and HOCl, making the leukocyte a major source of N-chlorotaurine. Although N-chlorotaurine exhibits bactericidal and fungicidal activity, it is less cytotoxic than hypochlorous acid[46,47]. In fact, the formation of N-chlorotaurine may protect the neutrophil itself against oxidative stress from excessive HOCl production[48]. Another important property of N-chlorotaurine is the ability to regulate the severity of the inflammatory response[40,49,50]. Several groups have shown that N-chlorotaurine downregulates the generation of inflammatory mediators, such as superoxide, nitric oxide, tumor necrosis factor-α, interleukin-6 and prostaglandin E_2[51-53]. It also inhibits the expression of chemokines involved in the recruitment of neutrophils into the lung during pulmonary inflammation[54]. Moreover, there is some evidence that taurine may reduce neutrophil activation and adherence to endothelial cells. These findings are supported by Raschke et al.[55], who found that taurine protects the heart against neutrophil-induced reperfusion injury, an effect involving oxidative stress. Therefore, one of the important functions of N-chlorotaurine and taurine is to restrict the cytotoxicity of the neutrophil, setting limitations on the amount of damage done during inflammation. This is likely the most important anti-oxidant action of taurine.

Taurine may also affect oxidative damage by limiting the availability of lipids for lipid peroxidation. Biological membranes are naturally arranged as bilayers. However, some of the phospholipids found in the membrane are capable of disrupting the bilayer structure. Phosphatidylethanolamine belongs to a group of disrupting phospholipids known as hexagonal formers. Interestingly, phosphatidylethanolamine can be converted into a bilayer former, phosphatidylcholine. Hamaguchi et al.[56] found that taurine inhibits the enzyme phospholipid N-methyltransferase, which catalyzes the conversion of phosphatidylethanolamine to phosphatidylcholine. Besides blocking the conversion of a hexagonal former into a bilayer former, taurine also affects the distribution of phospholipids within the membrane. Phosphatidylethanolamine is preferentially located on the inner bilayer (the cytosolic side) of the cell membrane, while phosphatidylcholine is preferentially found on the outer bilayer facing the extracellular space. By inhibiting the N-methyltransferase reaction, taurine prevents the movement of phospholipids from the inner bilayer to the outer bilayer. Taurine is also capable of forming an ionic interaction with the zwitterionic headgroups of certain phospholipids[65]. Because of these myriad of actions, it would not be surprising if taurine indirectly affects lipid peroxidation.

4. CONJUGATION AS A CYTOPROTECTIVE MECHANISM

One of the recognized functions of taurine is conjugation with bile acids, a reaction in which the amino group of taurine (a nucleophile) reacts with the carbonyl of the bile acid (an electrophile). The reaction of nucleophiles with electrophiles is very common in toxicology, as illustrated by acetaminophen toxicity. At therapeutic doses, only a small fraction of acetaminophen is oxidized by the cytochrome P450 system to form N-acetyl-p-benzoquinone-imine[57]. At these doses, glutathione is able to detoxify the reactive imine by forming a glutathione conjugate. However, the reactive imine intermediate accumulates after an overdose of acetaminophen. As a result, the glutathione defenses are overwhelmed and N-acetyl-p-benzoquinone-imine begins reacting with other nucleophiles, such as the amino group of proteins. The end result is severe hepatic injury, characterized by lipid peroxidation, mitochondrial dysfunction, disruption of calcium homeostasis and apoptosis. Waters et al.[58] found that taurine treatment protects against acetaminophen-induced hepatic injury, an effect they attributed to the prevention of lipid peroxidation and calcium overload. However, it is our contention that taurine might also detoxify N-acetyl-p-benzoquinone-imine by forming a taurine conjugate.

Another nucleophile-electrophile reaction involving taurine is the formation of a Schiff base. Several cytotoxic aldehydes appear to be detoxified by taurine through this reaction. Ogasawara et al.[59] found that taurine reacts with acetaldehyde, a metabolite of ethanol that is important in ethanol toxicity. Taurine also reacts with malondialdehyde, a product of lipid peroxidation, whose toxicity is linked in part to its modification of proteins[59]. However, one of the most important reactions involving a reactive carbonyl occurs in diabetes. Glucose and fructose can react with amino groups to form an 1-amino-1-deoxyketose Amadori product. The Amadori product can be further oxidized to form a highly reactive decarbonyl intermediate. The reaction of the reactive decarbonyl intermediate with amino groups leads to the formation of either pentosidine or advanced glycation end products[12,60]. This sequence of events is toxic to the cell. Therefore, it is not surprising that the reaction of taurine with the reactive carbonyl compounds has been implicated in the cytoprotective actions of taurine in the diabetic kidney[14].

5. IMPROVED CALCIUM METABOLISM AS A CYTOPROTECTIVE MECHANISM

Prevention of calcium overload was the first recognized cytoprotective action of taurine[1]. In their 1981 study, Kramer et al.[1] found that hearts exposed to a period of Ca^{2+} free perfusion followed by reperfusion with buffer containing Ca^{2+} underwent severe necrosis. The damage resulting from the Ca^{2+} paradox resulted from excessive accumulation of Ca^{2+} during the reperfusion period. Inclusion of taurine in the perfusion medium minimized the degree of cellular necrosis. In a subsequent study, Nakashima et al.[61] attributed to protection against the oxygen and calcium paradoxes of the hepatocyte to taurine-mediated reductions in Ca^{2+} influx. More recently, El Idrissi and Trenkner[62] and Chen et al.[63] reported that neuronal cells incubated with medium containing taurine exhibit a reduction in glutamate-induced excitotoxicity, an effect attributed to inhibition of Ca^{2+} influx via the Na^{+}/Ca^{2+} exchanger. Interestingly, the Na^{+}/Ca^{2+} exchanger has also been implicated in the effects of taurine depletion on doxorubicin cardiotoxicity. Harada et al.[64] found that both taurine deficiency and doxorubicin suppress the activity of the Na^{+}/Ca^{2+} exchanger, a transporter that functions in the cardiomyocyte to extrude Ca^{2+}. Since doxorubicin also facilitates the release of Ca^{2+} from intracellular stores, it dramatically increases $[Ca^{2+}]_i$. Taurine treatment improves Ca^{2+} homeostasis by facilitating the efflux of Ca^{2+} via the Na^{+}/Ca^{2+} exchanger. These effects of taurine can be traced in part to its membrane stabilizing activity[65]. High rates of phospholipid N-methylation are associated with a decrease in Na^{+}/Ca^{2+} exchanger activity. Therefore, taurine-mediated reductions in phospholipid N-methyltransferase activity enhance flux through the Na^{+}/Ca^{2+} exchanger and lower tissue Ca^{2+} content[56]. Taurine also promotes Ca^{2+} efflux via the Na^{+}/Ca^{2+} exchanger by increasing $[Ca^{2+}]_i$ in the vicinity of the exchanger[65]. The delivery of more Ca^{2+} to the Na^{+}/Ca^{2+} exchanger partially overcomes limitations arising from its high Km for Ca^{2+}.

Chronic osmotic stress and taurine treatment affect the cell, not only by acutely improving the osmotic balance, but also by affecting the activity of several key ion transporters. Among the transporters, whose flux is altered by osmotic stress and taurine treatment are the Na^{+}/Ca^{2+} exchanger, the ATP-sensitive K^{+} channel, the L-type Ca^{2+} channel and the fast Na^{+} channel[66]. Although not directly measured, there is every reason to believe that taurine also affects the activity of the osmotically sensitive Na^{+}/H^{+} exchanger. All of these taurine-sensitive transporters are key players in ischemia-reperfusion injury[67] and presumably in other forms of Ca^{2+}-dependent cytotoxicity.

Several mechanisms have been proposed for the effectiveness of various agents against ischemia-reperfusion injury. One of the most studied mechanisms is ischemic preconditioning. In 1986, Murry et al.[68] found that brief periods of ischemia confer protection against a prolonged period of sustained ischemia, a process that is now known as ischemic preconditioning. Mechanistic studies have revealed that ischemic preconditioning triggers a phosphorylation cascade that culminates in the activation of the mitochondrial ATP-sensitive K^+ channel. Although the role of the ATP-sensitive K^+ channel has not been established, inhibitors of the channel block the beneficial effects of a wide range of cardioprotective agents and protocols. There is reason to believe that taurine might also affect ischemic preconditioning. Han et al.[69] reported that intracellular taurine (20 mM) markedly depresses the activity of the cell membrane ATP-sensitive K^+ channel. Although the effect of taurine on the mitochondrial ATP-sensitive K^+ channel has not been studied, it is important to point out that the sarcolemmal and mitochondrial transporters share certain properties. Clearly, experiments examining the effect of taurine on the activity of the mitochondrial ATP-sensitive K^+ channel need to be performed.

Another cytoprotective mechanism is mediated by reductions in $[Ca^{2+}]_i$. Foremost among the agents that interfere with the accumulation of Ca^{2+} during ischemia-reperfusion are the Na^+/H^+ exchange inhibitors[67]. During ischemia the hydrolysis of ATP and the generation of certain metabolic intermediates lead to a reduction in pH_i. The Na^+/H^+ exchanger senses the decline in pH_i and promotes an exchange between extracellular Na^+ and intracellular H^+. Some of the Na^+ that enters the cell undergoes further exchange with extracellular Ca^{2+}. When influx of Ca^{2+} via the Na^+/Ca^{2+} exchanger is excessive, severe cellular damage ensues. Osmotic stress and taurine treatment appear to affect the activities of both the Na^+/H^+ and Na^+/Ca^{2+} exchangers. According to Earm et al.[70], cells exposed to medium containing taurine show enhanced Na^+/Ca^{2+} exchanger activity. Therefore, the effects of taurine on the Na^+/H^+ and Na^+/Ca^{2+} exchangers are unlikely to be cytoprotective. Nonetheless, there is every reason to believe that the combination of taurine treatment and a Na^+/H^+ exchange inhibitor will be more cytoprotective than either the Na^+/H^+ exchanger inhibitor alone or taurine alone.

6.   CONCLUSION

Taurine affects the two major causes of cellular toxicity, namely, Ca^{2+} overload and oxidative stress. Often, the beneficial effects of taurine have been attributed to both an improvement in oxidative stress and Ca^{2+} overload.

Nonetheless, it is important to recognize that taurine does not directly scavenge superoxide, hydrogen peroxide and superoxide although it directly scavenges HOCl in the presence of myeloperoxidase.

Four mechanisms may contribute to taurine-mediated reductions in oxidative stress. First, there is some evidence that taurine might upregulate the anti-oxidant defenses. Second, N-chlorotaurine suppresses the activity of the neutrophils, thereby reducing their ability to generate free radicals. Third, taurine may prevent Ca^{2+} overload, thereby minimizing free radical generation. Fourth, the major cause of taurine-mediated cytoprotection against certain xenobiotics is the formation of a taurine conjugate that is incapable of generating free radicals.

At least three mechanisms contribute to the modulation of Ca^{2+} movement by taurine. First, taurine indirectly alters the activity of the Na^+/Ca^{2+} exchanger. Second, as an osmolyte, taurine affects the activity of a number of key osmotically sensitive ion transporters. These transporters directly affect Na^+ and K^+ transport, which in turn alter Ca^{2+} transport. Third, taurine detoxifies specific xenobiotics that alter Ca^{2+} movement.

REFERENCES

1. Kramer, J.H., Chovan, J.P., and Schaffer, S.W., 1981, Effect of taurine on calcium paradox and ischemic heart failure. *Am. J. Physiol.* 240: H238-H246.
2. Hoffmann, E.K., and Dunham, P.B., 1995, Membrane mechanisms and intracellular signaling in cell volume regulation. In *International Review of Cytology*, vol. 161 (K.K. Jeon and J. Jarvik, eds.), Academic Press, San Diego/London, pp. 173-262.
3. Garcia-Dorado,D., and Oliveras, J., 1993, Myocardial oedema: a preventable cause of reperfusion injury? *Cardiovasc. Res.* 27: 1555-1563.
4. Trachtman, H., del Pizzo, R., and Sturman, J.A., 1990, Taurine and osmoregulation. III. Taurine deficiency protects against cerebral edema during acute hyponatremia. *Pediatr. Res.* 27: 85-88.
5. Allo, S.N., Bagby, L., and Schaffer, S.W., Taurine depletion, a novel mechanism for cardioprotection from regional ischemia. *Am. J. Physiol.* 273: H1956-H1961.
6. Okada, Y., and Maeno, E., 2001, Apoptosis, cell volume regulation and volume-regulatory chloride channels. *Comp. Biochem. Physiol. Part A* 130: 377-383.
7. Lang, F., Madlung, J., Siemen, D., Ellory, C., Lepple-Wienhues, A., and Gulbins, E., 2000, The involvement of caspases in the CD95 (Fas/Apo1)- but not swelling-induced cellular taurine release from Jurkat T-lymphocytes. *Pfluegers Arch.* 440: 93-99.
8. Lang, F., Busch, G.L., Ritter, M., Voelkl, H., Waldegger, S., Gulbins, E., and Haeussinger, D., 1998, Functional significance of cell volume regulatory mechanisms. *Physiol. Rev.* 78: 247-306.
9. Ozasa, H., and Gould., K.G., 1982, Protective effect of taurine from osmotic stress on chimpanzee spermatozoa. *Arch. Andrology* 9: 121-126.
10. Law, R.O., 1995, Taurine efflux and cell volume regulation in cerebral cortical slices during chronic hypernatremia. *Neurosci. Lett.* 185: 56-59.
11. Trachtman, H., Barbour, R., Sturman, J.A., and Finberg, L., 1988, Taurine and osmoregulation: taurine is a cerebral osmoprotective molecular in chronic hypernatremic dehydration. *Pediatr. Res.* 23: 35-39.

12. Hansen, S.H., 2001, The role of taurine in diabetes and the development of diabetic complications. *Diabetes Metab. Res. Rev.* 17: 330-346.

13. Pop-Busui, R., Sullivan, K.A., Van Huysen, C., Bayer, L., Cao, X., Towns, R., and Stevens, M.J., 2001, Depletion of taurine in experimental diabetic neuropathy: implications for nerve metabolic, vascular and function deficits. *Expt. Neurol.* 168: 259-272.

14. Trachtman, H., Futterweit, S., Maesaka, J., Ma, C., Valderrama, E., Fuchs, A., Tarectecan, A.A., Rao, P.S., Sturman, J.A., Boles, T.H., Fu, M-X, and Baynes, J, 1995, Taurine ameliorates chronic streptozotocin-induced diabetic nephropathy in rats. *Am. J. Physiol.* 269: F429-F438.

15. Ha, H., Yu, M-R, and Kim, K.H., 1999, Melatonin and taurine reduce early glomerulopathy in diabetic rats. *Free Radical Biol. Med.* 26: 944-950.

16. Obrosova, I.G., Fathallah, L., and Stevens, M.J., 2001, Taurine counteracts oxidative stress and nerve growth factor deficit in early experimental diabetic neurology. *Expt. Neurol.* 172: 211-219.

17. Wu, Q.D., Wang, J.H., Fennessey, F., Redmond, H.P., and Bouchier-Hayes, D., 1999, Taurine prevents high glucose-induced human vascular endothelial cell apoptosis. *Am. J. Physiol.* 277: C1229-C1238.

18. Pasantes-Morales, H., and Cruz, C., 1985, Taurine and hypotaurine inhibit light-induced lipid peroxidation and protect rod outer segment structure. *Brain Res.* 330: 154-157.

19. Pasantes-Morales, H., and Cruz, C., 1984, Protective effect of taurine and zinc on peroxidation-induced damage in photoreceptor outer segments. *J. Neurosci. Res.* 11: 303-311.

20. Banks, M.A., Porter, D.W., Martin, W.G., and Castranova, V., 1991, Ozone-induced lipid peroxidation and membrane leakage in isolated rat alveolar macrophages: protective effects of taurine. *J. Nutr. Biochem.* 2: 308-313.

21. Gordon, R.E., Shaked, A.A., and Solano, D.F., 1986, Taurine protects hamster bronchioles from acute NO_2-induced alterations. *Am. J. Pathol.* 125: 585-600, 1986.

22. Gurujeyalakshmi, G., Wang, Y., and Giri, S.N., 2000, Suppression of bleomycin-induced nitric oxide production in mice by taurine and niacin. *Nitric Oxide: Biol. Chem.* 4: 399-411.

23. Bhat, M., Rojanasakul, Y., Weber, S.L., Ma, J.Y.C., Castranova, V., Banks, D.E., and Ma, J.K.H., 1994, Fluoromicroscopic studies of bleomycin-induced intracellular oxidation in alveolar macrophages and its inhibition by taurine. *Environ. Health Perspect.* 102 (Suppl 10): 91-96.

24. Yan, C.C., and Huxtable, R.J., 1996, Effects of taurine and guanidinoethane sulfonate on toxicity of the pyrrolizidine alkaloid monocrotaline. *Biochem. Pharmacol.* 51: 321-329.

25. Wang, Q., Hollinger, M.A., and Giri, S.N., 1992, Attenuation of amiodarone-induced lung fibrosis and phospholipidosis in hamsters by taurine and/or niacin treatment. *J. Pharmacol. Expt. Therapeut.* 262: 127-132.

26. Ungemach, F.R., 1987, Pathobiochemical mechanisms of hepatocellular damage following lipid peroxidation. *Chem. Physics Lipids* 45: 171-205.

27. Waterfield, C.J., Mesquita, M., Parnham, P., and Timbrell, J.A., 1993, Taurine protects against the cytotoxicity of hydrazine, 1,4-naphthoquinone and carbon tetrachloride in isolated rat hepatocytes. *Biochem. Pharmacol.* 46: 589-595.

28. Timbrell, J.A., Seabra, V., and Waterfield, C.J., 1995, The *in vivo* and *in vitro* protective properties of taurine. *Gen. Pharmacol.* 26: 453-462.

29. Wu, C., Miyagawa, C., Kennedy, D.O., Yano, Y., Otani, S., Matsui-Yuasa, I., 1997, Involvement of polyamines in the protection of taurine against the cytotoxicity of hydrazine or carbon tetrachloride in isolated rat hepatocytes. *Chemico-Biol. Interactions* 103: 213-224.

30. Hwang, D.F., and Wang, L.C., 2001, Effect of taurine on toxicity of cadmium in rats. *Toxicology* 167: 173-180.

31. Balkan, J., Dogru-Abbasoglu, S., Kanbagli, O., Cevikbas, U., Aykac-Toker, G., and Uysal, M., 2001, Taurine has a protective effect against thioacetamide-induced liver cirrhosis by decreasing oxidative stress. *Human Expt. Toxicol.* 20: 251-254.

32. Hwang, D.F., Hour, J.L., and Cheng, H.M., 2000, Effect of taurine on toxicity of oxidized fish oil in rats. *Food Chem. Toxicol.* 38: 585-591.

33. Trachtman, H., del Pizzo, R., Futterweit, S., Levine, D., Rao, P.S., Valderrama, E., and Sturman, J.A., 1992, Taurine attenuates renal disease in chronic puromycin aminonucleoside nephropathy. *Am. J. Physiol.* 262: F117-F123.

34. Trachtman, H., Futterweit, S., and Bienkowski, R.S., 1993, Taurine prevents glucose-induced lipid peroxidaiton and increased collagen production in cultured rat mesangial cells. *Biochem., Biophys. Res. Commun.* 191: 759-765.

35. Eppler, B., and Dawson, R. Jr., 2002, Cytoprotective role of taurine in a renal epithelial cell culture model. *Biochem. Pharmacol.* 63: 1051-1060.

36. Hamaguchi, T., Azuma, J., Awata, N., Ohta, H., Takihara, K., Harada, H., Kishimoto, S., and Sperelakis, N., 1988, Reduction of doxorubicin-induced cardiotoxicity in mice by taurine. *Res.Commun. Chem. Pathol. Pharmacol.* 59: 21-30.

37. Ohta, H., Azuma, J., Awata, N., Hamaguchi, T., Tanaka, Y., Sawamura, A., Kishimoto, S., and Sperelakis, N., 1988, Mechanism of the protective action of taurine against isoprenaline induced myocardial damage. *Cardiovasc. Res.* 22: 407-413.

38. Aruoma, O.I., Halliwell, B., Hoey, B.M. and Butler, J., 1988, The antioxidant action of taurine, hypotaurine and their metabolic precursors. *Biochem. J.* 256: 251-255.

39. Fukuda, T., Ikejima, K., Hirose, M., Takei, Y., Watanabe, S., and Sato, N., 2000, Taurine preserves gap junctional intercellular communication in rat hepatocytes under oxidative stress. *J. Gastroenterol.* 35: 361-368.

40. Giri, S.N., Blaisdell, R., Rucker, R.B., Wang, Q., and Hyde, D.M., 1994, Amelioration of bleomycin-induced lung fibrosis in hamsters by dietary supplementation with taurine and niacin: biochemical mechanisms. *Environ. Health Perspect.* 102 (Suppl. 10): 137-148.

41. Vohra, B.P.S., and Hui, X., 2001, Taurine protects against carbon tetrachloride toxicity in the culture neurons and *in vivo. Arch. Physiol. Biochem.* 109: 90-94.

42. Nonaka, H., Tsujino, T., Watari, Y., Emoto, N., Yokoyama, M., 2001, Taurine preventthe decrease in expression and secretion of extracellular superoxide dismutase induced by homocysteine: amelioration of homocysteine-induced endoplasmic reticulum stress by taurine. *Circulation* 104: 1165-1170.

43. Mochizuki, H., Oda, H., and Yokogoshi, H., 2000, Dietary taurine alters ascorbic acid metabolism in rats fed diets containing polychlorinated biphenyls. *J. Nutr.* 130: 873-876.

44. Pitari, G., Dupre, S., Spirito, A., Antonini, G., and Amicarelli, F., 2000, Biochemical and ultrastructural alterations in rat after hyperoxic treatment: effect of taurine and hypotaurine. In *Taurine 4: Taurine and Excitable Tissues* (L. Della Corte, R.J. Huxtable, G. Sgaragli, and K.F. Tipton, eds.), Kluwer Academic/Plenum Publishers, New York/Boston/Dordrecht/London, Moscow, pp. 149-156.

45. Wright, C.E., Lin, T.T., Sturman, J.A., and Gaull, G.E., 1985, Taurine scavenges oxidized chlorine in biological systems. In *Taurine: Biological Actions and Clinical Perspectives* (S.S. Oja, L. Ahtee, P. Kontro and M.K. Paasonen, eds.), Alan R. Liss, Inc., New York, pp. 137-147.

46. Cantin, A.M., 1994, Taurine modulation of hypochlorous acid-induced lung epithelial cell injury in vitro: role of anion transport. *J. Clin. Invest.* 93: 606-614.

47. Nagl, M., Lass-Florl, C., Neher, A., Gunkel, A., and Gottardi, W., 2001, Enhanced fungicidal activity of N-chlorotaurine in nasal secretion. *J. Antimicrobial Chemotherapy* 47: 871-874.

48. Weiss, S.J., Klein, R., Slivka, A., and Wei, M., 1982, Chlorination of taurine by human neutrophil: evidence for hypochlorous acid generation. *J. Clin. Invest.* 70: 598-607.

49. Kato, S., Umeda, M., Takeeda, M., Kanatsu, K., and Takeuchi, K. (2002) Effect of taurine on ulcerogenic response and impaired ulcer healing induced by monochloramine in rat stomachs. *Aliment. Pharmacol. Ther.* 16 (Suppl. 2): 35-43.

50. Son, M. Kim, H.K., Kim. W.B., Yang, J., and Kim, B.K., 1996, Protective effect of taurine on indomethacin-induced gastric mucosal injury. *Arch. Pharm. Res.* 19: 85-90.

51. Schuller-Levis, G.B., Levis, W.R., Ammazzalorso, M., Nosrati, A., and Park, E., 1994, Mycobacterial lipoarabinomannan induced nitric oxide and tumor necrosis factor alpha production in a macrophage cell line: down regulation by taurine chloramines. *Infect. Immun.* 62: 4671-4674.

52. Quinn, M.R., Park, E., and Schuller-Levis, G., 1996, Taurine chloramines inhibits prostaglandin E$_2$ production in activated raw 264.7 cells by post-transcriptional effects on inducible cyclooxygenase expression. *Immunol. Lett.* 50: 185-188.

53. Marcinkiewicz, J., Grabowska, A., Bereta, J., Bryniarski, K., and Nowak, B., 1998, Taurine chloramines down-regulates the generation of murine neutrophil inflammatory mediators. *Immunopharmacology* 40: 27-38.

54. Liu, Y., and Quinn, M.R., 2002, Chemokine production by rat alveolar macrophages is inhibited by taurine chloramines. *Immunol. Lett.* 80: 27-32.

55. Raschke, P., Massoudy, P., and Becker, B.F., 1995, Taurine protects the heart from neutrophil-induced reperfusion injury. *Free Rad. Biol. Med.* 19: 461-471.

56. Hamaguchi, T., Azuma, J., and Schaffer, S., 1991, Interaction of taurine with methionine: inhibition of myocardial phospholipids methyltransferase. *J. Cardiovasc. Pharmacol.* 18: 224-230.

57. Klaassen, C.D., 1996, *Casarett and Doull's Toxicology: The basic science of poisons*, McGraw-Hill, New York /St. Louis /San Francisco /Auckland /Bogota/ Caracas /Lisbon /London /Madrid /Mexico City /Milan /Montreal /New Delhi /San Juan / Singapore /Sydney /Tokyo /Toronto.

58. Waters, E., Wang, J.H., Redmond, H.P., Wu, Q.D., Kay, E.l, and Bouchier-Hayes, D., 2001, Role of taurine in preventing acetaminophen-induced hepatic injury in the rat. *Am. J. Physiol.* 280: G1274-G1279.

59. Ogasawara, M., Nakamura, T., Koyama, I., Nemoto, M., and Yoshida, T., 1993, Reactivity of taurine with aldehydes and its physiological role. *Chem. Pharm. Bull.* 41: 2172-2175.

60. Grandhee, S.K., and Monnier, V.M., 1991, Mechanism of formation of the maillard protein cross-link pentosidine. *J. Biol. Chem.* 266: 11649-11653.

61. Nakashima, T., Seto, Y., Nakajima, T., Shima, T., Sakamoto, Y., Cho, N., Sano, A., Iwai, M., Kagawa, K., Okanoue, T., 1990, Calcium-associated cytoprotective effect of taurine on the calcium and oxygen paradoxes in isolated rat hepatocytes. *Liver* 10: 167-172.

62. El Idrissi, A., and Trenkner, E., 1999, Growth factors and taurine protect against excitotoxicity by stabilizing calcium homeostasis and energy metabolism. *J. Neurosci.* 19: 9459-9468.

63. Chen, W.Q., Jin, H., Nguyen, M., Carr, J., Lee, Y.J., Hsu, C.C., Faiman, M.D., Schloss, J.V., and Wu, J.Y., 2001, Role of taurine in regulation of intracellular calcium level and neuroprotective function in cultured neurons. *J. Neurosci. Res.* 66: 612-619.

64. Harada, H., Cusack, B.J., Olson, R.D., Stroo, W., Azuma, J., Hamaguchi, T., and Schaffer, S.W., 1990, Taurine deficiency and doxorubicin: interaction with the cardiac sarcolemmal calcium pump. *Biochem. Pharmacol.* 39: 745-751.

65. Schaffer, S.W., Azuma, J., and Madura, J.D., 1995, Mechanisms underlying taurine-mediated alterations in membrane function. *Amino Acids* 8: 231-246.

66. Schaffer, S., Takahashi, K., and Azuma, J., 2000, Role of osmoregulation in the actions of taurine. *Amino Acids* 19: 527-546.

67. Karmazyn, M., and Moffat, M.P., 1993, Role of Na$^+$/H$^+$ exchange in cardiac physiology and pathophysiology: mediation of myocardial reperfusion injury by the pH paradox. *Cardiovasc. Res.* 27: 915-924.

68. Murry, C.E., Jennings, R.B., and Reimer, K.A., 1986, Preconditioning with ischemia: a delay of lethal cell injury in ischemic myocardium. *Circulation* 74: 1124-1136.

69. Han, J., Kim, E., Ho, W-K, and Earm, Y.E., 1996, Blockade of the ATP-sensitive potassium channel by taurine in rabbit ventricular myocytes. *J. Mol. Cell. Cardiol.* 28: 2043, 2050.

70. Earm, Y.E., Ho, W.K., and So, I.S., 1993, Effect of taurine on the activation of background current in cardiac myocytes of the rabbit. In *Ionic Channels and Effect of Taurine on the Heart* (D. Noble, and Y. Earm, eds), Kluwer Academic Publishers, Boston, pp. 119-138.

71. Mozaffari, M.S., and Schaffer D., 2001, Taurine modulates arginine vasopressin-mediated regulation of renal function. *J. Cardiovasc. Pharmacol.* 37: 742-750.

Taurine, Analogues and Bone
A Growing Relationship

RAMESH C. GUPTA and SUNG-JIN KIM
Department of Pharmacology, School of Dentistry
Kyung Hee University, Seoul 130-701, Korea

1.   INTRODUCTION

The recent finding that bone tissue contains high amounts of taurine[1] has prompted the investigation of taurine's role in bone. This has resulted in the surge of reports yielding two major breakthroughs.   First, taurine stimulates bone formation.   Second, it inhibits bone resorption, thereby preventing alveolar bone loss. Apart from this, beneficial effects of taurine on fenestration, healing of wounds and following tooth extraction have been discussed. In a review related to osteoporosis, alveolar bone loss and drug development, Wyhn[2] argued that it is time to look beyond bisphosphonate to other drugs that might prevent the resorptive phase of alveolar bone loss; bisphosphonates are the only available remedy now and have been for many years. He suggested that any new agent must have the ability to block the production of inflammatory factors and to arrest the cause of osteoporosis. The recent finding that taurine inhibits bone resorption indicates that it and perhaps some of its analogues deserve further consideration as new therapeutic agents.   In this review, we first discuss taurine's effect on bone formation followed by a discussion on bone resorption.   Finally, we make the case that taurine and its analogues may serve as important alternatives to bisphosphonate therapy.

2.   TAURINE IN BONE TISSUE FORMATION

Bone provides both mechanical support to vital organs and protection to bone marrow. It also serves as a reservoir of calcium and phosphate ions for maintenance of serum homeostasis. Much of bone tissue contains matrix, and is composed of collagen fibers and non collagenous proteins. Bone tissue contains three types of cells, including osteoblasts, which are bone forming cells, and osteoclasts, which are bone resorption cells. The plasma membrane of the osteoblast is rich in alkaline phosphatase, the marker of bone formation. In their nuclei, osteoblasts also have several receptors for hormones such as estrogens and vitamin D3. To modulate osteoclast activity in the osteoblast, the osteoblast has receptors for various bone resorption stimulatory factors, such as parathyroid hormone, prostaglandin and the interleukins. The bone formation process involves matrix maturation, mineralization and measurement of bone formation. It is believed that regulation of osteoblast function occurs at three levels: (1) endocrine regulation involving the actions of hormones like parathyroid hormone. In the osteoblast, parathyroid hormone enhances ion and amino acid transport, regulates collagen synthesis, and stimulates matrix and alkaline phosphatase synthesis, (2) paracrine regulation of osteoblast activity in bone formation promotes the initial developmental stages of the embryo, long bone formation and fracture healing, and (3) autocrine regulation is involved in the remodeling process.

An effect of amino acids on matrix vesicle formation was first reported in 1986[3]. It was shown that a decrease in the level of free amino acids in the culture media was associated with a progressive decline in matrix vesicle formation, an effect reversed by increasing the levels of amino acids. A set of five amino acids, including taurine, was used in the study. Although taurine increased cellular alkaline phosphatase activity, the most effective combination was glutamate, alanine, serine, asparagine and taurine.

A significant study was carried out by Park et al.[4], who found that taurine stimulated alkaline phosphatase activity and collagen synthesis in osteoblast-like UMR-106 cells. Taurine also stimulated tyrosine phosphorylation of a number of cellular proteins. Taurine has little cytotoxicity over a concentration range of 0.01 to 10 mM, a concentration range in which it has been shown to stimulate alkaline phosphatase activity by 32-87%. The maximum increase in collagen synthesis (56%), as measured from hydroxyproline formation, was observed at 20 mM taurine, with concentrations below 1 mM having very little effect. To gain insight into the level of taurine involvement in osteoblast regulation, several experiments have been conducted. It was found that even at a concentration of 0.1 mM taurine stimulated tyrosine phosphorylation within 1 minute of a number of proteins (65, 53, 42 and 38 KDa). These stimulatory actions of taurine were blocked by MEK inhibitor, confirming the claim that taurine acts through ERK2.

3.   TAURINE AND OSTEOPOROSIS

Metabolic bone disease can be broadly classified as either osteoporosis or osteomalacia. In osteoporosis, there is both a decrease in bone mass and microarchitectural disruption. Osteoporosis has been correlated with alveolar bone loss. Alveolar bone resorption followed by loss of teeth is clinically the most important issue in human periodontitis. Certain gram negative bacteria, such as actinomycetemcomitans and porphyromonus gingivalis, are involved in the pathogenesis of these diseases[5]. Lipopolysaccharide is a constituent of the outer membrane of these bacteria and acts as a stimulator of bone resorption, with 1L-1α and PGE2 participating in lipopolysaccharide-induced resorption. These mechanistic views were also confirmed by Koide et al[6], who reported on taurine's inhibitory action of bone resorption and osteoclast formation. Taurine was able to suppress lipopolysaccharide induced bone resorption even at a concentration of 100 μg/ml. Such studies were successfully repeated in hamsters. In the future, taurine may be introduced as an effective agent in preventing inflammatory bone resorption in periodontal disease and may provide an alternative to bisphosphonate therapy.

4. TAURINE - BONE INTERACTION: A FUTURE VISION

Apart from the two actions mentioned above, taurine exerts a variety of other actions on bone that warrant further study. Taurine supplementation aids tissue repair in destructive diseases like periodontitis. In addition, taurine has the ability to enhance bone formation during wound healing and, along with chitosan[7], a natural healing agent, accelerate an early phase in healing. This may provide the logic for using taurine supplementation under these conditions. Apart from this, taurine along with precursors[8] cysteine and methionine have a protective effect in injury after tooth extraction. Taurine supplementation appears to be linked to rapid recovery, in other words taurine deficiency may delay the healing process. Since taurine deficiency causes abnormalities in retina and heart, in a similar manner it may cause abnormalities leading to metabolic bone disease. Such studies are required to provide a link between taurine deficiency and abnormalities in bone tissue development. Strikingly, taurine does have radioprotective properties. Chemotherapy and radiation lead to a reduction in plasma taurine levels, but taurine supplementation improves the survival rate of mice after total body irradiation[9].

5. TAURINE: EXERCISE AND BONE GAIN

Skeletal development in late adolescence and early adulthood is linked to regular physical activity. Exercise, along with a diet supplemented with minerals, amino acids and vitamins, may further enhance development. Taurine containing drinks stimulate cognitive performance, well being and physical activity[10]. Exercise also increases taurine levels, thus there may be a possible link between exercise, taurine content and bone development.

6.   TAURINE: A NATURAL PREVENTIVE AGENT FOR METABOLIC BONE DISEASES

Taurine has been mentioned as a preventive medicine[11] for heart, liver and many other diseases. Taurine's actions appear to be mediated via osmo-regulation, antioxidation and the host defenses. Since taurine stimulates bone formation and inhibits bone loss, an adequate reservoir of taurine may prevent, as well as correct, metabolic bone diseases, thereby helping to cope with advancing age.

7.   TAURINE: A POSSIBLE BONE METABOLISM MARKER

Several markers for bone formation and resorption are in use, but no single marker is available for both. As taurine has proved its effectiveness in both situations, studies are warranted to determine the taurine concentration in normal, developing and diseased state, providing information on the potential of taurine as a marker of bone metabolism.

8.   FURTHER AHEAD - BEYOND BISPHOSPHONATES; TAURINE ANALOGUES: POSSIBLE ALTERNATIVES

To prevent osteoporosis, a drug must arrest the cause of the disease. Bisphosphonates are presently indicated for this condition. Recently, amino bisphosphonates have been tested for their ability to prevent resorption of alveolar bone. In this regard, taurine, which is an amino sulphonate, may serve as a structural analogue of the amino bisphosphonates. Taurine analogues are also in use as anti-convulsants[12] and anti-alcohol agents[13]. Taurine analogues, which are more lipophilic, may constitute a new class of therapeutic agents for metabolic bone disease. Indeed, our laboratory has found that some of these analogues exhibit similar or even better bone formation activity than taurine.

9.   CONCLUSION

This update of taurine - bone interaction is unique and perhaps the first of its kind. Although little has been done to estimate the taurine content of bone tissue and study its regulation and transport, it is clear that taurine is an important dietary nutrient. More studies on the effects of dietary taurine on bone development are warranted. This may provide the link between taurine and metabolic bone diseases. Although the stimulatory effect of taurine on bone formation and on inhibition of bone resorption has been experimentally documented, more pharmacological studies are needed to establish taurine as a therapeutic agent and even as a clinical marker of bone metabolism. The importance of exercise-induced taurine accumulation also deserves further study. Finally, as taurine analogues are developed for other therapeutic uses, attention should be directed to using them as an alternative to bisphosphonates.

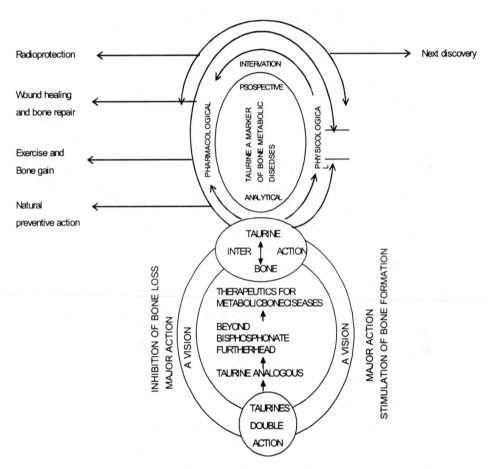

Fig. 1. A sign of growing relation. A holistic view of taurine-bone interaction and a future vision.

ACKNOWLEDGMENT

Dr. S.-J. Kim thanks KOSEF and KOFST for support through the 2000 Brain pool program. Dr. R. C. Gupta is thankful to KOSEF for a brain pool award and also to Nagaland University India for granting a leave of absence which enables him to undertake this work. Authors thank Ms. M.H. Kim for the excellent secretarial assistance.

REFERENCES

1. Lubec, B., Ya-hua, Z., Pertt, S., Pentti, T., Kitzmuller ,E. and Lubec, G. 1997, Distribution and disappearance of the radio labeled carbon derived from arginine and taurine in mouse. *Life Sci* 60: 2378-2381.
2. Wyhn, R.L.,2000, Osteoporosis, alveolar bone loss and drug development. *General Dentistry* 218-225.
3. Iishikawa, Y.L., Chin, J.E., Schalk, E.M. and Wuthier, R.E., 1986, Effect of amino acid levels on matrix vesicle formation by epiphyseal growth plate chondrocytes in primary culture. *J. Cellular physiology* 126 : 391-406.
4. Park, S., Kim, H. and Kim, S.J., 2001, Stimulation of ERK2 by taurine with enhanced alkaline phosphates activity and collagen synthesis in osteoblast like UMR-106 cells. *Biochem. Phamacology* 62: 1107-1111.
5. Ishihara, Y., Nishihara, T., Maki, E., Noguchi, T., and Koga, .T, 1991, Role of interleukim-1 and prostaglandin in invitro bone resorption induced by Actinobacillus actinomycetemcomitans lipopolysaccharid. *J. Periodont Res* 26: 155-160.
6. Koide, M., Okahashi, N., Tanaka, R., Kazuno, K., Shibasaki ,K.,, Yamazaki, Y., Kaneko, K., Veda, N., Ohguchi, M., Ishihara, Y., Noguchi, T., and Nishihara, T., 1999, Inhibition of experimental bone resorption and osteoclast formation and survival by 2-aminoe ethane sulphonic acid. *Arch. Oral Biol*, 44: 711-719.
7. Ozmeri, C., Ozcan, G., Haytacc, M., Alaaddinoglu, E.E., Sargon, M.F. and Senel, S., 2000, Chitosan film enriched with an antioxidant agent, taurine in fenestration defects. *J. Biomed Mater. Res* 51: 500-503.
8. Iito, R., Uchiyama, T., Ohno, H., and Kohda, T., 1985, Proposal of cascade action of taurine related sulphur amino acids, on cardiac and vascular sympathetic Nerves; protection and therapy of injuries after tooth extraction in Guinea pigs. *Prog. Desal in Clincal and Biological Res*. 179: 2155-224.
9. Desal, T.K., Maliakkol, J., Kinxie, J.L., Ehrinpreis, M.N., Luk, G.D. and Cejka, J., 1992, Taurine deficiency after intensive chemotherapy and or radiaion. *Am. J. Clin. Nutr* 55: 708-711.
10. Barthel, T., Mechau, D., Wehr, T., Schnittker, R., Liesen, H. and Weiss, M., 2001, Readiness potential in different states of physical activation and after ingestion of taurine and or cafferine containing drinks. *Amino acids*, 20; 63-73.
11. Birdsall, T.C., 1998, Therapeutic Application of Taurine; *Alt. Med. Rev* ; 3: 128-136.
12. Huxtable, R.J., and Nakagawa, K., 1985, The anticonvalsant actions of two taurine derivatives in genetic and chemically induced seizures, *Prog. in clinical and Biol. Res*. 179; 435-448.
13. Wilde, M.I. and Wagstaff, A.J., 1997, Acamprosate. A review of its pharmacology and clinical potential in the management of alcohol dependence after detoxification. *Drugs* 53; 1038-1053.

Anti-inflammatory Activities of Taurine Chloramine

Implication for Immunoregulation and Pathogenesis of
Rheumatoid Arthritis

EWA KONTNY[*], WŁODZIMIERZ MAŚLIŃSKI[*] and
JANUSZ MARCINKIEWICZ[#]
[*]Department of Pathophysiology & Immunology, Institute of Rheumatology, Warsaw, and
[#]Jagiellonian University Medical College, Cracow, Poland

1. INTRODUCTION

1.1 Initial Comments

Neutrophils recruited into the site of inflammation generate a large number of highly reactive oxidants, including hypochlorous acid (HOCl), produced by the myeloperoxidase-catalyzed oxidation of Cl^- by hydrogen peroxide[1]. Hypochlorous acid is the major neutrophil microbicidal agent but its excessive production leads to tissue damage. Oxidative tissue damage is thought to be of pathogenic significance in a large number of diseases, *e.g.* atherosclerosis, malignancy and rheumatoid arthritis. Taurine (Tau), a dominant free amino acid present in most mammalian tissues and in the cytosol of phagocytic cells at high (10-20 mM) concentration[2], acts as a major trap for HOCl. The reaction of Tau with HOCl forms the long-lived but weaker oxidant taurine chloramine (Tau-Cl) and thus reduces HOCl toxicity. On the other hand, there is growing evidence that Tau-Cl: (i) down-regulates the production of pro-inflammatory mediators by macrophages, neutrophils and other cells engaged in inflammatory response[3], (ii) regulates the function of dendritic cells[4], and (iii) enhances protein immunogenecity by chlorination[5]. This implies Tau-Cl to act as an important physiologic immunoregulatory factor maintaining the delicate balance between mounting an effective immune response, and minimizing the destruction of the tissue by the inflammatory cells.

Taurine 5: Beginning the 21st Century
Edited by Lombardini *et al.*, Kluwer Academic/Plenum Publishers, New York 2003

1.2   Generation and Metabolism of Tau-Cl

At physiological pH (7.4) HOCl reacts with Tau to form relatively stable taurine monochloramine (Tau-Cl), but at acidic pH, which may occur inside phagolysosome, the reaction of HOCl with Tau results in formation of mono- and dichloramine (Tau-Cl$_2$), that is more toxic and has much stronger bactericidal properties than Tau-Cl[6]. The physiologic concentrations of Tau-Cl are not known. *In vitro* data show that in the presence of a physiologically relevant (12.5 -15 mM) concentration of Tau activation of human neutrophils (~2x10^6 cell/ml) results in generation of ~ 100 nmol of Tau-Cl[7] that is further converted into sulphoacetaldehyde at a rate of 637 ± 18 pmol/h/ml[8]. Taking into account that the *in vivo* inflammatory site is heavily infiltrated with neutrophils (*e.g.* a large interstitial inflammatory site may contain as many as 25x10^6 neutrophils/ml)[1], it is conceivaible that local accumulation of Tau-Cl may easily reach the mM range. Consequently, the local sulphoacetaldehyde concentration is likely to reach up to 100 μM. Therefore, in our *in vitro* experiments we applied Tau, Tau-Cl and sulphoacetaldehyde at 50-500 μM concentrations that were also not cytotoxic.

2.   IMMUNOREGULATORY ACTIVITY OF TAU-CL IN MURINE SYSTEM

Tau-Cl inhibits the production of pro-inflammatory mediators by activated cells derived from various sources. *In vitro* the modulatory effect of Tau-Cl on murine inflammatory cells such as neutrophils, macrophages, bone-marrow-derived dendritic cells as well as on cell lines (RAW 264.7 macrophages, A-20-J – B cell antigen presenting cell line) has been shown. Tau-Cl inhibits the generation of reactive oxygen intermediates (ROI), nitric oxide (NO), TNF-α, IL-6, and PGE$_2$, all proinflammatory mediators, with similar IC$_{50}$ value, approximately 250 μM[3], but through different mechanisms: Tau-Cl inhibits transcription of the iNOS gene[9], supresses the translation of TNF-α mRNA[10], and affects the post-transcriptional regulation of COX-2 expression[11]. In general, the effect of Tau-Cl was reversible and observed when the compound was added at the time of cell activation. These results indicate that Tau-Cl not only protects our tissues, as a trap for highly toxic HOCl, but also, in an active way exerts anti-inflammatory properties. *In vivo*, however, the Tau-Cl impact on the cells present at the site of inflammation will depend on its local concentration and on the time of their interaction. In contrast to neutrophils and macrophages, which remain at the site of inflammation for hours and days, respectively, sentinel dendritic cells (DC) after maturation migrate out from a site of inflammation into the local lymph nodes. Nevertheless, one may speculate that Tau-Cl contributes to the modulation of DC functions in a paracrine fashion *in vivo*. Previously we have shown that Tau-Cl in a dose dependent manner inhibits the production of pro-inflammatory cytokines (TNF-α, IL-6, IL-12) by DC[4]. Moreover, Tau-Cl selectively modulates the ability of DC to induce the release of IL-2 (Th1 type cytokine) and IL-10 (Th2 type cytokine) from T cells[4]. Incubation of T cells with Tau-Cl suppressed their activity[5]. *In vivo*, the generation of Tau-Cl usually occurs at a distance from the lymphoid tissue in which T cells are initially activated by antigen. Thus, it is unlikely that the inhibitory action of Tau-Cl on T cells affects T cell priming. However, Tau-Cl may contribute to the regulation of the latter event by chlorination (oxidation) of protein antigens, which results in an increased immunogeneticity of modified antigen. We have shown that chlorination of ovalbumin by both HOCl, and to the lesser extent by Tau-Cl, facilitates processing and presentation of these proteins by APC[5]. Mechanisms are still unknown, but it may be an effect of increased susceptibility of chlorinated proteins to degradation by endoproteases[12].

In conclusion, it is apparent that Tau-Cl has pleiotropic effects on the inductive phase of the immune response. Therefore, it provides another molecular link between innate and acquired immunity. *In vitro* data suggest that in the microenvironment of the localized inflammatory response, Tau-Cl will exert predominantly a negative regulatory influence, inhibiting not only production of pro-inflammatory mediators but also inhibiting both T cell and APC function. However, interaction between antigen and Tau-Cl will result in an enhanced immunogenecity of the protein antigens, thus enhancing the initiation of a subsequent adaptive specific response.

3. RHEUMATOID ARTHRITIS

3.1 Pathogenesis and Therapeutic Targets

Rheumatoid arthritis (RA) is an autoimmune, chronic, inflammatory disease of uncertain etiology, characterized by hyperplasia of the synovial cell lining layer, synovitis and destruction of joint cartilage and bone. Although the mechanisms of RA initiation are obscure they may involve the innate immunity activators that trigger synoviocytes to synthesize cytokines and chemokines attracting blood cells (neutrophils, macrophages, T and B lymphocytes) into the joint, followed by inflammation and activation of the adaptive immune response specific to indefinite antigen(s) (either foreign or self)[13,14]. The destructive phase includes cartilage dissolution by locally produced proteolytic enzymes and bone erosion by osteoclasts. A growing body of evidence indicates that numerous cytokines of different cellular origin are overproduced in RA affected joints. Among them, TNF-α and IL-1, produced mostly by macrophage-like synoviocytes, are thought to be the key mediators of joint inflammation and destruction by inducing the synthesis of other pro-inflammatory cytokines (IL-6 and IL-8), tissue degrading metalloproteinases (MMPs), prostaglandins and NO in a variety of cell types, and by inhibiting the production of matrix components[15]. The targeting of TNF-α and IL-1 has became a strategic basis for developing therapies to treat RA[15]. Moreover, recently developed new anti-inflammatory drugs, the selective inhibitors of inducible cycloxygenase-2 (COX-2)[16] and the dual inhibitors of both 5-lipoxygenase and cyclooxygenases[17], represent another beneficial therapeutic tool in RA treatment.

3.2   Fibroblast-Like Synoviocytes: Pathogenic Functions and Their Inhibition by Tau-Cl

Recently increasing attention has been directed toward fibroblast-like synoviocytes (FLS), the cells of mesenchymal origin located in the joint intimal lining, as potential therapeutic targets for RA treatment. These cells participate in all phases of RA progression. It is proposed that synovial hyperplasia, an early event that precedes clinically apparent joint inflammation[14], may be a consequence of intractable proliferation and partial resistance of FLS to apoptosis[18,19]. Local FLS proliferation was demonstrated by showing the expression of PCNA (proliferating cell nuclear antigen), c-Myc, and nucleolar organizer region in these cells[18]. Importantly, RA FLS have many features of partial transformation: anchorage-independent growth, loss of contact inhibition, oncogene activation, monoclonal or oligoclonal expansion, detectable telomerase activity, and somatic gene (e.g. p53) mutation[18]. It is hypothesized that alteration in the structure or function of p53 gene might be the most important in RA FLS transformation and renders these cells resistant to apoptotic death[18]. Moreover, RA FLS are partly resistant to Fas-induced apoptosis, mostly due to elevated expression of anti-apoptotic molecules and protection given by growth factors and cytokines[19,20]. It is debatable whether RA FLS are irreversibly changed due to genetic alterations or are passively responding to the local environment. RA synovium is chronically exposed to oxidative stress and reactive oxygen and nitrogen species are possible causes of somatic mutations[18]. On the other hand, growth factors and cytokines known to support FLS proliferation are abundant in the inflamed RA joint[19,20]. Because RA FLS are also active participants of inflammatry response and the tissue destruction process it is obvious that inhibition of FLS growth is one of the main goals in the treatment of disease. We have previously demonstrated[21,22] that *in vitro* Tau-Cl inhibited with similar potency (IC$_{50}$ ≈ 300-400 μM) proliferation of RA

FLS triggered by bFGF, PDGF or TNF-α. By contrast, neither Tau[21,22] nor sulphoacetaldehyde (unpublished data) affected proliferation of these cells. Consistent with the inhibition of cell proliferation, we found that Tau-Cl down-regulated the stimuli-induced expression of PCNA in these cells[22]. The PCNA acts as a cofactor of DNA type δ polymerase, is expressed predominantly in the late G1 and S phases of cell cycle and is crucial for the initiation of DNA synthesis. The importance of this protein expression for RA FLS cell-cycle progression was proved using antisense oligonucleotides strategy[23]. It was also shown that the induction of a senescence gene, the p16 INK4a cyclin dependent kinase inhibitor, was capable of inhibiting RA FLS growth and ameliorated adjuvant arthritis[24]. Thus, the blocking of RA FLS cell-cycle progression seems to be an intriguing novel therapeutic approach for RA treatment, and interestingly Tau-Cl possesses such properties.

The engagement of FLS in the perpetuation of RA synovitis is well documented. These cells spontaneously produce IL-6 and angiogenic factors while macrophage-derived IL-1 and TNF-α further activate FLS and raise IL-6 production, trigger IL-8 and PGE$_2$ synthesis as well as the release of proteases[18]. Although IL-6 exerts both pro- and anti-inflammatory activities, its implication in RA pathogenesis is not questioned. IL-6 is primarily the product of FLS, supports proliferation of these cells, participates in the bone loss and in erosive process, contributes to systemic symptoms and is deleterious in animal models of antigen-induced arthritis[25]. IL-8, another product of activated FLS, participates in RA synovitis due to its chemotactic and angiogenic activities, and ability to activate macrophages and neutrophils[26]. We found Tau-Cl to be a potent inhibitor of both IL-6 (IC$_{50}$ ≈ 225 μM) and IL-8 (IC$_{50}$ ≈ 450 μM) production in RA FLS stimulated *in vitro* with either IL-1β or TNF-α[21]. Moreover, we revealed that Tau-Cl acts at the level of the transcription of the cytokine genes[27]. By contrast, neither Tau[21,27] nor sulphoacetaldehyde (unpublished data) were effective.

Prostaglandin E$_2$ (PGE$_2$) is another inflammatory mediator implicated in the inflammatory feature of RA[28]. Cyclooxygenase (COX), a key enzyme in prostaglandin (PGs) biosynthesis, is either constitutively expressed as a COX-1 isoform that generate PGs for homeostatic functions, or COX-2 isoenzyme which is induced by many mediators of inflammation, including IL-1[28]. When cultured *in vitro*, RA FLS express constitutively only COX-1 and spontaneously synthesize little amount of PGE$_2$ while stimulation of these cells with IL-1β significantly raises PGE$_2$ and triggers COX-2 protein expression[29]. We have also found, that both of these events are dose-dependently inhibited by Tau-Cl[29]. Neither Tau[29] nor sulphoacetaldehyde (unpublished data) affected the PGE$_2$ synthesis in IL-1β-stimulated cells. Thus, we report that in RA FLS Tau-Cl is a potent and selective inhibitor of COX-2-generated PGE$_2$ synthesis.

By contrast, in our preliminary experiments we failed to show any significant effect of Tau-Cl on both the spontaneous expression of MMP-2 and MMP-9 (gelatinases), and on the IL-1β-induced expression of MMP-1 (collagenase) and MMP-3 (stromelysin)(data not shown). Although this suggests that Tau-Cl does not inhibit RA FLS function related to tissue degradation, further studies are needed to support this observation.

In summary, our *in vitro* data demonstrate that Tau-Cl affects several pathogenic functions of RA FLS: (i) proliferation of these cells, that is a critical event in synovial hyperplasia, as well as the synthesis of pro-inflammatory (ii) cytokines (IL-6 and IL-8) and (iii) PGE_2.

Because monocyte/macrophage-derived pro-inflammatory cytokines are potent activators of RA FLS, we asked the question whether Tau-Cl is able to affect the synthesis of these cytokines and found that in peripheral blood mononuclear cells (PBMCs) of healthy volunteers stimulated *in vitro* with lipopolysaccharide (LPS) Tau-Cl inhibits the synthesis of IL-1β, IL-6, and TNF-α with different potency (IC_{50} values ≈ 300, 400, and 500 μM, respectively)[30]. This observation was further confirmed in PBMCs of RA patients with a short duration of the disease (≤ 2 years), but PBMCs of RA patients with more advanced disease (disease duration ≈ 6 years) were refractory to Tau-Cl inhibition of TNF-α synthesis[31]. Despite the difference, it is likely that Tau-Cl may also affect the pathogenic function of RA FLS indirectly, by reducing monocyte/macrophage-derived pro-inflammatory cytokine synthesis, at least in early RA.

3.3   The Mechanism of Anti-inflammatory Tau-Cl Action

Accumulating evidence proves that Tau-Cl acts at the level of either transcription or translation of genes encoding crucial mediators of the inflammatory response. Little is known however, what signal transduction pathways implicated in the expression of these genes are affected by Tau-Cl. Two transcription factors, the activator protein-1 (AP-1) and the nuclear factor κB (NFκB) are important regulators of gene expression in the setting of inflammation, including the chronic inflammatory response associated with RA[32]. Regulation of AP-1 and NFκB is complex but both of them are also under the control of an oxidoreductive mechanism (redox regulation)[33] and thus they are likely to be targets for a weak oxidant such as Tau-Cl. We reported that in RA FLS Tau-Cl down-regulates IL-1β-triggered DNA-binding activity of both NFκB and AP-1[27]. The mechanism of NFκB inhibition by Tau-Cl was further studied in detail by others. In Jurkat T-cell line Tau-Cl was shown to prevent NFκB activation by oxidation of IκBα inhibitor[34]. This may account for potent anti-inflammatory activity of Tau-Cl, for NFκB regulates a broad range of genes critical for inflammation (*e.g.* cytokines, growth factors, COX-2, adhesion molecules)[32].

AP-1 includes members of the Jun and Fos families of transcription factors and regulates production of (for example) cytokines and MMPs[32]. Transcription of c-fos gene is under control of mitogen-activated protein kinases (MAPK), including Erk (the extracellular signal-regulated kinases)[32]. Therefore, we investigated the effect of Tau-Cl on the expression of non-phosphorylated (enzymatically inactive) and phosphorylated (enzymatically active) forms of Erk in RA FLS (Fig.1). These cells show constitutive high expression of non-phosphorylated Erk 2, that did not change upon cell treatment. Spontaneous expression of phosphorylated Erk 1 and Erk 2 was weak or undetectable. Although the expression of the latter enzymatically active forms of Erk 1/Erk 2 was raised upon stimulation of the cells with either TNF-α or IL-1β, this was not affected by Tau-Cl. Therefore, we conclude that the other signaling events, but not Erk activity, are relevant to the inhibition of AP-1 activity by Tau-Cl.

Phosphorylated Erk1 (44 kDa) and Erk 2 (42 kDa)

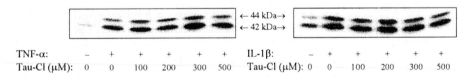

TNF-α:		−	+	+	+	+	+	IL-1β:		−	+	+	+	+	+
Tau-Cl (μM):	0	0	100	200	300	500		Tau-Cl (μM): 0	0	100	200	300	500

Non-phosphorylated Erk2

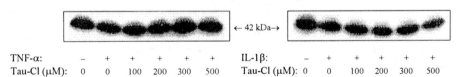

TNF-α:		−	+	+	+	+	+	IL-1β:		−	+	+	+	+	+
Tau-Cl (μM):	0	0	100	200	300	500		Tau-Cl (μM): 0	0	100	200	300	500

Figure 1. Tau-Cl does not affect the cytokine-triggered expression of phosphorylated (enzymatically active) forms of Erk1/Erk2. The representative Western blot of 5 performed.

3.4   Metabolism of Tau-Cl in RA

There are only limited data concerning the metabolism of Tau-Cl in RA. However, the elevated plasma levels of Tau[35] and hypertaurinuria[36] demonstrated in RA patients suggest the disturbed metabolism of Tau/Tau-Cl in this disease. There is little doubt that neutrophils, representing the major source of Tau-Cl, are implicated in RA pathogenesis[37]. These cells accumulate mostly in the joint synovial fluid where they exhibit features indicative of both: partial activation and functional "exhaustion"[37]. Consistently, our recent data suggest that neutrophils from synovial fluid of RA patients generate *in vitro* less Tau-Cl than their peripheral blood counterparts and neutrophils from healthy volunteers[38]. Therefore, it is likely that the impaired generation of Tau-Cl in RA joints provides the concentration of this compound below that required to exert anti-inflammatory action. However, further data concerning real Tau-Cl levels in RA affected joints are needed to prove this supposition.

3.5   Tau-Cl Application in the Treatment of Autoimmune and Chronic Inflammatory Diseases. A Perspective.

Tau-Cl shows significant anti-inflammatory and microbicidal properties along with lower toxicity, as compared with other oxidants[39,40]. This optimal compromise between microbicidal activity and high tolerability[39] suggests Tau-Cl as a useful anti-inflammatory and disinfectant agent for a special applications in human medicine. A recent clinical study shows that Tau-Cl could be considered as a therapeutic option (local application) in some infectious diseases. In contrast, very little is known about the ability of exogenous Tau-Cl to inhibit inflammatory response *in vivo*. However, there is evidence that Tau administration exerts anti-inflammatory effects *in vivo*. Taurine was reported to prevent or attenuate lung inflammation and the development of experimentally induced atherosclerosis in animals. It is unlikely that Tau acts *in vivo* as a passive scavenger of toxic HOCl. We suggest that the final effect of Tau may depend on local Tau-Cl formation. It has been reported that both MPO and hypochlorite-modified proteins are present in human atherosclerotic lesions[41]. It implicates that HOCl oxidation of LDL may contribute to atherogenesis. On the other hand, unexpectedly, increased atherosclerosis in MPO-deficient mice was observed[42]. Thus, MPO may play a dual role in atherosclerosis: deleterious *via* generation of toxic HOCl and protective *via* unknown mechanisms. In our opinion, these data together with the observation of the beneficial effects of Tau suggest a Tau-Cl contribution in this process. To confirm this hypothesis, the influence of Tau administration on the development of atherosclerosis in MPO-deficient mice should be tested.

Only recently, we have described the effect of *in vivo* Tau-Cl administration (subcutaneous daily injections) on the development of murine type II collagen induced arthritis (CIA)[43]. Tau-Cl treatment prior to disease onset, simultaneously with immunization, significantly delayed the appearance of arthritic symptoms. However, on the last day of the experiment, incidence and severity of arthritis was similar in mice treated with Tau-Cl and placebo mice. These results suggest that the effect of Tau-Cl was short term and reversible. One can speculate that systemic administration of Tau-Cl was ineffective either due to a very low intra-articular concentration of Tau-Cl or decomposition of Tau-Cl into ineffective metabolites. To improve the effectiveness of Tau-Cl therapy, further investigation should include local, intra-articular administration of more stable *in vivo* forms of Tau-Cl. Finally, in our opinion, local application of Tau-Cl could be more promising and effective than the systemic one.

4. CONCLUSION

Data from *in vitro* studies clearly demonstrate both anti-inflammatory and immunoregulatory properties of Tau-Cl. The activity of this compound is not species limited for its effects on murine and human cells are similar. Importantly, Tau-Cl influences the functions of a broad range of cell types (monocytes/macrophages, lymphocytes, dendritic cells, fibroblasts, *etc.*) that co-operate at a site of inflammation. Therefore, the contribution of this physiologic product of activated neutrophis to down-regulate the inflammatory response *in vivo* is strongly suggested. Consistently, the impaired generation of Tau-Cl may favour transformation of inflammatory response from the acute into the chronic phase, characteristic *e.g.* for RA. Due to bactericidal effect Tau-Cl has successfully been applied locally in some infectious diseases. In our opinion, it is rational to consider therapeutic application of exogenous Tau-Cl also in inflammatory diseases, including RA. However, further investigations in animal models are necessary to support this assumption more strongly.

ACKNOWLEDGEMENTS

The work from our laboratories was supported by grants from the State Committee for Scientific Research of Poland (No: 4P05A10419 and 4P05B01018), and from the Institute of Rheumatology, Warsaw, Poland.

REFERENCES

1. Weiss, S.J., 1989, Tissue destruction by neutrophils. *N. Engl. J. Medicine* 320: 365-376.
2. Learn, D.B., Fried, V.A., and Thomas, E.L., 1990, Taurine and hypotaurine content of human leukocytes. *J. Leukoc. Biology* 48:174-182.
3. Marcinkiewicz, J., 1997, Neutrophil chloramines: missing links between innate and acquired immunity. *Immunol. Today* 18:577-580.
4. Marcinkiewicz, J., Nowak, B., Grabowska, A., Bobek, M., Petrovska, L., and Chain, B., 1999, Regulation of murine dendritic cell functions in vitro by taurine chloramine, a major product of neutrophil myeloperoxidase-halide system. *Immunology* 98:371-378.
5. Marcinkiewicz, J., Grabowska, A., and Chain, B.M., 1998, Modulation of antigen-specific T-cell activation in vitro by taurine chloramine. *Immunology* 94:325-330.
6. Marquez, L.A., and Dunford, H.B., 1994, Chlorination of taurine by myeloperoxidase. *J. Biol. Chemistry.* 269:7950-7956.
7. Weiss, S.J., Klein, R., Slivka, A., and Wei, M., 1982, Chlorination of taurine by human neutrophils. *J. Clin. Investigations* 70:598-607.
8. Cunningham, C., Tipton, K.F., and Dixon, H.B.F., 1998, Conversion of taurine into N-chlorotaurine (taurine chloramine) and sulphoacetaldehyde in response to oxidative stress. *Biochem. Journal* 330:939-945.
9. Park, E., Schuller-Levis, G.,Jia, J.H., Ouinn, M.R., 1997, Preactivation exposure of RAW 264.7 cells to taurine chloramine attenuates subsequent production of nitric oxide and expression of iNOS mRNA. *J. Leukoc. Biology* 61:161-166.
10. Park, E., Schuller-Levis, G., Quinn, MR., 1995, Taurine chloramine inhibits production of nitric oxide and TNF-α in activated RAW 264.7 cells by mechanisms that involve transcriptional and translational events. *J. Immunology* 154:4778-4784.
11. Quinn, M.R., Park, E., Schuller-Levis, G., 1996, Taurine chloramine inhibits prostaglandin E2 production in activated RAW 264.7 cells by post-transcriptional effects on inducible cyclooxygenase expression. *Immunol. Letters* 50:185-188.
12. Marcinkiewicz, J., Olszowska, E., Olszowski, S., and Zgliczynski, J.M., 1991, Enhancement of immunogenic properties of ovalbumin as a result of its chlorination. *It. J. Biochemistry* 23, 1393-1395.
13. Arend, W.P., 2001, The innate immune system in rheumatoid arthritis. *Arthritis & Rheumatism* 44:2224-2234.
14. Firestein , G.S., and Zvaifler N.J., 2002, How important are T cells in chronic rheumatoid synovitis. II. T-cell-independent mechanisms from beginning to end. *Arthritis & Rheumatism* 46:298-308.
15. Carteron, N.L., 2000, Cytokines in rheumatoid arthritis: trials and tribulations. *Mol. Med. Today* 6:315-323.
16. Everts, B., Währbord, P., and Hedner, T., 2000, COX-2-specific inhibitors-the emergence of a new class of analgestic and anti-inflammatory drugs. *Clin. Rheumatology* 19:331-343.
17. Bertolini, A., Ottani, A., and Sandrini M., 2001, Dual-acting anti-inflammatory drugs: a reappraisal. *Pharmacol. Research* 44:437-450.
18. Yamanishi, Y., and Firestein, G.S., 2001, Pathogenesis of rheumatoid arthritis: the role of synoviocytes. *Rheum. Dis. Clin. North America* 27:355-371.
19. Müller-Ladner, U., Gay, R.E., Gay, S., 2000, Activation of synoviocytes. *Curr. Opin. Rheumatology* 12:186-194.
20. Kurowska, M., Rudnicka, W., Kontny, E., Janicka, I., Chorąży, M., Kowalczewski, J., Ziółkowska, M., Ferrari-Lacraz, S., Strom, T.B., and Maśliński, W., 2002, Fibroblast-like synoviocytes from rheumatoid arthritis patients express functional IL-15 receptor complex: endogenous IL-15 in autocrine fashion enhances cell proliferation and expression of Bcl-x_L and Bcl-2. *J. Immunology* 169:1760-1767.

21. Kontny, E., Grabowska, A., Kowalczewski, J., Kurowska, M., Janicka, I., Marcinkiewicz, J., and Maśliński,W., 1999, Taurine chloramine inhibition of cell proliferation and cytokine production by rheumatoid arthritis fibroblast-like synoviocytes. *Arthritis & Rheumatism* 42:2552-2560.
22. Kontny, E., Kurowska, M., Kowalczewski, J., Janicka, I., Marcinkiewicz, J., and Maśliński, W., 2001, The mechanism of taurine chloramine inhibition of fibroblast-like synoviocytes growth. *Amino Acids* 21:75.
23. Morita, Y., Kasjihara, N., Yamamura, M., Okamoto, H., Harada, S., Maeshima, Y., et.all., 1997, Inhibition of rheumatoid synovial fibroblast proliferation by antisense oligonucleotides targeting proliferating cell nuclear antigen messenger RNA. *Arthritis & Rheumatism* 40:1292-1297.
24. Taniguchi, K., Kohsaka, H., Inoue, N., Terada, Y., Ito, H., Hirokawa, K., Niyasaka, N., 1999, Induction of the p16INK4a senescence gene as a new therapeutic strategy for the treatment of rheumatoid arthritis. *Nature Medicine* 5:760-767.
25. Dinarello, Ch. A., and Moldawer, L.L., 2000, Interleukin-6 and its superfamily.In *Proinflammatory and anti-inflammatory cytokines in rheumatoid arthritis.* 2nd editions, Amgen Inc., pp:49-57.
26. Dinarello, Ch. A., and Moldawer, L.L., 2000, Chemokins and their receptors. In *Proinflammatory and anti-inflammatory cytokines in rheumatoid arthritis.* 2nd editions, Amgen Inc., pp:99-111.
27. Kontny, E., Szczepanska, K., Kowalczewski, J., Kurowska, M., Janicka, I., Marcinkiewicz, J., and Maśliński, W., 2000, The mechanism of taurine chloramine inhibition of cytokine (interleukin-6, interleukin-8) production by rheumatoid arthritis fibroblast-like synoviocytes. *Arthritis & Rheumatism* 43:2169-2177.
28. Amin, A.R., Attur, M., Abramson, S.B., 1999, Nitric oxide synthase and cyclooxygenases: distribution, regulation, and intervention in arthritis. *Curr. Opin. Rheumatology* 11:202-209.
29. Kontny, E., Rudnicka, W., Kowalczewski, J., Marcinkiewicz, J., Maśliński, W., 2002, Selective inhibition of COX-2-generated prostaglandin E_2 synthesis in rheumatoid arthritis synoviocytes by taurine chloramine. Submitted to *Arthritis & Rheumatism.*
30. Chorąży, M., Kontny, E., Marcinkiewicz, J., and Maśliński, W., 2002, Taurine chloramine modulates cytokine production by human peripheral blood mononuclear cells. *Amino Acids* 23:407-413.
31. Chorąży, M., Kontny, E., Rell-Bakalarska, M., Marcinkiewicz, J., and Maśliński, W., 2002, Taurine chloramine modulates cytokine production by peripheral blood mononuclear cells (PBMCs) from RA patients. *Ann. Rheum. Diseases* 61 (suppl.1): 270.
32. Firestein, G.S., and Manning, A.M., 1999, Signal transduction and transcription factors in rheumatic disease. *Arthritis & Rheumatism* 42:609-621.
33. Sen, C.K., Packer, L., 1996, Antioxidant and redox regulation of gene transcription. *FASEB Journal* 10:709-720.
34. Kanayama, A., Inoue, J., Sugita-Konishi, Y., Shimizu, M., and Miyamoto, Y., 2002, Oxidation of Ikappa Balpha at methionine 45 is one cause of taurine chloramine-induced inhibition of NF-kappa B activation. *Biol. Chemistry* 277:24049-24056.
35. Trang, L.E., Furst, P., Odeback, A.C., Lovgren, O.,1985, Plasma amino acids in rheumatoid arthritis. *Scand. J. Rheumatology* 14:393-402.
36. Rylance, H.J., 1969, Hypertaurinuria in rheumatoid arthritis. *Ann. Rheum. Diseases* 28:41-44.
37. Edwards, S.W., and Hallet, M.B., 1997, Seeing the wood for the threes: the forgotten role of neutrophils in rheumatoid arthritis. *Immunol. Today* 18:320-324.

38. Kontny, E., Wojtecka-Łukasik, E., Rell-Bakalarska, K., Dziewczopolski, W., Maśliński W., and Maśliński S., 2002, Impaired generation of taurine chloramine by synovial fluid neutrophils of rheumatoid arthritis patients. *Amino Acids* 23:415-418.
39. Nagl, M., Miller, B., Daxecker, F., Ulmer, H., and Gottardi, W., 1998, Tolerance of N-chlorotaurine, an endogenous antimicrobial agent, in the rabbit and human eye – a phase I clinical study. *J. Ocul. Pharmacol. Therapy* 14:283-290.
40. Marcinkiewicz, J., Chain, B., Nowak, B., Grabowska, A., Bryniarski, K., and Baran, S.J., 2000, Antimicrobial and cytotoxic activity of hypochlorous acid: interactions with taurine and nitrite. *Inflamm. Research* 49:280-289.
41. Hazell, L., Arnold, L., Flowers, D., Waeg, G., Malle, E., and Strocker, R., 1996, Presence of hypochlorite-modified proteins in human atherosclerotic lesions. *J. Clin. Investigations* 97:1535-1544.
42. Brennan, M.L., Anderson, M.M., Shih, D.M., et.al., 2001, Increased atherosclerosis in myeloperoxidase-deficient mice. *J. Clin. Investigations* 107:419-430.
43. Kwaśny-Krochin, B., Bobek, M., Kontny, E., Głuszko, P., Biedron, R., Chain, B.M., Maśliński, W., Marcinkiewicz, J., 2002, Effect of taurine chloramine, the product of activated neutrophils, on the development of collagen-induced arthritis in DBA 1/J mice. *Amino Acids* 23:419-426.

Taurine Chloramine Inhibits Production of Inflammatory Mediators and iNOS Gene Expression in Alveolar Macrophages; a Tale of Two Pathways: Part I, NF-κB Signaling

MICHAEL R. QUINN*, MADHABI BARUA, YONG LIU, and VALERIA SERBAN*

Laboratory of Molecular Cell Signaling, Department of Developmental Biochemistry, New York State Institute for Basic Research in Developmental Disabilities, and *Center for Developmental Neuroscience, Staten Island, NY, USA

1. INTRODUCTION

Taurine protects against tissue damage in a variety of models involving inflammation, especially the lung[1,2]. The mechanism of taurine protection is not well understood but the ability of taurine to attenuate the toxic effects of HOCl/OCl⁻ by formation of taurine chloramine (Tau-Cl) and its subsequent effects are thought to be important. Tau-Cl is formed by the action of the halide-dependent myeloperoxidase (MPO) system associated with neutrophils[3,4]. Considering the crucial role that alveolar macrophages play during pulmonary inflammatory events, we determined the effects of Tau-Cl on the production of NO, TNF-α, MCP-1, and MIP-2 in NR8383 cells. NR8383 cells are a cell line derived from rat alveolar macrophages.

Production of proinflammatory mediators is regulated primarily at the level of gene transcription. Since iNOS is considered a prototypical inflammatory gene in many species, we determined the effects of Tau-Cl on iNOS gene expression using NR8383 cells that were transiently transfected with a luciferase reporter gene driven by fragments of the rat iNOS gene promoter. Since the nuclear transcription factor NF-κB is involved in regulating the expression of iNOS, TNF-α, MCP-1, MIP-2, and most other inflammatory mediators[5-7], we evaluated the effects of Tau-Cl on the NF-κB signaling pathway.

2. MATERIALS AND METHODS

2.1 Materials

Polyclonal antibody to p65, p50, cRel, and IκB-α were obtained from Santa Cruz Biotechnology (Santa Cruz, CA). Antibody against phosphorylated IκB-α was from New England Biolabs (Beverly, MA). Culture media and rat recombinant IFN-γ were purchased from Gibco-BRL (Grand Island, NY). LPS W (*Escherichia coli* 0111:B4) from Difco Laboratories (Detroit, MI). All other chemicals were obtained from commercial sources and Tau-Cl was freshly synthesized on the day of use as previously described[8].

2.2 Cell Culture

NR8383 cells were grown as previously detailed[8,9]. Experiments were conducted in DMEM supplemented with 2% FBS, 1% penicillin, and 1% streptomycin. Cells were activated with LPS (1 μg/ml) and IFN-γ (10 U/ml). Tau-Cl was added at the time of activation. Cells were collected and nuclear protein extracts, cytosolic fractions, and cell lysates were prepared[8,9]. Electrophoretic mobility shift assays (EMSA), supershift assays, and Western blots were performed as previously described[8]. Media content of TNF-α, MCP-1, and MIP-2 were determined 24 hr after activation by ELISA (Biosource International, Camarillo, CA). Conditioned media NO$_2$ content was measured spectrophotometrically[8].

2.3 Transient Transfections and Reporter Gene Analyses

Plasmids containing 1.7 kb of the 5′-region and 133 bp of the first exon of the rat iNOS gene promoter, referred to as piNOS-Luc-3/2[10,11], and a truncated version, piNOS-Luc-3/1, were graciously provided by Drs. J. Pfeilschifter and W. Eberhardt. piNOS-Luc-3/1 contains the proximal 526 bp of 5′-region of the iNOS promoter and 133 bp of the first exon[12]. Both promoter fragments were subcloned into the pGL3 basic vector to control expression of a modified luciferase gene (Promega, Madison, WI). Transfections of NR8383 cells were performed using Superfect reagent

Qiagen, Valencia, CA). Transfections of NR8383 cells with phRG-TK which drives a synthetic *Renilla* luciferase (Promega) served as an internal control and was used to normalize reporter gene luciferase activity within each sample. Luciferase activities were measured using the dual luciferase reporter gene assay (Promega).

3. RESULTS AND DISCUSSION

Addition of Tau-Cl to NR8383 cells at the time of activation inhibited production of proinflammatory mediators (Table 1). Inhibition was dependent on the concentration of Tau-Cl, with 50% inhibition occurring in the presence of 0.5 mM Tau-Cl and 80-85% inhibition by 1.0 mM Tau-Cl. Production of proinflammatory mediators is inhibited by Tau-Cl in activated cells of various sources and it has been proposed that this action of Tau-Cl may represent a major mechanism by which taurine protects against tissue damage[1].

Table 1. Tau-Cl inhibits production of inflammatory mediators by NR8383 cells.

Inflammatory mediator	Tau-Cl Concentration, mM		
	0	0.5	1.0
NO_2^- (μM)	41 ± 2	20 ± 2	9 ± 1
TNF-α (ng/ml)	27 ± 5	12 ± 4	5 ± 2
MCP-1 (pg/ml)	292 ± 10	150 ± 20	52 ± 8
MIP-2 (ng/ml)	26 ± 1	13 ± 0.2	4 ± 0.2

Cells were exposed to Tau-Cl at the time of activation and the media content of each mediator was measured 24 hrs later. Values represent the x̄ ± S.D. of triplicate samples. Similar results were obtained in 3-10 additional independent experiments.

The temporal expression pattern for iNOS, TNF-α, MCP-1 and MIP-2 mRNAs was examined in NR8383 cells[8,9]. Tau-Cl exerted a profound inhibitory effect on message expression for these inflammatory mediators, especially for iNOS mRNA. It was suggested that early events in gene transcription may be particularly affected by Tau-Cl. For these reasons we evaluated the effects of Tau-Cl on transcription of the rat iNOS gene in NR8383 cells transfected with piNOS-Luc-3/2 and with a truncated version of the rat iNOS gene promoter region contained in piNOS-Luc-3/1 (Fig. 1).

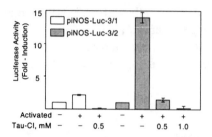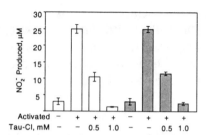

Figure 1. Tau-Cl inhibits expression of piNOS-Luc-3/1 and piNOS-Luc-3/2 in transiently transfected NR8383 cells. Cells were transfected for 3 hr before activation. Luciferase activities in cell lysates and NO_2^- media content were measured 24 hr after activation. Values represent $\bar{x} \pm$ SD of duplicate transfections. Similar results were obtained in 2 independent experiments.

The rat iNOS gene promoter consists of a proximal core promoter region containing an NF-κB binding site and a distal enhancer element containing a second NF-κB binding site along with binding sites for other transcription factors[13-15]. The potentiating effects of IFN-γ are mediated through IRF-1 and Stat-1α binding to cognate sites downstream of –526 bp. The results presented in Fig. 1 demonstrate that LPS + IFN-γ stimulates induction of the truncated iNOS gene promoter 2.2-fold above that of unactivated cells while that of the 1.7 kb iNOS gene promoter is stimulated 14-fold. Tau-Cl potently inhibited the induced expression of luciferase driven by both iNOS gene promoter fragments. Preliminary studies demonstrated that Tau-Cl does not inhibit the luciferase reaction (not shown). The inhibition of iNOS gene expression by Tau-Cl appears specific because expression of the *Renilla* luciferase activity driven by the basal expression of phRG-TK was unaffected by Tau-Cl, e.g., the relative light units expressed for activated cells transfected with piNOS-Luc-3/2 was 8.3 ± 1.6 vs 7.8 ± 0.6 for transfected NR8383 cells activated in the presence of 1.0 mM Tau-Cl. Thus, Tau-Cl did not inhibit protein synthesis or luciferase enzyme activity in general. Similar observations were made with cells transfected with piNOS-Luc-3/1. In addition, production of NO was not impaired by the conditions of transfection as shown in Fig. 1. These results suggest that the NF-κB signaling pathway is an important target for Tau-Cl.

Nuclear protein extracts of NR8383 cells were prepared to determine the kinetics of NF-κB binding by EMSA (Figure 2). The binding activity of NF-κB increased within 30 min of activation and continued to increase over the remaining 24 hr of activation. Nuclear protein extracts from cells activated in the presence of Tau-Cl (1.0 mM) exhibited greatly reduced NF-κB binding activity relative to that of activated cells, at all times examined. In addition, a second NF-κB complex was evident after 3, 6, and 24 hr of activation, but was barely detectable in the nuclear protein fraction of Tau-Cl treated cells.

Activation
Time (hrs)

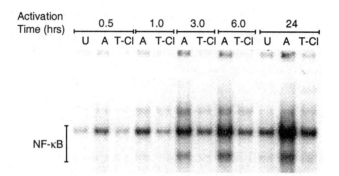

Figure 2. Kinetics of NF-κB nuclear protein binding in NR8383 cells. Cells were treated for the indicated times before preparing nuclear protein extracts for analyses by EMSA. U, unactivated cells; C, control activated cells; T-Cl, cells exposed to 1.0 mM Tau-Cl. Similar results were obtained in 3 additional independent experiments. (From ref. 13 with permission copyright 2001, The Amer. Assoc. Immunologists, Inc.).

Supershift EMSA analyses of nuclear protein extracts prepared from NR8383 cells activated for 1 hr demonstrated that the upper NF-κB band was a heterodimer containing p50 and p65 subunits (Fig. 3). This was evident because antibody against p50 and antibody against p65, but not cRel, supershifted the upper band. The lower NF-κB band in Figure 2 is a homodimer consisting of p50 subunits. Composition of the NF-κB complex in Tau-Cl treated cells was similar to that of controls (Fig. 3). These results indicate that Tau-Cl inhibits some earlier event in the activation of NF-κB because Western blot analyses of nuclear protein extracts confirmed reduced levels of p65 and p50 in cells activated in the presence of Tau-Cl[8]. In unstimulated cells NF-κB is sequestered in the cytoplasm as a complex with IκB. Upon stimulation, IκB-α is phosphorylated at serine residues 32 and 36, ubiquitinated, and degraded by the 26S proteasome. Degradation of IκB-α unmasks the nuclear localization sequence of NF-κB allowing translocation into the nucleus and binding to the promoter region of target genes[5-7].

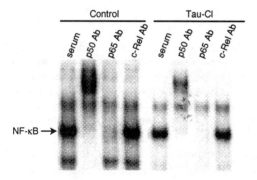

Figure 3. Characterization of NF-κB complexes was assessed by supershift assays using preimmune serum and antibodies against NF-κB proteins. Nuclear protein extracts were prepared 1 hr after activation. Similar results were obtained in 4 independent experiments.

The effects of Tau-Cl on NF-κB activation was evaluated by determining the temporal changes in cytosolic IκB (Fig. 4). Control cells exhibited a rapid disappearance of cytosolic IκB-α during the first 1 hr of activation with gradual reappearance occurring between 3 and 24 hr of activation (Fig. 4). Cells activated in the presence of Tau-Cl (1.0 mM) did not exhibit this transient change in cytoplasmic IκB. These results suggest that in Tau-Cl treated cells IκB was not being degraded and resynthesized, but was being retained as a complex with NF-κB in the cytoplasm.

As a prerequisite for NF-κB translocation to the nucleus, cytosolic IκB must be phosphorylated, ubiquitinated, and degraded by the 26S proteasome[5-7]. To address the effects of Tau-Cl on this event, we utilized MG-132, a peptide that blocks proteasome activity, thus allowing phosphorylated IκB to accumulate in the cytoplasm. Effects of Tau-Cl on phosphorylation of IκB-α were determined by using cells that were preincubated (1 hr, 37°C, 5% CO_2) with 10 μM MG-132 or vehicle (DMSO, 0.25% (v/v) final concentration) before activation with LPS + IFN-γ. Tau-Cl was added at the time of activation. Cell lysates were prepared 1 hr later and analyzed by Western blot using antibodies specific for IκB-α phosphorylated at serine residue 32 (Fig. 4).

Preparations from activated cells contained reduced amounts of unphosphorylated IκB-α and increased p-IκB-α relative to unactivated cells. Preincubation with MG-132 (10 μM) for 1 hr before activation greatly enhanced the p-IκB-α signal whereas cells activated in the presence of Tau-Cl (1 mM) accumulated only low amounts of p-IκB. Cells activated in the

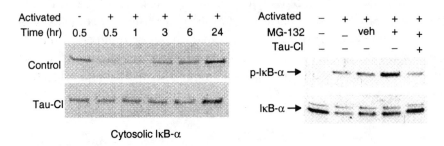

Figure 4. (Left panel*)* Tau-Cl inhibits degradation of IκB. Cytosolic fractions of NR8383 cells were analyzed by Western blot after various times of activation. (Right panel) Tau-Cl inhibits phosphorylation of IκB-α in NR8383 cells. Cells were exposed to media alone, MG-132, or to vehicle (DMSO) for 1 hr before addition of LPS + IFN-γ. Tau-Cl (1.0 mM) was added at the time of activation. Cell lysates were prepared 1 hr later and analyzed by Western blot. (From ref. 13 with permission, copyright 2001, The Amer. Assoc. Immunologists, Inc.).

presence of Tau-Cl maintained unphosphorylated IκB-α levels that were similar to those of unactivated controls. Further studies demonstrated that the activity of IKK, the multiprotein complex that phosphorylates IκB-α, is not directly inhibited by Tau-Cl. Rather, Tau-Cl inhibits upstream signaling pathways that activate IKK[8].

In summary, the present results demonstrate that Tau-Cl inhibits production of inflammatory mediators and expression of the iNOS gene. The transcriptional effects of Tau-Cl on proinflammatory gene expression is mediated, in part, by inhibition of NF-κB activation at some point upstream of IKK activation. Our results are concordant with the recent report that Tau-Cl inhibits activation of NF-κB in fibroblast-like synoviocytes[16].

ACKNOWLEDGEMENTS

The authors are grateful to Drs. J. Pfeilschifter and W. Eberhardt for kindly providing us with iNOS promoter fragments and appreciate the excellent secretarial assistance of Mrs. A. Parese. This work was supported by the OMRDD of New York and by NIH grants HL-49942 and NS40721.

REFERENCES

1. Quinn, M.R., and Schuller-Levis ,G.B., 1999, Taurine Chloramine, an Inhibitor of iNOS Expression and a Potential Modulator of Inflammation. In *Molecular and Cellular Biology of Nitric Oxide* (J. Laskin and D. Laskin, eds.), Marcel Dekker, Inc., New York, pp. 309-331.
2. Gurujeyalakshmi, G., Wang, Y, and Giri, S.N., 2000, Taurine and niacin block lung injury and fibrosis by down-regulating bleomycin-induced activation of transcription nuclear factor-kappa B in mice. *J. Pharmacol. Exp. Ther.* 293: 82-90.
3. Zgicznski, J.M., Stelmaszynska, T., Domanski, J., and Ostrowski, W., 1971, Chloramines as intermediates of oxidation reaction of amino acids by myeloperoxidase. *Biochim. Biophys. Acta* 235: 419-424.
4. Weiss, S.J.R., Klein, A., Slivka, A., and Wei, M., 1982, Chlorination of taurine by human neutrophils: evidence for hypochlorous acid generation. *J. Clin. Invest.* 70: 598-607.
5. Baeuerle, P.A., and Henkel, T., 1994, Function and activation of NF-κB in the immune system. *Annu. Rev. Immunol.* 12: 141-179.
6. Barnes, P.J., and Karin, M., 1997, Nuclear factor-κB, a pivotal transcription factor in chronic inflammatory diseases. *N. Engl. J. Med.* 336: 1066-1071.
7. Kopp, E., and Ghosh, S., 1995, NF-κB and Rel proteins in innate immunity. *Adv. Immunol.* 58: 1-27.
8. Barua, M., Liu, Y., and Quinn, M.R., 2001, Taurine chloramine inhibits nitric oxide synthase and TNF-α gene expression in activated alveolar macrophages: Decreased NF-κB activation and IκB kinase activity. *J. Immunol.* 167: 2275-2281.
9. Liu, Y., and Quinn, M.R., 2002, Chemokine production by rat alveolar macrophages is inhibited by taurine chloramine. *Immunol. Lett.* 80: 27-32.

10. Eberhardt, W., Kunz, D., Hummel, R., and Pfeilschifter, J., 1996, Molecular cloning of the rat inducible nitric oxide synthase gene promoter. *Biochem. Biophys. Res. Commun.* 223: 752-756.
11. Beck, K.-F., Eberhardt, W., Walpen, S., Apel, M., and Pfeilschifter, J., 1998, Potentiation of nitric oxide synthase expression by superoxide in interleukin 1β-stimulated rat mesangial cells. *FEBS Lett.* 435: 35-38.
12. Eberhardt, W., Plüss, C., Hummel, R., and Pfeilschifter, J., 1998, Molecular mechanisms of inducible nitric oxide synthase gene expression by IL-Iβ and cAMP in rat mesengial cells. *J. Immunol.* 160: 4961-4969.
13. Lowenstein, C.J., Alley, E.W., Raval, P., Snowman, A.M., Snyder, S.H., Russell, S.W., and Murphy, W.J., 1993, Macrophage nitric oxide synthase gene: two upstream regions mediate induction by interferon gamma and lipopolysaccharide. *Proc. Natl. Acad. Sci., U.S.A.* 90: 9730-9734.
14. Xie, Q.W., Whisnant, R, and Nathan, C., 1993, Promoter of the mouse gene encoding calcium-independent nitric oxide synthase confers inducibility by interferon-gamma and bacterial lipopolysaccharide. *J. Exp. Med.* 177: 1779-1784.
15. Murphy, W.J., 1999, Transcriptional Regulation of the Genes Encoding Nitric Oxide Synthase. In *Cellular and Molecular Biology of Nitric Oxide* (J.D. Laskin and D.L. Laskin, eds.), Marcel Dekker, Inc., New York, pp. 1-56.
16. Kontny, E., Szczepańska, K., Kowalczewski, J., Kurowska, M., Janicka, I., Marcinkiewicz, J., and Maśliński, W., 2000, The mechanism of taurine chloramine inhibition of cytokine (interleukin-6, interleukin-8) production by rheumatoid arthritis fibroblast-like synoviocytes. *Arthritis Rheum.* 43: 2169-2177.

Taurine Chloramine Inhibits Production of Inflammatory Mediators and iNOS Gene Expression in Alveolar Macrophages; a Tale of Two Pathways: Part II, IFN-γ Signaling Through JAK/Stat

MICHAEL R. QUINN[*], YONG LIU, MADHABI BARUA, and VALERIA SERBAN[*]

Laboratory of Molecular Cell Signaling, Department of Developmental Biochemistry, New York State Institute for Basic Research in Developmental Disabilities, and *Center for Developmental Neuroscience, Staten Island, NY, USA

1. INTRODUCTION

Taurine protects against tissue damage in a variety of models that share inflammation as a common pathogenic feature[1]. This has been particularly well documented in animal models of lung injury where prophylactic administration of taurine protects lung from damage induced by inhalation exposure to NO_2[2] or O_3[3] and by intratracheal instillation of bleomycin[4,5], amidarone[6], or paraquat[7]. Lung damage in these models is associated with an influx of neutrophils along with mononuclear leukocytes and production of inflammatory mediators, which contribute to the ensuant pulmonary damage.

The ability of taurine to attenuate the highly toxic effects of $HOCl/OCl^-$ produced by activated polymorphonuclear leukocytes, eosinophils, and basophils is thought to be important in the mechanism of taurine protection. Taurine reacts with $HOCl/OCl^-$ to form taurine monochloramine (Tau-Cl), an oxidant that is far less reactive and more stable than $HOCl/OCl^{-}$[8-10]. While the detoxification of $HOCl/OCl^-$ by Tau-Cl formation contributes to the protective effects of taurine, more recent studies suggest that Tau-Cl may be a significant biological effector molecule. This is supported by reports that Tau-Cl inhibits production of proinflammatory mediators by activated cells of various origin[1,11-14].

Alveolar macrophages play a critical role in propagating pulmonary lung inflammation[15,16] and Tau-Cl has been demonstrated to inhibit production of inflammatory mediators in alveolar macrophages activated with LPS and IFN-γ[17,18]. While LPS activates target genes through the NF-κB signaling pathway, IFN-γ induces expression of its target genes through JAK/Stat. Expression of iNOS protein is regulated primarily at the level of gene transcription and the promoter region of the iNOS gene contains several consensus sequence binding sites for IFN-γ activated transcription factors, i.e., Stat1α, IRF-1, and C/EBP-β[19-21]. For these reasons we evaluated the effects of Tau-Cl on IFN-γ signaling pathways focusing on IRF-1 and Stat1α in NR8383 cells, a cell line derived from rat alveolar macrophages. NR8383 cells retain many of the characteristics of alveolar macrophages and have been useful for molecular studies of alveolar macrophage function[22,23].

2.   MATERIALS AND METHODS

2.1   Materials

Rabbit polyclonal antibodies against the carboxy-terminal of IRF-1, Stat1α (p91), and Oct-1 were purchased from Santa Cruz Biotechnology (Santa Cruz, CA). Rabbit polyclonal antibody against phospho-Tyr 701 of Stat1 was from Cell Signaling Technology (Beverly, MA). Plasmid containing the cDNA probe for rat glyceraldehyde-3-phosphate dehydrogenase (GAPDH) was kindly provided by Dr. R. Dong (London University College). The cDNA probe used for hybridizing with IRF-1 mRNA contained a sequence corresponding to nucleotides 225 to 1204 of murine IRF-1 mRNA (GenBank accession number BC003821) prepared from pCMV-SPORT6, IMAGE clone 3600525 (American Type Culture Collection, Manassas, VA). Oligonucleotides were synthesized by Bioserve Biotechnologies (Laurel, MD). Tri-Reagent was purchased from Molecular Research Center (Cincinnati, OH), nylon (Nytran) membranes were from Schleicher & Schuell (Keen, NH), and [^{32}P]dCTP was obtained from Amersham Corporation (Arlington Heights, IL). Tau-Cl was prepared fresh on the day of use and all other materials were from commercial suppliers as previously described[17,18].

2.2   Cell Culture

NR8383 cells were obtained from the American Type Culture Collection (Manassas, VA). Cells were grown in flasks containing F-12 nutrient mixture (HAM) supplemented with 15% heat-inactivated FBS, 2 mM glutamine, 1% penicillin, and 1% streptomycin. Experiments were conducted in phenol red-free DMEM supplemented with 2% FBS and antibiotics. Cells were activated with LPS (1 µg/ml) and IFN-γ (10 U/ml) with or without the addition of Tau-Cl (1.0 mM) and media content of NO_2^- was routinely measured 24 hr later in parallel cultures using Griess reagent. Cells were collected at the times indicated and nuclear protein extracts, lysates, or total RNA extracts were prepared as previously described[17].

2.3   Electrophoretic Mobility Shift Assay (EMSA)

Double-stranded oligonucleotide probes were synthesized and labeled using the Klenow fragment of DNA polymerase and [^{32}P]dCTP. Probe for IRF-1 contained the iNOS specific IRF-1 binding sequence (underlined) 5'-CACTGTCAAT<u>ATTTCAC</u>TTTCATAATG-3', and the probe for Oct-1 contained the iNOS specific Oct-1 binding site motif, 5'-GACTTT<u>ATGCAAAA</u>CAGCTCT-3'. The Stat1α binding probe contained a consensus sequence for GAS as follows, 5'-GATGTA TT<u>TCCCAGAA</u>AGG AAC-3'.

Protein-DNA binding interactions were performed by incubating (22°C, 15 min) 3-5 µg of nuclear protein extract with 1 ng of [^{32}P] labeled (50,000 cpm) double stranded oligonucleotide probe in 20 mM Tris-HCl, pH 7.9, containing 50 mM NaCl, 2 mM $MgCl_2$, 1 mM EDTA, 10% glycerol, 0.1% Nonidet P-40, 1 mM DTT, 0.5% BSA, and 2 µg poly dI-dC in a final volume of 20 µl. For supershift assays antibodies were included in the above reaction mixture and incubated at 4°C for 90 min before addition of the radiolabeled probe, followed by incubation at 22°C for 15 min. The protein-oligonucleotide complexes were fractionated on 6% nondenaturing polyacrylamide gels run in 0.5 X TBE buffer. Gels were dried under vacuum and bands were visualized by autoradiography. Western blot and Northern blot analyses were conducted as previously described[17].

3.   RESULTS AND DISCUSSION

The transcription factor interferon response factor-1 (IRF-1) is essential for the induced expression of the iNOS gene[24,25]. IRF-1 is a major signaling protein mediating the effects of IFN-γ. In addition, IRF-1 is also the product of an IFN-γ stimulated gene[26]. The effects of Tau-Cl on the nuclear translocation and binding of IRF-1 to its cognate recognition site, ISRE, was determined by EMSA (Fig. 1). Two complexes of IRF-1/ISRE were detected in nuclear protein extracts prepared from NR8383 cells that were activated for 1 hr. Cells activated in the presence of Tau-Cl (1 mM) exhibited decreased formation of both complexes while unactivated cells did not exhibit IRF-1 binding to the ISRE probe. Preincubation of nuclear protein extracts with antibody against IRF-1 before addition of radiolabeled probe supershifted both IRF-1/ISRE complexes confirming IRF-1 as a constituent of both complexes (Fig. 1).

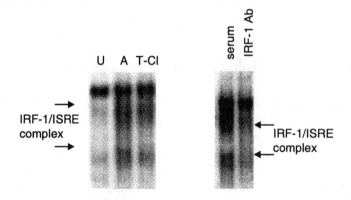

Figure 1. (Left panel) Cells were treated for 1 hr before preparing nuclear protein extracts. IRF-1 binding to ISRE was measured by EMSA. U, unactivated cells; A, activated control cells; T-Cl, cells activated in the presence of 1.0 mM Tau-Cl. (Right panel) Specificity of the IRF-1/ISRE binding interaction was assessed by supershift with IRF-1 antibody. Similar results were obtained in 2–3 additional independent experiments.

Western blot analysis of IRF-1 in nuclear protein extracts demonstrated that Tau-Cl treated cells contained less IRF-1 than that of control activated cells (Fig. 2). This suggested that Tau-Cl does not influence IRF-1 binding to ISRE, but inhibits the amount of IRF1 that translocates to the nucleus upon cell activation. Since IRF-1 is a protein that is induced in response to IFN-γ, expression of IRF-1 mRNA was evaluated by Northern blot (Fig. 2). Cells activated in the presence of Tau-Cl expressed lower levels of IRF-1 transcript than that of activated cells. The decreased expression of IRF-1 mRNA in

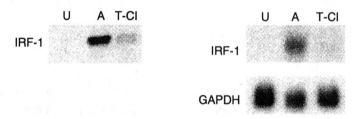

Figure 2. (Left panel) IRF-1 in nuclear protein extracts of NR8383 cells activated for 1 hr was measured by Western blot. Cells were treated as described in Figure 1. (Right panel) IRF-1 mRNA expression was measured by Northern blot in NR8383 cells activated for 30 min. Similar results were obtained in two additional independent experiments.

NR8383 cells activated in the presence of Tau-Cl accounts for the decreased nuclear protein content of IRF-1 and the observed effects of Tau-Cl in the IRF-1/ISRE EMSA.

IFN-γ stimulates expression of IRF-1, as well as other target genes, through the JAK/Stat pathway[26-28]. Upon interaction with the IFN-γ receptor in the plasma membrane, receptor subunits become phosphorylated and activate associated tyrosine kinases, JAK1 and JAK2. JAK 1 and 2 phosphorylate Tyr-701 of Stat1 allowing p-Stat1 to form homodimers, which migrate into the nucleus and bind to a GAS sequence motif on the promoter region of iNOS, IRF-1, and other IFN-γ target genes[24-28]. Analysis of Stat1α binding to a probe containing a GAS consensus sequence was performed using nuclear protein extracts of NR8383 cells activated in the absence or presence of Tau-Cl (1 mM) by EMSA (Fig. 3). The formation of Stat-1/GAS

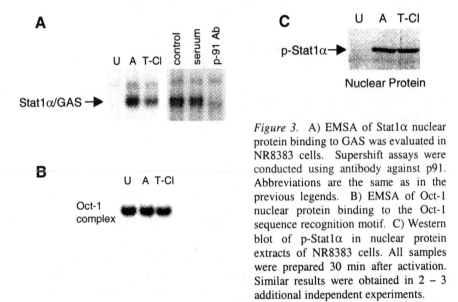

Figure 3. A) EMSA of Stat1α nuclear protein binding to GAS was evaluated in NR8383 cells. Supershift assays were conducted using antibody against p91. Abbreviations are the same as in the previous legends. B) EMSA of Oct-1 nuclear protein binding to the Oct-1 sequence recognition motif. C) Western blot of p-Stat1α in nuclear protein extracts of NR8383 cells. All samples were prepared 30 min after activation. Similar results were obtained in 2 – 3 additional independent experiments.

complex recovered from the nuclear protein extract was much less in Tau-Cl treated cells relative to that of controls. The identity of the complex was confirmed by the supershifting of the Stat1/GAS band in samples that were preincubated with Stat1 (p91) antibody (Fig. 3). As an example of a transcription factor that functions at the promoter region of the iNOS gene but is not part of the IFN-γ signaling pathway, we evaluated Oct-1 binding to its recognition sequence motif (Fig. 3). Tau-Cl did not affect Oct-1 binding in activated cells. Oct-1 is a transcription factor that resides in the nucleus and because of this, unactivated cells express levels of Oct-1 complex similar to that of activated cells[21].

Western blot analysis indicated similar amounts of p-Stat1 in the nuclear protein extracts of Tau-Cl treated cells and controls. This suggests that Tau-Cl does not interfere with upstream JAK/Stat signaling events mediating the effects of IFN-γ. Rather, Tau-Cl appears to somehow inhibit the binding of p-Stat1 homodimers to the GAS sequence motif. It is uncertain whether this is a direct effect of Tau-Cl or the result of some other regulatory step that influences the transactivation of IFN-γ target genes. For instance, there is a family of proteins, referred to as protein inhibitor of activated stat (PIAS), that function within the nucleus to inhibit p-Stat activation of IFN regulated genes[29]. PIAS1 is a member of this protein family that specifically binds p-Stat1α homodimers in the nucleus thus preventing binding of p-Stat1α to its cognate GAS site[30]. Perhaps Tau-Cl enhances this binding interaction to prevent the transcription of IFN-γ target genes.

ACKNOWLEDGEMENTS

The authors appreciate the excellent secretarial assistance of Mrs. A. Parese. This work was supported by the OMRDD of New York and by NIH grants HL-49942 and NS40721.

REFERENCES

1. Quinn, M.R., and Schuller-Levis ,G.B., 1999, Taurine Chloramine, an Inhibitor of iNOS Expression and a Potential Modulator of Inflammation. In *Molecular and Cellular Biology of Nitric Oxide* (J. Laskin and D. Laskin, eds.), Marcel Dekker, Inc., New York, pp. 309-331.
2. Gordon, R.E., Shaked, A.A., and Solano, D.F., 1986, Taurine protects hamster bronchioles from acute NO_2^- induced alterations. *Am. J. Pathol.* 125: 585-600.
3. Schuller-Levis, G.B., Gordon, R.E., Park, E., Pendino, K.J., and Laskin, D.L., 1995, Taurine protects rat bronchioles from acute ozone-induced lung inflammation and hyperplasia. *Exp. Lung Res.* 21: 877-888.
4. Gurujeyalakshmi, G., Wang, Y, and Giri, S.N., 2000, Taurine and niacin block lung injury and fibrosis by down-regulating bleomycin-induced activation of transcription nuclear factor-kappa B in mice. *J. Pharmacol. Exp. Ther.* 293: 82-90.
5. Gurujeyalakshmi, G., Wang, Y, and Giri, S.N., 2000, Suppression of bleomycin-induced nitric oxide production in mice by taurine and niacin. *Nitric Oxide* 4: 399-411.
6. Wang, Q., Hollinger, M.A., and Giri, S.N., 1992, Attenuation of amiodarone induced lung fibrosis and phospholipidosis in hamsters by taurine and/or niacin treatment. *J. Pharmacol. Exp. Ther.* 262: 127-132.
7. Izmi, K., Nagata, R., Motoya, T., Yamashita, Y., Hirokane, T., Naguta, T., Satoh, Y., Sawada, Y., Ishibashi, M., Yoshida, H., and Fukuda, T., 1989, Preventive effect of taurine against acute paraquat intoxication in beagles. *Jpn. J. Pharmacol.* 50: 229-233.
8. Zgicznski, J.M., Stelmaszynska, T., Domanski, J., and Ostrowski, W., 1971, Chloramines as intermediates of oxidation reaction of amino acids by myeloperoxidase. *Biochim. Biophys. Acta* 235: 419-424.

9. Weiss, S.J.R., Klein, A., Slivka, A., and Wei, M., 1982, Chlorination of taurine by human neutrophils: evidence for hypochlorous acid generation. *J. Clin. Invest.* 70: 598-607.

10. Grisham, M.B., Jefferson, M.M., Melon, D.F., and Thomas, E.L., 1984, Chlorination of endogenous amines by isolated neutrophils. *J. Biol. Chem.* 259: 10404-10413.

11. Marcinkiewicz, J., Grabowska, A., Bereta, J., and Stelmaszynska, T., 1995, Taurine chloramine, a product of activated neutrophils, inhibits *in vitro* the generation of nitric oxide and other macrophage inflammatory mediators. *J. Leukoc. Biol.* 58: 667-674.

12. Park, E., Alberti, J., Quinn, M.R., and Schuller-levis, G., 1998, Taurine chloramine inhibits the production of superoxide anion, IL-6 and IL-8 in activated human polymorphonuclear leukocytes. *Adv. Exp. Med. Biol.* 442: 177-182.

13. Liu, Y., Tonna-DeMasi, M., Park, E., Schuller-Levis, G., and Quinn, M.R., 1998, Taurine chloramine inhibits production of nitric oxide and prostaglandin E$_2$ in activated C6 glioma cells by suppressing inducible nitric oxide synthase and cyclooxygenase-2 expression. *Molec. Brain Res.* 59: 189-195.

14. Kontny, E., Grabowska, A., Kowalczewski, J., Kurowska, M., Janicka, I., Marcinkiewicz, J., and Maslinski, W., 1999, Taurine chloramine inhibition of cell proliferation and cytokine production by rheumatoid arthritis fibroblast-like synoviocytes. *Arthritis Rheum.* 42: 2552-2560.

15. Sakanashi, Y., Takeya, M., Yoshimura, T., Feng, L., Morioka, T., and Takahashi, K., 1994, Kinetics of macrophage subpopulations and expression of monocyte chemoattractant protein-1 (MCP-1) in bleomycin-induced lung injury of rats studied by a novel monoclonal antibody against rat MCP-1. *J. Leukoc. Biol.* 56: 741-750.

16. Lentsch, A.B., Czermak, B.J., Bless, N.M., Rooijen, N.V., and Ward, PA., 1999, Essential role of alveolar macrophages in intrapulmonary activation of NF-κB. *Am. J. Respir. Cell Mol. Biol.* 20: 692-698.

17. Barua, M., Liu, Y., and Quinn, M.R., 2001, Taurine chloramine inhibits nitric oxide synthase and TNF-α gene expression in activated alveolar macrophages: Decreased NF-κB activation and IκB kinase activity. *J. Immunol.* 167: 2275-2281.

18. Liu, Y., and Quinn, M.R., 2002, Chemokine production by rat alveolar macrophages is inhibited by taurine chloramine. *Immunol. Lett.* 80: 27-32.

19. Eberhardt, W., Kunz, D., Hummel, R., and Pfeilschifter, J., 1996, Molecular cloning of the rat inducible nitric oxide synthase gene promoter. *Biochem. Biophys. Res. Commun.* 223: 752-756.

20. Zhang, H., Chen, X., Teng, X., Snead, C., and Catravas, J.D., 1998, Molecular cloning and analysis of the rat inducible nitric oxide synthase gene promoter in aortic smooth muscle cells. *Biochem. Pharmacol.* 55: 1873-1880.

21. Murphy, W.J., 1999, Transcriptional Regulation of the Genes Encoding Nitric Oxide Synthase. In *Cellular and Molecular Biology of Nitric Oxide* (J.D. Laskin and D.L. Laskin, eds.), Marcel Dekker, Inc., New York, pp. 1-56.

22. Helmke, R.J., German, V.F., and Mangos, J.A., 1989, A continuous alveolar macrophage cell line: comparisons with freshly derived alveolar macrophages. *In Vitro Cell. Develop. Biol.* 25: 44-48.

23. Griscavage, J.M., Rogers, N.E., Sherman, M.P., and Ignarro, L.J., 1993, Inducible nitric oxide synthase from a rat alveolar macrophage cell line is inhibited by nitric oxide. *J. Immunol.* 151: 6329-6337.

24. Martin, E., Nathan, C., and Xie, Q., 1994, Role of interferon regulatory factor 1 in induction of nitric oxide synthase. *J. Exp. Med.* 180: 977-984.

25. Gao, J., Morrison, D.C., Parmely, T.J., Russell, S.W., and Murphy, W.J., 1997, An interferon-γ-activated site (GAS) is necessary for full expression of the mouse iNOS gene in response to interferon-γ and lipopolysaccharide. *J. Biol. Chem.* 272: 1226-1230.

26. Der, S.D., Zhou, A., Williams, B.R.G., and Silverman, R.H., 1998, Identification of genes differentially regulated by interferon α, β or γ using oligonucleotide arrays. *Proc. Natl. Acad. Sci. USA* 95: 15623-15628.

27. O'Shea, J., Gadina, M., and Schreiber, R.D., 2002, Cytokine signaling in 2002: New surprises in the Jak/Stat pathway. *Cell* 109: S121-S131.
28. Ramana, C.V., Gil, M.P., Schreiber, R.D., and Stark, R.D., 2002, Stat1-dependent and-independent pathways in IFN-γ-dependent signaling. *Trends Immunol.* 23: 96-101.
29. Starr, R., and Hilton, D.J., 1999, Negative regulation of the JAK/STAT pathway. *BioEssays* 21: 47-52.
30. Liu, B., Liao, J., Rao, X., Kushner, S.A., Chung, C.D., Chang, D.D., and Shuai, K., 1998, Inhibition of Stat1-mediated gene activation by PIAS1. *Proc. Natl. Acad. Sci. USA* 95: 10626-10631.

Production of Nitric Oxide by Activated Microglial Cells Is Inhibited by Taurine Chloramine

VALERIA SERBAN*†, YONG LIU, and MICHAEL R. QUINN*
Laboratory of Molecular Cell Signaling, Department of Developmental Biochemistry, New York State Institute for Basic Research in Developmental Disabilities, and *Center for Developmental Neuroscience, Staten Island, NY and †Samaritan Program, Temple University Hospital, Philadelphia, PA, USA

1. INTRODUCTION

Microglia are considered to be brain resident macrophages and along with astrocytes are the major immunoresponsive cells in the CNS[1,2]. When activated by bacterial endotoxin or cytokines, microglia respond rapidly by proliferating, changing morphology, and by producing proinflammatory cytokines and NO[2-4]. Although transient activation of microglia contributes to brain repair processes, chronic activation as occurs in CNS viral infections[5], AIDS dementia complex[6], Alzheimer's disease[7], multiple sclerosis[8,9], traumatic injury, and stroke, leads to neuronal cell death as a result of inflammation and oxidative stress[2,4,10]. Production of nitric oxide is of particular importance in the pathology of several CNS disorders because of the toxicity of its byproducts, e.g. peroxynitrite. The increased production of NO by activated microglia results primarily from increased expression of the iNOS gene. The therapeutic potential of downregulating activation of microglia and/or production of NO may be of significant clinical value in developing strategies for treatment of neurodegeneative diseases.

Recent studies with macrophages and various other immuno-responsive cells suggest that taurine chloramine (Tau-Cl) may function as a

Taurine 5: Beginning the 21st Century
Edited by Lombardini *et al.*, Kluwer Academic/Plenum Publishers, New York 2003

physiological modulator of the inflammatory response because Tau-Cl inhibits production of inflammatory mediators, including NO[11-15]. Formation of Tau-Cl is mediated by the activity of halide-dependent myeloperoxidase (MPO) and the presence of taurine[16,17]. MPO is associated primarily with neutrophils, but recent reports demonstrate induced expression of MPO in activated microglial cells[18,19] and thus provide a means of Tau-Cl formation in the CNS without breach of the blood-brain barrier. Although the inhibitory effects of Tau-Cl on production of inflammatory mediators by C6 glioma cells has been described[20,21], Tau-Cl modulation of NO production by activated microglia has not yet been reported.

2.   MATERIALS AND METHODS

2.1   Materials

Recombinant murine IFN-γ, HBSS, DMEM with or without phenol red, penicillin, streptomycin, and trypsin-EDTA were purchased from GIBCO-BRL (Grand Island, NY). Heat inactivated fetal bovine serum was from Gemini Bio-Products, Inc. (Calabasas, CA), and LPS W (Escherichia coli 0111:B4) was purchased from Difco Laboratories (Detroit, MI). Plasmids containing cDNA probe for iNOS, and glyceraldehyde-3-phosphate dehydrogenase (GAPDH) were graciously provided by Drs. C. Nathan, and R. Dong, respectively. All other materials were obtained from commercial sources and Tau-Cl was prepared fresh before use as previously described[20,21].

2.2   BV-2 Cells

BV-2 cells are an immortalized murine microglial cell line that retain the morphological and functional characteristics of microglia[22,23]. BV-2 cells were graciously provided to us by Dr. Michael McKinney (Mayo Clinic, Jacksonville, FL) with permission from Dr. Virginia Bocchini (Univ. Perugia, Italy). BV-2 cells were grown in DMEM with 10% FCS, 2 mM L-glutamine, and antibiotics at 37°C in a humidified 95%/5% mixture of air and CO_2. Cells were seeded and used 2 – 3 days later when they were 70 – 90% confluent. Media was changed at the initiation of experiments. Cells were washed twice in Hanks' balanced salt solution (HBSS), and experiments were conducted in serum free DMEM without phenol red. Cells were activated with LPS or IFN-γ, or a combination of both as indicated.

2.3   Primary Cell Cultures

Primary cultures of rat microglial cells were prepared from postnatal day one rat cortex using a modification of the procedure described by McCarthy and de Villis[24]. After brains were carefully stripped of meningeal membranes, the cortex was removed, minced and mechanically dispersed by passing through 18 and 21 gauge needles. The resulting cells were seeded in DMEM/F-12 media containing 15% FCS, 6 mM L-glutamine, and antibiotics, and cultured at 37°C in humidified air. Media was replaced the following day to remove floating cells (neurons) and was exchanged (~70%) every 2 – 3 days. After 9 – 11 days, nearly confluent (90 – 95%) cells were shaken for one hour at 120 rpm. Detached cells were collected, washed, and seeded in a 24-well plate. The medium was replaced after 2 hours to remove non-adherent cells, and the microglial cells were allowed to grow in DMEM containing 10% fetal bovine serum, 2 mM L-glutamine, and antibiotics for an additional 24 hours before initiating experiments. The medium was changed to serum free DMEM before cells were activated with 1 µg/ml LPS. Tau-Cl was added at the time of activation, and NO_2^- production was measured 24 hours later by quantification of nitrite, a metabolic endproduct of NO[20].

2.4   Northern Blot Analyses

Northern blot analyses were conducted as previously described[20]. Briefly, total RNA was extracted from cells using Tri-Reagent, size-fractioned by electrophoresis in 1% agarose-formaldehyde gel, transferred to Nytran membrane, and cross-linked to the membrane by UV irradiation. Blots were prehybridized in ExpressHyb Hybridization Solution (Clontech, Palo Alto, CA) for 1 hr (68°C) before hybridization with [^{32}P]dCTP random prime-labeled cDNA at 68°C for 16 – 18 hr. Blots were washed three times at room temperature in 2 X SSC containing 0.5 SDS followed by two washes at 50°C in 0.1 X SSC containing 0.1% SDS. Membranes were stripped of cDNA probe between sequential hybridizations. RNA hybridized with cDNA probe was visualized by autoradiography and analyzed by computer-assisted densitometry.

3.   RESULTS AND DISCUSSION

In order to establish optimal conditions for activation of BV-2 cells various concentrations of LPS were evaluated for eliciting NO production measured 24 hr later (Fig. 1A). Unactivated BV-2 cells did not produce detectable amounts of NO_2^- over the course of the experiments. Cells activated with LPS produced NO_2^- in a concentration dependent manner. A robust and consistent response (70 μM NO_2^-) was elicited by 1 μg/ml LPS, and 1 μg/ml LPS was selected for use in subsequent studies. BV-2 cells activated with IFN-γ alone, generated NO_2^- in the culture media in a fashion similar to that of LPS. Physiological concentrations of IFN-γ (100 – 200 U/ml) were sufficient to produce a robust response, 30–50 μM NO_2^- (Fig. 1B), and 100 U/ml was selected for further studies when using IFN-γ as the sole activator.

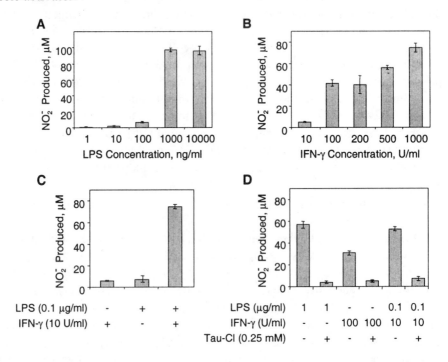

Figure 1. Activated BV-2 cells produce NO and NO production is inhibited by Tau-Cl. Cells were activated with the indicated concentration of LPS (A) or IFN-γ (B) and NO_2^- was measured 24 hr later. C) Cells were activated with LPS, IFN-γ, or with a combination of LPS + IFN-γ. D) Cells were activated with LPS, IFN-γ, or with LPS + IFN-γ and Tau-Cl was added at the time of activation. Values represent x̄ ± SD for triplicate samples. Similar results were obtained in 3-10 independent experiments.

These results demonstrate that BV-2 cells are somewhat unique in that they produce large amounts of NO in response to either LPS or IFN-γ when each activator is used alone. However, when lower concentrations of LPS (0.1 μg/ml) are used in combination with low levels of IFN-γ (10 U/ml) the activators exhibit a synergistic effect in eliciting production of NO from BV-2 cells (Fig. 1C). Effects of Tau-Cl on the production of NO by BV-2 cells were evaluated under each of the three activation paradigms (Fig. 1D).

Initial dose-response curves indicated 0.25 mM Tau-Cl to be an effective concentration for inhibiting NO production. Under each of the three activator conditions 0.25 mM Tau-Cl inhibited production of NO by approximately 90%. The viability of the cells was unaffected by Tau-Cl as determined by morphologic examination and trypan blue exclusion.

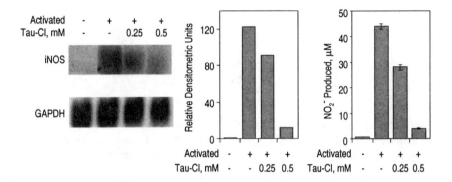

Figure 2. Tau-Cl inhibits expression of iNOS mRNA. Cells were activated with LPS in the presence of various Tau-Cl concentrations (0, 0.25, 0.5 mM). RNA was extracted from BV-2 cells 8 hr after activation and analyzed by Northern blot. Bands obtained by autoradiography were measured by computer-assisted densitometry. iNOS bands were normalized by GAPDH for each lane and expressed as Relative Densitometric Units. NO_2^- was measured 24 hr after activation in conditioned media of parallel cultures. Similar results were obtained in two independent experiments.

Since NO is produced by iNOS and iNOS is regulated primarily, but not exclusively, at the level of gene expression, total RNA extracts were prepared from BV-2 cells after 8 hr of activation and iNOS mRNA was measured by Northern blot (Fig. 2). Transcripts for GAPDH were used to normalize the amount of RNA applied to each lane. After 8 hr of activation with LPS(1 μg/ml) iNOS mRNA was robustly expressed in BV-2 cells. Unactivated cells did not express iNOS mRNA and did not produce detectable levels of NO_2^-. Cells activated in the presence of Tau-Cl express dose-dependent decreases in iNOS mRNA relative to that of control activated cells. Inhibition of iNOS mRNA expression by Tau-Cl accounted for the inhibition of NO_2^- production (Fig. 2). Western blot analyses of cells lysates revealed similar dose-dependent Tau-Cl inhibition of iNOS protein expression (not shown) in accordance with the Northern blot results. Since BV-2 cells are a cell line, the effects of Tau-Cl on NO production was evaluated in authentic microglial cells. Primary cultures of rat cortical microglial cells were utilized for these purposes. Tau-Cl dose-dependently inhibited production of NO by microglial cells activated with LPS (1 μg/ml) for 24 hr (Fig. 3).

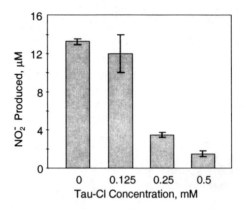

Figure 3. Tau-Cl inhibits production of NO_2^- by primary cultures of rat cortical microglial cells activated with LPS (1 µg/ml). Tau-Cl was added at the time of activation and NO_2^- was measured 24 hr later. Values represent x̄ ± SD of triplicate samples.

Several transcription factors serve as convergence points for many different signaling pathways that regulate expression of the iNOS gene[25]. Among these transcription factors NF-κB and IRF-1 are considered to be of primary importance. Although NF-κB is not thought to mediate effects of IFN-γ in most cell types, IFN-γ activates NF-κB in BV-2 cells[25] and apparently accomplishes this through a different protein kinase cascade than that utilized by LPS[27]. BV-2 cells will be useful for unraveling the signal transduction pathways that regulate iNOS expression in microglial cells as demonstrated by the effects of LPS, IFN-γ, and the synergistic effect of the combined use of LPS + IFN-γ on NO production. The results of these studies strengthen the idea that Tau-Cl has the potential to serve as a useful therapeutic agent for downregulating inflammatory events in the CNS.

ACKNOWLEDGEMENTS

The authors appreciate the excellent secretarial assistance of Mrs. A. Parese. This work was supported by the OMRDD of New York and by NIH grants HL-49942 and NS40721.

REFERENCES

1.   Kreutzberg, G.W., 1996, Microglia: a sensor for pathological events in the CNS. *Trends Neurosci.* 19: 312-318.
2.   Minghetti, L., and Levi, G., 1998, Microglia as effector cells in brain damage and repair: focus on prostanoids and nitric oxide. *Prog. Neurobiol.* 54: 99-125.

3. Ding, M., St. Pierre, B.A., Parkinson, J.F., Medberry, P., Wong, J.L., Rogers, N.E., Ignarro, L.J., and Merrill, J.E., 1997, Inducible nitric-oxide synthase and nitric oxide production in human fetal astrocytes and microglia. *J. Biol. Chem.* 272: 11327-11335.
4. González-Scarano, F., and Baltuch, G., 1999, Microglia as mediators of inflammatory and degenerative diseases. *Ann. Rev. Neurosci.* 22: 219-240.
5. Koprowski, H., Zheng, Y.M., Heber-Katz, E., Fraser, N., Rorke, L., Fu , Z.F., Hanlon, C., and Dietzschold, B., 1993, *In vivo* expression of inducible nitric oxide synthase in experimentally induced neurologic diseases. *Proc. Natl. Acad. Sci. USA* 90: 3024-3027.
6. Koka, P., He, K., Zack, J.A., Kitchen, S., Peacock, W., Fried, I., Tran, T., Yashar, S.S., and Merill, J.E., 1995, Human immunodeficiency virus 1 enveloped proteins induce interleukin 1, tumor necrosis factor α, and nitric oxide in glial cultures derived from fetal, neonatal, and adult human brain. *J. Exp. Med.* 182, 941-952.
7. Griffin, W.S., Sheng, J.G., Royston, M.C., Gentleman, S.M., McKenzie, J.E., Graham, D.I., Roberts, G.W., and Mrak, R.E., 1998, Glial-neuronal interactions in Alzheimer's disease: the potential role of a `cytokine cycle´ in disease progression. *Brain Pathol.* 8: 65-72.
8. Merrill, J.E., Ignarro, L.J., Sherman, M.P., Melinek, J., and Lane, T.E., 1993, Microglial cell cytotoxicity of oligodendrocytes is mediated through nitric oxide. *J. Immunol.* 151: 2132-2141.
9. Ding, M., Zhang, M., Wong, J.L., Rogers, N.E., Ignarro, L.J., and Voskuhl, R.R., 1998, Cutting Edge: Antisense knockdown of inducible nitric oxide synthase inhibits induction of experimental autoimmune encephalomyelitis in SJL/J mice. *J. Immunol.* 160: 2560-2564.
10. Mayer, A.M.S., 1998, Therapeutic implications of microglia activation by lipopolysaccharide and reactive oxygen species generation in septic shock and central nervous system pathologies: a review. *Medicine* 58: 377-385.
11. Park, E., Quinn, M.R., Wright, C.E., and Schuller-Levis, G., 1993, Taurine chloramine inhibits the synthesis of nitric oxide and the release of tumor necrosis factor in activated RAW 264.7 cells. *J. Leukoc. Biol.* 54: 119-124.
12. Marcinkiewicz, J., Grabowska, A., Bereta, J., and Stelmaszynska, T., 1995, Taurine chloramine, a product of activated neutrophils, inhibits *in vitro* the generation of nitric oxide and other macrophage inflammatory mediators. *J. Leukoc. Biol.* 58: 667-674.
13. Quinn, M.R., and Schuller-Levis ,G.B., 1999, Taurine Chloramine, an Inhibitor of iNOS Expression and a Potential Modulator of Inflammation. In *Molecular and Cellular Biology of Nitric Oxide* (J. Laskin and D. Laskin, eds.), Marcel Dekker, Inc., New York, pp. 309-331.
14. Kontny, E., Grabowska, A., Kowalczewski, J., Kurowska, M., Janicka, I., Marcinkiewicz, J., and Maslinski, W., 1999, Taurine chloramine inhibition of cell proliferation and cytokine production by rheumatoid arthritis fibroblast-like synoviocytes. *Arthritis Rheum.* 42: 2552-2560.
15. Barua, M., Liu, Y., and Quinn, M.R., 2001, Taurine chloramine inhibits nitric oxide synthase and TNF-α gene expression in activated alveolar macrophages: Decreased NF-κB activation and IκB kinase activity. *J. Immunol.* 167: 2275-2281.
16. Weiss, S.J.R., Klein, A., Slivka, A., and Wei, M., 1982, Chlorination of taurine by human neutrophils: evidence for hypochlorous acid generation. *J. Clin. Invest.* 70: 598-607.
17. Grisham, M.B., Jefferson, M.M., Melon, D.F., and Thomas, E.L., 1984, Chlorination of endogenous amines by isolated neutrophils. *J. Biol. Chem.* 259: 10404-10413.
18. Nagra, R.M., Becher, B., Tourtellotte, W.W., Antel, J.P., Gold, D., Paladino, T., Smith, R.A., Nelson, J.R., and Reynolds, W.F., 1997, Immunohistochemical and genetic evidence of myeloperoxidase involvement in multiple sclerosis. *J. Neuroimmunol.* 78: 97-107.

19. Reynolds, W.F., Rhees, J., Maciejewski, D., Paladino, T., Sieburg, H., Maki, R.A., and Massillon, E., 1999, Myeloperoxidase polymorphism is associated with gender specific risk for Alzheimer's disease. *Exp. Neurology* 155: 31-41.
20. Liu, T., Tonna-DeMasi, M., Park, E., Schuller-Levis, G., and Quinn, M.R., 1998, Taurine chloramine inhibits production of nitric oxide and prostaglandin E$_2$ in activated C6 glioma cells by suppressing inducible nitric oxide synthase and cyclooxygenase expression. *Molec. Brain Res.* 59: 189-195.
21. Liu, Y ., Schuller-Levis, G., and Quinn, M.R., 1999, Monocyte chemoattractant protein-1 and macrophage inflammatory protein-2 production is inhibited by taurine chloramine in rat C6 glioma cells. *Immunol. Lett.* 70: 9-14.
22. Blasi, E., Barluzzi, R., Bocchini, V., Mazzolla, R., and Bistoni, F., 1990, Immortalization of murine microglial cells by a v-*raf*/v-*myc* carrying retrovirus. *J. Neuroimmunol.* 27: 229-237.
23. Bocchini, V., Mazzolla, R., Barluzzi, R., Blasi, E., Sick, P., and Kettenmann, H., 1992, An immortalized cell line expresses properties of activated microglial cells. *J. Neurosci. Res.* 31: 616-621.
24. McCarthy, K.D., and de Villis, J., 1989, Preparation of separate astroglial and oligodendroglial cell cultures from rat cerebral tissue. *J. Cell Biol.* 85: 890-902.
25. Murphy, W.J., 1999, Transcriptional regulation of the genes encoding nitric oxide synthase. In *Cellular and Molecular Biology of Nitric Oxide* (J.D. Laskin and D.L. Laskin, eds.,), Marcel Dekker, Inc., New York, pp. 1-56.
26. Jana, M., Liu, X., Koka, S., Ghosh, S, Petro, T.M., and Pahan, K., 2001, Ligation of CD40 stimulates the induction of nitric-oxide synthase in microglial cells. *J. Biol. Chem.* 276: 44527-44533.
27. Han, I.-O., Kim, K.-W., Ryu, J.H., and Kim, W.-K., 2002, p38 Mitogen-activated protein kinase mediates lipopolysaccharide, not interferon-γ-induced nitric oxide synthase expression mouse BV2 microglial cells. *Neurosci. Lett.* 325: 9-12.

Production of Inflammatory Mediators by Activated C6 Cells Is Attenuated by Taurine Chloramine Inhibition of NF-κB Activation

YONG LIU, MADHABI BARUA, VALERIA SERBAN*, and MICHAEL R. QUINN*
Laboratory of Molecular Cell Signaling, Department of Developmental Biochemistry, New York State Institute for Basic Research in Developmental Disabilities, and *Center for Developmental Neuroscience, Staten Island, NY, USA

1.   INTRODUCTION

Taurine chloramine (Tau-Cl) is produced through the actions of a halide-dependent myeloperoxidase system (MPO) that is associated primarily with polymorphonuclear leukocytes (PMN)[1-3]. Tau-Cl may be synthesized directly by MPO within the intracellular environment of PMN[4], in part, because PMN intracellular taurine concentrations are in the range of 20 mM[5]. However, activated PMN secrete MPO and release it into the extracellular milieu upon death. Under extracellular conditions, where physiological taurine concentrations are usually low, MPO peroxidizes halide ions to produce HOCl/OCl$^-$, which causes indiscriminate cellular damage to tissues at the site of inflammation. Taurine chemically reacts with HOCl/OCl$^-$ to rapidly form Tau-Cl, a more stable, less reactive and more selective oxidant than HOCl/OCl$^-$. Administration of pharmacologic doses of taurine has been demonstrated to protect tissue from damage in several different paradigms where inflammation is likely to be elicited[6].

Results of recent studies suggest that Tau-Cl may be a significant immunomodulator at sites of inflammation because Tau-Cl inhibits production of proinflammatory mediators by activated macrophages[6-8]. Glial cells, similar to macrophages, may be activated by bacterial endotoxin and/or cytokines to produce nitric oxide (NO), PGE$_2$, and other proinflammatory mediators in the CNS[9-11]. Glial cell production of NO and PGE$_2$ contributes to neuronal cell death during CNS inflammatory events, e.g., CNS viral infections, and AIDS dementia complex[12-13]. Production of NO and PGE$_2$ in the CNS during ischemic stroke also leads to neuronal cell death within the area of the penumbra[14,15] and PMN are recruited into this zone within hours of infarct formation. Therefore, increasing levels of Tau-Cl by providing taurine to PMN associated MPO at the site of injury has potential therapeutic implications if Tau-Cl modulates production of inflammatory mediators in glia, similar to its effects in activated macrophages. For the present studies we used rat C6 glioma cells because they are derived from astrocytes and express many astrocytic properties in culture. Results demonstrate that Tau-Cl inhibits the production of NO, PGE$_2$, and the chemokines MCP-1 and MIP-2 by activated C6 cells. Tau-Cl also inhibits iNOS gene expression in primary cultures of rat cortical astrocytes. The effects of Tau-Cl appear to be mediated, at least in part, by interfering with NF-κB signaling.

2. MATERIALS AND METHODS

2.1 Materials

Recombinant rat IFN-γ, HBSS without Ca^{2+}, and MG^{2+}, HBSS without phenol red, RPMI-1640, penicillin, streptomycin, trypsin-EDTA, heat inactivated horse serum, and DMEM were purchased from GIBCO-BRL (Grand Island, NY). Heat inactivated fetal bovine serum was from Gemini Bio-Products, Inc. (Calabasas, CA), recombinant human TNF-α was obtained from Endogen, Inc. (Cambridge, MA), and LPS W (Escherichia coli 0111:B4) was purchased from Difco Laboratories (Detroit, MI). Plasmids containing cDNA probe for iNOS, COX-2, MCP-1, MIP-2, and glyceraldehyde-3-phosphate dehydrogenase (GAPDH) were graciously provided by Drs. C. Nathan, R. Dubois, T. Yoshimura, K.E. Driscoll, and R. Dong, respectively. All other materials were obtained from commercial sources and Tau-Cl was prepared fresh before use as previously described[16,17].

2.2 Cell Culture

Rat C6 glioma cells were purchased from the American Type Culture Collection and grown in RPMI-1640 containing 15% horse serum, 2% fetal calf serum, 2 mM L-glutamine, and antibiotics. Cells were passaged weekly and used at p41-56. Cells were seeded and allowed to grow to approximately 90% confluency over 2 - 3 days before initiating treatment. Treatment was initiated by washing the cells and replacing the media with HBSS containing various concentrations of Tau-Cl for 2 hr in a humidified CO_2 (37°, 5% CO_2) incubator. Cell exposure to Tau-Cl was terminated by washing the cells and replacing the media with DMEM (phenol red free) containing 4% fetal calf serum and antibiotics. Cells were then activated by adding LPS, 10 μg/ml; rIFN-γ, 50 U/ml; and rTNF-α, 50 ng/ml (activators) for times indicated. Highly enriched (98% GFAP positive) cultures of astrocytes were prepared from 1-day old rats according to the procedure of McCarthy and de Villis[18]. Astrocytes were grown to 90% confluency and experiments were conducted in DMEM containing 2% FCS and antibiotics. Cells treated with Tau-Cl as described for C6 cells and were activated with LPS (1 μg/ml) and IFN-γ (10 U/ml).

2.3   Inflammatory Mediator Assays

Media content of MCP-1, MIP-2, PGE_2, and NO_2^- was determined as previously described[16,17]. Cell lysates, nuclear protein extracts, cytosolic fractions, and total RNA extracts were prepared and Northern blots, Western blots, EMSA and supershift assays were conducted as previously described in full detail[16,19]. Expression of iNOS mRNA in astrocytes was measured by RT-PCR. The primers used for iNOS amplification were 2881F (5′-AGTGTCAGTGGCTTCCAG CTC-3′), corresponding to bases 2881-2901; and 3308R (5′-CCTCAACC TGCTCCTCACTCA-3′), complementary to bases 3308-3328 of rat inducible NOS cDNA sequence (genebank accession: U03699). Length of PCR product is 447bp. The primers used for internal control amplification were 553F (5′-ACAGTCCATGCCAT CACTGCC-3′), corresponding to bases 553-573; and 818R (5′-GCCTGCTTCACC ACCTTCTTG-3′), complementary to bases 798-818 of rat glyceraldehyde-3-phosphate dehydrogenase (GAPDH) cDNA sequence (genebank accession: M17701). Length of PCR product is 265bp.

3.   RESULTS AND DISCUSSION

Results presented in Fig. 1 summarize our previous findings[16,17]. Tau-Cl inhibits the production of NO, PGE_2, MCP-1, and MIP-2 by C6 cells measured 24 hrs after activation. Western blot analyses of cell lysate proteins indicate that expression of inducible COX-2 and iNOS is inhibited by Tau-Cl and account for the effects of Tau-Cl on PGE_2 and NO, their respective products of enzymatic activity. Northern blot analyses of the corresponding mRNAs are shown representing times of peak expression for each transcript. Results presented in Figure 1 illustrate an inhibitory effect of Tau-Cl on the production of inflammatory mediators that is coordinate with decreased

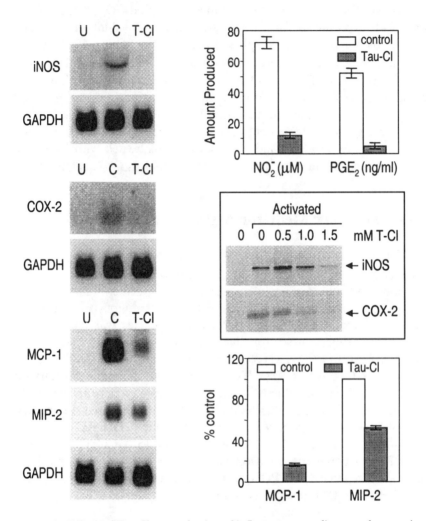

Figure 1. Effects of Tau-Cl on production of inflammatory mediators and expression of their respective genes in C6 cells. Media content of NO (μM), PGE_2 (ng/ml), MCP-1, and MIP-2 was measured 24 - 48 hr after activation. Control values for MCP-1 and MIP-2 were 28 ± 2 and 45 ± 2 ng/ml, respectively. Western blot analyses of iNOS and COX-2 were measured in cell lysates prepared 24 hr after activation. Northern blot analyses of iNOS gene expression was measured 8 hr after activation, all other transcripts were measured 4 hr after activation. U, unactivated; C, control activated; T-Cl, cells exposed to 1.5 mM Tau-Cl before activation. Similar results were obtained in 2 – 4 additional independent experiments.

expression of the respective genes. Inhibition of NO production and iNOS gene expression by Tau-Cl is also observed in primary cultures of rat cortical astrocytes as shown in Figure 2. Expression of COX-2, TNF-α, MCP-1 and MIP-2 genes also are inhibited by Tau-Cl in primary cultures of rat cortical astrocytes (manuscript in preparation). These results suggest that Tau-Cl inhibits an early event involved in the regulation of inflammatory gene expression. For these reasons, we evaluated the effects of Tau-Cl on the nuclear transcription factor, NF-κB.

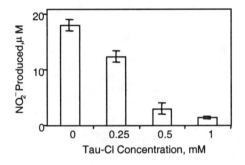
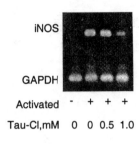

Figure 2. Effects of Tau-Cl on production of NO and expression of iNOS and GAPDH mRNA by primary cultures of rat cortical astrocytes measured 24 hr after activation. Similar results were obtained in 3 – 5 independent experiments.

NF-κB is present in the cytoplasm of most cells held in an inactive form as a complex bound to a member of a family of inhibitory proteins, IκB, which prevent NF-κB nuclear translocation. IκB interacts with NF-κB blocking its nuclear localization sequence (NLS). IκB-α is a critical component in the NF-κB signaling pathway and is the most studied of the IκB proteins[20,21]. Cell activation leads to rapid phosphorylation of IκB by a large (~ 800 kD) multisubunit IκB kinase (IKK) complex. IKK is the point of convergence for most NF-κB-activating stimuli, including LPS and cytokines, and is activated by phosphorylation. Phosphorylation of IκB-α at Ser-32 and Ser-36 residues by IKK elicits polyubiquitination by an ubiquitin ligase complex, and degradation of IκB by the 26S proteasome, which frees NF-κB[21]. The NLS of NF-κB is exposed and the NF-κB molecule interacts with the nuclear import machinery and translocates into the nucleus. Once in the nucleus, NF-κB binds to the regulatory elements (κB sites) of target genes. NF-κB is required for expression of TNF-α. IL-1β, IL-6, IP-10, COX-2, iNOS, MIP-2, and MCP-1 genes[22-24].

Results presented in Figure 3 illustrate the inhibitory effects of Tau-Cl on NF-κB activation as evaluated by EMSA of nuclear protein extracts prepared from C6 cells 30 min after activation. The identity of the radiolabeled complex was determined to be a p65/p50 diamer as indicated by supershift with p65 and p50 antibody. Antibody against c-Rel, another NF-κB family member protein, was without effect. Tau-Cl decreased activation of NF-κB but did not affect the composition of NF-κB that migrated into the nucleus. Analysis of cytosolic IκB-α by Western blot (Fig. 3) indicated that Tau-Cl treated cells maintained IκB-α in the cytosol presumably as a complex with NF-κB.

Y. Liu et al.

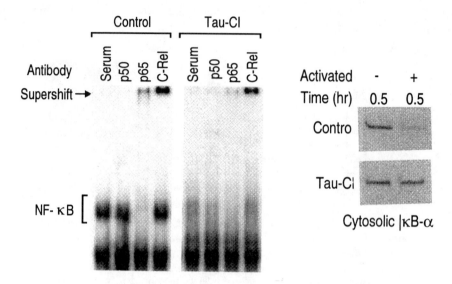

Figure 3. (Left panel) EMSA of NF-κB in nuclear protein extracts of C6 cells. Cells were exposed to 1.5 mM Tau-Cl or HBSS before activating cells and samples were prepared 30 min later. Supershifting of bands was examined using preimmune serum or antibody against p50, p65, or c-Rel. (Right panel) Western blot of IκB-α from cytosolic fraction of C6 cells prepared 30 min after activation. Cells were exposed to HBSS or to 1.5 mM Tau-Cl before activation.

The results of these studies demonstrate that Tau-Cl inhibits the production of inflammatory mediators by activated C6 glioma cells and the effects of Tau-Cl are recapitulated in primary cultures of rat cortical astrocytes. The inhibitory effects of Tau-Cl are mediated, in part, through the attenuation of NF-κB activation. Taken together, these results suggest that Tau-Cl has the potential to function as a negative regulator of inflammatory gene expression in the CNS.

ACKNOWLEDGEMENTS

The authors appreciate the excellent secretarial assistance of Mrs. A. Parese. This work was supported by the OMRDD of New York and by NIH grants HL-49942 and NS40721.

REFERENCES

1. Weiss, S.J., Klein, R., Slivka, A., and Wei, M., 1982, Chlorination of taurine by human neutrophils: evidence for hypochlorous acid generation. *J. Clin. Invest.* 70: 598-607.
2. Weiss, S.J., Lampert, M.B., and Test, S.T., 1983, Long-lived oxidants generated by human neutrophils: characterization and bioactivity. *Science* 222: 625-628.
3. Grisham, M.B., Jefferson, M.M., Melton, D.F., and Thomas E.L., 1984, Chlorination of endogeneous amines by isolated neutrophils. *J. Biol. Chem.* 259: 10404-10413.
4. Marquez, L.A., and Dunford, H.B., 1994, Chlorination of taurine by myeloperoxidase: Kinetic evidence for an enzyme-bound intermediate. *J. Biol. Chem.* 269: 7950-7956.
5. Learn, D.B., Fried, V.A., and Thomas, E.L., 1990, Taurine and hypotaurine contents of human leukocytes. *J. Leukoc. Biol.* 48: 174-182.
6. Quinn, M.R., and Schuller-Levis ,G.B., 1999, Taurine Chloramine, an Inhibitor of iNOS Expression and a Potential Modulator of Inflammation. In *Molecular and Cellular Biology of Nitric Oxide* (J. Laskin and D. Laskin, eds.), Marcel Dekker, Inc., New York, pp. 309-331.
7. Park, E., Quinn, M.R., Wright, C.E., and Schuller-Levis, G., 1993, Taurine chloramine inhibits the synthesis of nitric oxide and the release of tumor necrosis factor in activated RAW 264.7 cells. *J. Leukoc. Biol.* 54: 119-124.
8. Marcinkiewicz, J., Grabowska, A., Bereta, J., and Stelmaszynska, T., 1995, Taurine chloramine, a product of activated neutrophils, inhibits *in vitro* the generation of nitric oxide and other macrophage inflammatory mediators. *J. Leukoc. Biol.* 58: 667-674.
9. Galea, E., Feinstein, D.L., and Reis, D.L., 1992, Induction of calcium-independent nitric oxide synthase in primary rat glial cultures. *Proc. Natl. Acad. Sci. USA* 89: 10945-10949.
10. O'Banion, M.K., Miller, J.C., Chang, J.W., Kaplan, M.D., and Coleman, P.D., 1996, Interleukin-1β induces prostaglandin G/H synthase 2 (cyclooxygenase-2) in primary murine astrocyte cultures. *J. Neurochem.* 66: 2532-2540.
11. Ding, M., St. Pierre, B.A., Parkinson, J.F., Medberry, P., Wong, J.L., Rogers, N.E., Ignarro, L.J., and Merrill, J.E., 1997, Inducible nitric-oxide synthase and nitric oxide production in human fetal astrocytes and microglia. *J. Biol. Chem.* 272: 11327-11335.
12. Akaike, T., Weihe, E., Schaefer, M., Fu, Z.F., Zheng, Y.M., Vogel, W., Schmidt, H., Koprowski, H., and Dietzschold, B., 1995, Effect of neurotropic virus infection on neuronal and inducible nitric oxide synthase activity in rat brain. *J. Neuro. Virol.* 1: 118-125.
13. Koka, P., He, K., Zack, J.A., Kitchen, S., Peacock, W., Fried, I., Tran, T., Yashar, S.S., and Merill, J.E., 1995, Human immunodeficiency virus 1 envelope proteins induced interleukin 1, tumor necrosis factor α, and nitric oxide in glial cultures derived from fetal, neonatal, and adult human brain. *J. Exp. Med.* 182: 941-952.
14. Endoh, M., Maiese, K., and Wagner, J., 1994, Expression of the inducible form of nitric oxide synthase by reactive astrocytes after transient global ischemia. *Brain Res.* 651: 92-100.
15. Nogawa, S., Zhang, F., Ross, M.E., and Iadecola, C., 1997, Cyclooxygenase-2 gene expression in neurons contributes to ischemic brain damage. *J. Neurosci.* 17: 2746-2755.
16. Liu, Y., Tonna-DeMasi, M., Park, E., Schuller-Levis, G., and Quinn, M.R., 1998, Taurine chloramine inhibits production of nitric oxide and prostaglandin E_2 in activated C6 glioma cells by suppressing inducible nitric oxide synthase and cyclooxygenase-2 expression. *Molec. Brain Res.* 59: 189-195.

17. Liu, Y., Schuller-Levis, G., and Quinn, M.R., 1999, Monocyte chemoattractant protein-1 and macrophage inflammatory protein-2 production is inhibited by taurine chloramine in rat C6 glioma cells. *Immunol. Lett.* 70: 9-14.
18. McCarthy, K.D., and de Villis, J., 1989, Preparation of separate astroglial and oligodendroglial cell cultures from rat cerebral tissue. *J. Cell. Biol.* 85: 890-902.
19. Barua, M., Liu, Y., and Quinn, M.R., 2001, Taurine chloramine inhibits nitric oxide synthase and TNF-α gene expression in activated alveolar macrophages: Decreased NF-κB activation and IκB kinase activity. *J. Immunol.* 167: 2275-2281.
20. Barnes, P.J., and Karin, M., 1997, Nuclear factor-κB, a pivotal transcription factor in chronic inflammatory diseases. *N. Engl. J. Med.* 336: 1066-1071.
21. Karin, M., Ben-Neriah, Y., 2000, Phosphorylation meets ubiquitination: the control of NF-κB activity. *Annu. Rev. Immunol.* 18: 621-663.
22. Ueda, A., Ishigatsubo, Y., Okubo, T., and Yoshimura, T., 1997, Transcriptional regulation of the human monocyte chemoattractant proten-1 gene. *J. Biol. Chem.* 272: 31092-31099.
23. Newton, R., Kuitert, L.M.E., Bergmann, M., Adcock, I.M., and Barnes, P.J., 1997, Evidence for involvement of NF-κB in the transcriptional control of COX-2 gene expression by IL-1β. *Biochem. Biophys. Res. Commun.* 237: 28-32
24. Pahl, H.L., 1999, Activators and target genes of Rel/NF-κB transcription factors. *Oncogene* 18: 6853-6866.

Taurine Is Involved in Oxidation of IκBα at Met45
N-Halogenated Taurine and Anti-inflammatory Action

YUSEI MIYAMOTO[1], ATSUHIRO KANAYAMA[2], JUN-ICHIRO INOUE[3], YOSHIKO S. KONISHI[4], and MAKOTO SHIMIZU[2]

[1]*Department of Integrated Biosciences,*[2]*Department of Applied Biological Chemistry, and* [3]*Institute of Medical Science, University of Tokyo, Chiba and Tokyo, Japan, and* [4]*Division of Microbiology, National Institute of Health Sciences, Tokyo, Japan*

1.   INTRODUCTION

The neutrophil is a phagocyte and functions to kill invading microorganisms in the defense of the host[1]. Phagocytosis rapidly turns on the respiratory burst, resulting in release of superoxide anion radical ($Q_2 \cdot{}^-$) in the phagosomes as illustrated in Fig. 1. Superoxide anion radical is formed from oxygen (O_2) by one-electron reduction that is catalyzed by NADPH oxidase assembled for activation at the phagosomal membrane. Cytosolic NADPH is the electron donor for this reduction. Spontaneous dismutation converts $O_2 \cdot{}^-$ to hydrogen peroxide (H_2O_2) in phagosomes: $2O_2 \cdot{}^- + 2H^+ \rightarrow O_2 + H_2O_2$. Myeloperoxidase (MPO) also released into phagosomes catalyzes the production of hypochlorous acid (HClO) from H_2O_2 and chloride ion (Cl⁻). The final product, HClO, is a strong oxidant that most effectively kills microorganisms. However, reactive oxygen species (ROS) such as $O_2 \cdot{}^-$, H_2O_2 and HClO cause damage to surrounding tissues as observed in inflammation. If HClO infiltrates into neutrophils themselves, it impairs their proteins, lipids and genes, eventually leading to their cell death. Because taurine, the most abundant free amino acid in the neutrophil, has a high reactivity with HClO, taurine chloramine (TauCl) is scarcely synthesized when taurine encounters HClO. The cytotoxicity of TauCl is much less than that of HClO and hence the biological function of taurine has been thought to be antioxidation and

Taurine 5: Beginning the 21st Century

Edited by Lombardini *et al.*, Kluwer Academic/Plenum Publishers, New York 2003

detoxification in the neutrophil[2].

Phagocytosis triggers another cellular event, the activation of nuclear factor kappa B (NFκB) in the neutrophil[3]. This transcription factor plays a role in inflammation because the production of many proteins relating to inflammation such as cytokines depends on the activation of NFκB.

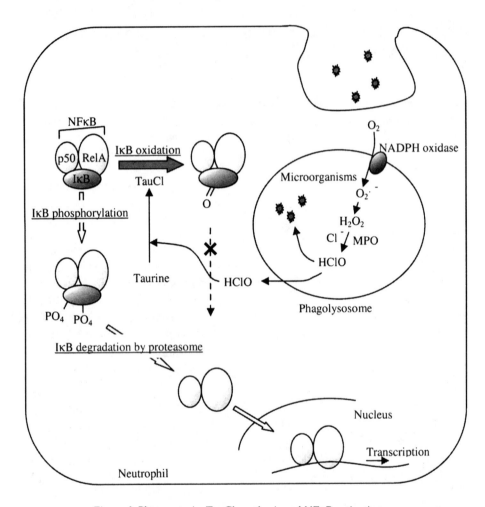

Figure 1. Phagocytosis, TauCl synthesis and NFκB activation.

Figure 1 depicts a cascade of the NFκB activation processes. In unstimulated cells, NFκB, a heterodimer of RelA and p50, is in its inactive state; it is associated with the inhibitory protein κB (IκB) and sequestered in the cytoplasm. Stimulation of cells activates IκB kinase (IKK) through an intracellular signal pathway. Activated IKK phosphorylates IκB. In the case of the α isoform of IκB (IκBα), Ser32 and Ser36 are phosphorylated, which leads to ubiquitination on Lys21 and Lys22. After ubiquitination, IκB is susceptive to proteolysis by a proteasome-dependent system. The degradation and dissociation of IκB enables NFκB to relocate to the nucleus and initiate gene expression. Recently, studies have shown that TauCl inhibits NFκB-dependent production of inflammatory cytokines[4,5]. However, the molecular mechanism of this inhibition is unclear.

2.   MODIFICATION OF IκBα by TauCl

2.1   Band Shift of IκBα and Inhibition of NFκB Activation

Proteolysis of IκB in response to stimulation is the crucial step in the activation of NFκB. Therefore, this study was initiated to research the effect of TauCl on IκB degradation using Jurkat T cells. Western blot analyses showed that treatment of cells with 1 mM TauCl for 1 hour generated two shifted bands of IκBα and also increased resistance to degradation triggered by stimulation with tumor necrosis factor α (TNFα). This modification of IκB by TauCl was specific to the α isoform. The decrease of NFκB nucleus translocation by a similar TauCl treatment was observed by an electrophoretic mobility shift assay. A reporter assay revealed that NFκB-dependent luciferase expression was decreased by TauCl in a dose dependent manner. These results indicate that inhibition of IκBα degradation by TauCl leads to inhibition of nuclear transfer of NFκB and consequent NFκB-dependent gene expression.

We examined the effects of TauCl treatment on IKK activity as well as NFκB activation triggered by overexpression of IKKβ. The IKK extracted from Jurkat cells treated with TauCl (1 mM and 1 hour) prior to TNFα stimulation was able to phosphorylate IκBα *in vitro*. When IKKβ was overexpressed in HEK293 cells, NFκB-dependent luciferase expression was increased. TauCl treatment inhibited two thirds of this increase. The results show that TauCl interacts with the signal pathway in downstream rather than upstream of IKK activation.

2.2   Oxidation of IκBα at Met45

When we first observed a band shift of IκBα caused by TauCl treatment, we assumed that TauCl might modify IκBα at IKK phosphorylation sites, Ser32/36, because the TauCl-induced band shift was similar in molecular size to that of phosphorylated IκBα. However, TauCl treatment generated a band shift of an IκBα mutant whose Ser32/36 were converted to alanine, indicating that modification sites of IκBα by TauCl are not at Ser32/36. Furthermore, alkaline phosphatase treatment (10 units and 30 min prior to electrophoresis) of the extract from TauCl-pretreated cells did not eliminate shifted bands of IκBα, showing that phosphorylation is not a cause of the TauCl-induced IκBα band shift. Cell extracts were first boiled to inactivate enzymes and then treated with TauCl but TauCl treatment still generated the band shift. All the results support an idea that modification of IκBα by TauCl is not enzymatic. We assumed that TauCl may chlorinate IκBα because TauCl has a high reactive chloride. IκBα was treated with radiolabeled ^{14}C-TauCl and Tau^{36}Cl *in vitro*. An imaging analysis revealed that the radioactivity was in the shifted band of IκBα treated with Tau^{36}Cl but not with ^{14}C-TauCl, indicating that TauCl chlorinates IκBα. Because the amino acid residue that is most sensitive to chlorination is tyrosine, an IκBα mutant whose 6 tyrosine residues were converted to phenylalanine (IκBα Y/F) was expressed in HEK293 cells and cells were treated with TauCl. However, TauCl treatment generated a band shift of the IκBα mutant. Furthermore, IκBα pretreated with TauCl was treated at 37°C for 30 min with acetic acid (pH 4) in the presence of a small amount of zinc but the shifted band did not disappear. The results indicate that TauCl chlorinates IκBα but this chlorination did not occurred at tyrosine residues. We carried out amino acid analysis of IκBα after TauCl treatment and found the decrease of tryptophan. Therefore, we performed a similar experiment with IκBα W/F (all three tryptophan sites were converted to phenylalanine) but a band shift of this mutant was still observed. Therefore, we assumed that chlorination of IκBα might not be essential in the band shift of IκBα.

To determine modification sites of IκBα by TauCl, three deletion mutants of IκBα 1-180, 1-248 and 1-298 were made and treated for 30 min at 37°C with 0.4 mM TauCl *in vitro* together with wild type IκBα (IκBα 1-317). Band shifts were generated in all three mutants. Then, two more mutants of IκBα 43-180 and 67-180 were made and treated with TauCl. A band shift was observed with IκBα 43-180 but not with IκBα 67-180, indicating that a modification site is located between 43rd and 66th amino acids of IκBα. To narrower down possible modification sites, IκBα 47-180 was made and tested. A similar treatment with TauCl did not generate a band shift, indicating that a modification site is located between 43rd and 46th amino acids of IκBα. Finally, we made three more deletion mutants of IκBα 44-180, 45-180 and 46-180 and repeated TauCl treatment. Mutants of IκBα 44-180 and 45-180 exhibited band shifts while IκBα 46-180 did not. These results indicate that a TauCl modification site is Met45 of IκBα and also indicate that if TauCl modification of IκBα always generates its band shift, a TauCl modification site does not exist from the 46th amino acid to the carboxyl terminus of IκBα.

A small peptide of IκBα 43-50 with acetylated N-terminal end, AcEQMLKELQ, was used to understand how IκBα Met45 is modified by TauCl. HPLC analysis was carried out using octadodecyl sulphate C18 column. TauCl treatment decreased the retention time of the peptide, indicating that TauCl treatment increases the hydrophilicity of the peptide. Furthermore, molecular weight analyses with matrix-assisted laser desorption/ionization time-of-flight mass spectrometry revealed that TauCl treatment increased the molecular weight of the peptide by a factor of approximately 16. Because the increased molecular weight of 16 is equal to that of one oxygen atom, the modification of the peptide by TauCl is estimated to be oxidation of IκBα Met45 to methionine sulfoxide.

To confirm that Met45 of IκBα is an amino acid residue that is modified by TauCl, two deletion mutants, IκBα 1-280 (wild type) and IκBα 1-280 (M45A), were expressed in Jurkat cells and their band shifts by TauCl treatment were studied. The IκBα 1-280 (M45A) degraded by TNFα-stimulation regardless of TauCl treatment while in the case of IκBα 1-280 (wild type) TauCl treatment generated its band shift and increased resistance to degradation initiated by TNFα-stimulation. The results indicate that IκBα is modified at Met45 with TauCl, which increases the resistance. However, a reporter assay revealed that switching the expression of IκBα 1-280 (M45A) from IκBα 1-280 (wild type) did not completely recover NFκB-dependent luciferase expression when cells were treated with TauCl. Though the low level of expression of IκBα 1-280 (M45A) and (wild type) in comparison with endogenous IκBα may partially account for failure in full recovery of luciferase expression, we could not conclude that oxidation of IκBα by TauCl occurs only at Met45.

2.3   HL60 Cells and Rat Neutrophils

Because HL-60 cells express MPO[6], HL-60 cells were used to determine if TauCl intracellularly synthesized can modify IκBα like TauCl extracellularly administered. TNFα stimulation caused IκBα degradation but did not after cell treatment with 100 μM H_2O_2 for 5 min. When HL-60 cells were pretreated for 30 min with a specific MPO inhibitor, 4-aminobenzoic hydrazide (ABAH), at 500 μM and subsequently treated with H_2O_2 in the presence of this inhibitor at the same concentration, the band shift of IκBα was not generated and IκBα degraded in response to TNFα stimulation. The nuclear transfer of NFκB was also initiated by TNFα stimulation and inhibited by H_2O_2 pretreatment. A similar cell treatment with ABAH recovered this TNFα-induced nucleus transfer of NFκB regardless of H_2O_2 treatment. The results suggest that intracellularly synthesized TauCl is able to modify IκBα and inhibit NFκB nuclear translocation.

To demonstrate that intracellular taurine really affects the NFκB-dependent production of inflammatory cytokine, the release of IL-8 was measured with rat neutrophils. The intracellular concentration of taurine in neutrophils was perturbed by osmolarily change in buffer. The release of IL-8 initiated by co-incubation with bacteria was significantly higher in taurine-reduced neutrophils than reloaded ones. Furthermore, when taurine-reloaded neutrophils were pretreated with 500 μM ABAH before and during incubation with bacteria, the release of IL-8 was significantly increased. Because neutrophils express MPO, intracellular taurine reacts with HClO produced during phagocytosis of bacteria to form TauCl. Intracellularly synthesized TauCl seems to work to decrease NFκB-dependent IL-8 production.

3.   DISCUSSION AND PROSPECT

In this study, we found that TauCl oxidizes IκBα at Met45 to generate its band shift and oxidized IκBα becomes resistant to proteolysis induced by stimulation, resulting in inhibition of NFκB activation. Similar band shift and inhibition of NFκB nuclear translocation were observed by H_2O_2 treatment of HL-60 cells expressing MPO. Secretion of IL-8 was regulated by intracellular taurine in rat neutrophils. Therefore, taurine may help suppress excessive inflammatory reaction in neutrophils. Taurine may also act as an intermediate regulator of two distinct events occurred in neutrophils during their phagocytosis of bacteria, synthesis of HClO and NFκB activation. Because MPO is released to the extracellular space of inflammatory site, TauCl may be possibly synthesized in the extracellular space, diffuse into surrounding tissues and inhibit NFκB activation. Therefore, we assume that taurine may suppress the inflammation in which ROS is generated to an excessive extent. However, we still have following questions: 1) Does TauCl specifically oxidize IκBα? (How about other proteins?); 2) Does TauCl oxidize another methionine residue of IκBα, besides Met45, because two shifted bands of IκBα were observed when cells were treated with TauCl?; 3) Which does oxidation of IκBα at Met45 disturb its phosphorylation or ubiquitination?; 4) How does chlorination of IκBα by TauCl take part in the inhibition of NFκB activation? We hope that we could answer these questions.

Recently, we studied effects of taurine bromamine (TauBr) on IκBα. A Western blot analysis revealed a similar band shift of IκBα as observed with TauCl (Fig. 2). This result indicates that taurine may have a similar function in eosinophils because eosinophils express eosinophil peroxidase that mediates the synthesis of hypobromous acid (HBrO) from H_2O_2 and bromide ion (Br⁻). Taurine reacts with HBrO to form TauBr that may inhibit NFκB activation like TauCl. It has been reported that orally administered taurine reduces the number of eosinophils accumulated in the bronchus of rat model of asthma[7]. Because eosinophils produce ROS causing inflammation, the authors concluded that taurine has beneficial effects on asthma. We think differently. If taurine inhibits NFκB activation via TauBr formation, this amino acid not only decreases the production of NFκB-dependent inflammatory cytokines but also leads eosinophils to apoptotic cell death, because expression of inhibitor of apoptosis proteins (IAP) depends on NFκB[8]. Therefore, the number of eosinophils was decreased by taurine administration as they observed. In the case of neutrophils, we can speculate a similar phenomenon. After phagocytosis and killing of bacteria, taurine reacts with excessive HClO to form TauCl and inhibits NFκB activation. Neutrophils become ready to fall into apoptosis. This means that TauCl-induced inhibition of NFκB activation promotes the removal of neutrophils that have performed their part from blood circulation. We speculate that N-halogenated taurine may lead cells to death by inhibition of NFκB activation though taurine itself helps protect cells against DNA fragmentation[9].

TauCl	−	+	−
TauBr	−	−	+

Figure 2. Effects of TauCl and TauBr on IκBα

ACKNOWLEDGEMENTS

This work was supported in part by a grant-in-aid for Scientific Research from the Ministry of Education, Science, Sports, and Culture of Japan and a research grant from Taisho Pharmaceutical Co., Ltd. (Y.M.).

REFERENCES

1. Hampton, M.B., Kettle, A.J., and Winterbourn, C.C., 1998, Inside the neutrophil phagosome: Oxidants, myeloperoxidase, and bacterial killing. *Blood* 92: 3007-3017.

2. Stapleton, P.P., O'Flaherty, L., Redmond, H.P., and Bouchier-Hayes, D.J., 1998, Host defense-a role for the amino acid taurine? *J. Parnter. Enteral. Nutr.* 22: 42-48.

3. McDonald, P.P., and Cassatella, M.A., 1997, Activation of transcription factor NF-κB by phagocytic stimuli in human neutrophils. *FEBS Lett.* 412: 538-586.

4. Marcinkiewicz, J., Grabowska, A., Bereta, J., Bryniarski, K., and Nowak, B., 1998, Taurine chloramine down-regulates the generation of murine neutrophil inflammatory mediators. *Immunopharmacology* 40: 27-38.

5. Kontny, E., Szczepańska, K., Kowalczewski, J., Kurowska, M., Janicka, I., Marcinkiewicz, J., and Maśliński, W., 2000, The mechanism of taurine chloramine inhibition of cytokine (interleukin-8) production by rheumatoid arthritis fibroblast-like synoviocytes. *Arthritis Rheum.* 43: 2169-2177.

6. Åbrink, M., Gobl, A.E., Huang, R., Nilsson, K., and Hellman, L., 1994, Human cell lines U-937, THP-1 and Mono Mac 6 represent relatively immature cells of the monocyte-macrophage cell lineage. *Leukemia* 8: 1579-1584.

7. Cortijo, J., Blesa, S., Martinez-Losa, M., Mata, M., Seda, E., Santagelo, F., and Morcillo, E.J., 2001, Effects of taurine on pulmonary responses to antigen in sensitized Brown-Norway rats. *Eur. J. Pharmacol.* 431: 111-117.

8. Chu, Z-L., McKinsey, T.A., Liu, L., Gentry, J.J., Malim, M.H., and Ballard, D.W., 1997, Suppression of tumor necrosis factor-induced cell death by inhibitory of apoptosis c-IAP is under NF-κB control. *Prot. Natl. Acad. Sci. USA* 94: 10057-10062.

9. Waters, E., Wang, J.H., Redmond, H.P., Wu, Q.D., Kay, E., and Bouchier-Hayes, D., 2001, Role of taurine in preventing acetaminophen-induced hepatic injury in the rat. *Am. J. Physiol. Gastrointest. Liver Physiol.* 280:G1274-G1279.

The Combined Treatment with Taurine and Niacin Blocks the Bleomycin-Induced Activation of Nuclear Factor-kB and Lung Fibrosis

SHRI N. GIRI

Department of Molecular Biosciences, School of Veterinary Medicine, University of California, Davis, California, USA

1.   INTRODUCTION

Interstitial lung fibrosis (ILF) is a potentially lethal and chronic response of the lung to injury resulting from a wide range of causes[1]. Regardless of the causes, ILF is invariably associated with fibrosing alveolitis characterized by inflammation, an excess number of fibroblats[2] and an absolute increase in lung collagen content, and abnormality in the ultrastructural appearance and spatial distribution of collagen types[3, 4]. A number of studies have documented that cytokines are released in the lungs of patients with pulmonary fibrosis and in the animal models of lung fibrosis. Inflammatory cells such as macrophages, lymphocytes and neutrophils play a key role in the production of a variety of cytokines and growth factors which regulate the proliferation, chemotactism and secretary activity of the fibroblasts. Activated macrophages in the inflamed lungs in response to bleomycin (BL) instillation, synthesize increased amounts of several proinflammatory and fibrogenic cytokines that mediate an enhanced fibroproliferative response[5-7]. It is commonly understood that not a single cytokine but a network of cytokines control the inflammatory processes[8].

In the BL-rodent model of lung fibrosis, proinflammatory cytokines such as IL-1, IL-6, and TNF-α are found initially in the airway and alveoli with subsequent increases in the expression of MCP-1 and MIP-1α in alveolar macrophages and airway epithelial cells[9, 10]. The chemoattractants, MCP-1 and MIP-1α facilitate recruitment and activation of specific subsets of lymphocytes, eosinophils, and macrophages in the alveolar space. These cells in turn secrete cytokines, like TGF-β which is capable of stimulating the proliferation of myofibroblasts and upregulating the procollagen gene expression. The macrophages, lymphocytes, and myofibroblasts present in the BL-induced established fibrotic lesions stimulate one another and thus initiate and maintain an excess collagen synthesis via a complex network of chemokines and other soluble proinflammatory substances[8].

A number of studies indicate that many cytokine genes are regulated in part by nuclear factor kB (NF-kB), which is normally sequestered in the cytoplasm as an inactive multiunit complex bound to an inhibitory protein, IkBα[11]. Several agents are found to activate this complex by causing phosphorylation and degradation of IkBα and thus allowing the translocation of the active dimmer of NF-kB into the nucleus, where it binds to the promoter region of genes such as IL-1α, Il-6 and TNF-α containing the NF-kB motif and thereafter, it stimulates the expression of these genes. The generation of reactive oxygen species (ROS) has been linked with NF-kB activation by a variety of stimuli[11]. One of the proposed mechanisms of BL-induced lung injury is its ability to generate ROS by binding to DNA/Fe^{2+} and forming a DNA/Fe^{2+}/BL complex. This complex undergoes redox cycling and generates ROS such as superoxide and hydroxyl radicals[12]. In the present study, we tested the hypothesis that an excess production of proinflmmatory cytokines in BL-induced lung injury results in response to activation of NF-kB and the beneficial effect of the combined treatment with taurine (T) and niacin (N) against BL-induced lung injury as reported in our earlier papers[13,14], resides in the ability of this treatment to suppress the NF-kB activation followed by decreases in the synthesis and release of proinflammatory and fibrogenic cytokines such as IL-1α, TNF-α, IL-6 and TGF-β.

2. METHODS

2.1 Animal Model

The BL- mouse model of lung fibrosis as previously described was used in the present study[15]. Briefly, all experiments were carried out in male C57BL/6 mice weighing 25-28 g (Simonsen, Gilroy, CA). Animals were caged in groups of four. They were first allowed to acclimatize for one week before starting the study. The mice had access to water and either pulverized Rodent Laboratory Chow 5001(Purina Mills, St. Louis, MO) or the same pulverized chow containing 2.5% niacin (w/w) and 1% taurine in water. Animals were randomly divided into four experimental groups: saline-instilled (SA) with a control diet (CD) and water (SA+CD); saline-instilled with taurine in water and niacin in diet (SA+TN); BL-instilled with the control diet and water (BL+CD); and BL-instilled with taurine in water and niacin in diet (BL+TN). The animals were fed these diets 3 days prior to intratracheal (IT) instillation and continuing for the entire duration of the study. After mice were anesthetized with ketamine and xylazine, either 50 μl sterile isotonic saline or 0.1 unit of BL sulphate in 50 μl saline per mouse was instilled by IT route.

2.2 BALF Collection and Lung Processing

The mice were sacrificed by an overdose of sodium pentobarbital at 1, 3, 5, 7, 14 and 21 days after IT instillation and bronchoalveolar lavage (BAL) was carried out as previously described [16]. After the lavage, the lungs were excised; freeze clamped and stored at -80°C for hydroxyproline assay. The BALF was centrifuged at 4°C for 10 min at 1500 rpm. The supernatant was gently aspirated and stored at -80°C until used for cytokine assays. In another set, the animals were killed by decapitation and their lungs were quickly removed, freeze clamped and dropped in liquid N_2 and later stored at -80°C until used for RNA extraction

2.3 Lung Hydroxyproline Content

Collagen deposition was estimated by determining the hydroxyproline content of the whole lung. The lung was excised, homogenized, and hydrolyzed in 6N HCl for 16-18 hr at 110°C. Hydroxyproline content was assessed by the colorimetric method of Woessner[17]. Data are expressed as μg of hydroxyproline/lung.

2.4 Measurement of IL-1α, TNF-α, IL-6, and TGF-β in BALF

IL-1α, TNF-α, IL-6, and TGF-β were assayed by specific enzyme-linked immunosorbent assay (ELISA) from Genzyme. The sensitivities of the assay for different cytokines were as follows: IL-1α, 15pg/ml; TNF-α, 15pg/ml; IL-6, ≤ 5pg/ml; and TGF-β, 50pg/ml.

2.5 RNA Extraction and RT-PCR

Total RNA from the lung was isolated using the RNeasy total RNA extraction protocol (Qiagen, Chatsworth, CA) according to the manufacture's description. The PCR primers for GAPDH, IL-1α, and TNF-α in message amplification and phenotyping analysis of cytokine mRNAs were obtained from Clontech Laboratories (Palo Alto, CA). The following 5'-primer and 3'-primer sequences were employed:

GAPDH: (5') primer 5' TGAAGGTCGGTGTGAACGGATTTGGC 3'
          (3') primer 5' CATGTAGGCCATGAGGTCCACCAC 3'
IL-1α: (5') primer 5' AAGATGTCCAACTTCACCTTCAAGGAGAGCCG 3'
          (3') primer 5' AGGTCGGTCTCACTACCTGTGATGAGTTTTGG 3'
TNF-α: (5') primer 5' TTCTGTCTACTGAACTTCGGGGTGATCGGTCC 3'
          (3') primer 5' GTATGAGATAGCAAATCGGCTGACGGTGTGGG 3'.

First strand cDNA synthesis was performed using an Advantage RT-PCR kit (Clontech, Palo Alto, CA). After the completion of the first-strand cDNA synthesis, the samples were diluted 1:5 and the PCR reactions were carried out as described previously[18].

2.6   Protein Assay and Tissue Protein Extraction

Total proteins extracted from the lung were determined by the coomassie blue-dye binding assay (Bio-Rad) and protein from lungs was extracted according to the technique described previously[19].

2.7   Electrophoretic Mobility Shift and Western Blot Assays

NF-kB activities in lung nuclei from mice in SA+CD, BL+CD and BL+TN groups were determined by electrophoretic mobility shift assays using the Promega gel shift assay system (Promega, Madison, WI). DNA binding activity was determined following incubation of 20 µg of tissue proteins at room temperature for 20 min with ^{32}P-labeled double stranded oligonucleotide containing the NF-kB (5'-AGTTGAGGGGACTTTCCCAGGC-3') binding motif as described previously[19]. The expression of IkBα protein in lungs in various groups was evaluated by Western Blot analysis as reported previously[19].

2.8   Statistics

Treatment - related differences were evaluated using a two way ANOVA, followed by pairwise comparisons using the Newman-Keuls test. Statistical significance was considered at the p values of ≤ 0.05.

3. RESULTS

3.1 Hydroxyproline Content of Lungs

The lung levels of hydroxyproline; a marker of collagen deposition, in whole lung homogenate was evaluated. There were significant increases in the lung hydroxyproline content in mice in BL+ CD groups at 14 and 21 days after BL instillation, compared with the mice in either saline control (SA+CD and SA+TN) groups; and treatment with taurine and niacin significantly attenuated these increases in BL+TN groups as compared to their respective BL + CD groups at both time points (Fig. 1).

3.2 Cytokine Levels in BALF

To test the hypothesis that treatment with taurine and niacin alters the cytokine release during the course of BL-induced lung fibrosis, we measured the levels of IL-1α, TNF-α, IL-6, and TGF-β in the BALF from mice in SA+CD, BL+CD and BL+TN groups (Fig.2). IL-1α protein levels in the BALF from BL-treated mice in BL+CD groups were increased significantly by 2- and 3- fold at 3 and 5 days as compared to the corresponding SA+CD control groups, respectively. Thereafter, the levels declined to the control levels and stayed that way for the remaining part of the study (Fig. 2A). Treatment with taurine and niacin prevented the BL-induced increases in IL-1α protein levels in the BALF from mice in BL+TN groups at 3 and 5 days and thereafter the protein levels returned to the control values. The TNF-α protein levels in the BALF from BL-treated mice in BL+CD groups continued to remain elevated from day 1 through day 21 as compared to the corresponding SA+CD control groups and significant increases occurred at all times except the initial two time points (Fig. 2B). Treatment with taurine and niacin decreased the BL-induced increases in the TNF-α levels almost at all time points but significant decreases occurred only at 7 and 21 days in BL + TN groups as compared to BL + CD groups at the corresponding times. The IL-6 levels were significantly elevated starting day 1 through day 5 in the BALF from mice in BL+CD groups as compared to mice in SA+CD groups at the corresponding times. Although, treatment with taurine and niacin decreased the IL-6 levels at these times in BL+ TN groups as compared to BL+ CD groups at the corresponding times but significant decrease between the two groups occurred only at day 5 (Fig.2C). As compared to SA+CD control groups, the TGF-β protein levels in BALF were significantly increased in BL+CD groups at the corresponding times by 3, 14, 3.4 and 7.9 fold at 5, 7, 14 and 21 days after IT instillation of BL, respectively (Fig. 2D). Treatment with taurine and niacin decreased the BL-induced increases in the TGF-β levels in BL+TN groups at these time points by 40%, 57%, 40% and 47%, as compared to BL+CD groups at the corresponding time points, respectively; and the decreases were significant at the last three time points (Fig. 2D).

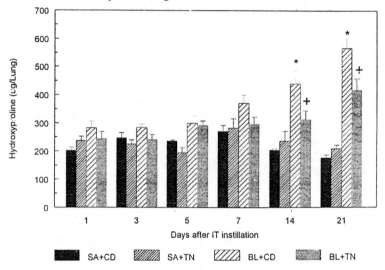

Figure 1. Effects of IT instillation of saline or BL (0.1 U/mouse) with or without 1% taurine in drinking water and 2.5% niacin (w/w) in diet on hydroxyproline content of mouse lungs at different times after instillation. See *Materials and Methods* for treatment details. Each value represents the mean ± S.E. of five animals. *, significantly higher ($P \leq .05$) than all other groups at the corresponding times, and +, significantly lower than the BL + CD group at the corresponding time.

3.3   Effects of Taurine and Niacin on IL-1α and TNF-α mRNAs Expression

RT-PCR analysis was carried out to evaluate the effects of treatment with taurine and niacin on the steady state level of IL-1α and TNF-α mRNAs in the lungs from mice in SA+CD, BL+CD and BL+TN groups. The increases in IL-1α mRNA levels occurred in BL+CD groups from day 1 through day 14 but significant increases were seen only at day 7 and day 14 as compared to SA+CD groups at the corresponding times (Fig. 3). Although, treatment with taurine and niacin decreased the message levels in BL+TN groups from day 3 through 14, significant decrease occurred only at day 7 as compared to mice in BL+CD groups at this time (Fig. 3). The TNF-α message levels were significantly increased in lungs from mice in BL+CD groups from day 1 through day 14 as compared to mice in SA + CD groups at the corresponding times and treatment with taurine and niacin significantly decreased the TNF-α mRNA in BL + TN group only at day 7 as compared to BL + CD group at this time (Fig. 4).

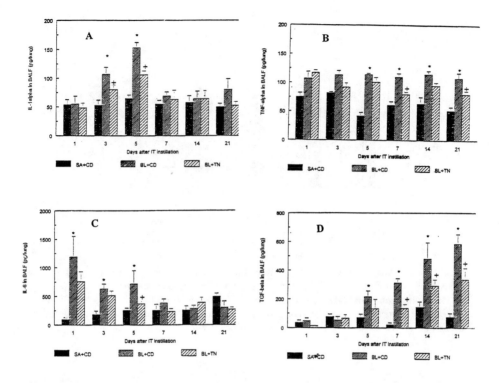

Figure 2. Effects of combined treatment with taurine and niacin on cytokine levels in the BALF of mice receiving IT instillation of either saline or BL at different times. See Materials and Methods for treatment details. Each value represents the mean ± S.E. of five animals. *, significantly higher ($P \leq .05$) than all other groups at the corresponding times, and +, significantly lower than the BL + CD groups at the corresponding times. A, IL-1α; B, TNF-α; C, IL-6; and D, TGF-ß.

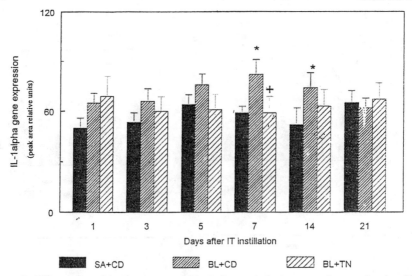

Figure 3. Effects of combined treatment with taurine and niacin on steady-state levels of IL-1 α mRNA in lungs of mice receiving IT instillation of either saline or BL at different times. RNA was isolated and RT-PCR was performed with primers specific for IL-1α and the housekeeping gene GAPDH. See *Materials and Methods* for treatment details. Each value represents the mean ±   S.E. of five animals. *, significantly higher ($P \leq .05$) than all other groups at the corresponding times, and +, significantly lower than the BL + CD group at the corresponding time.

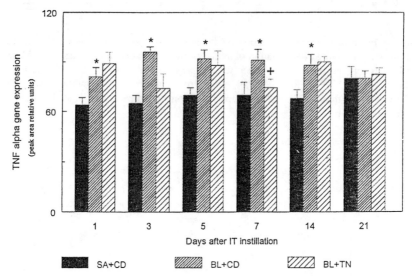

Figure 4. Effects of combined treatment with taurine and niacin on steady-state levels of TNF-α mRNA in lungs of mice receiving IT instillation of either saline or BL at different times. RNA was isolated and RT-PCR was performed with primers specific for TNF-α and the housekeeping gene GAPDH. See *Materials and Methods* for treatment details. Each value represents the mean ± S.E. of five animals. *, significantly higher ($P \leq .05$) than all other groups at the corresponding times, and +, significantly lower than the BL + CD group at the corresponding time.

3.4 Activation of NF-kB During BL-Induced Lung Injury and Fibrosis

The specificity of the NF-kB consensus oligonucleotide probe was first confirmed by experiments using nuclear extracts from whole lungs of mice at 5 days after BL instillation (Fig. 5). In DNA binding reactions, nuclear extract from lungs of BL-treated mice was incubated with only ^{32}P-labeled NF-kB consensus oligonucleotide probe and it showed typical binding to the labeled oligonucleotide (Fig. 5) (lanes 1&2). Competition with excess amount of unlabeled AP-2 oligonucleotide (100-fold) failed to reduce the binding of NF-kB (lane, 3), whereas competition with an excess amount of unlabeled NF-kB oligonucleotide (100-fold) completely prevented NF-kB binding to the labeled probe. Thus, it confirmed the signal specific to NF-kB (lane, 4). The DNA binding reaction using nuclear extracts from the lungs of mice in SA+CD control groups was also carried out in the same way (gel picture not shown).

1   2   3   4

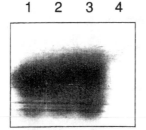

Figure 5. Demonstration of specificity of NF-kB consensus oligonucleotide gel shift assays. Binding reactions with nuclear extracts from whole lungs harvested 5 days after BL instillation. The extracts were incubated with 32P-labeled NF-kB oligonucleotide (lanes 1 and 2). Specificity of NF-kB complex formation was confirmed by competition experiments with a 100-fold excess of unlabeled AP-2 oligonucleotide (lane 3) or unlabeled NF-kB oligonucleotide (lane 4).

The kinetics of NF-kB activation in lungs from mice in SA+CD, BL+CD and BL+TN groups were determined by electrophoretic mobility shift assays using nuclear extracts from whole lung at various time points. The basal levels of NF-kB DNA binding activity was found in the lung nuclear extracts from mice in SA+CD groups (Fig. 6). However, the level of NF-kB activation in BL+CD groups was higher than that observed in SA+CD control groups. The nuclear localization of NF-kB in BL+CD groups was increased within 1 day after BL instillation, and it progressively increased and peaked by day 14, and remained high until day 21. The statistical analysis revealed that the levels of NF-kB activation in BL+ CD groups were higher at 1, 5,7,14 and 21 days than the SA+CD control groups at the corresponding times. The combined treatment with taurine and niacin inhibited the BL-induced increased nuclear localization of NF-kB in BL+TN

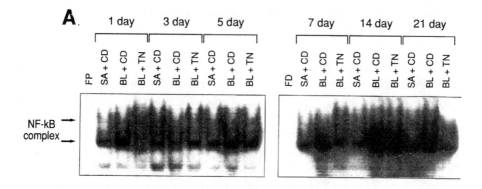

NF-kB
complex

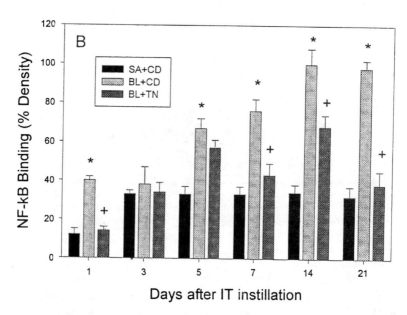

Days after IT instillation

Figure 6. Effects of combined treatment with taurine and niacin on the kinetics of NF-kB activation in lungs of mice receiving IT instillation of either saline or BL at different times. A, nuclear extracts were obtained from mouse lungs at the indicated time points and analyzed for NF-kB activity with a 22-mer double-stranded oligonucleotide probe containing a NF-kB site. Each lane represents lung nuclear extracts from the SA + CD, BL + CD, or BL + TN group. FP, free probe. Arrowhead represents NF-kB complex. This gel is a representative of three independent experiments. See *Materials and Methods* for treatment. B, time course of densitometric scanning of the band shift data as percentage of maximum of NF-kB binding activity. Each value represents the mean ± S.E. of three animals. *, significantly higher ($P \leq$ .05) than the corresponding SA + CD groups, and +, significantly lower than the corresponding BL + CD groups.

groups beginning day 1 through day 21; and significant inhibitions occurred at 1,7,14, and 21 days as compared to BL+CD groups at the corresponding times (Figs. 6A and 6B).

3.5 Preservation of IkBα Protein by Taurine and Niacin Treatment

The Western blot analysis of whole lung homogenates revealed that IkBα protein levels were elevated in BL+TN groups as compared to BL+CD and SA+CD groups (Fig.7). Although a marginal level of IkBα was detected in SA+CD groups, the IkBα protein levels were completely depleted in BL+CD groups indicating that NF-kB activation in BL+CD groups occurred via IkBα degradation. In contrast, treatment with taurine and niacin in BL+TN groups protected the IkBα degradation, which in turn, prevented the translocation of NF-kB from the cytoplasm to the nucleus.

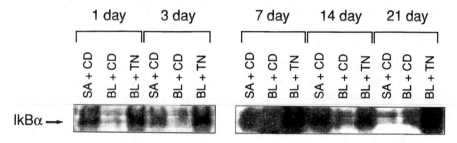

Figure 7. Effects of combined treatment with taurine and niacin on the IkBα protein expression in the lungs of mice receiving IT instillation of either saline or BL at different times. Western blot analysis shows a time-dependent increased preservation in the IkBα levels in BL + TN groups as opposed to depletion in BL + CD groups. This gel is representative of three independent experiments. See *Materials and Methods* for treatment details.

4. DISCUSSION

The data presented in this report demonstrate that taurine and niacin treatment significantly attenuated the BL-induced lung injury and fibrosis in mice as reflected by decreased content of the lung hydroxyproline, an index of fibrosis. It is interesting that taurine and niacin treatment suppressed the BL-induced increased levels of inflammatory and fibrogenic cytokines such as IL-1α, IL-6, TNF-α and TGF-β. These results are consistent with the results reported by other investigators that the release of proinflammatory cytokines plays a central role in the BL-induced lung fibrosis[5, 7, 20].

One of the widely accepted mechanisms for BL-induced lung injury is its ability to generate ROS[12]. The BL-induced generation of ROS will explain our finding of NF-kB activation in the lungs of mice in BL+CD groups. This will also explain the inhibitory effect of taurine and niacin (taurine in particular) on BL-induced NF-kB activation by scavenging ROS in BL+TN groups since taurine is known to scavenge ROS. It appears that the pathway for NF-kB activation involves the degradation of its regulatory protein IkBα because the amount of IkBα was barely detectable in BL+CD groups as opposed to complete preservation of this protein in BL+TN groups. The findings reported in this paper are consistent with the findings of other investigators since antioxidants are known to suppress cytokine gene activation via inhibiting the activation of NF-kB in vitro[21]. The measurement of cytokine levels in the BALF has provided valuable information in demonstrating the expression of early response (IL-1α, IL-6 and TNF-α) and late response (TGF-β) cytokines secondary to activation of NF-kB and this may be one of the underlying mechanisms for BL-induced lung inflammation and fibrosis.

It appears that the activation of NF-kB plays a critical role in cytokine-mediated inflammation by upregulating the transcription of a specific set of cytokine genes in response to IT instillation of BL in mice. However, it is not known as how the activation of NF-kB coordinates the differential production of these cytokines. The temporal sequence for production of various cytokines and their relative amounts in response to a fibrogenic dose of BL in mice are probably functions of interactions between NF-kB and other transcription factors as well as factors independent of NF-kB[22]. The activation of NF-kB *in vivo* is a necessary step in the production of early response cytokines such as IL-1α and TNF-α in BL-induced lung inflammation. It is known that taurine offers protection against oxidant induced lung injury by inhibiting production of nitric oxide and TNF-α, which are directly linked to tissue injury[23, 24]. Our findings are consistent with the findings of other *in vitro* and *in vivo* studies in which treatment with antioxidants such as N-acetyl cysteine[25], pyrrolidine dithiocarbamate[26] epigallocatechin-3-gallate[27] were found to block the activation of NF-kB by inhibiting the signal transduction-induced phosphorylation of IkBα. It has been demonstrated in several studies that activation of NF-kB can often be prevented by antioxidants and has led to the prevailing theory that NF-kB is an oxidant-sensitive transcription factor[28].

5.   CONCLUSION

We have provided evidence that the activation of transcription factor NF-kB plays a central role in BL-induced lung injury and fibrosis in mice; and the anti-inflammatory and antifibrotic effects of the combined treatment with taurine and niacin may reside in their ability to suppress the NF-kB activation by preserving the IkBα protein. The results of the present investigation have also uncovered a novel strategy to develop specific inhibitors of NF-kB activation which might prove to be therapeutically efficacious for the management of inflammation and fibrosis in general.

REFERENCES

1. Crouch, E. 1990, Pathophysiology of pulmonary fibrosis. *Am. J. Physiol.* 59:L159-184.
2. Chandler, D. B., Hyde, D.M., and Giri, S.N. 1983, Morphometric estimates of infiltrative cellular changes during the development of bleomycin-induced pulmonary fibrosis in hamsters. *Am. J. Pathol.* 112: 177-177.
3. Clark, J. G., Overton, J. E., Marino, B. A., Uitto, J., and Starcher, B. C. 1980, Collagen biosynthesis in bleomycin-induced pulmonary fibrosis in hamsters. *J. Lab.Clin. Med.* 96: 943-953.
4. Seyer, J. M.,Hutcheson, E. T., and kang, A. H. 1976, Collagen polymorphism in idiopathic chronic pulmonary fibrosis. *J. Clin. Invest.* 57: 1498-1507.
5. Scheule, R.K., Perkins, R.C., Hamilton, R., and Hollan, A. 1992, Bleomycin stimulation of cytokine secretion by the human alveolar macrophages. *Am. J. Physiol.* 262: 386-L391.
6. Khalil, N., and O'Conner, R. N. 1995, Cytokine regulation of pulmonary fibrosis: *In Pulmonary Fibrosis* (S.M. Phan and R.S. Thrall, eds.) Mercel Dekker, New York, Volume 80 pp. 627-646.
7. Piguet, P.F., Collart, M.A., Grau, G.E., Kapanchi, Y., and Vassali, P. 1989, Tumor necrosis factor/cachectin plays a key role in bleomycin-induced pneumopathy and fibrosis. *J. Exp. Med.* 170: 655-663.
8. Smith, R. E., Strieter, R. M., Phan, S. H., and Kunkel, S. L. 1996, C-C Chemokines: Novel mediators of the profibrotic inflammatory response to bleomycin challenge. *Am. J. Respir. Cell Mol. Biol.* 15: 693-702.
9. Piguet, P. F., and Vesin, C. 1994, Treatment of human recombinant soluble TNF soluble receptor of pulmonary fibrosis induced by bleomycin or silica in mice. *Eur. Respir. J.* 7: 515-518.
10. Smith, R.E., Strieter, R.M., Phan, S.H., Lukacs, N.W., Huffnagle, G.B., Wilke, C.A., Burdick, M.D., Lincoln, P., Evanoff, H., and Kunkel, S.L. 1994, Production and function of murine macrophage inflammatory protein-1α in bleomycin-induced lung injury. *J. Immunol.* 153: 4704-4712.
11. Bauerle, P.A., and Henkel, T. 1994, Function and activation of NF-kappa B in the immune system. *Ann. Rev. Immunol.* 12: 141-179.
12. Caspary, W.J., Lanzo, D.A., and Niziak, C. 1982, Effect of deoxyribonucleic acid on the production of reduced oxygen by bleomycin and iron. *Biochemistry.* 21: 334-338.
13. Wang, Q., Giri, S.N., Hyde, D.M., and Li, C. 1991, Amelioration of bleomycin-induced pulmonary fibrosis in hamsters by combined treatment with taurine and niacin. *Biochem. Pharmacol.* 42: 1115-1122.

14. Gurujeylakshmi, G., Iyer, S.N., Hollinger, M.A., and Giri, S.N. 1996, Procollagen gene expression is down-regulated by taurine and niacin at the transcriptional level in the bleomycin hamster model of lung fibrosis. *J. Pharmacol. Exp. Therap.* 277: 1152-1157.
15. Giri, S.N., Hyde, D.M., and Marfino, Jr., B. J. 1986 Ameliorating effect of murine interferon gamma on bleomycin-induced lung collagen fibrosis in mice. *Biochem. Med. Metab. Biol.* 36: 194-197.
16. Giri, S.N., Hollinger, M.A., and Schiedt, M.J. 1981, The effects of paraquat and superoxide dismutase on pulmonary vascular permeability and edema in mice. *Arch. Envin. Health.* 36:149-154.
17. Woessner, Jr., J.F. 1961, The determination of hydroxyproline in tissue and protein samples containing small proportions of this imino acid. *Anal. Biochem.* 93: 440-444.
18. Gurujeyalakshmi, G., Hollinger, M.A., and Giri, S.N. 1998, Regulation of transforming growth factor-β_1 mRNA expression by taurine and niacin in the bleomycin hamster model of lung fibrosis. *Am. J. Respir. Cell Mol. Biol.* 18: 334-342.
19. Gurujeylakshmi, G., Wang, Y., and Giri, S.N. 2000, Taurine and niacin block lung injury and fibrosis by down-regulating bleomycin-induced activation of transcription nuclear factor –kB in mice. *J. Pharmacol. Exp. Ther.* 293: 82-90.
20. Phan, S.H., and Kukel, S.L. 1992, Lung cytokine production in bleomycin-induced pulmonary fibrosis. *Exp. Lung. Res.* 18: 29-43.
21. Collins, T., Read, M.A., Neish, A.S., Whitley, M.J., Thanos, D., and Maniatis, T. 1995, Transcriptional regulation of endothelial cell adhesion molecules: NF-kappa B and cytokine-inducible enhancers. *FASEB J.* 9: 899-909.
22. Blackwell, T.S., and Chrisman, J.W. 1997, The role of nuclear factor-kB in cytokine gene regulation. *Am. J. Respir. Cell Mol. Biol.* 17:3-9.
23. Schuller-Levis, G., Quinn, M.R., Wright, C., and Park, E. 1994a, Taurine protects against oxidant-induced lung injury: Possible mechanism(s) of action. *Adv. Exp. Med. Biol.* 359: 31-39.
24. Schuller-Levis, G., Gordon, R.E., Park, E., Pendino, K.J., and Laskin, D.L. 1994b, Taurine protects rat bronchioles from acute ozone-induced lung inflammation and hyperplasia. *Exp. Lung Res.* 21: 877-888.
25. Leff, J.A., Wilke, C.P., Hybertson, B.M., Shanley, P.F., Beehler, C.J., and Repine, J.E.1993, Postinsult treatment with N-acetylcysteine decreases IL-1 induced neutrophil influx and lung leak in rats. *Am. J. Physiol.* 265: L501-L506.
26. Nathens, A.B., Bitar, R., Davreux, C., Bujard, M., Marshall, J.C., Dackiw, A.P.B., Watson, R.W.G., and Rotstein, O.D. 1997, Pyrrolidine dithiocarbamate attenuates endotoxin-induced acute lung injury. *Am. J. Respir. Cell Mol. Biol.* 17: 608-616.
27. Li, Y.L., and Lin, J.K. 1997, (-)-Epigallocatechin-3-gallate blocks the induction of nitric oxide synthase by down-regulating lipopolysaccharide-induced activity of transcription factor nuclear factor-kB. *Mol. Pharmacol.* 52: 465-472.
28. Sun, Y., and Oberley, L.W. 1996, Redox regulation of transcriptional activators. *Free Radic. Biol. Med.* 21: 335-348.

Taurine Reduces Lung Inflammation and Fibrosis Caused By Bleomycin

GEORGIA B. SCHULLER-LEVIS*, RONALD E. GORDON#,
CHUANHUA WANG* AND EUNKYUE PARK*

Department of Immunology, New York State Institute for Basic Research in Developmental Disabilities, Staten Island, New York, USA and #Department of Pathology, Mount Sinai Medical School, New York, New York, USA

1.   INTRODUCTION

Taurine has been shown to protect against lung injury induced by various oxidants such as ozone, nitrogen dioxide, amiodarone and paraquat and in combination with niacin to protect against bleomycin induced lung injury[1-6]. The protective effects of taurine may be explained by its detoxification of hypochlorous acid, a potent oxidant which can cause extensive tissue damage[7]. However, recent evidence from our laboratories[8-10] and others[11, 12] demonstrate that taurine chloramine (Tau-Cl), produced from highly toxic hypochlorous acid and taurine, is thought to be pivotal in regulating inflammation. Our previous in vitro data demonstrate that Tau-Cl inhibits proinflammatory mediators such as NO and TNF-α secreted from lipopolysaccharide (LPS) and gamma interferon activated murine macrophages[8, 9].

Bleomycin treatment results in dysregulated matrix remodeling, leading to thickened alveolar walls, alveolar collapse, and scarring[13]. Fibrosis culminates in the overproduction of interstitial collagen. Data from this study show that fibrosis is strikingly absent and inflammation is reduced in the lung of rats pretreated with taurine in the drinking water ten days prior to bleomycin instillation.

2.   Materials and Methods

Sprague-Dawley rats (Taconic Farm, German Town, NY) were kept in Thoren ventilated racks and cages for protection from air-borne infection and given water or water with 5% taurine for 10 days prior to installation of 1U bleomycin *i.t* with Nembutal anesthesia (Abbott Labs, North Chicago, Ill.). Animals were divided into the following 4 groups with at least 4 rats per group: saline and water (SW), bleomycin and water (BW), saline and taurine (ST), and bleomycin and taurine (BT). Taurine supplementation continued until the rats were sacrificed. The amount of water consumed and the weights of the animals were monitored daily. There were no differences in weigh between groups. The control groups (ST and SW) were injected *i.t.* with saline. Light microscopy was performed on lung tissues perfused with 4% paraformaldehyde in *situ* in 30 cm water gravity. The specimen for electron microscopy was fixed additionally with a mixture of 4% paraformaldehyde and 1% glutaraldehyde after perfusion with 4% paraformaldehyde. Immunohistochemistry was performed by the modified method described previously[4]. ICAM was detected by immunogold staining as previously described[14].

3.   Results and Discussion

At day 28, the BW group developed extensive fibrosis in the presence of infiltrated monocytes and fibroblasts as well as thickening of the alveolar walls (Fig. 1A). The BT group showed fewer inflammatory infiltrates and significantly less fibrotic foci compared to the BW group (Fig. 1B). Our previous studies demonstrated that Tau-Cl significantly decreased in vitro production of both NO and TNF-α in murine macrophages and RAW 264.7 cells, a macrophage-like cell line[7, 8]. Thus, the expression of NOS and TNF-α in the lungs of bleomycin treated rats was examined (Fig. 2). By LM immunochemistry, NOS staining was evident both in the airways and in alveolar macrophages on day 3 and 7 in the BW group (Fig. 3A&B). Day 3 demonstrated minimal staining in airway epithelium for NOS in the BT group (Fig. 3C), but was absent at Day 7 indicating protection of NOS induced lung damage in the taurine treated group (Fig. 3D). TNF-α was present in both the BW and BT group through day 28 but was more abundant in the BW group than the BT group (Fig. 4A&B). Taurine treatment diminished TNF-α expression in non- ciliated Clara cells, macrophages and fibroblasts at all time points (Fig. 4C).

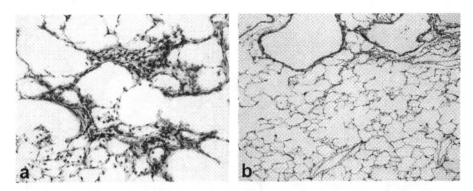

Figure 1. LM of an area of lung taken from a BW animal 28 days after treatment. It is possible to see the formation of fibrosis and many interstitial and alveolar infiltrative cells, primarily macrophages **A**. **B** is from a BT animal at the same time point showing no evidence of fibrosis and only a few macrophages. Sections were stained with hemtoxylin and eosin. A 200X B 100X.

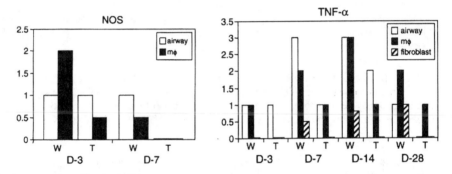

Figure 2. Rats were treated with water only (**W**) or with 5% taurine (**T**) in the drinking water 10 days prior to i.t. bleomycin. Airways cell ( ) are non-ciliated Clara cells. 0-4.5 on axis represents the degree of positive staining. **D** is the day of sacrifice.

While observing the lung specimens from each of the groups and time points by transmission electron microscopy, it was determined that the migratory inflammatory cells, neutrophils and monocytes were seen in the capillaries and in air spaces within the first 48 hours of the BW (Fig. 5A&B) and BT groups. The differences between these two groups appear to be that the number of inflammatory cells are significantly reduced in the animals treated with taurine. On days 1 through 3 the predominate inflammatory cell was the neutrophil, even in animals sacrificed at later time points. However, the most important finding was that the association between the early and later time points in the development of fibrosis in the BW group and not in the BT group was the adherence of these inflammatory cells, especially the

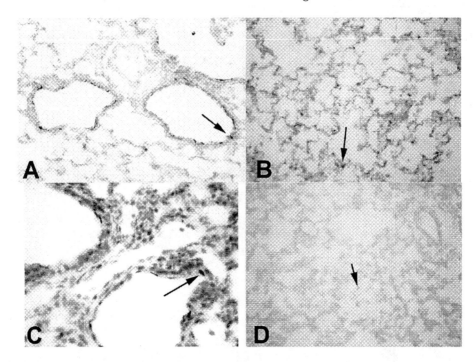

Figure 3. LM of lung from a BW animal **A.** compared to that of lung from a BT animal **C.** 3 days after treatment when the NOS upregulation was maximal. It is possible to see significant labelling in airway cells (arrow) and infiltrative cells (arrow) from the BW lung **B** and much less labelling in a corresponding area from a BT lung **D.** A 200X B 100X, C 200X, D 100X.

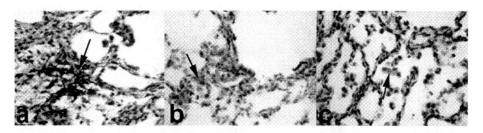

Figure 4. LM of lung from a BW animal **A&B.** compared to that of lung from a BT animal **C.** 14 days after treatment when the TNF-α upregulation was maximal. It is possible to see significant labeling (arrow) in areas of fibrosis and in infiltrative cells from the BW lung and much less labeling (arrow) in a corresponding area from a BT lung. A 200X B 200X. C.200X.

macrophages, to the type I pneumocytes after migration from vascular compartment (Fig. 5C, 6).

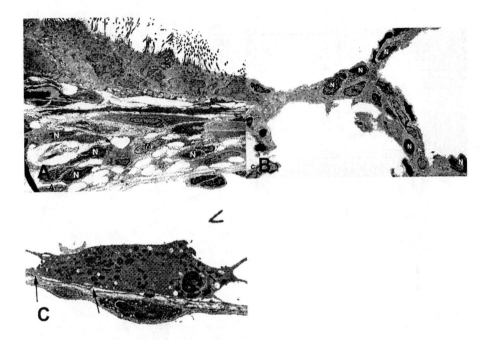

Figure 5. As seen in this TEM of bronchiole **A** and alveolar septa **B** from the lung of BW animal sacrificed 24-48 hours after treatment, there is a significant inflammatory infiltrate that includes neutrophils (N) and macrophages(M). Macrophages are strongly adherent with multiple points of contact(arrows) to type I pneumocytes **C**. Sections were stained with uranyl acetate and lead citrate. A 2000X B 2000 X, C 5000X.

ICAM is normally found in small quantities on the surface and within small vesicles of type I pneumocytes (Fig. 7C). More ICAM on type I pneumocyte was detected by EM immunogold staining (Figs. 7A&B) in the BW group compared to the BT group, more leukocytes adhered to type I pneumocytes, primarily at the tips of the pneumocyte microvilli. The damage in the BW group was more severe than in the BT group. ICAM staining correlated directly with lung damage, indicating that adhesion molecules induced by bleomycin in the BW group play an important role in lung injury.

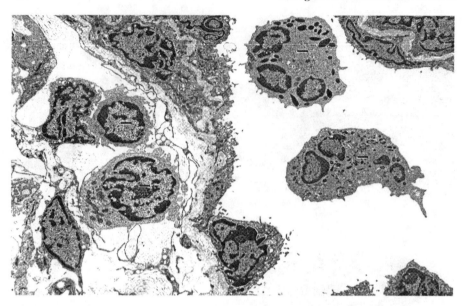

Figure 6. Alveolar septa from the lung of BT animal sacrificed 24-48 hours after treatment, there is an inflammatory infiltrate that includes infiltrative inflammatory cells(I). In this case the migrating cells once in the airspace stay rounded and few if any contacts with type I pneumocytes. Sections were stained with uranyl acetate and lead citrate. 5000X

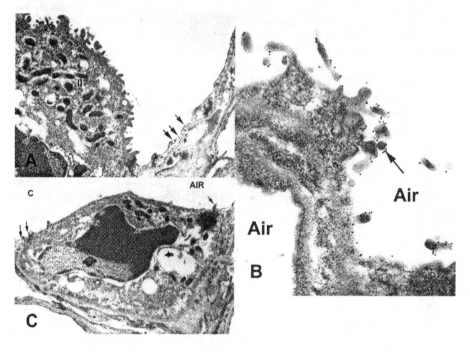

Figure 7. The lung tissue from a BW animal treated with immunogold to identify sites of ICAM presentation exhibit significant labeling(arrows) of the microvilli on the surface of type I pneumocytes and no labelling on type II cells **A**. **B** is an area from a BT lung exhibiting significantly less labeling(arrows) of the surface of type I pneumocyte, equivalent to the labelling of a lung **C** from an untreated animal. A 7000X B 7000X, C 13,000X.

Our hypothesis is that supplemental taurine in the drinking water increases the available taurine both sytemmically and at the site of inflammation. Leukocytes capable of generating H_2O and Cl^- via the MPO pathway, have intracellular concentrations of taurine from 20-50mM. Moreover, extracellular taurine concentrations in physiologic fluid range from 50-100 mM[17]. Thus, Tau-Cl, a stable oxidant, can be produced at the site of inflammation and modulate proinflammatory cytokine production. Specifically, the reduction of NO which is known to cause tissue damage in bleomycin treated mice[18, 19], and TNF-α which is known to up-regulate ICAM production[20], leads to reduced inflammation and the consequent fibrosis. Those cells which do enter the lung in the BT group do not appear to "stick and stay" leading to an absence of fibrosis. These data indicate taurine may provide a useful prophylatic approach to bleomycin induced lung fibrosis.

ACKNOWLEDGEMENTS

The authors are grateful to Vanessa DeBello for secretarial assistance and Joseph Samet for expertise with the photographic portions of this work. This work was supported, in part, by the Office of Mental Retardation and Developmental Disabilities of New York State and by U.S. Public Health Service Research Grant Award HL-49942 from the National Institutes of Health.

REFERENCES

1. Wang, Q., Giri, S.N., Hyde, D.M. and Li, C., 1991, Amelioration of bleomycin-induced pulmonary fibrosis in hamsters by combined treatment with taurine and niacin. Biochem Pharmacol. 42:1115-1122.
2. Wang, Q., Giri, S.N., Hyde, D.M. and Nakashima, J.M., 1989, Effects of taurine on bleomycin induced lung fibrosis in hamsters. Proc. Soc. Exp. Biol. Med. 190: 330-338.
3. Gordon, R.E. and Heller, R.F., 1992, Taurine protection of lungs in hamster models of oxidant injury; a morphological time study of paraquat and bleomycin treatment. In Taurine, edited by J. B. Lombardini, Plenum Press, New York. 319-328.
4. Schuller-Levis, G.B., Gordon, R.E. , Park, E., Pendino, K.J. and Laskin, D.L., 1995, Taurine protects rat bronchioles from acute ozone-induced lung inflammation and hyperplasia. Exp.Lung Res. 21:877-888.

5. Wang, Q.J., Hollinger, M.A., Giri, S.N., 1992, Attenuation of amiodarone-induced lung fibrosis and phopholipidosis in hamsters by taurine and/or niacin treatment. J. Pharmacol. Exp. Ther. 262:127-132.
6. Gould, R.E., Shaked, A.A., Solano, D.F., 1986, Taurine protects hamster bronchiols from acute NO_2 induced alteration. Am. J. Pathol. 125:585-600.
7. Lambert, M.B. and Weiss, S.J., 1983, The chlorinating potential of the human monocyte. Blood, 62: 645-651.
8. Park, E., Quinn, M. R., Wright C. and Schuller-Levis, G.B., 1993, Taurine chloramine inhibits the synthesis of nitric oxide and the release of tumor necrosis factor alpha in RAW 264.7 cells. J. Leukoc. Biol. 54: 119-124.
9. Park, E., Schuller-Levis, G.B. and Quinn, M. R., 1995, Taurine chloramine inhibits production of nitric oxide and TNF-α in activated RAW 264.7 cells by mechanisms that induce transcriptional and translational events. J. Immunol. 154:4778-4784.
10. Kim, C., Park, E., Quinn M.R. and Schuller-Levis G.B., 1996, The production of superoxide anion and nitric oxide by cultured murine leukocytes and the accumulation of TNF-α in conditioned media is inhibited by taurine chloramine. Immunopharmacol. 34:89-95.
11. Marcinkiewicz, J., Grabowska, A., Bereta, J., Stelmaszynska, T., 1995, Taurine chloramine, a product of activated neutrophils, inhibits in vitro the generation of nitric oxide and other macrophage inflammatory mediators. J. Leukoc. Biol. 58:667-674.
12. Kontny, E., Grabowska, A., Kowalczewski, J., Kurowska, M., Janicka, I., Marcinkiewicz, J., Maslinskki, W., 1999, Taurine chloramine inhibition of cell proliferation and cytokine production by rheumatoid arthritis fibroblast-like synoviocytes. Arthritis Rheum. 42:2552-2560.
13. Adamson, I.Y., and Bowden, D.H., 1974, The pathogenesis of bleomycin-induced pulmonary fibrosis in mice. Am. J. Pathol. 77:185-197.
14. Marin, M.L., Hardy, M.A., Gordon, R.E., Reemtsma, K., and Benvenisty, A.I., 1990, Immunomodulation of vascular endothelium III effects of UVB irradiation on vein allograft rejection. J. Vasc. Surg. 11:103-111
15. Green, T.R., Fellman, J.H., Eicher, A.L. and Pratt, K.L., 1991, Antioxidant role and subcellular location of hypotaurine and taurine in human neutrophils. Biochim. Biophys Acta. 1073: 91-97.
16. Fukuda, K., Hirai, Y., Yoshida, H., Hakajima, T., Usii, T., 1982, Free amino acid content of lymphocytes and granulocytes compared. Clin. Chem. 28, 1758-1761.
17. Vanhee, D., Delneste, Y., Lassalle, P., Gosset, P., Joseph, M., Tonnel, A.B., 1994, Modulations of endothial cell adhesion molecule expression in a situation of chronic inflammatory stimulation. Cell Immumol. 155;2:446-456.
18. Huot, A. and Hacker, M., 1990, Role of reactive nitrogen intermediate production in alveolar macrophage mediated cytostatic activity induced by bleomycin lung damage in rats. Cancer Res. 50:7863-7866.
19. Gurujeylakshimi, G., Wang, Y. and Giri, S.N., 2000, Suppression of bleomycin-induced nitric oxide production in mice by taurine and niacin. Nitric Oxide. 4:399-411.

Taurine and the Lung
Which Role in Asthma?

FRANCESCO SANTANGELO[*], JULIO CORTIJO[#] and ESTEBAN MORCILLO[#]

Preclinical Development, R&D, Zambon Group, Bresso, Milan, Italy and #Department of Pharmacology, Faculty of Medicine, University of Valencia, Spain

1.   INTRODUCTION

Taurine (TAU) is the most abundant free amino acid in many tissues and in particular in proinflammatory cells like polymorphonuclear leukocytes and tissues exposed to elevated levels of oxidants[1]. Furthermore, orally administered TAU has been reported to reduce lung oxidant damage from exposure to ozone, nitrogen dioxide, paraquat, amiodarone and bleomycin in animal models[2-5].

To our knowledge, the activity of taurine on experimental models of asthma has not yet been reported even though taurine administration has been reported to benefit asthmatics[6].

Asthma is a chronic inflammatory disease characterised by reversible bronchial obstruction and airway hyper-reactivity, eosinophil accumulation and plasma exudation. There is increasing clinical and experimental evidence that excessive production of reactive oxygen species and defective endogenous anti-oxidant defence mechanisms may be present in asthma[7]. In addition, TAU levels in bronchoalveolar lavage fluid and airway secretions have been reported to be altered in asthma[8]. Thus, the objective of the present work has been the study of the effects of orally administered taurine on antigen-induced bronchoconstriction, airways hyper-reactivity, eosinophilia, and airway microvascular leakage in actively sensitised Brown-Norway rats. We also determined the levels of taurine in bronchoalveolar lavage fluid. With the aim of investigating the potential mechanism of action, we also evaluated the effect of TAU and taurine chloramine (TAU-Cl) on superoxide anion generation by human eosinophils.

2.   METHODS AND RESULTS

2.1   Isolation of Human Eosinophils and Superoxide Anion Generation

Fresh anticoagulated human blood from healthy donors was diluted with PBS and aliquots were layered onto a Percoll solution in the magnetic cell separation system (MACS)[9], allowing for the preparation of eosinophils of >95% purity and >97% viability.

Release of superoxide from human eosinophils was measured as the superoxide dismutase (SOD)-inhibitable reduction of ferricytochrome c using a modified microassay[10].

2.2   Sensitisation Procedure of Animals

Male Brown-Norway rats weighing 220-300 g were used. The animals were actively sensitised by intraperitoneal injection of ovalbumin (OA) and Al(OH)$_3$ as previously outlined[11]. Experimental groups:

(i) Untreated groups: **Group A**: negative control group, sensitized animals receiving drug vehicle and exposed to aerosol saline; **Group B**: positive control group, sensitized animals subsequently exposed to aerosol antigen and receiving drug vehicle; (ii) Treated groups: **Group C1:** sensitized animals treated with TAU 1 mmol/kg (i.e. 125 mg/kg bw); **Group C2:** sensitized animals treated with TAU 3 mmol/kg (i.e. 375 mg/kg bw). Treated animals received oral TAU (1 or 3 mmol/kg) for 7 days before antigen challenge.

2.3   Effects of Taurine and Taurine Chloramine on Superoxide Anion Generation of Human Eosinophils

TAU-Cl decreased superoxide generation by fMLP (N-formylmethionyl-leucyl-phenylalanine, PMA (phorbol 12-myristate), and SOZ (serum opsonized zymosan) with similar potencies (-log IC$_{50}$ = 3.64±0.12, 3.69±0.13 and 3.89±0.07, respectively) while taurine was largely ineffective (Fig. 1).

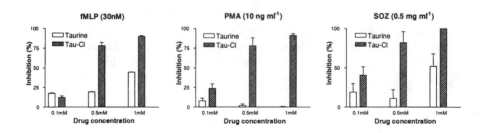

Figure 1. Inhibition by Taurine and Tau-Cl of superoxide release by human eosinophils.

2.4 Effects of Taurine on Antigen-Induced Acute Bronchoconstriction

Challenge of sensitized untreated animals with aerosol antigen provoked an acute rise in lung resistance reaching a peak in about 2-3 min. Taurine (1 or 3 mmol/kg) failed to significantly inhibit antigen-induced bronchoconstriction. The increase above baseline was $142\pm23\%$, $133\pm9\%$, and $127\pm13\%$ in the control and taurine treated groups, respectively ($P>0.05$ among groups) (data not shown).

2.5 Effects of Taurine on Antigen-Induced Hyperreactivity

Intravenous administration of 5-HT ($6.25-100\ \mu g\ kg^{-1}$) 24 h after saline (group A; negative control) or antigen (group B; positive control) exposure resulted in a dose-dependent increase in airway resistance of sensitised untreated rats. Statistical differences between the groups, i.e. 'hyper-reactivity', was obtained at concentrations ranging from 6.25 to 25 $\mu g/kg$ (Fig. 2). When the dose-response curve for 5-HT was reproduced in the taurine treated groups exposed to antigen, a decreased responsiveness (*i.e.* an anti-hyper-reactivity effect) was apparent at a concentration of taurine of 1 mmol kg^{-1} and 3 mmol kg^{-1} for group C1 and C2, respectively.

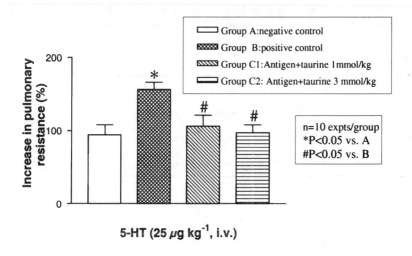

Figure 2. Airways hyperreactivity to 5-HT in sensitized Brown-Norway rats.

2.6 Effects of Taurine on Evans Blue Levels in Bronchoalveolar Lavage Fluid

An antigen challenge increased the concentration of Evans blue in bronchoalveolar lavage fluid. This increase was reduced by TAU at the two concentrations examined.

Table 1. EB in BALF (ng ml^{-1})

Group A	632±105		n=7
Group B	1241±155	significant vs. A	n=8
Group C1	742±76	significant vs. B	n=5
Group C2	477±196	significant vs. B	n=5

2.7 Taurine Levels in Bronchoalveolar Lavage Fluid

Taurine levels in bronchoalveolar lavage fluid of antigen challenged, untreated rats were higher than the levels observed in saline-exposed rats. The taurine levels in bronchoalveolar lavage fluid from antigen-challenged rats treated with taurine were also higher than the values in saline-exposed rats but not significantly higher than the taurine levels found in antigen-exposed, untreated rats. No significant correlation was found between taurine values in bronchoalveolar lavage fluid and either the airways responsiveness to 5-HT or neutrophil and eosinophil number.

Table 2. Taurine levels in BALF (nmol ml^{-1})

Group A	2.32±0.22		n=5
Group B	3.87±0.50	significant vs. A	n=8
Group C1	4.35±0.46	significant vs. A	n=9
Group C2	5.05±0.66	significant vs. A	n=9

2.8 Effects of Taurine on Antigen-Induced Increased Cellularity in Bronchoalveolar Lavage Fluid

Bronchoalveolar lavage was carried out 24 h after saline or antigen exposure of sensitized rats (Fig. 3). In rats exposed to antigen there was a significant increase in the total number of cells in the bronchoalveolar lavage fluid. The increase in eosinophil number was reduced by taurine while no inhibitory effect was produced on the other cell types (neutrophils, lymphocytes and macrophages).

Table 3. Total cell count in BALF (x10^6/ml)

Group A	0.093±0.0086	n=10
Group B	0.603±0.083	significant vs. A n=12
Group C1	0.409±0.071	significant vs. A n=13
Group C2	0.393±0.069	significant vs. A n=13

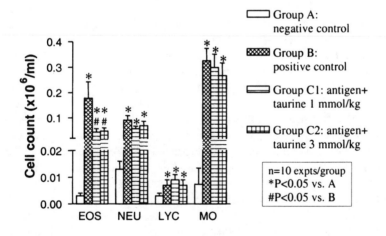

Figure 3. Differential cell count in BALF of sensitized BN rats.

3.   DISCUSSION

Taurine is an amino acid endowed with antioxidant, anti-inflammatory and immunomodulatory properties[1] that reduce oxidant-induced lung damage in a variety of experimental models[2-5]. In this study, we have extended these observations to show the anti-hyperreactivity and anti-inflammatory effects of oral taurine in an established experimental model of allergic asthma.

TAU was effective in reducing airways hyperreactivity to 5-HT, as well as eosinophil count in bronchoalveolar lavage fluid. To our knowledge, this is the first report in the literature on this anti-hyperreactivity effect of taurine. The mechanisms underlying hyper-reactivity to spasmogens in airways have not been fully elucidated. It is possible that the beneficial effect of taurine may be related to its anti-oxidant properties. In fact, there is experimental evidence that the activation of the redox-sensitive transcription factor NF-κB is required for the expression of various inflammatory genes in this experimental model of asthma[12]. Although we have not studied the effect of taurine on NF-κB activation, it has been recently shown that oral taurine in combination with niacin blocked the activation of NF-κB in a murine model of pulmonary fibrosis induced by oxidant stress[5]. In addition to reducing the number of inflammatory cells, taurine appears to lessen the oxidant burden by diminishing the generation of superoxide anion and other cytotoxic mediators by these cells[3,13]. It could also protect airway cells from oxidative damage as demonstrated for other antioxidants[14].

It is known that TAU-Cl depresses the release of inflammatory mediators by neutrophils and macrophages and modulates the activation of T-cells[15]. Our data on human eosinophils further support the role played by TAU-Cl in regulating the function of these cells. It is possible that, at least partially, our findings are mediated by an increase in TAU-Cl levels. Consistent with the findings on cell extravasation in airways, we found that taurine was able to reduce airway microvascular leakage in response to antigen provocation. However, the acute increase in airway resistance that follows the antigen challenge was not reduced by taurine. Antigen-mediated bronchoconstriction is related to the immediate release of mediators from mast cells degranulating in the airways[16] (Advenier et al., 1979). Thus, the amino acid was not effective in inhibiting mediator release from mast cells. This result is consistent with the inability of taurine (2.4 mmol kg^{-1}, i.v.) to inhibit antigen-induced bronchospasm of sensitized guinea-pigs [17].

The levels of TAU found in the bronchoalveolar lavage fluid of rats not challenged with antigen are within the range of levels reported in a previous study using a different rat strain[18]. Antigen exposure of sensitized rats resulted in an increase in taurine levels of bronchoalveolar lavage fluid 24 h post-challenge. This observation is consistent with the observed increase in TAU levels found in bronchoalveolar lavage fluid from asthmatic patients[8] although animal models of asthma are not fully equivalent to the human disease[19]. TAU levels in airway secretions were also augmented in patients with chronic bronchitis and cystic fibrosis[20], however, a negative correlation was found between taurine levels in sputum and respiratory parameters in cystic fibrosis[21]. Whether the increase in taurine content of bronchoalveolar lavage fluid is due simply to its release from damaged cells or is a compensatory rise in response to oxidant stress as described for other endogenous antioxidants[22] has not been established. Treatment with taurine further augmented taurine levels in bronchoalveolar lavage fluid compared to untreated rats but this increase failed to reach statistical significance.

In conclusion, the results from this study demonstrate that oral treatment with taurine produced anti-hyperreactivity and anti-inflammatory effects in an animal model of allergic asthma but the precise mechanisms involved in the effect remain to be fully ascertained.

NOTES

More details of some reported data have been published in ref. 23

REFERENCES

1. Stapleton, P.P., O'Flaherty, L., Redmon, H.P., Bouchier-Hayes, D.J., 1998, Host defense – a role for the amino acid taurine? *J. Parenter. Enteral Nutr.* 23: 366-367.
2. Schuller-Levis, G.B., Gordon, R.E., Park, E., Pendino, K.J., Laskin, D.L., 1995, Taurine protects rat bronchioles from acute ozone-induced lung inflammation and hyperplasia. *Exp. Lung Res.* 21: 877-888.
3. Timbrell, J.A., Seabra, V., Waterfield, C.J., 1995, The in vivo and in vitro protective properties of taurine. *Gen. Pharmacol.* 26: 453-462.
4. Gordon, R.E., Park, E., Laskin, D., Schuller-Levis, G.B., 1998, Taurine protects rat bronchioles from acute ozone exposure: a freeze fracture and electron microscopic study. *Exp. Lung Res.* 24: 659-674.
5. Gurujeyalakshmi, G., Wang, Y., Giri, S.N., 2000, Taurine and niacin block lung injury and fibrosis by down-regulating bleomycin-induced activation of transcription nuclear factor-κB in mice. *J. Pharmacol. Exp. Ther.* 293: 82-90.
6. Covarrubias, J., 1994, Taurine and the lung: Pharmacological intervention by aerosol route. *Adv. Exp. Med. Biol.* 359: 413-417.
7. Barnes, P.J., 2000, Reactive oxygen species in asthma. *Eur. Respir. Rev.* 10: 240-243.
8. Hofford, J.M., Milakofsky, L., Pell, S., Fish, J.E., Peters, S.P., Pollice, M., Vogel, W.H., 1997, Levels of amino acids and related compounds in bronchoalveolar lavage fluids of asthmatic patients. *Am. J. Respir. Crit. Care Med.* 155: 432-435.
9. Hatzelmann A, Tenor H, Schudt C., 1995, Differential effects of non-selective phosphodiesterase inhibitors on human eosinophil functions. *Br. J. Pharmacol* 114: 821-831
10. Cortijo J, Villagrasa V, Pons R, Berto L, Martí-Cabrera M, Martinez-Losa M, Domenech T, Beleta J, Morcillo EJ., 1999, Bronchodilator and anti-inflammatory activities of glaucine: *In vitro* studies in human airway smooth muscle and polymorphonuclear leukocytes. *Br. J. Pharmacol* 127: 1641-1651
11. Elwood, W., Barnes, P.J., Chung, K.F., 1992, Airway hyperresponsiveness is associated with inflammatory cell infiltration in allergic Brown-Norway rats. *Int. Arch. Allergy Immunol.* 99: 91-97.

12. Liu, S.F., Haddad, E.B., Adcock, I., Salmon, M., Koto, H., Gilbey, T., Barnes, P.J., Chung, K.F., 1997, Inducible nitric oxide synthase after sensitization and allergen challenge of Brown Norway rat lung. *Br. J. Pharmacol.* 121: 1241-1246.

13. Schuller-Levis, G.B., Sturman, J.A., 1992, Activation of alveolar leukocytes isolated from cats fed taurine-free diets. *Adv. Exp. Med. Biol.* 315: 83-90.

14. Cortijo, J., Martí-Cabrera, M., García de la Asunción, J., Pallardó, F.V., Esteras, A., Bruseghini, L., Viña, J., Morcillo, E.J., 1999, Contraction of human airways by oxidative stress. Protection by *N*-acetylcysteine. *Free Rad. Biol. Med.* 27 : 392-400.

15. Marcinkiewicz, J., Grabowska, A., Chain, B.M., 1998, Modulation of antigen-specific T-cell activation in vitro by taurine chloramine. *Immunology* 94: 325-330.

16. Advenier, C., Mallard, B., Santais, M.C., Ruff, F., 1979, The effects of metiamide and H$_1$ receptor blocking agents on anaphylactic response in guinea-pigs. *Agents Actions* 9: 467-473.

17. Kurachi, M., Hongoh, K., Watanabe, A., Aihara, H., 1987, Suppression of bronchial response to platelet activating factor following taurine administration. *Adv. Exp. Med. Biol.* 217: 189-198.

18. Cortijo, J., Cerdá-Nicolás, M., Serrano, A., Bioque, G., Estrela, J.M., Santangelo, F., Esteras, A., Llombart-Bosch, A., Morcillo, E.J., 2001, Preventive effect of oral N-acetycysteine on bleomycin-induced lung injury. *Eur. J. Resp.* 17: 1228-1235

19. Coleman, R.A., 1999, Current animal models are not predictive for clinical asthma. *Pulm. Pharmacol. Ther.* 12: 87-89.

20. Cantin, A.M., 1994, Taurine modulation of hypochlorous acid-induced lung epithelial cell injury in vitro. Role of anion transport. *J. Clin. Invest.* 93: 606-614.

21. Witko-Sarsat, V., Delacourt, C., Rabier, D., Bardet, J., Nguyen, A.T., Descamps-Latscha, B., 1995, Neutrophil-derived long-lived oxidants in cystic fibrosis sputum. *Am. J. Respir. Crit. Care Med.* 152: 1910-1916.

22. Smith, L.J., Houston, M., Anderson, J., 1993, Increased levels of glutathione in bronchoalveolar lavage fluid from patients with asthma. *Am. Rev. Respir. Dis.* 147: 1461-1464.

23. Cortijo J., Blesa S., Martinez-Losa M., Mata M., Seda E., Santangelo F., Morcillo E.J. 2001, Effects of taurine on pulmonary responses to antigen in sensitized Brown-Norway rat. Eur. J. Pharmacol. 413: 111-117

Effect of Taurine and Other Antioxidants on the Growth of Colon Carcinoma Cells in the Presence of Doxorubicin or Vinblastine in Hypoxic or in Ambient Oxygen Conditions

Effect of Antioxidants on the Action of Antineoplastic Drugs in MDR and non-MDR Cells

C. WERSINGER, G. REBEL and I. LELONG-REBEL
UPR 9003 du CNRS - Institut de Recherche Contre les Cancers de l'Appareil Digestif - Hôpitaux Universitaires. BP 426 - F 67091 Strasbourg, France

1. INTRODUCTION

Clinical cancer treatments are limited by the emergence of cumulative dose-dependent deleterious side effects of chemotherapeutic agents, some which are irreversible. Many of these side effects involve oxidative stress originating with the generation of reactive oxygen species (ROS) by chemotherapeutic agents, such as the anthracyclines[1,2].

Some of the well-known antioxidants exhibit interesting protective effects against the cardiotoxicity of the anthracyclines. These include carnitine[3], niacine[4], taurine[5,6], N-acetyl-L-cysteine[4,7] and vitamin E and some of its analogues[7,8]. However, the effects of these antioxidants on the cytotoxicity of the anthracyclines towards cancer cells have never been addressed. The results obtained with doxorubicin have also been compared to those obtained with combination therapy using antioxidants with vinblastine as the antineoplastic drug. In tumors most of the cells are hypoxic, the intratumoral oxygen pressure varying between 0.1% O_2 in the necrotic part of the tumor to 10% O_2 for cells close to intratumoral blood vessels[9-11]. Though it is well documented that hypoxia notably affects cell physiology, no studies have addressed the possible alteration of antioxidant properties under conditions reproducing tumor oxygenation status.

2.   MATERIALS AND METHODS

2.1   Cell Cultures

LoVo cell lines, derived from a supraclavicular metastasis of a human colon carcinoma were used. The parental non-MDR LoVo cell line and its multidrug resistant (MDR) counterpart (LoVo-Dox) were obtained from Pharmacia & Upjohn (Milan, Italy). From the parental drug sensitive cell line two variants were selected, which differed in morphology and degree of differentiation, the LoVo-S "fusoid" type (LoVo-f) and the LoVo-S "small cells" type (LoVo-sc), respectively. Cell lines were grown as monolayer cultures in DMEM/HamF12 supplemented with a mixture of 50U/ml penicillin and 50μg/ml streptomycin, 5% (v/v) heat inactivated selected FCS, 365mg/l L-glutamine and 55mg/l sodium pyruvate (1mM final concentration). To avoid reversion of the MDR phenotype, LoVo-Dox culture medium was supplemented with 0.4μg/ml doxorubicin. Doxorubicin and vinblastine were added as DMSO solutions.

Drug resistance characteristics of the cell lines were as follows under standard culture conditions: IC50 for doxorubicin: LoVo-f 3.58±0.15ng/ml, LoVo-sc 5.82±0.19ng/ml, LoVo-Dox 1.41±0.22μg/ml and IC50 for vinblastine: LoVo-f 2.84±0.12ng/ml, LoVo-sc 4.71±0.18ng/ml, LoVo-Dox 1.28±0.25μg/ml.

Absence of mycoplasma contamination was routinely checked using Hoechst 33258. Cell proliferation was determined by Crystal violet stain[12].

2.2   Effect of Antioxidants on Doxorubicin or Vinblastine Induced Inhibition of Cell Proliferation

Approximately 400 cells were seeded in 35 mm Petri dishes (Corning 430165) in 2 ml of DMEM/Ham F12 + 5% (v/v) FCS + 0.4μg/ml doxorubicin for LoVo-Dox cells. Twelve hours later, the culture medium was removed and replaced by 2ml of DMEM/Ham F12 + 5% FCS containing or not one of the antioxidants (2mM), in presence or absence of either vinblastine (VBL) or doxorubicin (Dox) (VBL or Dox concentrations used : 15ng/ml for LoVo-f and LoVo-sc cells, and 1μg/ml for LoVo-Dox cells, respectively, were chosen in relation to the respective sensitivities to the drugs). Culture media were changed every other day. Cells were grown either in a 95%air:5%CO_2 atmosphere (20%O_2) or in a Sanyo multigas incubator in 5%O_2:5%CO_2:90%N_2 atmosphere (hypoxia). As cell attachment was slower at 5% O_2 than at 20% O_2, cells were allowed to attach at 5% O_2 for 12 h and then switched to 20% O_2 or maintained at 5% O_2. Supplementation of the media with 2mM antioxidant did not change significantly the medium osmolarity (1% increase).

3.   RESULTS

3.1   Effect of Oxygen Pressure on Doxorubicin or Vinblastine-Induced Inhibition of LoVo Cell Proliferation.

Table 1 shows that under similar culture conditions the growth rates of the two LoVo sensitive cells were quite similar whereas growth of LoVo-Dox was slower than that of the sensitive counterparts.

Table 1. Effect of pO_2 on doxorubicin and vinblastine inhibition of cell growth

Cell line	Culture conditions	Cell growth (OD 600 nmX10^3)			Percentage		
		Day 0	Day 8 at 20% O_2	Day 8 at 5% O_2	I	II	III
LoVo-f	control	99±5	2844±69	1763±42	62		
	doxorubicin		454±37	687±42	151	16	61
	vinblastine		426±23	233±21	55	15	13
LoVo-sc	control	97±3	3014±76	1959±125	65		
	doxorubicin		589±48	827±58	140	19	42
	vinblastine		499±24	280±20	56	17	14
LoVo-Dox	control	94±5	1344±73	1080±75	80		
	doxorubicin		135±3	147±2	109	10	14
	vinblastine		108±9	92±5	85	8	8

Cells were grown as indicated in Materials and Methods. Percentage: (I) growth at 5% O_2 compared to 20% O_2; growth in presence of drugs compared to control at 20% O_2 (II) or at 5% O_2 (III), respectively.

Growing cells at 5% O_2 instead of 20% O_2 significantly reduced the proliferation of the three variants, the inhibition observed being higher for the chemosensitive cells (about 40%) than for the LoVo-Dox cells (about 20%). Doxorubicin and vinblastine strongly inhibited growth of sensitive cells grown at 20% O_2. Inhibition by doxorubicin but not inhibition by vinblastine was notably reduced when these cells were grown at 5% O_2. By contrast, sensitivity of chemoresistant cells towards both chemotherapeutic agents was only weakly affected by pO_2.

3.2.   Effect of Antioxidants on Doxorubicin Toxicity.

With the exception of N-acetylcysteine, none of the tested antioxidants affected the growth of the three cell lines. Whereas N-acetylcysteine slightly inhibited (15%) LoVo-sc growth, the growth of LoVo-f and of LoVo-Dox was increased (about 20%) by the compound when cultures were grown at 20% O_2. At 5% O_2, none of the antioxidants affected cell growth.

Doxorubicin was found to be a potent inhibitor of growth of the three cell lines (Table 2). This inhibition was not modified by taurine, carnitine, niacin or N-acetylcysteine. Surprisingly, trolox notably reversed the cytotoxicity of the anthracyclines, the effect nearly complete under hypoxic conditions. Vinblastine also inhibited LoVo cell growth. Taurine, trolox, carnitine, niacin and N-acetylcysteine had no effect on vinblastine cytotoxicity.

Table 2. Effect of antioxidants on doxorubicin and vinblastine cytotoxicity in LoVo cells

Cell line	% O_2	Anticancer drug	Antioxidant (2 mM)	Cell growth (% control) on day		
				4	8	12
LoVo-f and LoVo sc	20	vinblastine	±	48	15	2
	5	vinblastine	±	49	14	2
	20	doxorubicin	± , except trolox	58	20	2
		doxorubicin	+ trolox	90	67	53
	5	doxorubicin	± , except trolox	79	14	5
		doxorubicin	+ trolox	99	95	89
LoVo-Dox	20	vinblastine	±	52	10	2
	5	vinblastine	±	46	9	1
	20	doxorubicin	± , except trolox	52	9	4
		doxorubicin	+ trolox	91	65	50
	5	doxorubicin	± , except trolox	70	14	5
		doxorubicin	+ trolox	99	96	90

Cells were grown as indicated in Materials and Methods, in absence or in presence of either taurine, carnitine, niacine, N-acetylcysteine, or trolox.

4.   DISCUSSION

Growth of the three LoVo cell lines was notably reduced when cultured in a hypoxic environment instead of in air:5% CO_2 (20% O_2). This growth reduction was less important for the chemoresistant LoVo-Dox cells than for the two chemosensitive -f and -sc variants. The growth of the two sensitive lines in the presence of doxorubicin was affected much more when cultured in 20% O_2 than in hypoxic medium. By contrast, this anticancer agent similarly affected the growth of the MDR cells at the two oxygen pressures. Vinblastine was a potent inhibitor of growth of the three LoVo lines, an effect independent of the oxygen pressure. The observed difference between the cytotoxic effects of the two drugs is certainly related to their mechanism of action and their metabolism. In contrast to vinblastine, doxorubicin by its chemical structure, is involved in a redox-system which produces a semiquinone free radical that in the presence of O_2 enables the formation of reactive oxygen species[13]. It is clear that ROS production will be more efficient in a well oxygenated condition than in hypoxia, accounting for the greater sensitivity of cells cultured in air:5% CO_2 (20% O_2). Our results are not in accordance with the data obtained with

EMT6 or with Ehrlich cells[14,15]. However, these latter cells were tested in a N_2/CO_2 atmosphere for a one hour period and not at 5% O_2: 5% CO_2:N_2 for 12 days, as we did. The absence of effect of oxygen level on LoVo-Dox could be related to a decrease of ROS formation as previously observed in other MDR cells[13,16]. From the various antioxidant tested, none of them affected the cytotoxicity of vinblastine, whose mechanism of action does not involve ROS. Trolox strongly inhibited doxorubicin toxicity in all the LoVo lines. Vitamin E and certainly its hydrophilic derivative trolox are chain breaking compounds, inhibiting the formation of ROS[17,18]. The fairly complete inhibition of the doxorubicin effect by trolox observed in hypoxic cells (98% inhibition) could therefore be related to the inhibition of ROS formation under these conditions. The absence of an effect of the other antioxidants is related to their respective mechanisms of action: carnitine increases mitochondrial fatty-acid transport and ß-oxidation and subsequently increases ATP levels; niacin increases NAD concentration thereby activating various repair mechanisms.

Because the deleterious effects of the anthracyclines toward cardiac cells and kidney cells are decreased or abrogated *in vitro* or in rodent animals by numerous antioxidants, we focused on the effects of taurine. This amino acid was shown to suppress the cardiotoxicity of doxorubicin in mice [5]. Numerous studies report on the benefit of using taurine in cardiology, mainly in promoting a fast recovery after heart failure[19]. Taurine is not metabolized through the cytochrome P450 system, with the only known catabolism being the formation of isethionic acid, a neuroprotector generated from taurine chloramine by neutrophils and by macrophages[20]. Taurine regulates the intracellular calcium concentration[21] probably one of the most important factors responsible for anthracycline induced cardiopathy[22,23]. Taurine exhibits no side effects at high concentration. Our results show that in contrast to trolox, taurine does not affect the cytotoxicity of doxorubicin against tumoral MDR and non-MDR cells, which differ with respect to taurine transport[24]. Together, our data suggest that taurine could be a promising molecule in decreasing the side effects of anticancer drugs that promote free radical formation.

Comparison of the effects of trolox, N-acetylcysteine and taurine indicate that hydroxy free radicals are the main intermediary compounds involved in doxorubicin toxicity towards LoVo tumourous cells. Although N-acetylcysteine is known to increase the levels of reduced gluthatione[25] it has no protective effect on doxorubicin treated cells. Altogether, our data indicate that the superoxide anion produced by the anthracycline redox cycle[13] is probably directly converted into hydroxy free radicals, the final ROS produced from doxorubicin[13], but is not dismutated into hydrogen peroxide, which would be eliminated by gluthatione peroxidase. Taurine is a scavenger of reactive aldehydes[26] generated by doxorubicin[27]. Since taurine does not protect LoVo cells against doxorubicin cytotoxicity, reactive aldehydes do not appear to play a crucial role in the cytotoxic mechanism of the anthracyclines. Our results highlight the role of the oxygenation status of tissues in cancer therapy. Moreover, they emphasize the importance of testing pharmacological parameters in pO_2 conditions as close as possible to the *in vivo* situation(s) of the tumorous cell.

ACKNOWLEDGEMENTS

This work was supported by grants from the "Ligue Nationale contre le Cancer - Comite du Bas-Rhin" and from the "Fondation pour la Recherche Medicale".

REFERENCES

1. Singal, P.K., Li, T., Kumar, D., Danelisen, I. and Iliskovic, N., 2000, Adriamycin-induced heart failure: mechanism and modulation. *Mol.Cell.Biochem.* 207:77-85.
2. Barbey, M.M., Fels, L.M., Soose, M., Poelstra, K., Gwinne, W., Bakker, W. and Stolte, H., 1989, Adriamycin affects glomerular renal function : evidence for the involvment of oxygen radicals. *Free Rad. Res. Comms.* 7: 195-203.
3. Andrieu-Abadie, N., Jaffrezou, J.P., Hatem, S., Laurent, G., Levade, T. and Mercadier, J.J., 1999, L-carnitine prevents doxorubicin-induced apoptosis of cardiac myocytes: role of inhibition of ceramide generation. *FASEB J.* 13:1501-1510.
4. Schmitt-Graff, A. and Scheulen, M.E., 1986, Prevention of adriamycin cardiotoxicity by niacin, isocitrate or N-acetyl-cysteine in mice. A morphological study. *Pathol. Res. Pract.* 181:168-174.
5. Hamaguchi, T., Azuma, J., Awata, N., Ohta, H., Takihara, K., Harrada, H., Kishimoto, S. and Sperelakis, N., 1988, Reduction of doxorubicin induced cardiotoxicity in mice by taurine. *Res. Comm. Chem. Pathol. Pharmacol.* 59:21-30.
6. Hamaguchi, T., Azuma, J., Harada, H., Takahashi, K., Kishimoto, S. and Schaffer, S., 1989, Protective effect of taurine against doxorubicin-induced cardiotoxicity in perfused chick hearts. *Pharmacol. Res.* 21:729-734.
7. Venditti, P., Balestrieri, M., De Leo, T. and Di Meo, S., 1999, Free radical involvement in doxorubicin-induced electrophysiological alterations in rat papillary muscle fibers. *Cardiovasc Res.* 38:695-702.
8. DeAtley, S.M., Aksenov, M.Y., Aksenova, M.V., Harris, B., Hadley, R., Cole Harper, P., Carney, J.M. and Butterfield, D.A., 1999, Antioxidants protect against reactive oxygen species associated with adriamycin-treated cardiomyocytes. *Cancer Lett.* 136:41-46.
9. Vaupel, P., Kallinowski, F., and Okunieff, P., 1989, Blood flow, oxygen and nutrient supply and metabolic microenvironment of human tumors: a review. Cancer Res. 49, 6449-6465.
10. Dewhirst, M., Ong, E., Klitzman, B., Secomb, T.W., Vinuya, R.Z., Dodge, R., Brizel, D. and Gross, J.F., 1992, Perivascular oxygen tensions in a transplantable mammary tumor growing in a dorsal flap window chamber. *Radiation Res.* 130:171-182.
11. Adam, M.F., Dorie, M.J., Brown, J.M., 1978, Oxygen tension measureemnt of tumors growing in mice. Int. J. radiation Oncology Biol. Phys. 45:171-180.
12. Lelong, I.H. and Rebel, G., 1998, pH drift of "physiological buffers" and culture media used for cell incubation during *in vitro* studies. *J. Pharmacol. Toxicol. Methods* 39:203-210.
13. Sinha, B.K. and Mimnaugh, E.G., 1990, Free radicals and anticancer drug resistance: oxygen free radicals in the mechanism of drug toxicity and resistance by certain tumors. *Free Radical Biol. Med.* 8: 567-581.

14. Teicher, B.A., Lazo, J.S. and Sartorelli, A.C., 1981, Classification of antineoplastic agents by their selective toxicity towards oxygenated and hypoxic tumor cells. *Cancer Res.* 41: 73-81.
15. Gupta, V. and Costanzi, J.J., 1987, Role of hypoxia in anticancer drug induced cytotoxicity for Ehrlich ascites cells. *Cancer Res.* 47: 2401-2412.
16. Benchekroun, M.N., Sinha, B.K. and Robert, J., 1993, Doxorubicin-induced oxygen free radical formation in sensitive and doxorubicin-resistant variants of rat glioblastoma cell lines. *FEBS Lett.* 326:302-305.
17. Burton, G.W. and Ingold, K.U., 1989, Vitamin E as an in vitro and in vivo antioxidant. *Ann. N.Y. Acad. Sci.* 570:7-22.
18. Miura, T., Muraoka, S. and Ogiso, T., 1993, Inhibition of hydroxyl radical-induced protein damages by trolox. *Biochem. Mol. Biol. Int.* 31:125-133.
19. Azuma, J., Sawamura, A., and Awata, N., 1992, Uselfulness of taurine in chronic congestive heart failure and its prospective application. *Jpn. Circ. J.* 56:95-99.
20. Cunningham, C.; Tipton, K.F. and Dixon, H.B.; 1998, Conversion of taurine into N-chlorotaurine (taurine chloramine) and sulphoacetaldehyde in response to oxidative stress. *Biochem. J.* 330: 939-945.
21. Huxtable, R.J. 1992, Physiological actions of taurine. Physiol. Rev. 72: 101-163.
22. Kapelko, V.I., Williams, C.P., Gutstein, D.E. and Morgan, J.P., 1996, Abnormal myocardial calcium handling in the early stage of adriamycin cardiopathy. *Arch. Physiol. Biochem.* 104: 185-191.
23. Azuma, J., Hamaguchi, T., Ohta, H., Takihara, K., Awata, N., Sawamura, A. and Harada, H., 1987, Calcium overload-induced myocardial damage caused by isoproterenol and by adriamycin: possible role of taurine in its prevention. *Adv. Exp. Med. Biol.* 217: 167-179.
24. Wersinger, C., Lelong-Rebel, I.H., and Rebel, G. 2000, Detailed study of the different taurine uptake systems in colon LoVo MDR and non MDR cell lines. *Amino Acids*, 19:667-685.
25. Ferrari, G., Yan, C.Y. and Greene, L.A., 1995, N-acetylcysteine (D-and L-stereoisomers) prevents apoptotic death of neuronal cells. *J. Neurosci.* 15: 2857-2866.
26. Ogasawara, M., Nakamura, T., Koyama, I., Nemoto, M. and Yoshioa, T., 1993, Reactivity of taurine with aldehydes and its physiological role. *Chem. Pharm. Bull.* 41:2172-2175.
27. Benchekroun, M.N., and Robert, J.,1993, Measurement of doxorubicin induced lipid peroxidation under conditions that determine cytotoxicity in cultured tumor cells. *Anal. Biochem.* 201: 326-330.

Part 6:

Taurine: Retina and the Brain

The Role of Taurine in Cerebral Ischemia: Studies in Transient Forebrain Ischemia and Embolic Focal Ischemia in Rodents

ASHFAQ SHUAIB MD FRCPC FAHA
Department of Medicine,
University of Alberta, Edmonton, Canada

ABSTRACT

Sudden cessation of blood flow to the brain results in a series of events that either result in rapid loss of brain cells or delayed neuronal injury in certain vulnerable regions of the brain. Research over the last three decades has allowed for a better understanding of how neurons and other brain cells die from the effects of ischemia and hypoxia in the central nervous system. Excitatory and inhibitory neurotransmitters exist in a very precise balance for normal function of the brain. Ischemia very rapidly disrupts this balance resulting in a rapid build-up of excitatory neurotransmitters, especially glutamate in the extracellular space. The increased glutamate together with energy loss opens a number of different types of calcium and sodium channels resulting in the build-up of these ions in neurons, leading to cellular dysfunction and death. While most ischemia research has focused on antagonism of excitatory amino acids, there are some reports on enhancement and amplification of inhibitory responses in focal and global ischemia. The majority of work relates to potentiation of GABA, either endogenous or through GABA potentiating medications. Taurine has neuroinhibitory properties and may also have potential for neuroprotection in cerebral ischemia. This present review focuses on the role of taurine as a neuroprotective agent, possibly acting through several different inhibitory mechanisms. Taurine may inhibit neurotransmitter release and may result in normal intracellular osmolality. In transient global ischemia in gerbils, we studied *in vivo* microdialysis of amino acids before, during and after ischemia. We were able to show that taurine resulted in attenuation of glutamate during ischemia (however did not reach significance). In similar experiments, neuronal damage was assessed in the hippocampus. Our results show 48% damage in taurine treated animals, 60% in alanine treated animals and 69% in control groups (trend towards protection but again did not reach significance) Focal ischemia was induced by embolizing a thrombus into the distal internal carotid artery and origin of the middle cerebral artery. Again, in studies where we compared taurine to a placebo treated animal, there was no significant decrease in the amount of damage with taurine.

There are reports in the literature that taurine may attenuate neuronal injury during ischemia. Our studies in two models of cerebral ischemia in rodents did not reveal neuronal protection. It is possible that higher doses or possibly prolonged use of taurine may show better results. Taurine may also potentially offer additive protective effects when used in combination with thrombolysis or other neuroprotective agents. Further studies are necessary to better understand the potential for taurine as a neuroprotective agent in cerebral ischemia.

1.   INTRODUCTION

Neuronal function requires a high continuous supply of blood to the nervous system. Neuronal activity may cease to function immediately if the blood flow decreases to below 25% of normal values [1]. With further decrease in blood flow, there is depolarization of neurons, increase in extracellular glutamate and rapid cell death [1-3]. Almost immediately after cessation of blood flow, there is massive depolarization of neuronal membranes producing a several fold increase of glutamate in the extracellular fluid [1,3-5]. Glutamate, acting as an excitatory neurotransmitter, allows for entry of calcium into post-synaptic neurons. Most of this glutamate-mediated excitatory transmission occurs at the ionotrophic glutamate receptors such as the NMDA (n-methyl-d-aspartate), AMPA (alpha amino-3 hydroxy-5 methyl-4 isoxazolepropionic acid) or kinate receptors [1,6]. The calcium influx mediates the rapid neurotoxicity associated with ischemia while sodium influx may contribute to cell swelling and more delayed neuronal damage. The concentration of free calcium in the cytoplasm of a resting neuron is extremely low (approximately 100 nM) [7-10] as compared to extracellular concentrations in the range of 1-2 Mm [10]. Cerebral ischemia and the resulting energy failure leads to intracellular accumulation of calcium from a variety of sources including activation of the receptors mentioned above, release from intracellular stores and through activation of voltage-gated calcium channels [8,10,11]. Prolonged elevation of intracellular calcium promotes a host of processes that result in damage to cellular and nuclear proteins, enzymes and mitochondrial function [10,11].

Neuronal damage may result in necrosis or apoptosis, depending on the severity and duration of the insult [1-3,6,12]. This 'necrosis' is seen most often with focal ischemia in the region of the core (areas with the most severely reduced blood flow). In contrast, apoptosis may be seen when the 'transient global ischemia' is reversed quickly (less than 5 minutes in gerbils and 10 minutes in rat models of ischemia) and has been most widely studied in the hippocampus in models of transient forebrain ischemia [13]. Interestingly, the CA-1 neurons in the hippocampus may survive the insult and remain functional for up to 48 hours after a transient cessation of blood flow to the brain, only to subsequently show destruction of over 90% of neurons. This 'delayed ischemic damage' with a window of 24-48 hours has allowed many researchers to study a host of neuroprotective strategies including glutamate antagonists [14], calcium receptor blockers, free radical scavengers [15], blockage of inflammatory response induced by ischemia [16] and amplification of the inhibitory mechanisms to prevent cell loss [17]. The remainder of this review will focus on inhibitory mechanisms with particular attention to taurine as a strategy for neuroprotection after ischemia.

2. INHIBITORY MECHANISMS AND NEURONAL PROTECTION OF CEREBRAL ISCHEMIA

Most information on the effects of ischemia on neurotransmitters comes from *in vivo* microdialysis techniques during cerebral ischemia [18-22]. There is consistent and sustained release of glutamate, serotonin, noradrenaline, glycine and GABA at the onset of cerebral ischemia [20,21,23]. Taurine levels also increase early after induction of transient forebrain ischemia [24,25]. These levels gradually return to normal after circulation has been re-established. A number of inhibitory neurotransmitters may be potentially protective if their activity can be enhanced. For example, the noradrenergic pathway from locus ceruleus to the hippocampus is an inhibitory pathway. Ablation of this pathway prior to transient forebrain ischemia results in more pronounced hippocampal damage [26]. It has also been shown that infusion of noradrenaline during ischemia, mimicking an enhanced activation from locus ceruleus can be neuroprotective [27]. Serotenergic import from the raphe nuclei may also regulate hippocampal damage during ischemia. Serotonin agonists have been shown to be neuroprotective in transient forebrain ischemia in several studies [28-32]. Other neurotransmitters that may also influence the neuronal damage include acetylcoline [33-35], adenosine [35-38] and GABA [17]. Several medications that enhance GABA activation have been shown to be neuroprotective [39-43]. Table 1 lists inhibitory mechanisms with potential neuroprotective effects during ischemia. In addition to neurotransmitters with inhibitory potential, conditions such as hypothermia may also inhibit the release of excitatory amino acids and therefore "inhibit" ischemic damage.

Table 1: Inhibitory mechanisms that may have potential neuroprotective effects during ischemia.

A: Presynaptic

1.	Hypothermia [53]
2.	Hypothyroidism [54]
3.	GABA 'b' receptor activation [55]
4.	Adenosine [56,57]
5.	Calcium (N-channel blocker)[58]
6.	Sodium channel blocker [59-61]
7.	De-efferentation [62-64]
8.	Others (limited data) taurine, alcohol, acetylcholine, pre-conditioning

B: Post-synaptic

1.	GABA 'a' receptor [41]
2.	Serotonin [28,65]
3.	Norepinephrine [27]
4.	Insulin (?)
5.	Adrenalectomy (?)
6.	Hypothermia (?)
7.	Stimulation of metabotropic receptors (?)

The "?" denotes that the mechanism postulated but no definitive studies.

3.   TAURINE AND CEREBRAL ISCHEMIA

Taurine is an inhibitory amino acid, which has osmoregulatory, neuromodulatory and possibly neuroprotective effects in neuronal tissue [17,44-47]. Taurine is one of the most abundant amino acids in the central nervous system. In the neuronal tissue, taurine is utilized to resist intracellular hypoosmalirity and maintain a normal intracellular balance (0.3-0.5 mM) [46]. Due to changes in the glucose consumption, taurine may also influence glutamic acid production, which in turn is dependent upon glucose metabolism.

The role of taurine in intracellular water regulation is envisaged as follows: Taurine and water are released from neurons and are taken in by glia (water gain). The glial cells synthesize glutamine from glutamate and glutamine together with water, is subsequently transported into the blood. The glial cells therefore become hyperosmotic relative to the extracellular fluid. Taurine is then released by the glial cells and taken up by neurons (no osmotic effect). Hence, taurine maintains cell osmolarity and glucose influx and therefore maintains intracranial water balance in ischemic conditions. This may possibly attenuate some of the effects of the ischemic insult. Taurine also affects glutamate by influencing calcium concentrations and regulating extra- and intracellular glutamate. It may also act as a membrane stabilizer linked to chloride channels and may have effects on the high affinity sodium-dependent uptake of calcium in neurons, glial cells and the blood brain barrier. Taurine appears to be continuously redistributed between closely opposing structural elements. It is not metabolized in the central nervous system and has a slow cerebral turnover thus very little taurine is lost from the brain [46].

4. EXPERIMENTAL STUDIES WITH TAURINE IN GLOBAL AND FOCAL ISCHEMIA

4.1 Global Ischemia:

Our laboratory has utilized the Mongolian gerbil for the study of transient forebrain ischemia [48,49]. The gerbil has an incomplete circle of Willis, allowing for severe forebrain ischemia with transient occlusion of the internal carotid arteries. Transient occlusion for 5 minutes results in 70-90% death of the CA1 region of the hippocampus. For studies with taurine, a dose of 25 mM/kg/hr was continuously infused through an infusion pump for 7 days, beginning 4 days prior to transient ischemia. There were two types of control groups for comparison. Alanine has no inhibitory effects and was infused for the same duration as taurine. This was used as a 'negative-control' to ensure that the use of pumps or insertion of a micro-tubing into the hippocampus did not have protective or adverse effects. A second group of control animals was not treated with any medication. In our experiments, we utilized 8 animals per group (total 24 animals). Using stereotactic techniques, taurine and alanine were administered intracranially into the CA1 region of the hippocampus using micro-osmotic pumps (alzet, alza corp., Palo Alto, Model 1007D). During ischemia, *in vivo* microdialysis techniques were utilized to collect amino acids prior to, during ischemia and immediately after the insult [45]. Seven days after the onset of transient ischemia, the animals were euthanized and the brain dissected and the hippocampus evaluated for neuronal damage. Cerebral in-vivo microdialysis results showed that glutamate concentration was elevated to a lesser extent in the taurine treated animals. This difference, however, did not reach statistical significance. The glutamate concentrations in the taurine and alanine treated groups is shown in Figure 1. Histological assessment showed that the mean percentage of neuronal damage was greater in the control groups compared

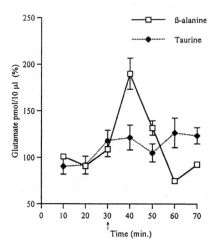

Figure 1: This figure shows the extracellular glutamate levels in animals treated with taurine or alanine. There was a trend towards lower levels of glutamate in the extracellular space with the use of taurine but this did not reach significance

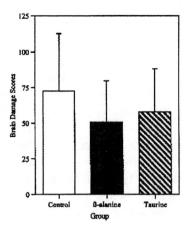

Figure 2: This figure shows the extent of neuronal injury in the hippocampal CA1 region in the taurine, alanine and vehicle treated animals. There was no difference in the extent of damage in the three groups.

to the alanine or taurine treated group. This is in Figure 2. The difference in the 3 groups was, however, not statistically significant. There appeared to be a trend towards a decrease in the extent of damage in the ipsilateral hippocampus where the taurine was infused by the alzet pumps (48% damage) but this too was not significantly lower.

4.2   Focal Ischemia:

Embolic focal cerebral ischemia was induced in anesthetized Wistar rats with the technique detailed in previous reports [50-52]. Twenty Wistar rats were randomly divided into two groups. The extent of neuronal injury was compared to vehicle treated animals. Animals treated with taurine received the medication intra-peritoneally (i.p) at a dose of 100 mg per kg at the onset of focal ischemia. Vehicle treated animals received i.p saline at the time of ischemia. All medications were delivered in a blinded manner.

All data in this study were expressed as mean ± SD. Statistical analysis of more than two groups of animals was performed with one-way analysis of variance (ANOVA), with subsequent individual comparisons using Scheffe's test. The difference was considered significant when *p* value <0.05.

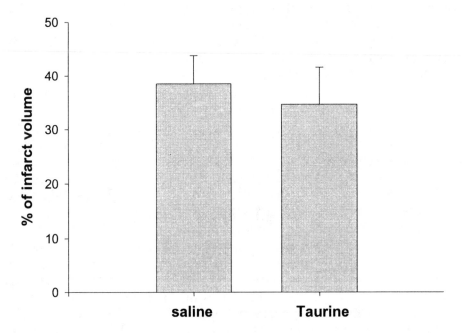

Figure 3: This figure shows the volume of infarction in the MCA territory in animals treated with taurine compared to vehicle treated animals. There is no significant difference in the extent of neuronal damage in the two groups.

All animals survived the 72 hours of required time after surgery. There were no adverse effects noted in any of the animals in both groups. Recovery from anesthesia was similar in both groups. Hemiplegia, followed by gradual recovery was also similar in the two groups. The extent of cerebral infarction in the active and placebo-treated groups is shown in Figure 3. While there was a trend towards protection with taurine, it did not reach significance. The damage was equally severe in the cerebral cortex and striatum in both groups.

5.   DISCUSSION

This review focuses on the mechanisms of ischemia, the role of inhibitory mechanisms in cerebral ischemia and in particular, the role of taurine in cerebral ischemia. Our research in models of global transient ischemia and focal ischemia has shown that, in the doses used, taurine does not appear to have neuroprotective effects. There was, however, a trend towards protection in transient global ischemia in gerbils. There are several reasons why neuroprotection was not evident in our models. The doses and the route of administration may not have been adequate in the models. Higher doses or prolonged therapy may therefore offer better outcomes. It is also possible that the insults in both focal and global ischemia were very severe and thus did not allow for assessment of moderate neuroprotection. Also, as a neuromodulator and stabilizer of membranes, it is possible that combining taurine with other neuroprotective or thrombolytic agents may offer additive or synergistic effects. Future studies may look at such combinations. It is important to note that taurine did not have any significant side effects and therefore may be an ideal combination agent. Under experimental ischemia, the concentration of taurine increases in extracellular space in the central nervous system [45;46]. Such increases have been seen in most regions of the brain, including the rat striatum [45] and rabbit's cerebral cortex, after transient forebrain ischemia. It has also been shown to be increased in the spinal cord after aortic occlusion in rabbits [17]. Multiple mechanisms for taurine release include calcium-dependent excitotoxic release from nerve terminals, reversal of the function of sodium-dependent membrane transporters and possibly through facilitated diffusion [17].

There are reports of neuroprotection with taurine in models of ischemia and excitotoxicity. Taurine protects against neuronal cell damage from increased concentrations of excitatory amino acids. It also may protect against metabolic cascades invoked by ischemia and hypoxemia. Ischemia induced calcium influx can also be decreased with treatment with taurine.

Taurine containing neurons have been shown to be resistant to the effects of forebrain and global ischemia as induced by the 4 vessel ischemia model in rats [17]. The discrepancy between our results and previous publications in the literature may be due to several reasons. Our models of focal and global ischemia result in severe injury and may not reveal 'moderate' protection with taurine that may be evident with models that deliver mild to moderate insults. The dose or route of administration of taurine in our studies may not have been sufficient to allow for best results to be evident. Our experiments with focal and transient global ischemia revealed that there was a trend towards protection in animals treated with taurine. Also, importantly, there were no side-effects noted with the use of the active drug. We believe that a combination of taurine with tissue plasminogen activator or other thrombolytic agents may offer potential synergy and should be tested in future studies.

6. SUMMARY

Taurine is an important inhibitory nerve transmitter that is available in abundance in the central nervous system. Its osmoregulatory and inhibitory properties may counteract the effects of excitotoxicity during cerebral ischemia. Sufficient reports in the literature indicate that in cell culture and some models of ischemia, taurine may have neuroprotective effects. We have studied the effects of infusion of taurine before and during ischemia in models of transient forebrain and focal ischemia. Whereas there appears to be a trend towards protection, our research indicates that this does not reach significance. Several factors may contribute to this lack of efficacy in our models, including the very severe nature of the model and possibly lower dosage of taurine that may not have been therapeutically optimal. Future studies utilizing increasing taurine concentrations in the central nervous system or in combination with other neuroprotective and thrombolytic agents may allow for a better study of this important neurotransmitter in cerebral ischemia. Taurine use offers a safe and potentially useful option in the management of cerebral ischemia but it requires further study.

REFERENCES

1. Sundi, TM, Grani, WC, Garcia, JH. *Journal of Neurosurgery* 1969; 37:311-322.
2. Hossmann, K. In: Ginsberg, M, Bogousslavsky ,J, editors. Cerebrovascular disease: pathophysiology, diagnosis, and management. Malden, Massachusetts: Blackwell Science, 1998: 193-216.

3. Heiss, WD, Rosner, G. *Annals of Neurology* 1983; 14:294-301.
4. Crowell, RM, Olsson ,Y, Katzo, I, Ommaya, A. *Stroke* 1970; 1:439-448.
5. DeGirolami, U, Crowell, RM, Marcoux ,FW. *Journal of Neuropathological Experimental Neurology* 1984; 43:57-71.
6. Gladstone, DJ, Black, SE, Hakim, AM. Stroke 23, 2123-2136. 2002.
7. Erecinska, M, Silver, IA. *Can J Physiol Pharmacol* 1992; 70 Suppl.:S190-S193.
8. Siesjo, BK, Bengtsson, F. *J Cerebral Blood Flow and Metabolism* 1989; 9:126-140.
9. Siesjo, BK. *Journal of Cerebral Blood Flow and Metabolism* 1981; 1:155-185.
10. Siesjo, BK, Zhao, Q, Pahlmark, K. *Ann Thorac Surg* 1995; 59:1316-1320.
11. Choi, DW. *Journal of Neuroscience* 1990; 10:2493-2501.
12. Partridge, LD, Müller, TH, Swandulla, D. *Brain Res Rev* 1994; 19:319-325.
13. Green, AR, Cross, AJ. In: Green AR, Cross AL, editors. Neuroprotective agents and cerebral ischemia. New York: Academic Press, 1997: 47-68.
14. Yang, Y, Li Q, Shuaib, A. Neuropharmacology 39, 881-888. 2000.
15. Yang, Y, Li ,Q, Shuaib, A. Experimental Neurology 163, 39-45. 2000.
16. Bowes, MP, Rothlein, R, Fagen, SC. Neurology 45, 815-819. 2001.
17. Shuaib, A, Breker-Klassen, MM. Neuroscience & Biobehavioral Reviews 21, 219-226. 1997.
18. Siddiqui, MM, Shuaib, A. Methods 23, 83-94. 2001.
19. Shuaib, A, Todd, KG, Baker, GB, Methods 23, 1-2. 2001.
20. Globus, MYT, Busto, R, Martinez, E, Valdés, I, Dietrich, WD, Ginsberg, MD. *Journal of Neurochemistry* 1991; 57:470-478.
21. Mitani, A, Andou, Y, Kataoka, K. *Neuroscience* 1992; 48:307-313.
22. Kanthan, R, Shuaib, A, Griebel, A, Miyashita, H. *Stroke* 1995; 26:870-873.
23. Mizui, T, Kinouchi, H, Chan, PH. *Am J Physiol Heart Circ Physiol* 1992; 262:H313-H317.
24. Menéndez, N, Solís, JM, Herreras, O, Galarreta, M, Conejero, C, Martín del Río, R. *Eur J Neurosci* 1993; 5:1273-1279.
25. Puka, M, Lehmann, A. *J Neurosci Res* 1994; 37:641-646.
26. Nishino, K, Lin, C-S, Morse, JK, Davis, JN. *Neuroscience* 1991; 43:361-367.
27. Koide, T, Wieloch, TW, Siesjo, BK. *Journal of Cerebral Blood Flow and Metabolism* 1986; 6:559-565.
28. Prehn, JHM, Backhauss, C, et al. *Eur J Pharmacol* 1991; 203:213-222.
29. Lance, JW. *Lancet* 1992; 339:1207-1209.
30. Bode-Greuel, KM, Klisch, J, Horvath, E, Glaser, T, Traber, J. *Stroke* 1990; 21(suppl 4):164-166.
31. Bielenberg, GW, Burkhardt, M. *Eur J Pharmacol* 1990; 183:1953-1954.
32. Azmitia, EC. *J Comparitive Neurology* 1981; 203:737-743.
33. Beley, A, Bertrand, N, Beley, P. *Neurochemistry Research* 1991; 16:555-561.
34. Bertrand, N, Bralet, J, Beley, A. *Neurochemistry Research* 1992; 17:321-325.
35. Kato, H, Araki, T, Hara, H, Kogure, K. *Brain Research* 1991; 553:33-38.
36. Obata, T, Hosokawa, H, Yamanaka, Y. *Comp Biochem Physiol [C]* 1993; 106C:629-634.
37. Dux, E, Schubert, P, Kreutzberg, GW. *Journal of Cerebral Blood Flow and Metabolism* 1992; 12:520-524.
38. O'Regan, MH, Simpson, RE, Perkins, LM, Phillis, JW. *Neurosci Lett* 1992; 138:169-172.
39. Lust, WD, Assaf, HM, Ricci, AJ, Ratcheson, R, Sternau, LL. *Metabolic Brain Disease* 1988; 3:287-292.
40. Sternau, LL, Lust, WD, Ricci, AJ, Ratcheson, R. *Stroke* 1989; 20:281-287.
41. Hallmayer, D, Hossman, KA, Meis, G. *Acta Neuropath* 1985; 68:27-31.
42. Shuaib, A, Hasan, S, Howlett, W, Ijaz, S. *Canadian J Neurol Sci* 1992; 19:263.
43. Shuaib, A, Ijaz, S, Hasan, S, Kalra, J. *Brain Research* 1992; 590:13-17.

44. Kamisaki, Y, Maeda, K, Ishimura, M, Omura, H, Itoh, T. *Brain Research* 1993; 627:181-185.
45. Khan, S, Benigesh, A, Todd, KG, Miyashita, H, Shuaib ANeurochem Res , 217-223. 2000.
46. Saranaari, P, Oja, SS. Amino Acids 1□, 509-526. 2002.
47. Lombardini, JB. Review: In: Lombardini JB, editor. Taurine. New York: Pleum Press, 1992: 245-251.
48. Owen, AJ, Ijaz, S., Miyashita, H, Wishart, T, Howlett, W, Shuaib, A. *Brain Res* 1997; 770(1-2):115-122.
49. Shuaib, A, Ijaz, MS, Miyashita, H, Hussain, S, Kanthan R. *Experimental Neurology* 1997; 147(2):311-315.
50. Yang, Y, Li, Q, Wang, CX, Jeerakathil, T, Shuaib, A. Neuroreport 11, 2307-2311. 2000.
51. Wang, CX, Todd, K, Yang, Y, Gordon, T, Shuaib, A. Journal of Cerebral Blood Flow and Metabolism 21, 413-421. 2000.
52. Wang, CX, Yang, ,Y Shuai,b A. Brain Research Reviews , 115-120. 2001.
53. Ginsberg, MD, Sternau, LL, Globus, MYT, Dietrich, WD, Busto,R. *Cerebrovasc Brain Metab Rev* 1992; 4:189-225.
54. Kaila, K. *Prog Neurobiol* 1994; 42:489-537.
55. Lal, S, Shuaib, A, Ijaz, S. *Neurochemistry Research* 1995; 20:115-119.
56. Hayward NJ, Hitchcott PK, McNight AT, Woodruff GN. British Journal of Pharmacology 110, 62P. 1993. Ref Type: Abstract
57. Heron, A, Lasbennes, F, Seylaz, J. *Brain Research* 1993; 608:27-32.
58. Buchan, AM, Gertler, SZ, Li, H, Xue D, Huang, Z-G, Chaundy, KE et al. *Journal of Cerebral Blood Flow and Metabolism* 1994; 14:903-910.
59. Smith, SE, Meldrum BS. *Stroke* 1992; 23:861-864.
60. Shuaib, A, Mahmood, R, Wishart, T, *Brain Research* 1995; 702:199-206.61.Graham, JL, Smith, SE, Chapman, AG, Meldrum, BS. British Journal of Pharmacology 112, 278P. 1994.
62. Jorjensen, MB, Johansen, FF, Diemer, NH. *Acta Neuropathol (Berl)* 1987; 73:189-194.
63. Onodera, H, Sato, G, Kogura, K. *Neurosci Letters* 1986; 68:169-174.
64. Globus, MY-T, Ginsburg, MD, Dietrich, WD, *Neurosci Lett* 1987; 251:256.
65. Nagai, Y, Narumi, S, Kakihana, M, *Arch Gerontol Geriatr* 1989; 8:273-289.

Studies on Taurine Efflux from the Rat Cerebral Cortex During Exposure to Hyposmotic, High K$^+$ and Ouabain-Containing aCSF

JOHN W. PHILLIS[a] and MICHAEL H. O'REGAN[b]

[a]Department of Physiology, Wayne State University School of Medicine, 540 E. Canfield, Detroit, MI 48201, USA and [b]Department of Biomedical Sciences, School of Dentistry, University of Detroit Mercy, 8200 W. Outer Drive, P.O. Box 19900, Detroit, MI 48219, USA

1.   INTRODUCTION

Cerebral ischemia leads to an accumulation in the extracellular space of several amino acids including the excitatory compounds, glutamate and aspartate, and inhibitory agents, γ-aminobutyric acid (GABA), glycine and taurine. There has been considerable interest in gaining an understanding of the mechanisms involved in ischemia-evoked release of these agents and the role that they may play in eliciting or protecting against ischemic-injuries.

Taurine is one of the most abundant free amino acids in the mammalian brain, where it appears to function both as a modulator of synaptic activity and also plays an important role in the regulation of intracellular osmolarity[1-3]. All cell types in the central nervous system appear to contain taurine[2]. Tissue concentrations of taurine are in the millimolar range within most of the brain and its synthesizing enzyme, cysteine sulfinate decarboxylate, has been localized in glial cells from the cerebellum[4]. Two Na$^+$-dependent, high affinity, transporters for taurine have been identified in brain tissue[5,6] and it is also a weak substrate for the GAT2 and GAT3 GABA transporters[7,8]. Barakat et al.[9] have demonstrated an ability of the Na$^+$-dependent taurine uptake in Bergman glia to reverse following intracellular perfusion of taurine together with a small induced depolarization. Inhibitors of taurine transport include alanine and guanidinoethane sulfonate, both of which are transported in place of taurine[2].

2.   NEUROREGULATORY ACTIONS OF TAURINE

Depressant effects of taurine on the firing of neurons in the spinal cord were initially observed in the early 1960's[10,11]. Taurine has subsequently been shown to induce hyperpolarization and inhibit the firing of neurons at many levels of the CNS[2]. Depression of neuronal firing induced by taurine is blocked by both strychnine and bicuculline in the medulla and spinal cord[12,13] suggesting that it may act as an agonist at both glycine and GABA receptor sites. On the basis of their studies with the selective taurine antagonist 6-amino-methyl-3-methyl-4H,1,2,4-benzothiadizine-1,1-dioxide   (TAG), Okamoto et al.[14] proposed taurine as an inhibitor neurotransmitter in cerebellar stellate interneurons, hyperpolarizing Purkinje cell dendrites.

In addition to this role as a neuromodulator and potential neurotransmitter, taurine appears to be important in the regulation of vasopressin release by osmoregulatory neurons in the supraoptic nucleus. Astrocytes in this nucleus accumulate high concentrations of taurine and hyposmotic stimulation of isolated nuclei elicits a rapid, reversible and highly sensitive release of taurine through volume sensitive anion channels from these cells, which is independent of extracellular Ca^{2+} and Na^+ and does not involve the taurine transporter[15-17]. Taurine released by glial cells activates inhibitory glycine receptors on magnocellular neurons in the supraoptic nucleus, thereby reducing the release of vasopressin from the axonal terminals of these neurons in the neurohypophysis[15]. By this action, taurine appears to complement the intrinsic osmosensitivity of supraoptic neurons, demonstrating a novel role for supraoptic glial cells in the neuroendocrine regulatory loop.

Inhibitory control of the neuroendocrine function of supraoptic neurons may also occur at the terminals of these neurons in the neurohypophysis. Taurine is highly concentrated in pituicytes, the glial cell equivalent of the neurohypophysis, and is released by hypotonic stimuli. Hyposmotic stimuli reduced   depolarization-evoked   vasopressin   release   from   isolated neurohypophyses, an inhibition which was prevented by strychnine[18].

3.   OSMOREGULATORY ACTIONS OF TAURINE: INDIVIDUAL CELLS

Taurine is considered to be a major contributor to volume regulation by individual neurons and glia in the central nervous system. Its release during hyposmotically-induced swelling of both neurons and glia has been demonstrated in culture preparations[19,20], brain slices[3], and *in vivo*[1,21,22]. Taurine efflux induced by hyposmotic stimulation is dependent on cell swelling and not on changes in ionic concentration; it is also sodium and temperature independent[23]. Swelling-induced taurine release appears to be independent of Ca^{2+}, being unaffected by Ca^{2+}-free media or by EGTA and BAPTA and is insensitive to verapamil and diltiazem[23]. The direction of the taurine flux can be reversed by increasing the extracellular concentration of taurine, demonstrating that it flows passively in the direction of its concentration gradient, excluding the involvement of an Na^{+}-dependent membrane taurine transporter[24].

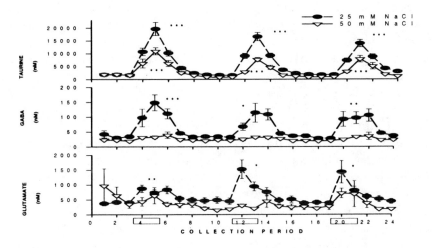

Figure 1. Time course of taurine, GABA and glutamate release from the rat cerebral cortex into continuously superperfused artificial cerebrospinal fluid upon three successive hyposmotic stimulation series. Boxed areas on the X-axis represent the three hyposmotic challenges that were conducted (collections 4 and 5; 12 and 13; 20 and 21). During all other collections, the aCSF was is osmotic. Each collection period represents 3 minutes. A 40 minute rest period (no samples were collected) was allowed before collection 9 and collection 16. aCSF made hyposmotic by reducing the NaCl concentration from 135 mM to 25 mM (filled circles; n = 13). aCSF made hyposmotic by reducing the NaCl concentration from 135 mM to 50 mM (open triangles; n = 5). Data are presented as mean ± SEM. Statistical significance determined by a two-tailed Student's t-test. Peak release was compared to the immediately preceding basal collection. *p<0.05; **p<0.01; ***p<0.001.

Hyposmotically-evoked taurine efflux is antagonized by a number of Cl⁻ channel and Cl⁻ transport inhibitors, including SITS, DIDS, NPPB, niflumic acid, polyunsaturated free fatty acids, dideoxyforskolin (DDF) and tamoxifen[22,24,25]. These results suggest that taurine efflux involves its diffusion through volume-sensitive anion channels, which are thought to be permeable to organic osmolytes including amino acids and polyalcohols. However, *in vivo* experiments have demonstrated that taurine release can be observed at levels of hypotonicity that do not elicit the efflux of other amino acids, which only appear with more pronounced reductions in osmoticity (Fig. 1)[1,22]. Furthermore in experiments with a variety of Cl[-]-channel blockers, taurine was the only amino acid to have its efflux significantly inhibited by all of the blockers[22]. The reduced sensitivity to anion channel blockers of the hyposmotically-induced release of glutamate, aspartate, glycine, GABA and phosphoethanolamine observed in this study, suggests that their release may involve other mechanisms, including depolarization. Cultured astrocytes experience a large depolarization upon hyposmotic stress[26]. Thus, whereas hyposmotic release of taurine may be primarily driven by osmotically induced swelling, that of the other amino acids may be secondary to the depolarization which accompanies the hyposmotic stimulus, with reversal of Na[+]-dependent plasma membrane transporters[27,28]. This result contrasted with the effects of anion channel blockers (SITS, NPPB, dipyridamole, tamoxifen) on ischemia-evoked release of amino acids, including taurine, which were significantly attenuated by these agents[29,30]. A likely explanation for this difference is that during a hyposmotic stress the swelling is due to the entry of water, rather than of Na[+] and Cl[-]. However, during ischemia, swelling is a consequence of the massive entry of Na[+] and Cl[-], together with osmotically attracted water. Thus, one of the reasons why anion channel blockers are neuroprotective during cerebral ischemia may be that they inhibit swelling by blocking the entry of Cl[-] into the cells[31]. Indeed, it has been demonstrated that DIDS, NPPB, and L-644-711 decrease KCl-induced astrocytic swelling[32], presumably by blocking Cl[-] entry.

Franco et al.[33] have recently provided evidence for two mechanisms of osmolyte release from hippocampal slices. ^{3}H-taurine release was inhibited by NPPB, niflumic acid and the tyrphostins AG18, AG879 and AG112, suggesting a tyrosine kinase-mediated mechanism. The release of ^{3}H-GABA and ^{3}H-D-aspartate occurred by a different mechanism, characterized by insensitivity to NPPB, niflumic acid and tyrphostins, with inhibition by chelerythrine and potentiation by phorbol 12-myristate, 13-acetate and cytochalasin E, suggesting an involvement of protein kinase C and disruption of the actin cytoskeleton.

Significant attenuation of taurine release from ischemic hippocampal slices in the presence of SITS or DIDS has also been reported by Saransaari and Oja (1998). These authors, however, expressed doubts regarding the role of taurine in brain cell volume regulation, as the reductions in osmolarity required to elicit a measurable release have generally exceeded those that would be encountered *in vivo* conditions[3].

Most the above studies of taurine efflux have involved responses to large, abrupt, decreases in external osmolarity, in contrast to the small and gradual reductions that occur in pathophysiological situations during which cells maintain their normal size over a large range of osmolarities. This phenomenon, termed isovolumetric or isovolumic regulation (IVR), has been studied by Franco et al.[35] who measured K^+ and amino acid efflux from hippocampal slices in response to gradual and continuous reductions (2.5 mOsm/min) in external osmolarity. With gradual hyposmotic changes no increase in cell water content was observed, even when the solution at the end of the experiment was 50% hyposmotic. The gradual decrease in osmolarity elicited a significant efflux of $[^3H]$-taurine and D-$[^3H]$-aspartate with a threshold of -5 mOsm. Niflumic acid significantly attenuated IVR-evoked $[^3H]$-taurine release. When hippocampal slices were loaded with $[^{86}]Rb$ and exposed to IVR conditions there was no significant increase in $[^{86}]Rb$, nor was there any decrease in the K^+ levels in the slices. These results indicate that when faced with gradual reductions in the external osmolarity, brain cells may respond by releasing organic, rather than inorganic, osmolytes, with loss of taurine being the primary amino acid contribution. Other organic osmolytes, including creatine, myo-inositol, soritol and N-acetyl-aspartate could also contribute to counteracting the loss of external osmolarity.

Nagelhus et al.[36] have argued for a role of taurine in osmoregulation at specific sites based on their observation that a decrease in plasma osmolarity triggered a shift of taurine from Purkinje cells to adjacent glia in the cerebellum, thus allowing the neurons to maintain their volume at the expense of the glia.

4. MECHANISMS UNDERLYING OSMOSENSITIVE RELEASE OF TAURINE

The observation that taurine efflux is sensitive to Cl^- channel blockers led to the proposal that its transport pathway was an anion channel-like molecule[37]. The channel is perceived as being an outwardly rectifying volume sensitive Cl^- channel (VSCC) permeable to monovalent anions as well as organic anions[38,39]. The intracellular signals involved in the regulation of these swelling-activated anion channels are still relatively poorly understood. Although hyposmotic swelling is associated with an increase in intracellular Ca^{2+} levels, Cl^- currents and osmosensitive taurine fluxes appear to be largely independent of extracellular Ca^{2+}, although a low level of intracellular Ca^{2+} appears to be necessary[40]. It has been postulated that activation of PLA_2 with the formation of arachidonic acid metabolites may regulate osmolyte efflux and the RVD response in ascites cells[41] and there is evidence that cultured astrocytes exhibit an increase in phosphoinositide hydrolysis during hyposmotic stress which can be blocked by the phospholipase C inhibitor U73122[42]. However, in an *in vivo* study[43], the non-selective PLA_2 inhibitor 4-bromophenacyl bromide and DEDA, a selective inhibitor of secretory PLA_2' failed to affect hyposmotically evoked taurine efflux from the rat cerebral cortex. $AACOCF_3$, a selective inhibitor of the cytosolic Ca^{2+}-dependent PLA_2 did significantly decrease taurine release, whereas U73122, a PLC inhibitor, was without effect. The effect of $AACOCF_3$ is consistent with a previous finding that this agent inhibited swelling-induced arachidonic acid release from CP100 neuroblastoma cells and prevented taurine and Cl^- fluxes[44].

Protein kinase C has been implicated in the release of amino acids from the rat cerebral cortex where the PKC inhibitor, chelerythrine, significantly attenuated glutamate and aspartate release during ischemia, and activation of PKC by phorbol 12-myristate, 13-acetate (PMA) enhanced glutamate release[45]. Protein kinases/phosphatases may also play a role in volume regulation[45]. In cultured astrocytes PKC activation with PMA facilitated hyposmotically induced swelling[42]. In studies on the *in vivo* rat cerebral cortex inhibition of PKC with chelerythrine significantly and selectively attenuated the hyposmotically induced release of taurine. Stimulation of PKC with phorbol 12,13 didecanoate (but not PMA) significantly enhanced taurine release as well as that of aspartate, glutamate and glycine[43]. Although these findings support the hypothesis that PKC plays a role in osmolyte efflux during hyposmotic stress[47], other investigators have failed to observe changes in osmosensitive taurine release from *in vitro* cerebellar granule cells following inhibition of PKC by H-7, H-8 or Go 6976 or its activation by PMA[48]. Chelerythrine also failed to inhibit hyposmotically-evoked taurine release from hippocampal slices[33].

An involvement of PKA in hyposmotic taurine release from the rat cerebral cortex was evaluated using the PKA stimulator, forskolin[43]. At a concentration of 100 µM (but not at 10 µM) forskolin enhanced hyposmotically induced taurine efflux suggesting a role for this kinase in swelling-activated taurine release.

A role for protein tyrosine kinases (PTK) in volume-regulated signaling is apparent from the numerous forms of this enzyme, including p125[FAK], p38, JNK, p56[lck], p76[syk] and ERK1/ERK2 which are phosphorylated during swelling[49]. Protein tyrosine phosphorylation appears to be important in the activation and modulation of osmosensitive taurine release from cerebellar granule cells as it was markedly inhibited by tryphostine A23 and potentiated by the tyrosine phosphatase inhibitor ortho-vandate[48]. The MAP kinases ERK1/ERK2 and p38 are activated during swelling and RVD but have no effect on taurine release[33,48]. The same lack of correlation was observed for the p38 kinase[50].

5. HIGH K⁺- AND OUABAIN-INDUCED SWELLING AND AMINO ACID EFFLUX

In a further study, the discrepancy between the action of DIDS on hyposmotic- (only taurine and glutamate release attenuated) and ischemia- (taurine, glutamate, aspartate, glycine, GABA, phosphoethanolamine-release attenuated) evoked amino acid release was examined. Two possible contributions to ischemia-evoked swelling and amino acid release, namely inhibition of Na^+/K^+ ATPase activity due to loss of ATP, and elevated extracellular K^+ were simulated by topical application of either ouabain or 75 mM K^+ to the rat cerebral cortex. The effect of DIDS was then compared to its effect on hyposmotically- and ischemia-evoked amino acid release[51].

Ouabain increased superfusates levels of taurine, GABA, glutamate, aspartate and phosphoethanolamine. DIDS significantly attenuated the release of all these amino acids with the exception of glutamate, in sharp contrast to the selective inhibition by DIDS of hyposmotically-evoked taurine and glutamate efflux. The discrepancy between the ability of DIDS to inhibit hyposmotic- but not ouabain-induced glutamate efflux may be because ouabain treatment would have substantially elevated extracellular K^+, initiating a reversal of the glutamate transporter. K^+ reversal of astrocyte glutamate transport has been described[52].

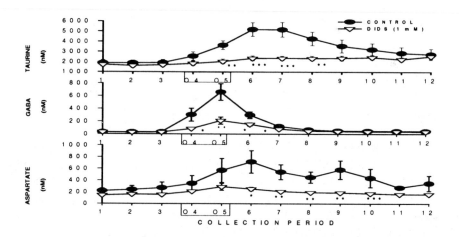

Figure 2. Time course of ouabain-induced taurine, GABA and aspartate release from the superfused rat cerebral cortex. Boxed areas on the X-axis represent the collections in which ouabain (30 μM) was present in the perfusing artificial cerebrospinal fluid. Values are expressed as mean ± SEM. Release without 4,4'-diisothiocyanatostilbene-2,2'-disulfonic acid (DIDS) (n = 12) was compared to release in the presence of DIDS (n = 16) with a two-tailed Student's t-test. *p<0.05; **p<0.01; ***p<0.001.

Extracellular levels of K$^+$ can rise up to 75 mM after 5 min of cerebral ischemia in rats[53]. High extracellular K$^+$ can then lead to depolarization and swelling via passive Donnan forces. Exposure to 75 mM K$^+$ induced a rapid exocytotic release of this amino acid, which would not have been a significant release factor for taurine. A similar lack of release of glutamate during exposure to high K$^+$ from the *in vivo* rat cerebral cortex was observed in earlier experiments[54] and may have been a consequence of continued uptake of glutamate by astrocytes[52].

The ability of DIDS to block the ischemia-evoked efflux of several amino acids, but only of taurine and glutamate release from hyposmotically swollen cells, may be a consequence of the ability of this agent to inhibit ischemia-induced swelling by blocking Cl$^-$ channels and thus water influx. K$^+$-induced swelling is inhibited in low Cl$^-$ solutions[55] and anion channel blockers, including DIDS, inhibit K$^+$-induced astrocyte swelling[32]. Thus, a reduction in swelling may represent a primary mechanism by which anion channel blockers attenuate ischemia-evoked amino acid release. Hyposmotic swelling, which requires only the influx of water into cells, would not be affected by Cl$^-$ channel blockers, but these could inhibit taurine efflux via volume-sensitive anion channels.

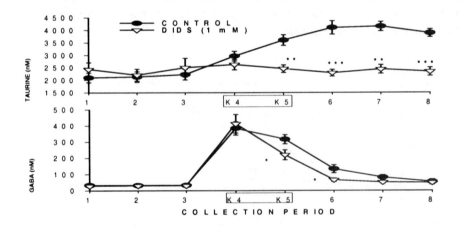

Figure 3. Time course of K$^+$-induced taurine and γ-aminobutyric acid (GABA) release from the superfused rat cerebral cortex. Boxed areas on the X-axis represent the collections in which K$^+$ (75 mM) was present in the perfusing artificial cerebrospinal fluid. Values are expressed as mean ± SEM. Release without 4,4'-diiosothiocyanatostilbene-2,2'-disulfonic acid (DIDS) (n = 20) was compared to release in the presence of DIDS (n = 20) with a two-tailed Student's t-test. *p<0.05; **p<0.01; ***p<0.001.

6.   CONCLUSION

Taurine is present at high concentrations in neurons and glial cells of the central nervous system, where it plays important roles in neuromodulation, possibly neurotransmission, and volume regulation. Swelling of brain cells as a consequence of depolarization, fluctuations in osmolarity, and during ischemia, initiates an efflux of inorganic and organic osmolytes with compensating reductions in cell volume. Taurine efflux has been demonstrated to make a significant contribution to such "regulatory volume decrease (RVD)". This is especially so when the hyposmolarity is of a gradual onset, evoking an "isovolumic regulation" with little increase in cell volume. Taurine efflux is proposed to be a consequence of the opening of outwardly rectifying, volume sensitive, Cl⁻ channels (VSCC) through which this amino acid can diffuse along its concentration gradient. The opening of such channels appears to be independent of an influx of extracellular Ca^{2+}. Agents known to block plasma membrane Cl⁻ channels inhibit both the RVD and taurine release. There is evidence that protein kinases, including protein tyrosine kinases, and also phospholipases are involved in the efflux of taurine from swollen cells. Another interesting, novel, indication of taurine's role in plasma osmolarity regulation is recent evidence that taurine release from swollen astrocytes in the supraoptic nucleus can inhibit the firing of magnocellular neurons, reducing vasopressin (antidiuretic hormone) secretion from the terminals of these neurons in the neurohypophysis.

ACKNOWLEDGEMENTS

Work from our laboratory was supported by USPHS: NINDS No. NS26912-09. The contributions to these studies of Dr. A.Y. Estevez is gratefully acknowledged.

REFERENCES

1.   Solis, J.M., Herranz, A.S., Herreras, O., Lerma, J., and Martin Del Rio, R., 1988, Does taurine act as an osmoregulatory substance in the rat brain? *Neurosci. Lett.* 91: 53-58.
2.   Huxtable, R.J., 1989, Taurine in the central nervous system and the mammalian actions of taurine. *Progr. Neurobiol.* 32: 471-533.
3.   Oja, S.S., and Saransarri, P., 1996, Taurine as osmoregulator and neuromodulator in the brain. *Metab. Brain Dis.* 11: 153-164.

4.   Almarghini, K., Remy, A., and Tappaz, M., 1991, Immunocytochemistry of the taurine biosynthesis enzyme, cyteine sulfinate decarboxylase, in the cerebellum: evidence for a glial localization. *Neuroscience* 43: 111-119.

5.   Liu, Q.R., López-Corcuera, B., Nelson, H., Mandiyan, S., and Nelson, N., 1992, Cloning and expression of a cDNA encoding the transporter of taurine and beta-alanine in mouse brain. *Proc. Natl. Acad. Sci. USA* 89: 12145-12149.

6.   Smith, K.E., Borden, L.A., Wang, C-H.D., Hartig, P.R., Branchek, T.A., and Weinshank, R.L., 1992, Cloning and expression of a high affinity taurine transporter from rat brain. *Molec. Pharmacol.* 42: 563-569.

7.   Liu, Q.R., López-Corcuera, B., Mandiyan, S., and Nelson, H., 1993, Molecular characterization of four pharmacologically distinct α-aminobutyric acid transporter in mouse brain. *J. Biol. Chem.* 268: 2106-2112.

8.   Borden, L.A., 1996, GABA transporter heterogeneity: pharmacology and cellular localization. *Neurochem. Int.* 29: 335-356.

9.   Barakat, L., Wang, D., and Bordey, A., 2002, Carrier-mediated uptake and release of taurine from Bergmann glia in rat cerebellar slices. *J. Physiol. (Lond.)* 541: 753-767.

10. Curtis, D.R., Phillis, J.W., and Watkins, J.C., 1961, Actions of amino-acids on the isolated hemisected spinal cord of the toad. *Br. J. Pharmacol.* 16: 262-283.

11. Curtis, D.R., and Watkins, J.C., 1965, The pharmacology of amino acids related to γ-aminobutyric acid. *Pharmacol. Rev.* 17: 347-391.

12. Haas, H.L., and Hösli, L., 1973, The depression of brain stem neurons by taurine and its interaction with strychnine and bicuculline. *Brain Res.* 52: 399-402.

13. Curtis, D.R., Hösli, H., and Johnston, G.A.R., 1968, A pharmacological study of the depression of spinal neurons by glycine and related amino acids. *Expl. Brain Res.* 6: 1-8.

14. Okamoto, K., Kimura, H., and Sakai, Y., 1983, Evidence for taurine as an inhibitory neurotransmitter in cerebellar stellate interneurons: selective antagonism by TAG (6-amino-methyl-3-methyl-4H,1,2,4-benzothiadizine-1,1-dioxide). *Brain Res.* 265: 163-168.

15. Hussy, N., Deleuze, C., Desarménien, M.G., and Moos, F.C., 2000, Osmotic regulation of neuronal activity: a new role for taurine and glial cells in a hypothalamic neuroendocrine structure. *Progr. Neurobiol.* 62: 113-134.

16. Bres, V., Hurbin, A., Duvoid, A., Orcel, H., Moos, F.C., Rabie, A., and Hussy, N., 2000, Pharmacological characterization of volume-sensitive, taurine permeable anion channels in rat supraoptic glial cells. *Br. J. Pharmacol.* 130: 1976-1982.

17. Voisin, D.L., and Bourque, C.W., 2002, Integration of sodium and osmosensory signals in vasopressin neurons. *TINS* 25: 199-205.

18. Hussy, N., Bres, V., Rochette, M., Duvoid, A., Alonso, G., Dayanithi, G., and Moos, F.C., 2001, Osmoregulation of vasopressin secretion via activation of neurohypophysial nerve terminals glycine receptors by glial taurine. *J. Neurosci.* 21: 7110-7116.

19. Pasantes-Morales, H., Alavez, S., Sánchez-Olea, R., and Morán, J., 1993, Contribution of organic and inorganic osmolytes to volume regulation in rat brain cells in culture. *Neurochem. Res.* 18: 445-452.

20. Kimelberg, H.K., Goderie, S.K., Higman, S., Pang, S., and Waniewski, R.A., 1990, Swelling-induced release of glutamate, aspartate, and taurine from astrocyte cultures. *J. Neurosci.* 10: 1583-1591.

21. Lehman, A., 1989, Effect of microdialysis-perfusion with aniosmotic medium on extracellular amino acids in the rat hippocampus and skeletal muscle. *J. Neurochem.* 53: 525-535.

22. Estevez, A.Y., O'Regan, M.H., Song, D., and Phillis, J.W., 1999, Effects of anion channel blockers on hyposmotically-induced amino acid release from the *in vivo* rat cerebral cortex. *Neurochem Res.* 24: 447-452.

23. Pasantes-Morales, H., and Schousboe, A., 1997, Role of taurine in osmoregulation: mechanisms and functional implications. *Amino Acids* 12: 281-292.

24. Pasantes-Morales, H., 1996, Volume regulation in brain cells: cellular and molecular mechanisms. *Met. Brain Dis.* 11: 187-204.

25. Jackson, P.S., and Strange, K., 1993, Volume-sensitive anion channels mediate swelling-activated inositol and taurine efflux. *Am. J. Physiol.* 265: C1489-C1500.

26. Kimelberg, H.K., and Kettenmann, H., 1990, Swelling-induced changes in electrophysiological properties of cultured astrocytes and oligodendrocytes. I. Effects on membrane potentials, input impedance and cell-cell coupling. *Brain Res.* 529: 255-261.

27. Phillis, J.W., Smith-Barbour, M., Perkins, L.M., and O'Regan, M.H., 1994, Characterization of glutamate, aspartate and GABA release from ischemic rat cerebral cortex. *Brain Res. Bull.* 34: 457-466.

28. Phillis, J.W., Ren, J., and O'Regan, M.H., 2000, Transporter reversal as a mechanism of glutamate release from the ischemic rat cerebral cortex: Studies with DL-threo-β-benzyloxyaspartate. *Brain Res.* 868: 105-112.

29. Phillis, J.W., Song, D., and O'Regan, M.H., 1997, Inhibition by anion channel blockers of ischemia-evoked release of excitotoxic and other amino acids from rat cerebral cortex. *Brain Res.* 758: 9-16.

30. Phillis, J.W., Song, D., and O'Regan, M.H., 1998, Tamoxifen, a chloride channel blocker, reduces glutamate and aspartate release from the ischemic cerebral cortex. *Brain Res.* 780: 352-355.

31. Kohut, J.J., Bednar, M.M., Kimelberg, H.K., McAuliffe, T.L., and Gross, C.E., 1992, Reduction in ischemic brain injury in rabbits by the anion transport inhibitor L-644,711. *Stroke* 23: 93-97.

32. Rutledge, E.M., Aschner, M., and Kimelberg, H.K., 1998. Pharmacological characterization of swelling-induced D-[H^3] aspartate release from primary astrocyte cultures. *Am. J. Physiol.* 274: C1511-C1520.

33. Franco, R., Torres-Marquez, M.E., and Pasantes-Morales, H., 2001, Evidence for two mechanisms of amino acid osmolyte release from hippocampal slices. *Pflugers Archiv. Eur. J. Physiol.* 442: 791-800.

34. Saransaari, P., and Oja, S.S., 1998, Mechanisms of ischemia-induced taurine release in mouse hippocampal slices. *Brain Res.* 807: 118-124.

35. Franco, R., Quesada, O., and Pasantes-Morales, H., 2000, Efflux of osmolyte amino acids during isovolumic regulation in hippocampal slices. *J. Neurosci. Res.* 61: 701-711.

36. Nagelhus, E.A., Lehmann, A., and Ottersen, O.P., 1993, Neuronal-glial exchange of taurine during hypo-osmotic stress: a combined immunocytochemical and biochemical analysis in rat cerebellar cortex. *Neuroscience* 54: 615-631.

37. Strange, K., and Jackson, P.S., 1995, Swelling-activated organic osmolyte efflux: a new role for anion channels. *Kidney Int.* 48: 994-1003.

38. Kirk, K., 1997, Swelling-activated organic osmolyte channels. *J. Membr. Biol.* 158: 1-16.

39. Okada, Y., 1997, Volume expansion-sensing outward-rectifier Cl-channel: fresh start to the molecular identity and volume sensor. *Am. J. Physiol.* 273: C755-C789.

40. Szucs, G., Heinke, S., Droogmans, G., and Nilius, B., 1996, Activation of volume-sensitive chloride current in vascular endothelial cells requires a permissive intracellular Ca^{2+} concentration. *Pflügers Arch. Eur.* 431: 467-469.

41. Lambert, I.H., and Hoffman, E.K., 1993, Regulation of taurine transport in Erlich ascites tumor cells. *J. Membr. Biol.* 131: 67-79.

42. Bender, A.S., Neary, J.T., Blicharska, J., Norenberg, L.-O.B., and Norenberg, M.D., 1992, Role of calmodulin and protein kinase C in astrocytic cell volume regulation. *J. Neurochem.* 58: 1874-1882.

43. Estevez, A.Y., O'Regan, M.H., Song, D., and Phillis, J.W., 1999, Hyposmotically induced amino acid release from the rat cerebral cortex: role of phospholipases and protein kinases. *Brain Res.* 844: 1-9.
44. Basavappa, S., Pedersen, S.F., Jorgensen, N.K., Ellory, J.C., and Hoffmann, E.K., 1998, Swelling-induced arachidonic acid release via the 85-kDa $cPLA_{(2)}$ in human neuroblastoma cells. *J. Neurophysiol.* 79: 1441-1449.
45. Phillis, J.W., and O'Regan, M.H., 1996, Mechanisms of glutamate and aspartate release in the ischemic rat cerebral cortex. *Brain Res.* 730: 150-164.
46. Grinstein, S., Furuya, W., and Bianchini, L., 1992, Protein kinases, phosphatases, and the control of cell volume. *News Physiol. Sci.* 7: 232-236.
47. Strange, K., Morrison, R., Shrode, L., and Putnam, R., 1993, Mechanism and regulation of swelling-activated inositol efflux in brain glial cells. *Am. J. Physiol.* 265: C244-C256.
48. Morales-Mulia, S., Cardina, V., Torres-Marquez, M.E., Crevenna, A., and Pasantes-Morales, H., 2001, Influence of protein kinases on the osmosensitive release of taurine from cerebellar granule cells. *Neurochem. Int.* 38: 153-161.
49. Pasantes-Morales, H., Franco, R., Torres-Marquez, M.E., Hernandez-Fonseca, K., and Ortega, A., 2000, Amino acid osmolytes in regulatory volume decrease and isovolumetric regulation in brain cells: contribution and mechanisms. *Cell. Physiol. Biochem.* 10: 361-370.
50. Tilly, B.C., Gaestel, M., Engel, K., Edixhoven, M.J., and de Jonge, H.R., 1996, Hypo-osmotic cell swelling activates the p38 MAP kinase signaling cascade. *FEBS Lett.* 395: 133-136.
51. Estevez, A.Y., Song, D., Phillis, J.W., and O'Regan, M.H., 2000, Effect of the anion channel blocker DIDS on ouabain- and high K^+-induced release of amino acids from rat cerebral cortex. *Brain Res. Bull.* 52: 45-50.
52. Longuemare, M.C., Rose, C.R., Farrell, K., Ransom, B.R., Waxman, S.G., and Swanson, R.A., 1999, K^+-induced reversal of astrocyte glutamate uptake is limited by compensatory changes in intracellular Na^+. *Neuroscience* 93: 285-292.
53. Hansen, A.J., and Zeuthen, T., 1981, Extracellular ion concentrations during spreading depression and ischemia in the rat cerebral cortex. *Acta Physiol. Scand.* 113: 437-489.
54. Phillis, J.W., Perkins, L.M., and O'Regan, M.H., 1993, Potassium-evoked efflux of transmitter amino acids and purines from rat cerebral cortex. *Brain Res. Bull.* 31: 347-552.
55. Schousboe, A., Morán, J., Pasantes-Morales, H., 1990, Potassium-stimulated release of taurine from cultured cerebellar granule neurons is associated with cell swelling. *J. Neurosci. Res.* 27: 71-77.

Interactions of Taurine and Adenosine in the Mouse Hippocampus in Normoxia and Ischemia

PIRJO SARANSAARI[*] and SIMO S. OJA[*,#]
[*]Tampere Brain Research Center, Medical School, FIN-33014 University of Tampere and
[#]Department of Clinical Physiology, Tampere University Hospital, Box 2000, FIN-33521
Tampere, Finland

1.   INTRODUCTION

Adenosine acts as an inhibitory neuromodulator in the brain through three types of receptors, namely A_1, A_2, (two subtypes A_{2A} and A_{2B}), and A_3[1]. By means of presynaptic A_1 receptors adenosine inhibits neurotransmitter release mainly in excitatory synapses, reducing thus neuronal excitability[2]. Postsynaptically, adenosine hyperpolarizes the postsynaptic membrane through A_1 receptors by increasing Ca^{2+} -dependent K^+ conductance[1]. Depolarization with K^+, electrical stimulation and various agents, including ionotropic glutamate receptor agonists, induce adenosine release[2]. Furthermore, adenosine is released during hypoxia and ischemia[3], acting then as an endogenous neuroprotectant against cerebral ischemia and excitotoxic neuronal damage[4]. Similar neuroprotective effects have also been assigned to taurine, a putative inhibitory neuromodulator[5], which is also released from nervous tissue under various cell-damaging conditions[6]. Being particularly enriched in the brain of immature animals, taurine has been thought to have a special function, being essential for the development and survival of neural cells[7,8]. We now characterized interactions in the release of taurine and adenosine from hippocampal slices from developing (7-day-old) and adult (3-month-old) mice in normal conditions and in ischemia, using a superfusion system[7].

Taurine 5: Beginning the 21st Century
Edited by Lombardini *et al.*, Kluwer Academic/Plenum Publishers, New York 2003   445

2. MODULATION OF TAURINE RELEASE BY ADENOSINE IN NORMOXIA

The basal release of [³H]taurine from hippocampal slices from 7-day-old mice under normoxic conditions was significantly enhanced in the presence of the adenosine receptor A_1 agonists N^6-cyclohexyladenosine (CHA) (1 µM) and R(-)N^6-(2-phenylisopropyl)adenosine (R-PIA) (0.1 µM). The release increased with superfusion time, as shown with R-PIA as an example (Fig. 1A).

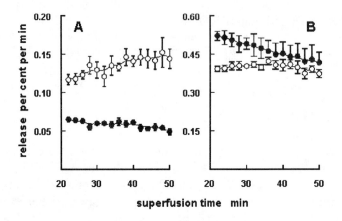

Figure 1. Effects of adenosine receptor agonist R-PIA (0.1 µM) (-O-) on basal taurine release (-●-) in hippocampal slices from 7-day-old (A) and 3-month-old (B) mice. The slices were superfused with the drugs from the beginning of superfusions. The results are mean values ± SEM of 4 independent experiments.

The effect of R-PIA was abolished by the A_1 receptor antagonist 8-cyclopentyl-1,3-dipropylxanthine (DPCPX) (1.0 µM) (Fig. 2). On the other hand, both CHA and R-PIA reduced the initial release (k_1, 20-30 min) of taurine in adults (Fig. 1B), this effect being antagonized by DPCPX (Fig. 2). The A_2 receptor agonist 2-p-(2-carboxyethyl)phenylamino-5'-N-ethyl-carboxamino adenosinehydrochloride (CGS 21680) (10 µM) also reduced the initial taurine release in adults (Fig. 2). The A_2 receptor antagonist 3,7-dimethyl-1-propargylxanthine (DMPX) (10 µM) had no marked effect (Fig. 2). The adenosine transport inhibitor dipyridamole (10 µM) slightly reduced the basal taurine release in adults and enhanced it in developing mice (Fig. 2). Adenosine receptors exist throughout the brain and both types A_1 and A_2 are expressed in the hippocampus[9], already during early brain development[10]. Both A_1[11] and A_2 receptors[12] modulate neuronal excitability in the CA1 area of the hippocampus.

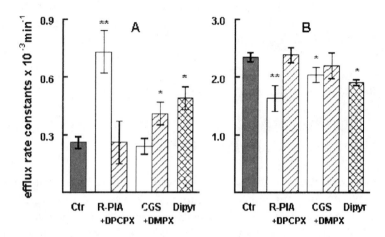

Figure 2. Effects of adenosine compounds on basal taurine release (ctr) from hippocampal slices from 7-day-old (A) and 3-month-old (B) mice. The results are efflux rate constants k_2 (A) and k_1 (B) ± SEM, calculated for the superfusion intervals of 20-30 min and 34-50 min, respectively. Number of independent experiments 4-8. Significance of differences from corresponding controls: *$p < 0.05$, **$p < 0.01$.

While presynaptic A_1 receptors are mainly involved in the attenuation of transmitter release[13], the A_{2A} receptors mediate the psychomotor effects of adenosine agonists and antagonists and interact with the functions of other transmitters and modulators[14]. Accordingly, the A_{2A} receptors have been shown both to reduce and to enhance transmitter release[15]. The adenosine receptors could thus modulate amino acid release in a positive or negative manner, depending on the receptor type activated. Both excitatory and inhibitory effects of adenosine receptor agonists and antagonists were also now discernible in hippocampal taurine release. Moreover, the effects were not alike in the mature and immature brain. The actions of adenosinergic compounds on taurine release have previously been studied only in vivo, using microdialysis and a few selected adenosinergic compounds. The present depression of basal taurine release in the adult hippocampus by the selective agonists CHA and R-PIA, however, indicates the involvement of adenosine A_1 receptors in the regulation of release. This inference is corroborated by the antagonism of DPCPX, an antagonist of A_1 receptors. The same type of adenosine receptors seem also to be operative in the immature hippocampus, though their effects were now opposite, A_1 receptor agonists enhancing and the antagonist DPCPX reducing the release.

The significance of adenosine A_{2A} receptors for hippocampal taurine release seems to be minor, since only the basal release was slightly reduced in adults by the specific A_{2A} agonist CGS 21680, which effect was not altered by the antagonist DMPX. The K[+]-stimulated release of taurine is not modified by adenosinergic compounds[16], indicating that in normoxia only the basal release of taurine in the hippocampus is subjected to adenosine receptor activation.

3.   MODULATION OF TAURINE RELEASE BY ADENOSINE IN ISCHEMIA

Under ischemic conditions the basal taurine release is markedly enhanced in both mature and immature hippocampus[6]. Of the adenosine receptor A_1 compounds, only 8-phenyltheophylline (8-PT) was effective in 7-day-old mice, inhibiting the basal ischemia-induced release. In the adult hippocampus this release was potentiated by CHA and 8-PT. The effect of CHA was antagonized by DPCPX[16]. In addition, when the slices were subjected to the drugs from the beginning of superfusion, the K[+]-stimulated release from 34 to 50 min was enhanced in the presence of CHA, 8-PT and DMPX (Fig. 3). The enhancement by CHA was again blocked by DPCPX. CGS 21680 does not affect taurine release in ischemia[16], but the transport inhibitor dipyridamole stimulated the K[+]-stimulated release in adults (Fig. 3).

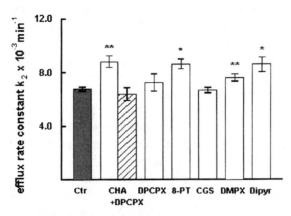

Figure 3. Effects of adenosine compounds on K[+]-stimulated taurine release (ctr) from hippocampal slices from 3-month-old mice. The results are means ± SEM of 4-8 independent experiments. Significance of differences from the control: *p<0.05, **p<0.01.

The markedly enhanced release of taurine in ischemia in both immature and mature hippocampus is partly mediated by Na^+-dependent transporters and partly by diffusion through anion channels[17]. In the developing hippocampus the adenosinergic compounds had no marked effects on the basal or K^+-stimulated taurine release. This may signify that the potentiation induced by ischemia was already maximal. On the other hand, in adults adenosine A_1 receptor activation was still able to potentiate taurine release in a receptor-mediated manner. Adenosine A_1 receptor agonists have been shown to protect neuronal cells against ischemic damage in vivo by depressing both the basal and the K^+-evoked release of excitatory amino acids[18]. The elevation of taurine levels together with the depression of excitatory amino acid release caused by adenosine A_1 receptor activation could be benefical under ischemic conditions, providing protection against excitotoxicity in the adult hippocampus. Such an effect could also be fortified by means of taurine release enhanced by the A_2 receptor antagonist DMPX.

4. MODULATION OF ADENOSINE RELEASE BY TAURINE

When 0.1, 1.0 or 10.0 mM taurine was added to the superfusion medium at 30 min, there was no change in unstimulated adenosine release at either age, but the K^+-stimulated release was significantly ($p<0.01$) reduced in the immature hippocampus, the efflux rate constant k_2 (32-40 min) decreasing from $1.14 \pm 0.08 \times 10^{-3}$ min^{-1} (mean $\pm$ SEM, n=7) to $0.79 \pm 0.04 \times 10^{-3}$ min^{-1} (mean $\pm$ SEM, n=5). When 0.1-10 mM taurine was present from the beginning of superfusion the basal and K^+-stimulated releases were unaltered in the adult hippocampus (data not shown), whereas in 7-day-old mice both releases were reduced (Fig. 4A), the effect being concentration-dependent[19]. This suppression of K^+-stimulated adenosine release by taurine in the immature hippocampus is a controversial finding. The present experiments cannot decisively elucidate the mechanism of this inhibition. However, under normal incubation conditions a part of taurine molecules penetrates during K^+ depolarization through plasma membranes by means of anion channels in the immature hippocampus[20], thereby hindering adenosine passage through this route. In the immature hippocampus. taurine release is mainly potentiated by A_1 adenosine receptors. Moreover, the NMDA-stimulated taurine release is also modulated by adenosine receptors only in the developing hippocampus, the enhancing effect being again mediated by the A_1 receptors[21].

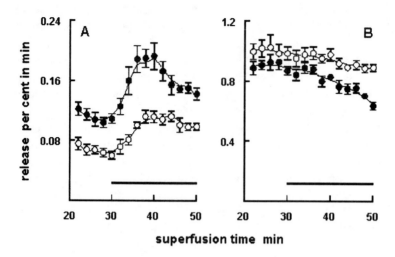

Figure 4. K$^+$-stimulated release of [^{3}H]adenosine (-●-) from hippocampal slices of 7-day-old mice in the presence of 10 mM (-O-) taurine in normal conditions (A) and in ischemia (B). Taurine was added at the beginning of superfusion and K$^+$ (50 mM) at 30 min, as indicated by the bar. The results are mean values ± SEM of 4-8 independent experiments.

The inhibitory effect of taurine on adenosine release could thus constitute some kind of a feedback control in the immature hippocampus, regulating the balance between inhibitory and excitatory currents. On the contrary, the ischemia-induced adenosine release is potentiated by 10.0 mM taurine in the developing hippocampus (Fig. 4B), this enhancement being more pronounced in adults (Saransaari and Oja, unpublished results). The simultaneous increase of both inhibitory agents, taurine and adenosine, in the extracellular space could be beneficial to nervous tissue, inhibiting neuronal excitability and thus exerting neuroprotective effects.

ACKNOWLEDGEMENTS

The skillful technical assistance of Mrs Irma Rantamaa, Mrs Oili Pääkkönen and Mrs Sari Luokkala and the financial support of the Medical Research Fund of Tampere University Hospital are gratefully acknowledged.

REFERENCES

1.   Cunha, R.A., 2001, Adenosine as a neuromodulator and as a homeostatic regulator in the nervous system: different roles, different sources and different receptors. *Neurochem. Int.* 38:107-125.

2. Latini, S., and Pedata, F., 2001, Adenosine in the central nervous system: release mechanisms and extracellular concentrations. *J. Neurochem.* 79:463-484.
3. Rudolphi, K.A., Schubert, P., Parkinson, F.E., and Fredholm, B.B., 1992, Adenosine and brain ischemia. *Cerebrovasc. Brain Metab. Rev.* 4:346-369.
4. Deckert, J., and Gleiter, C.H., 1994, Adenosine – an endogenous neuroprotective metabolite and neuromodulator. *J. Neural Transm.* 43:Suppl., 23-31.
5. Oja, S.S., and Saransaari, P., 1996, Taurine as osmoregulator and neuromodulator in the brain. *Metab. Brain Res.* 11, 153-164.
6. Saransaari, P., and Oja, S.S., 2000, Taurine and neural cell damage. *Amino Acids* 19:509-526.
7. Kontro, P., and Oja, S.S., 1987, Taurine and GABA release from mouse cerebral cortex slices: potassium stimulation releases more taurine than GABA from developing brain. *Dev. Brain Res.* 37:277-291.
8. Sturman, J.A., 1993, Taurine in development. *Physiol. Rev.* 73:119-147.
9. Cunha, R.A., Johansson, B., Van der Ploeg, I., Sebastiao, A.M., Ribeiro, J.A., and Fredholm, B.B., 1994, Evidence for functionally important adenosine A_{2A} receptors in the rat hippocampus. *Brain Res.* 649, 208-216.
10. Psarropoulou, C., Kostopoulos, G., and Haas, H.L., 1990, An electrophysiological study of the ontogenesis of adenosine receptors in the CA1 area of rat hippocampus. *Dev. Brain Res.* 55:147-150.
11. Sebastiao, A.M., Stone, T.W., and Ribeiro, J.A., 1990, The inhibitory adenosine receptor at the neuromuscular junction and hippocampus of the rat: antagonism by 1,3,8-substituted xanthines. *Br. J. Pharmac.* 101:453-459.
12. Sebastiao, A.M., and Ribeiro, J.A., 1992, Evidence for the presence of excitatory A_2 adenosine receptors in the rat hippocampus. *Neurosci. Lett.* 138:41-44.
13. Fredholm, B.B., and Dunwiddie, T.V., 1988, How does adenosine inhibit transmitter release? *Trends Pharmac. Sci.* 9:130-134.
14. Ribeiro, J.A., 1999, Adenosine A_{2A} receptor interactions with receptors for other neurotransmitters and neuromodulators. *Eur. J. Pharmac.* 375:101-113.
15. Corsi, C., Melani, A., Bianchi, L., Pepeu, G., and Pedata, F., 1999, Striatal A_{2A} adenosine receptors differentially regulate spontaneous and K^+-evoked glutamate release in vivo in young and aged rats. *Neuroreport* 10:687-691.
16. Saransaari, P., and Oja, S.S., 2000, Modulation of the ischemia-induced taurine release by adenosine receptors in the developing and adult mouse hippocampus. *Neuroscience* 97:425-430.
17. Saransaari, P., and Oja, S.S., 1998, Mechanisms of ischemia-induced taurine release in mouse hippocampal slices. *Brain Res.* 807:118-124.
18. Goda, H., Ooboshi, H., Nakane, H., Ibyashi, S., Sadoshima, S., and Fujishima, M., 1998, Modulation of ischemia-evoked release of excitatory and inhibitory amino acids by adenosine A_1 receptor agonist. *Eur. J. Pharmac.* 357:149-155.
19. Saransaari, P., and Oja, S.S., 2002, Mechanisms of adenosine release in the developing and adult mouse hippocampus. *Neurochem Res.* 27:911-918.
20. Saransaari, P., and Oja, S.S., 1999, Characteristics of ischemia-induced taurine release in the developing mouse hippocampus. *Neuroscience* 94:949-954.
21. Saransaari, P., and Oja, S.S., 2002, Characterization of N-methyl-D-aspartate-evoked taurine release in the developing and adult mouse hippocampus. *Amino Acids* DOI 10.1007/s00726-002-0310-z.

Involvement of Nitric Oxide in Ischemia-Evoked Taurine Release in the Mouse Hippocampus

PIRJO SARANSAARI[*] and SIMO S. OJA[*,#]
[*]Tampere Brain Research Center, Medical School, FIN-33014 University of Tampere and
[#]Department of Clinical Physiology, Tampere University Hospital, Box 2000, FIN-33521
Tampere, Finland

1.   INTRODUCTION

The novel type of neurotransmitter/neuromodulator nitric oxide (NO) is linked to the activation of the N-methyl-D-aspartate (NMDA) class of glutamate receptors and has been shown to modify transmitter release in the brain[1]. The synaptically released glutamate activates postsynaptic NMDA receptors and depolarizes the cell, allowing Ca^{2+} to enter through the receptor-linked ion channels, activating the Ca^{2+}-dependent enzyme, NO synthase (NOS)[2]. The NO formed may either act intercellularly or diffuse out of the cell and act extracellularly at the soluble guanylyl cyclase, enhancing the formation of 3',5'-cyclic guanosine monophosphate (cGMP)[3]. NO affects neurotransmitter transport in different brain areas, e.g., by evoking the release of monoamines[4], GABA[5] and glutamate[6]. On the other hand, inhibitory effects have been observed in dopamine[7], GABA[8] and glutamate release[9], indicating that the actions of NO are apparently complex.

The inhibitory amino acid taurine has been thought to function as a regulator of neuronal activity, particularly in the immature brain[10,11]. The levels of taurine in vivo are increased after forebrain ischemia in the hippocampus[12]. Taurine release in vitro is also markedly potentiated under various cell-damaging conditions, including ischemia, in both developing and adult mice[13-17]. Moreover we have recently demonstrated that NO-generating agents are able to modulate taurine release in both the immature and the mature hippocampus, suggesting an involvement of taurine in NO-mediated processes in this brain region under normal conditions[18]. We now characterized the effects of different NO-generating compounds hydroxylamine (HA), sodium nitroprusside (SNP) and S-nitroso-N-acetylpenicillamine (SNAP) on [³H]taurine release in mouse hippocampal slices from developing and adult mice under ischemic conditions. Ischemic conditions in vitro were achieved by superfusing slices in a nitrogen atmosphere without glucose[13].

2.   EFFECTS OF NO-GENERATING COMPOUNDS

The ischemia-induced release of taurine in the immature hippocampus was markedly increased when the NO-generating compounds SNP (1.0 mM), SNAP (1.0 mM) and HA (5.0 mM) were applied at the beginning of superfusions, the effects being fairly similar in magnitude (Table 1). In the adults only SNP and SNAP were effective. The 1.0 mM SNP-induced release at both ages studied was reduced by 0.1 mM nitroarginine (L-NNA), an inhibitor of NOS, and by 0.01 mM 1H-(1,2,4)oxadiazolo(4,3-a)quinoxalin-1-one (ODQ), the inhibitor of NO-activated guanylyl cyclase (Fig. 1A, B, Table 1).

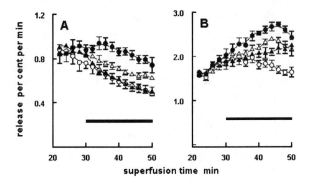

Figure 1. Effects of sodium nitroprusside (SNP) on the release of taurine from hippocampal slices from 7-day-old (A) and 3-month-old (B) mice under ischemic conditions. The slices were preloaded with 10 μM [³H]taurine for 30 min and then superfused under ischemic conditions (-O-, control) and with 1.0 mM SNP (-●-). Nitroarginine (L-NNA, 0.1 mM) (-▲-) and 1H-[1,2,4]oxadiazolo[4,3-a]quinoxalin-1-one (ODQ, 10 μM) (-△-) were added to medium at 30 min as indicated by the bar. Mean values ± SEM of 4-8 independent experiments are shown.

The SNAP-enhanced release was also reduced by L-NNA and ODQ (Table 1). HA potentiated the ischemia-induced taurine release only in developing mice. This effect was not altered by L-NNA or ODQ and 0.1 mM NMDA enhanced the release (Table 1).

Table 1. Effects of NO-generating compounds on taurine release in mouse hippocampal slices in ischemia

	Efflux rate constants k_2 x 10^{-3} min^{-1}			
Concentration mM	7-day-old		3-month-old	
Ischemia (basal)	2.49 ± 0.15	(7)	7.56 ± 0.42	(10)
SNP 1.0 (control)	3.66 ± 0.19	(8)	11.06 ± 0.25	(4)
+ L-NNA 1.0	2.54 ± 0.20**	(8)	9.15 ± 0.45*	(7)
+ ODQ 0.01	2.95 ± 0.05*	(4)	9.69 ± 0.25*	(4)
+ NMDA 0.1	3.72 ± 0.56	(4)	11.31 ± 0.74	(4)
SNAP 1.0 (control)	3.71 ± 0.19	(8)	11.71 ± 0.50	(7)
+ L-NNA 1.0	2.46 ± 0.07*	(4)	9.68 ± 0.56*	(8)
+ ODQ 0.01	2.86 ± 0.09*	(4)	9.59 ± 0.44*	(4)
+ NMDA 0.1	3.38 ± 0.11	(4)	11.60 ± 0.74	(4)
HA 5.0 (control)	3.25 ± 0.06	(4)	7.52 ± 0.34	(9)
+ L-NNA 1.0	3.29 ± 0.15	(8)	7.50 ± 0.60	(4)
+ ODQ 0.01	3.38 ± 0.18	(8)	8.38 ± 0.53	(4)
+ NMDA 0.1	3.71 ± 0.10**	(4)	8.14 ± 0.32	(4)

Ischemia (N_2, no glucose) and the NO-generating compounds were applied at the beginning of superfusion and the drugs at 30 min. The results are the efflux rate constants k_2 x10^{-3} min^{-1} (± SEM) for the time interval of 34-50 min with the number of independent experiments in parenthesis. The effects of SNP, SNAP and HA were significantly ($p<0.01$) enhanced in comparison to the basal release except for HA in adults. Significance of differences from the controls: *$p<0.05$, **$p<0.01$. Abbreviations: SNP, sodium nitroprusside; SNAP, S-nitroso-N-acetylpenicillamine; HA, hydroxylamine; L-NNA, nitroarginine; ODQ, 1H-(1,2,4)oxadiazolo(4,3-a)quinoxalin-1-one; NMDA, N-methyl-D-aspartate.

The retrograde messenger NO is apparently involved in ischemia-induced taurine release, since the NO-generating agents were able to enhance taurine release in ischemia at both ages studied, in a manner similar to that in under normal conditions[18]. Results at different effector concentrations have differed only quantitatively[18], but the effects of NO agents depend markedly on the duration and timing of the exposure and are not identical with different NO donors. They produce NO by different mechanisms. SNP and SNAP form NO extracellularly by spontaneous dissociation[1]. HA, in turn, can be regarded as an intracellular generator of NO, since it penetrates easily into cells and is broken down by a catalase-dependent reaction[19]. Its effects might be expected to mimic an increase in intracellular NO production. The present effects with exogenous NO donors in hippocampal slices may hence stem from alterations in the redox state of the NMDA receptor-channel complex brought about by extracellularly produced NO[20] and/or from actions of intracellular NO on the soluble guanylyl cyclase. Moreover, SNP may also have other effects owing to its Fe^{2+} and cyanide moieties[20].

The marked potentiating effects of SNP and SNAP on taurine release were antagonized by the NOS inhibitor L-NNA. Another NOS inhibitor, 7-nitroindazole (7-NINA), has also been effective in both immature and mature hippocampus[21]. Inhibition of NO production is thus able to reduce ischemic taurine release. Cerebral ischemia has profound effects on the biosynthetic pathway of NO, affecting both the synthesis of NO and the expression of genes encoding NOS, the increasing NO production being then accompanied by the upregulation of both NOS activity and the NOS gene[22]. The SNP- and SNAP-induced taurine releases were further reduced by the selective inhibitor of soluble guanylyl cyclase, ODQ, which drug has been shown to prevent potently and selectively the elevation of extracellular cGMP induced by SNP, SNAP or NMDA in the rat hippocampus and cerebellum, without affecting NOS activity[23]. In ischemia, the extracellular levels of cGMP have been found to be increased[24], showing that the NO/cGMP pathway also operates in this situation. Ischemia is thus able to evoke taurine release by a mechanism probably involving the soluble guanylyl cyclase in both developing and adult hippocampus.

3. EFFECTS OF cGMP

When the hippocampal slices were superfused with 10 μM ODQ the release was potentiated at both ages, as shown in Fig. 2A for developing mice. Moreover, the cGMP-phosphodiesterase inhibitor 2-(2-propyloxyphenyl)-8-azapurin-6-one (zaprinast) (0.1 mM) reduced the ischemia-induced release in both developing (Fig. 2A) and adult mice (data not shown). [³H]Taurine release was enhanced when slices from 7-day-old mice were superfused under ischemic conditions with the cGMP analog, 8-Br-cGMP (0.1 mM), from the beginning onwards (Fig. 2B). The analog had no effect on taurine release in adult mice (data not shown). The release stimulated by 8-Br-cGMP was unaffected by ODQ (10 μM) (Fig. 2B). The K^+-stimulated (50 mM) release was unaffected when the slices were superfused with 0.1 mM 8-Br-cGMP in ischemia (data not shown).

The phospodiesterase inhibitor zaprinast has been shown to increase cGMP levels in several brain preparations, including hippocampal slices[25-27]. Superfusing the slices with zaprinast or the membrane-permeable analog of cGMP, 8-Br-cGMP, has potentiated taurine release under normal conditions[28]. Increased cGMP levels thus induce taurine release, particularly in the immature hippocampus in normoxia, whereas the results in ischemia are the opposite.

Zaprinast and NO are known to inhibit synaptic transmission in area CA_1 of the rat hippocampus[25] and simultaneously to reduce glutamate release[26,27]. ODQ is known to reduce cGMP production targeted by NO, without interfering with any steps leading to NO formation[23].

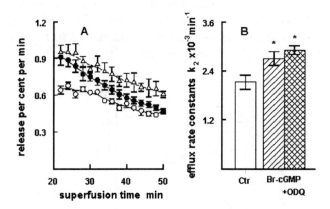

Figure 2. (A) Effects of 0.1 mM zaprinast (-O-) and 10 µM ODQ (-△-) on basal taurine release (-●-) from hippocampal slices from 7-day-old mice in ischemia. The drugs were applied at the beginning of superfusion. Results are mean values (±SEM) of 4-12 independent experiments. (B) Effect of 0.1 mM 8-Br-cGMP on taurine release from hippocampal slices from 7-day-old mice in ischemia. Control experiments (Ctr) and the effect of 10 µM 1H-[1,2,4]oxadiazolo[4,3-a]quinoxalin-1-one (ODQ) on the 8-Br-cGMP-evoked release. Results are the efflux rate constants k_2 (±SEM) for the time interval of 34-50 min. Number of independent experiments 4-8. Statistical significance of the effect of 8-Br-cGMP-: *p<0.05.

However, superfusion of slices in normoxia with 10 µM ODQ has failed to affect taurine release in 7-day-old mice and has enhanced the release in adults[28], as now in ischemia at both ages. This would appear to be at variance with the results obtained with zaprinast. One reason here might be that ODQ prevents the elevation of extracellular cGMP induced by NO generators[29] and we did not use any NO donors in these experiments. On the other hand, ODQ itself or a decrease in NO production could enhance hippocampal taurine release. This occurred now in ischemia.

4.   NO-INDUCED RELEASE AND NMDA

The ischemia-induced taurine release was significantly (p<0.05) potentiated by 0.1 mM NMDA applied at the beginning of superfusion. This stimulated release was reduced by SNP, SNAP (both 1.0 mM) and L-NNA (0.1 mM) in developing mice, while HA, SNP and L-NNA were inhibitory in adults (Fig. 3). The ionotropic glutamate receptor agonists, particularly NMDA, potentiate taurine release from hippocampal slices from developing and adult mice in a receptor-mediated manner in normoxia[30] and ischemia[31]. NMDA-evoked taurine release from both developing and adult mice has been found to be modified by the NO donors in normoxia[15]. The activation of NMDA receptors is known to induce NO synthesis in the hippocampus[32], whereas the endogenous production of NO is responsible for a sustained blockade of NMDA receptor activity[33].

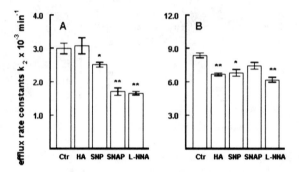

Figure 3. Effects of N-methyl-D-aspartate (NMDA) on the ischemia-induced taurine release from hippocampal slices from 7-day-old (A) and 3-month-old (B) mice. The slices were superfused under ischemic conditions with 0.1 mM NMDA (Ctr) from the beginning of superfusion. Five mM hydroxylamine (HA), 1.0 mM sodium nitroprusside (SNP), 1.0 mM S-nitroso-N-acetylpenicillamine (SNAP), and 0.1 mM nitroarginine (L-NNA) were added at 30 min. The graph shows the efflux rate constants ± SEM (k_2 x 10^{-3} min^{-1}) for the time interval of 34-50 min. Number of independent experiments 4-8. Significance of differences from the control: *p< 0.05, **p<0.01.

The NO-generating agents, particularly SNP, diminish NMDA-mediated changes in intracellular Ca^{2+} in neurons, independently of the redox state of NMDA receptors and cGMP[34], and block glutamate accumulation[35]. The NMDA-evoked release of taurine is markedly reduced by the NO donors in normoxia, especially by SNP in the adult hippocampus[15]. Furthermore, it has recently been confirmed by in vivo microdialysis of the rat cerebral cortex that the NMDA-induced taurine release is mediated via the NO cascade[36].

In ischemia, both NO-producing and -inhibiting compounds again reduced the NMDA-activated release, the results being essentially similar to those in normoxia[15]. The inhibition of NO synthesis by L-NNA and 7-NINA thus also inhibits the NMDA-evoked taurine release in ischemia. The location and timing of NO synthesis are apparently important for the effects of NO to manifest themselves, as suggested by Izumi and coworkers[37]. It appears that the overall degree of neuronal excitability may depend on the concentration of NO and subsequent increases or decreases in the release of inhibitory (taurine and GABA) and excitatory amino acids.

5.   CONCLUSIONS

Ischemia-induced taurine release is modulated by both ionotropic[31] and metabotropic[38] glutamate receptors in both the immature and the mature hippocampus. This release is also partly mediated by Na^+-requiring transporters and anion channels[14,16]. Furthermore, the NO-generating compounds in association with cGMP are able to modulate hippocampal taurine release in both adult and developing mice, suggesting the involvement of the NO/cGMP pathway. This could be a part of the neuroprotective properties of taurine, being important particularly under cell-damaging conditions. An increase in cGMP levels also depresses glutamate release[26,27] and together with a simultaneous increase in taurine release both mechanisms could aid in preventing excitotoxicity.

ACKNOWLEDGEMENTS

The skillful technical assistance of Mrs Oili Pääkkönen, Mrs Irma Rantamaa and Mrs Sari Luokkala and the financial support of the Medical Research Fund of Tampere University Hospital and the Academy of Finland are gratefully acknowledged.

REFERENCES

1.   Schuman, E.R., and Madison, D.V., 1994, Nitric oxide and synaptic function. *Annu. Rev. Neurosci.* 17:153-183.
2.   Bredt, S.D., Ferris, C.D., and Snyder, S.H., 1992, Nitric oxide synthase regulatory sites. Phosphorylation by cyclic AMP-dependent protein kinase, protein kinase C, and calcium/calmodulin protein kinase; identification of flavin and calmodulin binding sites. *J. Biol. Chem.* 267:10976-10981.
3.   Knowles, R.G., Palacios, M., Palmer, R.M.J., and Moncada, S., 1989, Formation of nitric oxide from L-arginine in the central nervous system: transduction mechanism for stimulation of the soluble guanylate cyclase. *Proc. Natl. Acad. Sci. USA* 86:5159-5162.
4.   Kaehler, S.T., Singewald, N., Sinner, C., and Philippu, A., 1999, Nitric oxide modulates the release of serotonin in the rat hypothalamus. *Brain Res.* 835:346-349.
5.   Ohkuma, S., Narihara, H., Katsura, M., Hasegawa, T., Kuriyama, K., 1995, Nitric oxide induced [^{3}H]GABA release from cerebral cortical neurons is mediated by peroxynitrite. *J. Neurochem.* 65:1109-1114.

6.   Segieth, J., Getting, S.J., Biggs, C.S., and Whitton, P.S., 1995, Nitric oxide regulates excitatory amino acid release in a biphasic manner in freely moving rats. *Neurosci. Lett.* 200:101-104.

7.   Guevara-Guzman, R., Emson, P.C., and Kendrick, K.M., 1994, Modulation of in vivo striatal transmitter release by nitric oxide and cyclic GMP. *J. Neurochem.* 62:807-810.

8.   Getting, S.J., Segieth, J., Ahmad, S., Biggs, C.S., and Whitton, P.S., 1996, Biphasic modulation of GABA release by nitric oxide in the hippocampus of freely moving rats in vivo. *Brain Res.* 717:196-199.

9.   Kamisaki, Y., Maeda, K., Ishimura, M., Omura, H., Moriwaki, Y., and Itoh, T., 1994, NO enhancement by nitric oxide of glutamate release from P2 and P3 synaptosomes of rat hippocampus. *Brain Res.* 644:128-134.

10.   Kontro, P., and Oja, S.S., 1987, Taurine and GABA release from mouse cerebral cortex slices: potassium stimulation releases more taurine than GABA from developing brain. *Devl. Brain Res.* 37:277-291.

11.   Sturman, J. A., 1993, Taurine in development, *Physiol. Rev.* 73:119-147.

12.   Ooboshi, H., Sadoshima, S., Yao, H., Ibayashi, S., Matsumoto, T., Uchimura H., and Fujishima, M., 1995, Ischemia-induced release of amino acids in the rat hippocampus of aged hypertensive rats. *J. Cereb. Blood Flow Metab.* 15:227-234.

13.   Saransaari, P., and Oja, S.S., 1997, Enhanced taurine release in cell-damaging conditions in the developing and ageing mouse hippocampus. *Neuroscience* 79:847-854.

14.   Saransaari, P., and Oja, S.S., 1998, Mechanisms of ischemia induced taurine release in mouse hippocampal slices. *Brain Res.* 807:118-124.

15.   Saransaari, P., and Oja, S.S., 1998, Release of endogenous glutamate, aspartate, GABA and taurine from hippocampal slices from adult and developing mice in cell-damaging conditions. *Neurochem. Res.* 23:567-574.

16.   Saransaari, P. and Oja, S.S., 1999, Characteristics of ischemia-induced taurine release in the developing mouse hippocampus. *Neuroscience* 94:949-954.

17.   Saransaari, P., and Oja, S.S., 2000, Taurine and neural cell damage. *Amino Acids* 19:509-526.

18.   Saransaari, P., and Oja, S.S., 1999, Taurine release modified by nitric oxide-generating compounds in the developing and adult mouse hippocampus. *Neuroscience* 89:1103-1111.

19.   DeMaster, E.G., Raij, L., Archer, S.L., and Weir, E.K., 1989, Hydroxylamine is a vasorelaxant and a possible intermediate in the oxidative conversion of L-arginine to nitric oxide. *Biochem. Biophys. Res. Commun.* 163:527-533.

20.   Lei, S.Z., Pan, Z.-H., Aggarwal, S.K., Chen, H.-S. V., Hartman, J., Sucher, N.J., and Lipton, S.A., 1992, Effect of nitric oxide production on the redox modulatory site of the NMDA receptor-channel complex. *Neuron* 8:1087-1099.

21.   Saransaari, P., and Oja S.S., 2002, Ischemia-induced taurine release is modified by nitric oxide-generating compounds in slices from the developing and adult mouse hippocampus. *Neurochem. Res.* 27, 395-402.

22.   Sorrenti, V., Di Giacomo, C., Campisi, A., Perez-Polo, J. R., and Vanella, A. 1999, Nitric oxide synthetase activity in cerebral post-ischemic reperfusion and effects of L-N^{G}-nitroarginine and 7-nitroindazole on the survival. *Neurochem. Res.* 24:861-866.

23.   Garthwaite, J., Southam, E., Boulton, C.L., Nielsen, E.B., Schmidt, K., and Mayer, B., 1995, Potent and selective inhibition of nitric oxide-sensitive guanylyl cyclase by 1H-[1,2,4]oxadiazolo[4,3-a]quinoxalin-1-one. *Molec. Pharmac.* 48:184-188.

24.   Fedele, E., and Raiteri, M., 1999, In vivo studies of the cerebral glutamate receptor/NO/cGMP pathway. *Prog. Neurobiol.* 58:89-120.

25. Boulton, C.L., Irving, A.J., Southam, E., Potier, B., Garthwaite, J., and Collingridge, G.L., 1994, The nitric oxide-cyclic GMP pathway and synaptic depression in rat hippocampal slices. *Eur. J. Neurosci.* 6, 1528-1535.
26. Sistiaga, A., Miras-Portugal, T., and Sánchez-Prieto, J., 1997, Modulation of glutamate release by a nitric oxide/cyclic GMP-dependent pathway. *Eur. J. Pharmac.* 321, 247-257.
27. Sequeira, S.M., Carvalho, A.P., and Carvalho C.M., 1999, Both protein kinase G dependent and independent mechanisms are involved in the modulation of glutamate release by nitric oxide in rat hippocampal nerve terminals. *Neurosci Lett.* 261, 39-32.
28. Saransaari, P., and Oja S.S., 2002, Taurine release in the developing and adult mouse hippocampus: involvement of cyclic guanosine monophosphate. *Neurochem Res.* 27, 15-20.
29. Trabace, L., and Kendrick, K.M., 2000, Nitric oxide can differently modulate striatal neurotransmitter concentrations via soluble guanylate cyclase and peroxynitrite formation. *J. Neurochem.* 75:1664-1674.
30. Oja, S.S., and Saransaari, P., 2000, Modulation of taurine release by glutamate receptors and nitric oxide. *Prog. Neurobiol.* 62:407-425.
31. Saransaari, P., and Oja, S.S., 1997, Glutamate-agonist-evoked taurine release from the adult and developing mouse hippocampus in cell-damaging conditions. *Amino Acids* 13:323-335.
32. Garthwaite, J., 1991, Glutamate, nitric oxide and cell-cell signalling in the nervous system. *Trends Neurosci.* 14:60-67.
33. Manzoni, O., Prezeau, L., Marin, P., Deshanger, S., Bockaert, J., and Fagni, L., 1992, Nitric oxide-induced blockade of NMDA receptors. *Neuron* 8:653-662.
34. Omerovic, A., Leonard, J. P., and Kelso, S. R., 1994, Effects of nitroprusside and redox reagents on NMDA receptors expressed in Xenopus oocytes. *Molec. Brain Res.* 22:89-96.
35. Oh, S., and McCaslin, P.P., 1995, The iron component of sodium nitroprusside blocks NMDA-induced glutamate accumulation and intracellular Ca^{2+} elevation. *Neurochem. Res.* 20:779-784.
36. Scheller, D., Korte, M., Szathmary S., and Tegtmeier, F., 2000, Cerebral taurine release mechanisms in vivo: pharmacological investigations in rats using microdialysis for proof of principle. *Neurochem. Res.* 25:801-807.
37. Izumi, Y., Clifford, D.B., and Zorumski, C.F., 1992, Inhibition of long-term potentiation by NMDA-mediated nitric oxide release. *Science* 257:1273-1276.
38. Saransaari, P., and Oja, S.S., 2000, Involvement of metabotropic glutamate receptors in ischemia-induced taurine release. *Neurochem. Res.* 25:1067-1072.

Effect of Ammonia on Taurine Transport in C6 Glioma Cells

MAGDALENA ZIELINSKA*, BARBARA ZABLOCKA# and
JAN ALBRECHT*

*Department of Neurotoxicology and #Laboratory of Molecular Biology, Medical Research Centre,
Polish Academy of Sciences, Warsaw, Poland

1.   INTRODUCTION

Taurine (Tau) in the CNS has been suggested to serve as an osmoregulator, inhibitory neuromodulator or neurotransmitter [1]. Suppositions regarding the functions of Tau have been associated with its readiness to become released from astrocytes or neurons in response to a plethora of pathological and physiological stimuli [2], also including pathologic concentrations of ammonia [3,4]. Two distinct mechanisms have been shown to underlie Tau transport across the plasma membrane: a) opening or activation of volume-operated anion channels (VSOAC), which subserves Tau efflux, b) activation of a specific taurine carrier - TauT, often associated with increased expression of TauT mRNA [5,6].

Previous *in vitro* studies have shown that ammonia at pathophysiologically relevant, low milimolar (1-10 mM) concentrations causes a massive efflux of radiolabelled or endogenous Tau from different CNS preparations *in vitro* [3,7,8], and *in vivo* [4]. The results have indicated that in contrast to the efflux evoked by hypotonic or high-potassium isotonic media that mimic hyposmotic shock, the ammonia-induced release, both in the acute and prolonged paradigm, is poorly correlated with cell volume changes and inconsistently susceptible to blocking the anion channels. Moreover, to our knowledge, the effects of ammonia on Tau uptake have not been investigated. Here we carried out a systematic analysis of ammonia-induced Tau uptake and efflux in cultured C6 glioma cells, with a focus on the relative contributions of carrier-mediated transport and passage through ion channels. The cells were treated either briefly (10 min) or for an extended period of time (24h) with 10 mM ammonium chloride, and the paradigms will be referred to as short-term ("st") and long-term ("lt").

Taurine 5: Beginning the 21st Century
Edited by Lombardini *et al.*, Kluwer Academic/Plenum Publishers, New York 2003   463

2.   MATERIAL AND METHODS

Cell Culture

Rat astrocytoma C6 cells (ATCC, passages 70-80) were grown in a Dulbecco's modified Eagle's medium (DMEM) containing 10% (v/v) fetal calf serum (Gibco) and supplemented with penicillin (50 U/ml) and streptomycin (50 U/ml) at 37 °C as described earlier [9].

HPLC of Tau

Neutralized perchloric acid extracts of the cells were analyzed by reverse-phase HPLC with fluorometric detection, after derivatization in a timed reaction with o-phthaldehyde plus mercaptoethanol, as previously described [3].

Tau Uptake Assay

A procedure previously described for Gln uptake was followed [9], except that the incubation mixtures contained 0.1 µCi/ml L-[1,2-^{3}H]Tau and unlabeled Tau in a 0.0025 mM to 0.150 mM concentration range; incubations were carried out for 10 min at 37 °C.

Tau Efflux Assay

Cultured cells were incubated for 10 min at 37 °C in the uptake medium. The medium contained 2.5 µM Tau and 0.1 µCi/ml L- [1,2-^{3}H]Tau. The cells were washed four times with nonradioactive medium. At two-minute intervals, the medium was removed and fresh medium was added. The procedure was carried out for 20 min. The radioactivities of collected fractions and lysates were measured. Measurement of Tau content in 20 min washout served to assess endogenous Tau efflux.

RT-PCR Analysis

Total RNA was isolated from C6 cells using TRIzol Reagent (Gibco/BRL) and total RNA (5 µg) was reverse transcribed using the Superscript II and oligo (dT)$_{12-18}$ primers (Gibco/BRL). Each cDNA sample (1 µl) was then amplified by PCR using the rat primers for TauT mRNA [10] and GAPDH mRNA. After 30 cycles of amplification (1 min at 94°C, 1 min at 65°C, and 1 min at 72°C) the PCR products were resolved on 1.2 % agarose gel.

Statistical evaluation

Statistical analysis of the data was performed using one way analysis of variance followed by the Dunnet's multiple comparisons test.

3.   RESULTS

Kinetic analysis of the uptake revealed its sodium and chloride dependence (Fig. 1), whereas competition experiments showed that the uptake is inhibited by model competitors hypotaurine, β-alanine and GES, in accord with the Tau transport characteristics observed in many different tissues, and also in oocytes transfected with a purified TauT mRNA [6,10]. Lt ammonia treatment decreased the Vmax of Tau uptake by ~25%, and the decrease was partly reversed 24h later ("lt+24h"), while st treatment remained without effect on the uptake kinetics (Fig. 1). Both st and lt ammonia treatment increased the efflux of newly loaded and endogenous Tau (Figs. 2 and 3). While the control efflux and that evoked by st ammonia treatment was unaffected by the absence of Na^+ or presence of GES in the medium, that evoked by lt ammonia treatment showed a distinct Na^+-dependent and GES- sensitive component (Fig. 2A,B). The control efflux of Tau was reduced by ~35% in the presence of an anion channel blocker, NA, whereas the efflux evoked by st and lt ammonia treatment was not sensitive to NA (Fig. 2C). A 65% above control increase of TauT mRNA expression relative to the expression of GAPDH mRNA was noted in lt ammonia-treated cells, and the increase retreated 24h after medium depletion of ammonia (Fig. 4), which coincided with the return of Tau uptake to the control level. Of note, in contrast to the efflux from control cells, the efflux from cells after recovery was insensitive to NA (Fig. 2C).

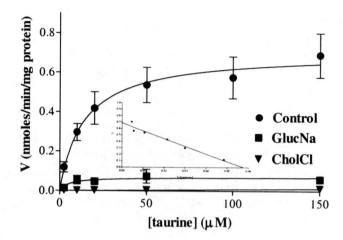

	Control	st	lt	lt+24h
V_{MAX} (nmoles/min/ mg protein)	0.70 ± 0.03	0.66 ± 0.03	0.50 ± 0.04*	0.61 ± 0.02*†
K_M (μM)	14.02 ± 2.16	12.66 ± 2.45	11.74 ± 3.70	11.57 ± 1.24

Competitors:	% of Control
none	98.0 ± 3.0 (n=4)
hypotaurine	5.5 ± 1.8 (n=4)
β-alanine	15.8 ± 1.7 (n=4)
GES	18.5 ± 0.3 (n=4)
Gln	97.1 ± 2.4 (n=4)

Figure 1. Upper panel: Michaelis-Menten (inset: Eadie-Hofstee) plot of Tau uptake in C6 cells. Medium panel: kinetic parameters of Tau uptake in control media ("Control"), following st and lt ammonia treatment and at 24h recovery after lt treatment ("lt+24h"). Bottom panel: inhibition of Tau uptake by competitors at 50 times excess. (mean ± SD, n=3-12); *p<0.05 vs Control, †p<0.05 vs lt.

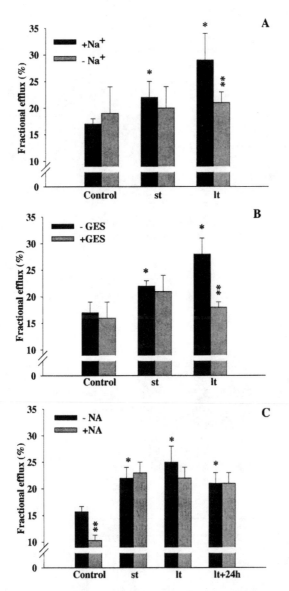

Figure 2. Fractional efflux of [³H] Tau from cultured C6 glioma cells grown in standard conditions ("Control"), following st and lt ammonia treatment and at 24h recovery after lt treatment ("lt+24h") in the presence or absence of sodium ions (**A**), Tau transport inhibitor guanidino ethane sulfonate (GES; 500 µM) (**B**) and an anion channel inhibitor, niflumic acid (NA; 200 µM) (**C**). (mean ± SD, n=4-8); *p<0.05 vs Control; **p<0.05 vs Control minus respective medium change.

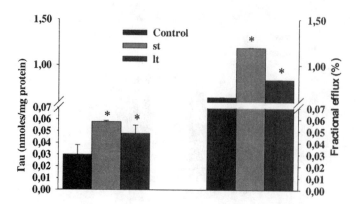

Figure 3. Efflux of endogenous Tau from cultured C6 glioma cells grown in standard conditions ("Control"), and following 10 min ("st") or 24h ("lt") ammonia treatment. (mean ± SD, n=5); *p<0.05 vs Control.

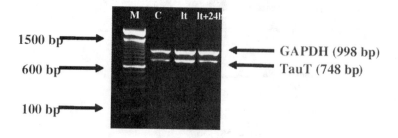

	TauT/GAPDH
Control	0.79 ± 0.09
lt	1.30 ± 0.10*
lt+24h	0.68 ± 0.10

Figure 4. Expression of TauT mRNA in cultured C6 glioma cells grown in standard conditions ("Control") or treated with 10 mM ammonium chloride for 24h ("lt") and cells treated with 10 mM ammonia for 24h in which ammonia was withdrawn for 24h ("lt+24h"). Below: relative expression of TauT mRNA vs GAPDH mRNA; (mean ± SD, n=6); *p<0.05 vs Control.

4.   SUMMARY AND CONCLUSIONS

-    Both short-term and prolonged treatment with ammonia stimulate Tau efflux from C6 glioma cells, which confirms earlier observations with different CNS preparations. In addition, prolonged, but not short-term treatment reversibly decreases Tau uptake. The results are consistent with *in vivo* observations showing robust extracellular accumulation of Tau in the CNS under hyperammonemic conditions [4].

-    Enhancement of Tau efflux by short term ammonia treatment is associated with its passage via a channel (or channels) which in contrast to the release from control cells is (are) not inhibited by NA. However, the increased Tau efflux does not appear to involve active Tau transport. Insensitivity to NA distinguishes ammonia-dependent Tau efflux from C6 cells from that reported in other CNS-derived preparations.

-    Tau release evoked by prolonged ammonia treatment likewise shows a NA-insensitive component, but also a component associated with activation of Tau transport in a reverse mode. Increased outward transport of Tau may be associated with transiently increased expression of TauT mRNA. Mutual relation of the two components, but also the identity of the NA-insensitive efflux route with the anion channels so far described remains to be analyzed.

ACKNOWLEDGEMENTS

Supported by SCSR grant no 6P05A00321.

REFERENCES

1.   Huxtable, R.J., 1992, Physiological actions of taurine. *Physiol. Rev.* 72:L 101-142.
2.   Saransaari, P., Oja, S.S., 2000, Taurine and neural cell damage. *Amino Acids* 19: 509-526.
3.   Zielińska, M., Hilgier, W., Law, R.O., Goryński, Pl Albrecht, J., 1999, Effects of ammonia in vitro on endogenous taurine efflux and cell volume in rat cerebrocortical minislices: influence of inhibitors of volume-sensitive amino acid transport. *Neuroscience* 91: 631-638.
4.   Zielińska, M., Hilgier, W., Borkowska, H.D., Oja,, S.S., Saransaari, P., Goryński, P., Albrecht, J., 2002, Ammonia-induced extracellular accumulationof taurine in the rat striatum *in vivo*: role of inotropic glutamate receptors, *Neurochem. Res.*, 27: 37-42.
5.   Pasantes-Morales, H., 1996, Volume regulation in brain cells: cellular and molecular mechanisms. *Metab. Brain Dis.* 11: 187-204.
6.   Bitoun, J., Tappaz, M., 2000, Gene expression of the transporters and biosynthetic enzymes of the osmolytes in astrocyte primary cultures exposed to hyperosmotic conditions. *Glia* 32: 165-176.

7. Albrecht, J., Bender, A.S., Norenberg, M.D., 1994, Ammonia stimulates the release of taurine from cultured astrocytes. *Brain Res.* 660: 228-232.

8. Faff-Michalak, L., Reichenbach, A., Dettmer, D., Kellner, K., Albrecht, J., 1994, K$^+$-hypoosmolarity-, and NH$_4$$^+$-induced taurine release from cultured rabbit Müller cells: role of Na$^+$ and Cl$^-$ ions and relation to cell volume changes. *Glia* 10: 114-120.

9. Dolińska, M., Dybel, A., Hilgier, W., Zielińska, M., Zablocka, B., Bużańska, L., Albrecht, J., 2001, Glutamine transport in C6 glioma cells: substrate specificity and modulation in a glutamine deprived culture medium. *J. Neurosci. Res.* 66: 959-966.

10. Smith, K.E., Borden, L.A., Wang,C.-H.D., Hartig, P.R., Branchek, T.a., Weinshank, R.L., 1992, Cloning and expression of a high-affinity taurine transporter from rat brain. *Mol. Pharmacol.* 42: 563-569.

Taurine and Hypotaurine Dynamics in Activated C6 Glioma

The Effects of Lipopolysaccharide (LPS) and Taurine Administration on Intracellular Hypotaurine and Taurine Dynamics in C6 Glioma

JOHN DOMINY, JR and RALPH DAWSON, JR
Department of Pharmacodynamics, University of Florida, Gainesville, Florida 32610, USA

1.   INTRODUCTION

Our lab has previously shown that there is a significant decrease in taurine content with ageing. Relative to their younger conspecifics, aged rats show appreciable declines in the taurine levels of serum[1] and peripheral tissues[2]. The brain would also appear to be affected by this trend as lower taurine concentrations have been noted in areas as diverse as the cortex, cerebellum, hippocampus, hypothalamic nuclei, and striatum[3,4,5].

The loss of taurine with senescence could have a profound impact on brain physiology. The central nervous system (CNS) uses taurine as a workhorse for a wide array of homeostatic processes: regulation of cell volume[6], scavenging of reactive oxygen species[7], modulation of innate immune response[8], and inhibitory neurotransmission[9]. Deprivation of adequate amounts of taurine, therefore, could reduce the ageing brain's capacity to buffet everyday wear and tear and render it more susceptible to pathological sequelae such as stroke, Alzheimer's, and Parkinson's disease.

Although the mechanism for taurine's decline in the CNS is not known, one possibility is reduced biosynthetic capacity brought on by enhanced immunological activity. Astrocytes have been shown to synthesize taurine *de novo* from cysteine[10,11] and are thus suspected to contribute to the brain's total taurine pool. Additionally, astroglia play an important role in exaggerating proinflammatory environments within the ageing brain[12]. Interestingly, septic infection, which is characterized by an inflammatory immune response, is known to affect cysteine catabolism and produce declines in the level of taurine in peripheral tissues[13]. The effects of sepsis on CNS taurine and its synthesis, however, remain largely unexplored.

Because of their potential to serve as both participants in immune response and producers of taurine, this study was designed to evaluated how LPS-induced activation of astrocytes would affect the levels of taurine and its biosynthetic precursor, hypotaurine. Using cultures of C6 glioma as a model for CNS astrocytes, a time-based exposure paradigm was employed to determine the effects of two commonly used doses of LPS (1 μg/ml and 10 μg/ml). Our hypothesis was that LPS-induced activation would cause a decrease in intracellular taurine stores, which should be ameliorated by supplementation with 1 mM taurine—a plasma concentration of taurine that is easily attainable by oral supplementation.

2.   MATERIALS AND METHODS

2.1   Basic Cell Culture

C6 glioma, passages 60-80, were cultured in Dulbecco's Modified Eagle Medium (DMEM; Gibco BRL, Grand Island, NY) supplemented with 10% fetal bovine serum (FBS; Gibco BRL, Grand Island, NY), 20 U/ml penicillin G (Fisher, Atlanta, GA) and 20 μg/ml streptomycin sulfate (Fisher, Atlanta, GA). Cultures were maintained in an incubator containing a humidified atmosphere enriched with 5% CO_2 and maintained at 37 °C.

2.2   Experimental Paradigm

All experiments employed flat bottom 24 well plates seeded at a density of 2×10^5 cells/ml. Cultures were permitted 48 h of incubation prior to experimental treatments. Forty-eight hours after seeding, wells were washed twice with HBSS and quickly refilled with medium. Wells were then dividing up into six treatment groups, with each group consisting of 4 wells. The treatments evaluated were 10 μg/ml LPS (Sigma, St. Louis, MO), 1 μg/ml LPS, 1 mM taurine (Sigma, St. Louis, MO), 10 μg/ml LPS with 1mM taurine, and 1 μg/ml LPS with 1mM taurine. These reagents were delivered in a vehicle of HBSS at an initial concentration ten times the final treatment concentration. Control cultures received only vehicle. After the appropriate experimental treatments were administered, plates were allowed to incubate under conditions earlier described for 1 h, 24 h, or 48 h.

At the end of the incubation period, medium was removed and frozen at -20°C until preparation for HPLC analysis. Wells were washed twice with ice-cold phosphate buffered saline and monolayers extracted with 0.2 M perchloric acid (PCA). Suspensions were homogenized and an aliquot removed for protein determination as outlined by the Bradford method[14] using bovine serum albumin as a standard. The remaining homogenate was microcentrifuged to pellet suspended protein. The resulting PCA supernatant was collected and frozen at -20°C until preparation for amino acid analysis. Amino acid content was assayed by HPLC coupled with electrochemical detection of o-pthaldehyde (OPA) derivatized samples as previously described[2].

2.3 Statistical Analysis

All data are expressed as means±standard deviation. Experiments were setup such that each treatment group consisted of 4 wells (N=4) and all experiments were independently replicated. To determine whether differences amongst various treatment groups were significant, analysis of variance (ANOVA) with Neumen-Keuls multiple comparison test was used with the aid of Prism (GraphPad Software, version 2.01). Differences among experimental groups were considered significant only if $P<0.05$. Data analysis was always performed on untransformed data, although for the purposes of presentation some results are expressed in terms of percent control.

3. RESULTS

3.1 The Effects of LPS-Induced Activation

Incubation of glioma in the presence of either 1 μg/ml or 10 μg/ml LPS for 1, 24, or 48 h produced no significant perturbations in intracellular taurine content (Table 1). Exposure to LPS had no effect on cell viability (data not shown).

Over the course of 48 h, exposure to LPS alone significantly ($P<0.05$) reduced the hypotaurine content of glioma cultures by 40% (Table 1). This effect was not dose-dependent. In both LPS-treated and control cultures, the hypotaurine content of the surrounding medium was below the limits of detection, <0.5 nmoles/well (data not shown).

Incubation with 1 mM taurine resulted in a time-dependent increase in the intracellular levels of this amino acid (Table 1). Taurine uptake was significantly ($P<0.01$) enhanced following activation with LPS for 48 h. This effect did not differ significantly between the two doses of LPS evaluated.

3.2 The Effects of Exogenous Taurine Exposure

Incubation with 1mM taurine produced a time-dependent decrease in the level of intracellular hypotaurine that first appeared after 24 h of incubation (Table 1). An additional 24 h of incubation further diminished the intracellular

Table 1: The effects of LPS and/or taurine on the intracellular levels of taurine and hypotaurine C6 glioma. Each experimental time point was replicated twice. Key: [a]=P<0.05 vs time-matched control, [b]=P<0.01 vs time-matched control, [c]=P<0.001 vs time-matched control, [d]=P<0.01 vs time-matched taurine group, [e]=P<0.001 vs time-matched taurine group, ND= below the limits of detection (<0.5 nmoles/well)

Length of incubation		Control (nmoles/mg)	LPS (10 µg/ml) (nmoles/mg)	LPS (1 µg/ml) (nmoles/mg)	Taurine (nmoles/mg)	Taurine/LPS (10 µg/ml) (nmoles/mg)	Taurine/LPS (1 µg/ml) (nmoles/mg)	DMEM + 10% FBS
1 h	Taurine	59.38±34.79	33.46±13.04	55.14±33.73	79.00±16.41	78.83±21.27	94.21±36.24	11.00 µM
	Hypotaurine	33.28±15.07	20.63±5.51	32.82±18.75	32.29±13.54	30.43±12.43	42.86±14.32	2.30 µM
24 h	Taurine	91.92±8.26	92.03±6.23	98.02±11.6	265.1±45.45 [c]	248.3 ±31.69 [c]	260.6±25.96 [c]	8.82 µM
	Hypotaurine	39.00±5.14	36.92±4.54	32.05±4.52	7.70±2.48 [c]	5.43±1.55 [c]	6.54±1.37 [c]	1.05 µM
48 h	Taurine	112.3±21.81	93.76±16.59	100.60±15.70	726.40±198.80 [c]	1867.00±834.1 [c,d]	1812±681.7 [c,d]	13.40 µM
	Hypotaurine	83.99±15.86	51.13±21.79 [a]	53.07±12.64 [b]	ND	ND	ND	2.11 µM

hypotaurine pool to below the limits of detection. Exposure to LPS had no significant additive effect on taurine's ability to depress intracellular hypotaurine.

Cultures exposed to taurine for 48 h did have detectable amounts of hypotaurine in the surrounding medium. When the total amount of extracellular hypotaurine is summed with the amount present intracellularly, however, an 80% discrepancy still exists between these cultures and time-matched control (Fig 1).

Incubation with other organic osmolytes, such as 1 mM myo-inositol, alanine, or glutathione, did not produce any significant change in hypotaurine (data not shown). Similar results, however, were obtained after 8 days of incubation with the taurine analogs guanidinoethane sulfonic acid and β-alanine (data not shown).

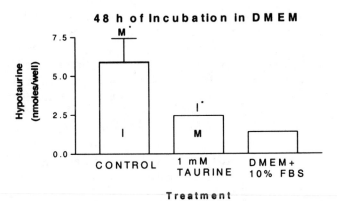

Figure 1. A breakdown of the total amount of hypotaurine present intracellularly (I) and extracellularly (M) in C6 glioma cultured for 48 h in the presence or absence of taurine. *=below the limits of detection (<0.5 nmoles/well)

4. DISCUSSION

The brain exhibits a marked decrease in taurine content with advanced age[5]. The underlying cause of this reduction is unknown. Decrements in the plasma pool of taurine available for uptake is one possibility. Another is diminished capacity for endogenous CNS biosynthesis. Why would this change with age? One possible explanation is the brain's exposure to exaggerated immune activity, a hallmark of ageing. Because pro-inflammatory conditions, such as sepsis, are known to induce drastic alterations in the cysteine catabolism of peripheral tissues[13], it is possible that similar changes could also occur in the brain. To investigate this hypothesis, C6 glioma were used as a model for CNS astrocytes to evaluate the effects of LPS-induced activation on intracellular levels of taurine and hypotaurine in the presence or absence of 1 mM taurine supplementation.

4.1   LPS-Induced Depletion of Hypotaurine

Results obtained from exposure to LPS indicate that immunological activation of glioma produces a significant depletion of intracellular hypotaurine. This effect would not appear to be media-specific, as C6 glioma cultured in Ham's F10 demonstrated significant ($P<0.05$) decreases in hypotaurine after 24 h of endotoxic activation (data not shown).   Nor is it specific to transformed astrocytes; preliminary studies show a 25% decrease in the intracellular hypotaurine pool of neonatal Wistar rat primary astrocyte cultures after 48 h of activation in DMEM.   While the mechanism cannot be directly determined by this study, there are two plausible explanations.   The first is a decrease in the availability of cysteine.   As mentioned earlier, sepsis results in profound perturbations in *in vivo* cysteine catabolic flux[13].   In particular, sepsis causes an elevated demand for cysteine as a consequence of significantly enhanced glutathione synthesis.   Because hypotaurine production is dependent upon adequate levels of cysteine to yield cysteine sulfinic acid (CSA), redistribution of intracellular cysteine stores towards glutathione could deplete hypotaurine levels by substrate attrition.   The earlier manifestation of a decrease in hypotaurine in Ham's F10 hints at this possibility, as Ham's F10 contains lower concentrations of sulfur amino acid precursors.   Additional experiments that culture LPS-activated cells with buthionine sulfoximine, a potent inhibitor of γ-glutamylcysteinyl synthase, will be needed to address this possibility.

Alternatively, LPS could be exerting an effect directly on cysteine sulfinic acid decarboxylase (CSAD).   As the enzyme that synthesizes hypotaurine from CSA, changes in either CSAD expression or activity should produce fluctuations in the hypotaurine pool.   Among the factors proven to affect CSAD activity is its state of phosphorylation.   In neuronally-derived CSAD, activity deacreases following dephosphorylation by phosphatase 2C and increases after phosphorylation by protein kinase C (PKC)[16].   Activation of the CD14 receptor by LPS is known to activate PKC in astrocytes[17].   PKC activation, however, would be expected to cause an increase in intracellular hypotaurine.   Because this is not seen, it is possible that some of the more distally related events associated with LPS activation could be at work.   Included in this list are mobilization of NF-κβ[18] and changes in the redox state[19].   Little research has been previously conducted on the direct impact of these molecular changes on cysteine catabolism, but they are certainly worth investigating.

Another explanation is that LPS is promoting the conversion of hypotaurine to taurine.   Because hypotaurine is an efficient scavenger of hydroxyl radicals[7], it is possible that the pro-oxidant environment created by LPS favors the non-enzymatic conversion of hypotaurine to taurine.   This is unlikely, however, as LPS alone caused no significant change in intracellular taurine.   In fact, it is uncertain if these cells are capable of synthesizing taurine from hypotaurine as 1) hypotaurine levels increased with incubation while taurine remained unchanged and 2) exposure to 1 mM hypotaurine for 24 h produced a significant ($P<0.001$) increase in intracellular hypotaurine with no accompanying increase in taurine synthesis (data not shown).

4.2   LPS-Induced Uptake of Taurine

LPS in the presence of 1 mM taurine resulted in significant (P<0.01) increases in taurine uptake with 48 h of incubation. Similar results have been obtained in primary astrocyte cultures[17]. This effect was attributed to LPS-induced production of tumor necrosis factor alpha (TNF-α). Although TNF-α was not measured in this study, it has been shown that LPS-induced activation does not increase the level of TNF-α transcription in C6 glioma[21]. Unlike primary astrocyte cultures, LPS did not significantly increase nitrite levels, used as an index of NO synthesis, in these cells (data not shown). LPS could therefore be operating independently of TNF-α/NO to either increase transporter expression or attenuate taurine-induced decreases in transporter activity. It is known that PMA-induced activation of PKC significantly down-regulates activity of the glial taurine transporter.[22] Curiously, LPS mobilizes PKC activity in glia but does not decrease taurine transport. Whether this is due to PKC isotype specificity or molecular events independent of PKC is uncertain. Whatever the mechanism of action, it is nevertheless intriguing that LPS produces an increase in the intracellular accumulation of taurine only with 48 h of exposure. Taking into consideration that taurine has many cytoprotective properties[23], this may be a defensive strategy adopted by glia to ward off some of the deleterious effects associated with chronic sepsis.

Additionally, it should be noted that the increase in transporter activity accompanying chronic exposure to LPS lends indirect support to the hypothesis that LPS depletes intracellular hypotaurine by biosynthetic means. Because of its capacity to indiscriminately transport β-amino acids, elevated taurine transporter activity should produce a corresponding increase in intracellular hypotaurine sequestration, which was not observed.

4.3   Depletion Of Hypotaurine By Taurine

Uptake of taurine was associated with a time-dependent decrease in intracellular hypotaurine. The simplest explanation for this observation is heteroexchange of intracellular hypotaurine for extracellular taurine by the taurine transporter. This finding has been previously reported for cultures of lymphoid and myeloid-tumor cells[24]. As shown in Fig 1, however, heteroexchange does not entirely explain the hypotaurine depletion in C6 glioma.

Alternatively, taurine could be inhibiting hypotaurine synthesis. Gene array data from our lab have revealed that C6 glioma express the genes for CDO and CSAD (data not shown). High levels of intracellular taurine may function in an inhibitory feedback fashion to diminish the CSAD pathway throughput mediated by these enzymes. Primary astrocytes, however, exhibited no change in the transcription of these enzymes when grown in medium supplemented with 500 µM taurine[25]. Nevertheless, post-transcriptional modulation of CDO and/or CSAD expression cannot be discounted. Chronic taurine supplementation in aged Fischer 344 rats, for instance, produced a significant decrease in liver CSAD activity while protein levels for this enzyme remained unaffected[26]. In addition, neurons cultured in 25mM taurine showed significant decreases in CSAD activity as a consequence of reduced phosphorylation by PKC[16]. These findings merit further investigation into possible taurine-induced changes in kinetic properties of the enzymes involved in hypotaurine synthesis, particularly in light of the widespread use of taurine as a dietary supplement.

5.   CONCLUSIONS

Inflammation could be an important contributor to the ageing-induced decrease in brain taurine by reducing the glial hypotaurine pool. While taurine supplementation can replenish diminished taurine stores in aged rats,[5] it may have the unanticipated result of reducing endogenous hypotaurine. Further studies are required to clarify both the mechanisms and significance of these two effects.

REFERENCES

1.   Wallace, D.R. and Dawson, R. Jr., 1990, Decreased plasma taurine in aged rats. *Gerontol.* 36: 19-27.
2.   Dawson, R. and Wallace, D.R., 1992, Taurine content in tissues from aged Fischer 344 rats. *Age* 15: 73-81.
3.   Palkovits, M., Banay-Schwartz, M., and Lajtha, A., 1990, Taurine levels in brain nuclei of young adult and aging rats. In: *Taurine: Functional Neurochemistry, Physiology, and Cardiology.* Wiley Liss, Inc. 45-51
4.   Strolin Benedetti, M.S., Russo, A., Marrari, P., and Dostert, P., 1990. The effects of lifelong treatment with MAO inhibitors on amino acid levels in rat brain. *J. Neural Transm.* 2: 239-248.
5.   Dawson, R. Jr., Liu, S., Eppler, B., and Patterson, T., 1999, Effects of dietary taurine supplementation or deprivation in aged male Fischer 344 rats. *Mech. Age. Dev.* 107: 73-91.
6.   H. Pasantes-Morales, S. Alavez, R. Sanchez Oela, and J. Moran 1993, Contribution of Organic and Inorganic Osmolytes to Volume Regulation in Rat Brain Cells in Culture.*Neurochem. Res.*, 18: 445-452.
7.   Aruoma, O., Halliwell, B., Hoey, B., and Butler, J., 1988, The antioxidant action of taurine, hypotaurine and their metabolic precursors. *Biochem. J.*, 265: 251-255.
8.   Marcinkiewicz, J., Grabowska, A., Bereta, J., and Stelmaszynska, T., 1995, Taurine chloramine, a product of activated neutrophils, inhibits in vitro the generation of nitric oxide and other macrophage inflammatory mediators. *J. Leukoc. Biol.*, 58 667-674.
9.   Huxtable, R.J., 1989, Taurine in the central nervous system and the mammalian actions of taurine. *Prog. Neurobiol.*, 32: 471-533.
10. Beetsch, J.W. and Olson, J.E., 1998, Taurine synthesis and cysteine metabolism in cultured rat astrocytes: effects of hyperosmotic culture. *Am. J. Physiol.*, 274: C866-C874.

11. Brand, A., Leibfritz, D., Hamprecht, B., and Dringen, R., 1998, Metabolism of cysteine in astroglial cells: synthesis of hypotaurine and taurine. *J. Neurochem.*, 71:827-832.
12. Kyrkanides, S., O'Banion, M.K., Whitely, P.E., Daeschner, J.C., and Olschowka, J.A., 2001, Enhanced glial activation and expression of specific CNS inflammation-related molecules in aged versus young rats following cortical stab injury. *J Neuroimmunol.* 119: 269-277.
13. Malmezat, T., Breuille, D., Pouyet, C., Mirand, P.P., and Obled, C., 1998, Metabolism of cysteine is modified during the acute phase of sepsis in rats. *J. Nutrition* 128: 97-105.
14. Bradford, M.M., 1976, A rapid and sensitive method for the quantification of microgram quantities of protein utilizing the principle of protein-dye binding. *Ann. Biochem.* 72: 248-254.
15. Franceschi, C., Bonafe, M., Valensin, S., Olivieri, F., De Luca, M., Ottaviani, E., De Benedictis, G., 2001, Inflamm-aging. an evolutionary perspective on immunosenescence. *Ann. N Y Acad. Sci.* 908: 244-254.
16. Wu, J., Tang, X.W., Schloss, J.V., and Faiman M.D., 1997, Regulation of taurine biosynthesis and its physiological significance in the brain. *J. Neurosci.* 17: 6947-6951.
17. Cheng, C.C., Wang, J.K., Chen, W.C., and Lin, S.B., 1998, Protein kinase C eta mediates lipopolysaccharide-induced nitric-oxide synthase expression in primary astrocytes. *J. Biol. Chem.* 273: 19424-19430.
18. Nishiya, T., Uehara, T., and Nomura, Y., 1995, Herbimycin A suppresses NF-κB activation and tyronsine phosphorylation of JAK2 and the subsequent induction of nitric oxide synthase in C6 glioma cells. *FEBS Lett.* 371: 333-336.
19. Liu, Y., Tonna-DeMasi, M., Park, E., Schuller-Levis, G., Quinn, M.R., 1998. Taurine chloramine inhibits production of nitric oxide and prostaglandin E_2 in activated C6 glioma cells by suppressing inducible nitric oxide synthase and cyclooxygenase-2 expression. *Mol. Brain Res.* 5: 189-195.
20. Chang, R.C.C., Stadlin, A., and Tsang, D., 2001, Effects of tumor necrosis factor alpha on taurine uptake in cultured rat astrocytes. *Neurochem. Int.* 38: 249-254.
21. Furman, I., Baudet, C., and Brachet, P. 1996, Differential expression of M-CSF, LIF, and TNF-alpha genes in normal and malignant rat glial cells: regulation by lipopolysaccharide and vitamin D. *J. Neurosci. Res.* 46: 360-366.
22. Tchoumkeu-Nzouessa, G.C. and Rebel, G., 1996, Activation of protein kinase C down-regulates glial but not neuronal uptake. *Neurosci. Lett.* 206: 61-64.
23. Huxtable, R.J., 1992, Physiological actions of taurine. *Physiol.. Rev.*, 72: 101-163.
24. Learn, D.B., Fried, V.A., and Thomas, E.L., 1990. Taurine and hypotaurine content of human leukocytes. *J Leukoc. Biol.* 48: 174-182.
25. Bitoun, M. and Tappaz, M., 2000, Taurine down-regulates basal and osmolarity-induced gene expression of its transporter, but not the gene expression of its biosynthetic enzymes in astrocyte primary cultures. *J. Neurochem.* 75: 919-924.
26. Eppler, B. and Dawson, R. Jr, 2001, Dietary taurine manipulations in aged male Fischer 344 rat tissue: taurine concentration, taurine biosynthesis, and oxidative markers. *Biochem. Pharm.* 62: 29-39

The Anti-Craving Taurine Derivative Acamprosate
Failure to Extinguish Morphine Conditioned Place Preference

M. FOSTER OLIVE
Ernest Gallo Clinic and Research Center, Department of Neurology, University of California at San Francisco, California, USA

1.   INTRODUCTION

Acamprosate (calcium acetylhomotaurinate) is a taurine derivative that is used as a pharmacotherapeutic for the treatment of alcoholism. In both animals and humans, acamprosate reduces alcohol intake and the propensity to relapse during abstinence. Acamprosate has therefore been called an "anti-craving" compound[1].

The neurochemical mechanisms of action of acamprosate remain unclear. Initially, acamprosate was thought to mimic the actions of the inhibitory neurotransmitter γ-aminobutyric acid (GABA), possibly due to its structural similarity to taurine. However, biochemical and electrophysiological studies have revealed that acamprosate more consistently modulates the function of the N-methyl-D-aspartate (NMDA) subtype of glutamate receptors as well as voltage-gated calcium channels[1-3].

1.1   Acamprosate-Morphine Interactions

Although acamprosate is used exclusively for the treatment of alcoholism, very few studies have examined the ability of acamprosate to modulate the actions of other drugs of abuse, such as opiates. Of these, however, acamprosate appears to modulate some of the actions of morphine. For example, Spanagel and colleagues[4] demonstrated that acamprosate attenuated the sensitised locomotor and mesolimbic dopamine responses to repeated morphine treatment in rats. Acamprosate was also demonstrated to reduce the induction of tolerance and physical dependence to morphine in mice[5], and attenuate the conditioned place aversion produced by naloxone-precipitated morphine withdrawal[6]. These data provide evidence that acamprosate can modulate the behavioural and biochemical effects of opiates such as morphine.

1.2   The Conditioned Place Preference Paradigm

One of the most widely used paradigms to examine the rewarding properties of drugs of abuse is the conditioned place preference (CPP) paradigm. A typical CPP experiment utilizes an apparatus in which two compartments are connected by a guillotine-type door. Each compartment has unique visual and tactile cues (i.e., the colour of the walls and the type of flooring). Occasionally, olfactory cues unique to each environment are also utilized.

During the conditioning phase, an experimental animal is injected with a moderate dose of the drug of interest (i.e., morphine) and immediately confined to one of the two chambers for a short period of time (i.e., 15 min). Then, on alternating days, the same animal is injected with a control solution (i.e., saline) and confined to the other chamber for the same amount of time. This process is repeated several times over a 1-2 week period in order for the rodent to learn to associate the unique environment with the subjective effects of the drug.

Following conditioning, the animal is placed in a small centre compartment joining the two chambers and allowed to freely move between the chambers for a set period of time (i.e., 30 min). Photobeams mounted near the floor monitor locomotor activity and time spent in each chamber during the test period. A conditioned place preference occurs when the animal spends significantly more time in the drug-paired chamber than the saline-paired chamber. A conditioned place aversion occurs when the animal spends significantly more time in the saline-paired chamber than the drug-paired chamber. It has been demonstrated that all drugs that are abused by humans will produce CPP in rodents, presumably due to the rewarding (i.e., pleasurable) subjective effects of the drug.

The CPP paradigm not only measures the rewarding effects of drugs of abuse, but can also be utilized to measure various aspects of relapse. For example, following the initial test period, the ability of the animal to retain CPP for a given drug will diminish (i.e., extinguish) over a period of time (i.e., 2-4 weeks). However, it has been demonstrated that animals will reinstate CPP following extinction if exposed to stress or a single administration (i.e., priming injection) of the drug. This reinstatement of CPP has been proposed to be a model of relapse to drug-seeking behaviour.

An ideal property of a pharmacotherapeutic for the treatment of any type of addiction would be the ability to inhibit not only the drive to self-administer the drug, but also inhibit the drive to seek drug-associated environmental cues (i.e., secondary reinforcers). Thus, a successful addiction therapeutic might facilitate the extinction of CPP for a given drug.

2.   EFFECTS OF ACAMPROSATE ON THE EXTINCTION OF MORPHINE CPP

The present study was conducted to determine if acamprosate, an anti-craving taurine derivative used in the treatment of alcoholism, would extinguish

morphine CPP. Male C57BL/6J mice (20-25 g, Jackson Laboratories) were
housed individually under a 12:12 light-dark cycle (lights-on at 0600 h) with *ad
libitum* access to food and water. Mice were given morphine (5 mg/kg i.p.) and
placed immediately into one of two conditioning chambers for 15 min. On the
following day, mice were administered saline i.p. and placed immediately into the
other conditioning compartment for 15 min. After this alternation between
morphine and saline conditioning was performed for 4 days (one session per
day), animals were allowed a 2-day conditioning-free period, and then given 4
additional conditioning sessions. On the day following the last conditioning
period, mice were placed into a centre compartment and allowed access to both
conditioning chambers for 30 min (Test Day 1). The time spent in each chamber
was measured with electronically with photobeams placed 1.5 cm apart.

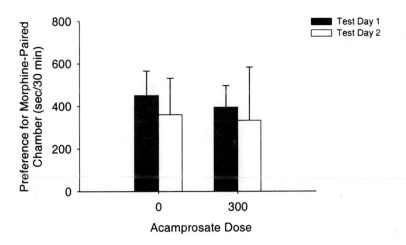

Figure 1. Repeated acamprosate fails to extinguish morphine CPP. Following the establishment of
morphine CPP on Test Day 1 (black bars), mice were treated i.p. with saline (0 mg/kg) or
acamprosate (300 mg/kg) once per day for 10 days, and then re-tested for morphine CPP (Test Day
2, open bars). No significant differences between test days or acamprosate treatment was observed
(p>0.05). n=8 per group.

Following Test Day 1, mice were treated i.p. with saline (0 mg/kg) or
acamprosate (300 mg/kg, Estechpharma, Seoul, Korea) once daily for 10 days.
Thirty minutes following the last saline or acamprosate injection, mice were re-
tested for morphine CPP (Test Day 2).

As seen in Figure 1, repeated acamprosate treatment failed to extinguish
orphine CPP. No differences were observed between Test Day 1 and Test Day 2
for either saline or acamprosate-treated animals.

3. CONCLUSIONS

Our findings indicate that the taurine derivative acamprosate does not influence the rewarding effects of morphine, as measured by the CPP paradigm. These findings are consistent with those of earlier reports in which acamprosate failed to reduce heroin self-administration[4] or the discriminative stimulus effects of morphine[7]. Thus, these data predict that acamprosate would have limited use in the treatment of opiate addiction.

ACKNOWLEDGEMENTS

This work was supported by funds from the State of California for medical research on alcohol and substance abuse through the University of California at San Francisco. The author wishes to thank Drs. Patricia Janak, Joseph Kim and Clyde Hodge for helpful comments and discussions.

REFERENCES

1. Spanagel, R., and Zieglgansberger, W., 1997, Anti-craving compounds for ethanol: new pharmacological tools to study addictive processes. *Trends Pharmacol. Sci.* 18:54-59.
2. Littleton, J., 1995, Acamprosate in alcohol dependence: how does it work? *Addiction* 90:1179-1188.
3. Olive, M.F., 2002, Interactions between taurine and ethanol in the central nervous system. *Amino Acids* 23:345-357.
4. Spanagel, R., Sillaber, I., Zieglgansberger, W., Corrigall, W.A., Stewart, J., Shaham, Y.,1998, Acamprosate suppresses the expression of morphine-induced sensitization in rats but does not affect heroin self-administration or relapse induced by heroin or stress. *Psychopharmacology* 139:391-401.
5. Sepulveda, J., Ortega, A., Zapata, G., Contreras, E., 2002, Acamprosate decreases the induction of tolerance and physical dependence in morphine-treated mice. *Eur. J. Pharmacol.* 445:87-91.
6. Kratzer, U., Schmidt, W.J., 1998, The anti-craving drug acamprosate inhibits the conditioned place aversion induced by naloxone-precipitated morphine withdrawal in rats. *Neurosci. Lett.* 252:53-56.
7. Pascucci, T., Cioli, I., Pisetzky, F., Dupre, S., Spirito, A., Nencini, P., 1999, Acamprosate does not antagonise the discriminative stimulus properties of amphetamine and morphine in rats. *Pharmacol. Res.* 40:333-338.

Ethanol-Induced Taurine Efflux
Low Dose Effects and High Temporal Resolution

ANTHONY SMITH[*], CHRISTOPHER J. WATSON[#], ROBERT T. KENNEDY[#] and JOANNA PERIS[*,#]

[*]Department of Pharmacodynamics, University of Florida, Gainesville, Florida USA
[#]Department of Chemistry, University of Michigan, Ann Arbor, Michigan, USA

1. INTRODUCTION

Taurine is one of the most abundant and versatile amino acids located in the CNS. Taurine has been shown to be an antioxidant[1] and to be involved in osmoregulation[2-5], calcium modulation[6-8] and neurotransmission[9]. As a neurotransmitter, taurine has been shown to increase chloride ion uptake in a manner similar to GABA[10] and is competitively inhibited by muscimol binding[11]. In addition, taurine enhances benzodiazepine binding to the $GABA_A$ receptor[12]. Taurine has also been shown to inhibit the NMDA receptor[13] and is likely to be the endogenous ligand for the glycine receptor in some areas of the brain[14]. These effects of taurine are similar to the action of ethanol as an agonist at the $GABA_A$ and glycine receptors, and an antagonist at the NMDA receptor[15-16].

Taurine has been shown to modify ethanol's effects such as enhancing ethanol-induced sleep time[17], decreasing locomotor stimulation caused by 1g/kg ethanol, yet increasing locomotor activity at 2 g/kg[18]. A similar pattern can be observed with conditioned taste aversion[19] (CTA). CTA is enhanced by taurine at low doses of ethanol (0.8 g/kg) yet taurine completely blocks CTA above 1 g/kg. Conversely, ethanol has been shown to affect taurine efflux in the brain. *In vivo* microdialysis studies have shown that an acute injection of ethanol increases taurine efflux in the nucleus accumbens[20-21], hippocampus, frontal cortex[22] and amyglada[23]. Genetic selection for ethanol preference also influences taurine efflux. For example, rats bred to be highly sensitive to ethanol displayed a reduced ethanol-induced taurine efflux in the nucleus accumbens compared to the low-alcohol sensitive rats[24]. Sardinian ethanol-preferring rats exhibited a reduced ethanol-induced taurine efflux in comparison to Sardinian ethanol-non-preferring rats[25]. However, these studies only measured taurine efflux after sedative doses of ethanol and not the lower doses associated with self-administration of ethanol. In addition, these studies employed traditional HPLC-based detection of amino acids in microdialysate samples collected every 15-20 minutes. This technique fails to resolve any fast neurotransmitter changes that might occur. Recently, by coupling microdialysis on-line to capillary electrophoresis with laser-induced fluorescence (CE-LIF), temporal resolution of multiple analytes has been improved[26-27]. The end result is higher resolution (seconds) with a greater capacity to measure a multitude of amino acids. The present microdialysis study was conducted to determine ethanol-induced taurine efflux in the nucleus accumbens after low doses (less than 1 g/kg body weight) of ethanol such that might be self-administered in a voluntary manner by rats.

2.   DOSE-RESPONSE RELATIONSHIP OF INJECTED ETHANOL

Male Sprague-Dawley rats (Harlan, Indianapolis, IN) weighing 250-350g were housed with *ad libitum* access to food and water throughout the experiment unless otherwise indicated. All procedures were conducted in strict adherence to the *National Institute of Health Guide for the Care and Use of Laboratory Animals*. One week prior to the experiment, each rat was surgically implanted with a guide cannula into the left side of Nac (+1.6 anteroposterior, +1.7 lateral, -6.2 dorsoventral).

On the day of the experiment, each rat was lightly anesthetized with halothane while the dialysis probe (o.d. 270 μm; active length 2 mm; cellulose membrane, 13,000 molecular weight cut-off) was inserted. The probe was perfused with artificial cerebrospinal fluid (ACSF; 145 mM NaCl, 2.8 mM KCl, 1.2 mM $MgSO_4$, 1.2 mM $CaCl_2$,) at 1 μL/min. After a 90-min equilibration period, saline (1 ml/kg) was injected i.p. and samples were taken every 10 sec for 60 minutes. Subsequent injections of ethanol (20% w/v in saline, i.p.) occurred in the following order: 0.5 g/kg, 1 g/kg, and 2 g/kg. After each injection, samples were taken every 10 sec for 60 minutes. After the experiment, probe placement was verified histologically and only those subjects with accurate placement were included in the analysis.

The amino acid concentrations in dialysate samples were determined using on-line CE-LIF as described previously[26]. Briefly, amino acids were derivatized on-line with a 1μL/min stream of *o*-pthalaldehyde/β-

mercaptoethanol solution with a reaction time of approximately 1.5 min. Samples of derivatized dialysate were injected onto the separation capillary (10 cm long, 10 μm ID, 150 μm OD fused silica) using a flow-gate interface. A 1 mL/min cross-flow of electrophoresis buffer (40 mM borate, 0.9 mM hydroxypropyl β cyclodextran, pH 9.5) was applied to the 30 μm gap between the reaction capillary and the separation capillary. To inject a plug of sample onto the separation capillary, the cross-flow was stopped for 1 s allowing the dialysate to fill the gap between capillaries and the voltage was raised to 2 kV for 100 ms. After injection, the gating cross-flow was resumed to wash excess dialysate to waste, and the separation voltage was ramped to 20 kV over 500 ms. The flow-gate was held at ground to isolate the animal from the voltage dropped across the separation capillary.

The derivatized amino acids were detected using LIF in a sheath-flow detector cell. The outlet of the separation capillary was inserted into a 200 μm² ID quartz cuvette. Grinding the outlet to a point reduced the dead volume at the outlet of the separation capillary. Sheath buffer was siphoned around the outside of the separation capillary by gravity. Analytes migrated off the end of the separation capillary with a laminar flow profile which reduces the background signal caused by laser scatter. Fluorescence was excited using the 351 nm line of an argon-ion laser focused onto the effluent stream. Emission was filtered through a 450 ± 25 nm longpass filter and detected by a photomultiplier tube. Current from the photomultiplier tube was amplified and filtered (10 ms rise-time) and then sampled digitally using a National Instruments board controlled by Labview-based software. A calibration curve was generated either before or after the experiment by placing the tip of the dialysis probe in three different concentrations of a standard solution containing taurine, glutamate, GABA, aspartate, dopamine, serine and glycine and then plotting peak height versus concentration.

The ethanol-induced efflux of taurine from the nucleus accumbens was dose-dependent as shown in the responses from a typical animal in Figure 1. A small increase in taurine levels was observed during the 15 min after 0.5 g/kg ethanol compared to either pre-injection basal levels or the effects of saline injection. A larger increase in taurine levels was observed during the 5 min after 1 g/kg ethanol which approached basal levels during the rest of the sampling period. The 2 g/kg ethanol dose increased taurine to 175% above baseline within the first 5 minutes after ethanol injection which then decreased after 20 minutes to about 40% above basal levels where they remained elevated for the duration of sampling. Levels of GABA and glutamate in microdialysate were not affected by any of the doses tested while levels of glycine were decreased by ethanol in a dose-dependent manner (data not shown).

Rat #1 - Dec. 12

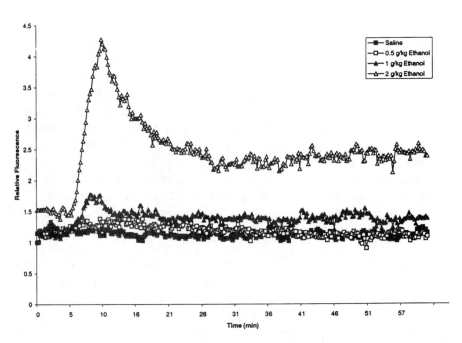

Figure 1. The effects of ethanol injections on taurine levels in microdialysate from the nucleus accumbens of a typical Sprague-Dawley rat. Microdialysate was sampled for one hour after saline or 0.5, 1.0 or 2.0 g/kg i.p. ethanol injections given in a cumulative fashion. The injection occurred at the start of each 1-hr sampling period but there is a 5 min delay for dialysate to travel from the probe tip to the flow-gate. Data are expressed as the relative fluorescence (in arbitrary units) detected at the end of the capillary electrophoresis column.

The mean maximum deviations of taurine levels from baseline for each rat and the mean plateau taurine levels 23-30 min after injection are shown in Figure 2. A significant dose-related increase in the maximum taurine levels occurred 10-15 min after ethanol injection (5-10 min corrected for dead volume). The ability of ethanol to cause a longer lasting increase in taurine levels was only significant for the 2 g/kg ethanol dose. Thus, rapid on-line detection of amino acids in microdialysate of rats allows reliable measurement of small magnitude changes in taurine efflux after sub-intoxicating doses of ethanol. Additionally, the ability to measure changes in less than a minute indicate two different phases of taurine efflux.

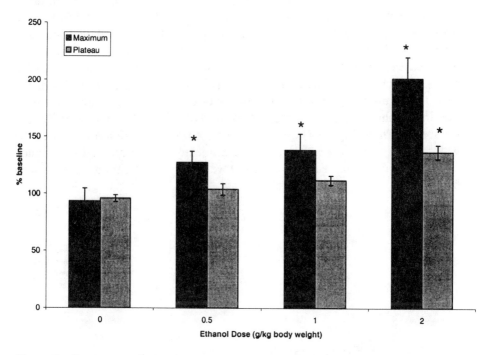

Figure 2. Data were calculated as a percent of the first twenty samples (pre-injection). Maximum indicates the largest change from baseline for each subject for each treatment. The mean latency for this maximum was 13.8±2.9, 14.9±5.5 and 11.1±1.7 min for 0.5, 1 and 2 g/kg ethanol, respectively. Plateau indicates the mean % of baseline 25 to 30 minutes after injection. Separate ANOVAs indicated an effect of dose for both maximum (F(3,18)=15.6; p<0.001) and plateau (F(3,18)=9.0; p<0.001). Shown are means ± SEM for N = 7.

3. SELF-ADMINISTERED ETHANOL

To induce voluntary ethanol consumption, ethanol-naïve rats were water-deprived for 24 hr and then given access to 0.2% w/v saccharin for 30 min. After this point, rats (no longer water-deprived) were given daily 30 min access to water and 0.2% w/v saccharin in separate graduated sipper tubes. Every 4 days, the ethanol concentration was gradually increased from 0 to 10% while the saccharin concentration was decreased to 0.1%. For the remainder of the experiment, rats had concurrent access to the 10% ethanol: 0.1% saccharin solution and water for 30 min/day at approximately 1300 h.

After rats attained stable daily levels of voluntary ethanol consumption (0.72±0.24 g/kg /30 min for three months), they underwent surgery for implantation of guide cannula as described above. After a one-week recovery period, during which they were ethanol deprived, rats were implanted with a microdialysis probe at 900 h. Baseline samples were taken for 3 hrs while the animal had food and water *ad libitum*. At 1300 h, the water bottle was replaced with the two sipper tubes containing water and 10% ethanol: 0.1% saccharin. The volume of each fluid consumed was recorded each minute. Of the five animals tested, two consumed 0.1 g/kg ethanol which did not alter taurine levels (Figure 3) and one rat did not drink. Two rats drank 0.7 g/kg ethanol, a majority of which was consumed within the first 5 min of access. These rats showed a mean 30% increase in taurine for about 30 min (Figure 3). After one hour, the tubes were replaced with the water bottle. Two hours later, rats were injected with 1 g/kg ethanol which caused a biphasic increase in taurine levels (Figure 3) similar to that seen previously.

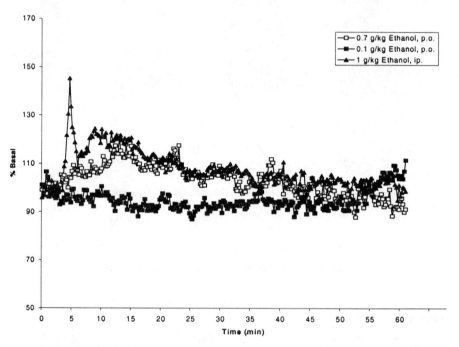

Figure 3. Taurine levels in nucleus accumbens after self-administration of 0.1 g/kg ethanol (N = 2) or 0.7 g/kg ethanol (N = 2) or injection of 1 g/kg ethanol i.p (N = 4). Data are expressed as a percent of the first twenty samples for each treatment (basal).

4.   CONCLUSIONS

Our results confirm previous findings using conventional microdialysis with HPLC detection (20 minute samples) that 1 and 2 g/kg ethanol injections increase taurine levels in the nucleus accumbens in a dose-dependent manner[20-21]. In the present study utilizing CE-LIF, we have demonstrated an earlier onset (within 7 minutes) of large magnitude, dose-dependent increases in taurine levels as well as a longer lasting, lower magnitude increase. In addition, the 0.5 g/kg dose elicited a statistically significant increase in taurine levels which has not been reported previously.

The utilization of CE-LIF enabled for the first time the observation of the effects of a clinically significant self-administered "drinking bout" on taurine levels in the nucleus accumbens. Most importantly, the time resolution of CE-LIF will allow better understanding of amino acid and catecholamine changes associated with the appetitive and consummatory phases of ethanol self-administration.

The exact mechanism of ethanol-induced taurine release and the role this efflux plays in regulating dopamine, glutamate and GABA transmission in nucleus accumbens has yet to be fully elucidated. The mechanism of ethanol-induced taurine efflux would appear to have two components: a large, fast response sensitive to lower doses and a slower response to higher doses. The versatile nature of taurine makes the reason for this release less clear. It could involve taurine's function as an osmolyte, however, the effects of the low doses of ethanol would argue against this. It could involve taurine's role as a neuromodulator. In either case, taurine could modulate many of ethanol's effects in the CNS either at neurotransmitter receptors or via altering neurotransmitter efflux. It is this property of taurine that could provide novel insight into possible therapies for alcoholism.

ACKNOWLEDGEMENTS

This work was supported by Public Health Service grants NS 38476 and AA 07561.

REFERENCES

1. Nakamori, K., Koyama, I., Nakamura, T., Nemoto, M., Yoshida, T., Umeda, M., Inoue, K., 1993, Quantitative evaluation of the effectiveness of taurine in protecting the ocular surface against oxidant. *Chem Pharm Bull* (Tokyo) 41: 335-8.
2. Pasantes-Morales, H., Moran, J., Schousboe, A., 1990, Volume-sensitive release of taurine from cultured astrocytes: properties and mechanism. *Glia* 3: 427-32.
3. Pasantes-Morales, H., Martin del Rio, R., 1990, Taurine and mechanisms of cell volume regulation. *Prog Clin Biol Res* 351: 317-28.
4. Solis, J.M., Herranz, A.S., Herreras, O., Lerma, J., Martin del Rio, R., 1988, Does taurine act as an osmoregulatory substance in the rat brain? *Neurosci Lett* 91(1): 53-8.
5. Hussy, N., Deleuze, C., Desarmenien, M.G., Moos, F.C., 2000, Osmotic regulation of neuronal activity: a new role for taurine and glial cells in a hypothalamic neuroendocrine structure. *Prog Neurobiol* 62: 113-34.

6. Li, Y.P. and Lombardini, J.B., 1991, Taurine inhibits protein kinase C-catalyzed phosphorylation of specific proteins in a rat cortical P2 fraction. *J Neurochem* 56: 1747-53.

7. Huxtable, R.J., and Sebring, L.A., 1987, Modulation by taurine of calcium binding to phospholipid vesicles and cardiac sarcolemma. *Proc West Pharmacol Soc* 30: 153-5.

8. Sawamura, A., Azuma, J., Awata, N., Harada, H., Kishimoto, S., 1990, Modulation of cardiac Ca++ current by taurine. *Prog Clin Biol Res* 351: 207-15.

9. Huxtable, R.J., 1989, Taurine in the central nervous system and the mammalian actions of taurine. *Prog Neurobiol* 32: 471-533.

10. Oja, S.S., Korpi, E.R., and Saransaari, P., 1990, Modification of chloride flux across brain membranes by inhibitory amino acids in developing and adult mice. *Neurochem Res* 15: 797-804.

11. Bureau, M.H., Olsen, R.W., 1991, Taurine acts on a subclass of $GABA_A$ receptors in mammalian brain in vitro. *Eur J Pharmacol* 207: 9-16.

12. Quinn, M.R., Miller, C.L., 1992, Taurine allosterically modulates flunitrazepam binding to synaptic membranes. *J Neurosci Res* 33: 136-41.

13. Kurachi, M., Yoshihara, K., Aihara, H., 1983, Effect of taurine on depolarizations induced by L-glutamate and other excitatory amino acids in the isolated spinal cord of the frog. *Jpn J Pharmacol* 33: 1247-54.

14. Mori, M., Gahwiler, B.H., Gerber, U., 2002, Beta-alanine and taurine as endogenous agonists at glycine receptors in rat hippocampus in vitro. *J Physiol.*, 539: 191-200.

15. Lovinger, D.M., White, G., and Weight, F.F., 1989, Ethanol inhibits NMDA-activated ion current in hippocampal neurons. *Science* 243: 1721-4.

16. Heidbreder C., De Witte, P., 1993, Ethanol differentially affects extracellular monoamines and GABA in the nucleus accumbens. *Pharmacol Biochem Behav* 46: 477-81.

17. Ferko, A.B., Bobyock, E., 1988, Effect of taurine on ethanol-induced sleep time in mice genetically bred for differences in ethanol sensitivity. *Pharmacol Biochem Behav* 31: 667-73.

18. Aragon, C.M., Trudeau, L.E., and Amit, Z., 1992, Effect of taurine on ethanol-induced changes in open-field locomotor activity. *Psychopharmacology* 107: 337-40.

19. Aragon, C.M., and Amit, Z., 1993, *Taurine and ethanol-induced conditioned taste aversion. Pharmacol Biochem Behav* 44: 263-6.

20. Dahchour, A., Quertemont, E., and De Witte, P., 1994, Acute ethanol increases taurine but neither glutamate nor GABA in the nucleus accumbens of male rats: a microdialysis study. *Alcohol Alcohol* 29: 485-7.

21. Dahchour, A., Quertemont, E., and De Witte, P., 1996, Taurine increases in the nucleus accumbens microdialysate after acute ethanol administration to naive and chronically alcoholised rats. *Brain Res* 735: 9-19.

22. Dahchour, A., and De Witte, P., 1999,Effect of repeated ethanol withdrawal on glutamate microdialysate in the hippocampus. *Alcohol Clin Exp Res* 23: 1698-703.

23. Quertemont, E., de Neuville, J., and De Witte, P., 1998, Changes in the amygdala amino acid microdialysate after conditioning with a cue associated with ethanol. *Psychopharmacology* (Berl) 139: 71-8.

24. Quertemont, E., Linotte, S., and De Witte, P., 2002, Differential taurine responsiveness to ethanol in high- and low-alcohol sensitive rats: a brain microdialysis study. *Eur J Pharmacol* 444: 143-50.

25. Quertemont, E., Lallemand, F., Colombo, G., De Witte, P., 2000, Taurine and ethanol preference: a microdialysis study using Sardinian alcohol-preferring and non-preferring rats. *Eur Neuropsychopharmacol*, 10: 377-83.

26. Bowser, M.T. and Kennedy, R.T., 2001, *In vivo* monitoring of amine neurotransmitters using microdialysis with on-line capillary electrophoresis. *Electrophoresis* 22: 3668-76.

27. Lada, M.W, Vickroy, T.W., Kennedy, R.T.,1998, Evidence for neuronal origin and metabotropic receptor-mediated regulation of extracellular glutamate and aspartate in rat striatum *in vivo* following electrical stimulation of the prefrontal cortex. *J Neurochem*, 70: 617-25.

Tyrosine Kinases and Taurine Release

Signaling Events and Amino Acid Release Under Hyposmotic and Ischemic Conditions in the Chicken Retina

L.D. OCHOA-DE LA PAZ* and R.A. LEZAMA*

*Department of Biophysics, Institute of Cell Physiology, National University of Mexico, Mexico City, Mexico

1.   INTRODUCTION

Amino acids are part of the pool of organic osmolytes contributing to volume regulation after hyposmotic swelling in most animal cells, including brain and retina[1-5]. Swelling occurs under ischemic conditions consequent to energy failure, which impairs the Na/K ATPase and leads to an increase in intracellular Na^+, followed by Cl^- influx and osmotically obligated water. The subsequent depolarization results in excessive extracellular K^+ which is removed by uptake, and is also followed by Cl^- and water[6]. While an active response to volume correction is known to occur following hyposmotic swelling, it is unclear whether ischemic swelling activates a similar volume regulatory mechanism.

Volume regulation after hyposmotic swelling is accomplished by the active extrusion of K^+, Cl^- and organic molecules, but such mechanisms may not operate under ischemic conditions, since the main osmolytes K^+ and Cl^- are causal of swelling. Under these conditions, amino acids may contribute importantly to cell volume adjustment or at least, to attenuate the magnitude of the swelling. Among amino acid osmolytes, taurine plays a key role, particularly in the retina where it is found in very high concentrations. Taurine has often been considered as representative of the organic osmolytes in terms of mechanisms of release, but this similarity has to be taken with caution, since marked differences are found between taurine and other amino acids, at least in brain cell preparations.

For volume regulation to occur after swelling, a number of steps have to be accomplished: first, a volume sensor indicates the change in cell volume, which in turn activates a signaling cascade transducing this information to the effectors, *i.e.*, osmolytes in charge of volume correction. Finally, inactivation of all these reactions takes place, once the volume sensor has detected the return to the original cell volume. Most of our present knowledge of this chain of events refers to the mechanisms of osmolyte efflux, while the signaling cascade has only just started to be explored. No information is yet available on the nature of the volume sensor. Most studies on volume regulation have been restricted so far to the hyposmotic swelling model, but considerably less is known about the ischemic model.

In the present study in the retina, a model of ischemia, consisting of high extracellular $[K^+]$ and ouabaine, was used to determine if volume regulation occurs and, if so, whether the release of amino acids contributes to this process. Efflux of labeled taurine, GABA and glutamate (traced as D-aspartate) was characterized in hyposmotic and ischemic swelling, stressing the fact that amino acids serve both as osmolytes and neurotransmitters. Attempts were made to identify the elements of the signaling cascade in the two conditions of swelling.

1.1   Volume Regulation and Amino Acid Efflux

A rapid increase in cell volume in the retina is observed following exposure to both hyposmotic and ischemic conditions, but whereas in the first condition volume regulation immediately starts, in the ischemic situation volume recovery is only partial. Both conditions elicit a rapid efflux of taurine, glutamate and GABA, the three most abundant amino acids in the retina [7]. Since these amino acids may function as synaptic transmitters, they may respond to depolarization rather than to swelling, since both hyposmolarity and ischemic conditions are highly depolarizing stimuli. To discriminate between these two possibilities, amino acid release was measured in a medium made hyperosmolar with raffinose. In this condition the efflux of taurine and GABA is essentially abolished while that of glutamate is less affected (Fig. 1).

The osmosensitive release of taurine is consistently sensitive to Cl⁻ channel blockers, and this is also true for the efflux evoked by ischemia. In contrast, glutamate efflux exhibits a much lower sensitivity to these agents, stressing the difference with a typical osmolyte behavior.

2.   Transducing Signal.   Role of Tyrosine Kinases

Our studies have focussed on the influence of tyrosine kinases as signaling elements for transducing the changes in cell volume in amino acid osmolyte fluxes. This is based on the modulatory effect of these kinases and the volume-sensitive Cl⁻ channel. We have previously reported the influence of tyrosine kinase manipulation on hyposmotic taurine efflux in cerebellar granule neurons and hippocampal·slices, and a similar influence has been recently found in fibroblasts[8-10]. There is only one report about an effect of tyrosine kinase blockers in ischemic conditions in the rat heart[11]. Table 1 shows that taurine release evoked by hyposmotic or ischemic conditions in the retina is markedly reduced by general tyrosine kinase blockers, tyrophostine A-23 and genistein, while glutamate fluxes are markedly less influenced by this treatment, suggesting either different stimuli for release and/or different pathways modulated by other types of kinases.

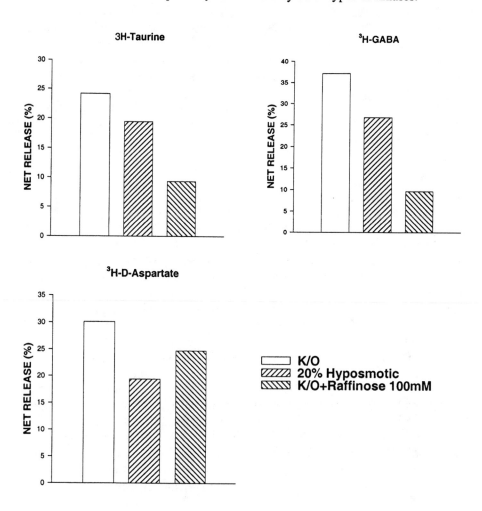

Figure 1. Release of ³H-taurine, ³H-GABA and ³H-D-Aspartate from retina evoked by hyposmotic or ischemic (K/O) conditions. Retinas were loaded with the labeled amino acids, washed and superfused with KH control medium. The K/O medium was made hyperosmotic with raffinose. Data represent net release, i.e. the percentage radioactivity released during the superfusion period with the experimental media (7 min) minus the basal release.

Table 1. Effect of tyrosine kinases and PI3-K blockers on ^{3}H-Taurine, ^{3}H-GABA and ^{3}H-D-Aspartate release in ischemic and hyposmotic conditions.

	Ischemia			20% Hyposmotic		
	^{3}H-Taurine (%)	^{3}H-GABA (%)	^{3}H-D-spart (%)	^{3}H-Taurine (%)	^{3}H-GABA (%)	^{3}H-D-Aspartate (%)
Control	24.8± 1.86	35.2 ± 3.6	31.8 ± 3.3	18.9 ± 2.6	26.3 ± 4.3	29.5 ± 1.8
Tyrphostine 25μM	2.5 ± 0.8**	2.6 ± 1.9**	34.5± 3.0	1.8 ± 0.2**	3.4 ± 1.2**	27.1 ± 2.2
Genistein 25μM	3.8 ± 0.6**	19.5± 3.0**	32.7 ± 3.4	1.9 ± 0.5**	25.0 ± 1.1	189± 1.4*
Wortmannin 100nM	9.5± 1.1**	14.2± 2.8*	28.9 ± 2.9	4.8± 3.5**	5.2 ± 2.0	30.9 ± 2.7

Data represent net release, i.e., the percentage radioactivity release during the superfusion period with the experimental media (7min) minus the basal release. Means ± SE, n=4-6. **P< 0.001 *P > 0.02 vs control (Student test t-test)

While these results underline the influence of tyrosine kinases on swelling-related amino acid release, the identity of the tyrosine kinases has not yet been established. In the retina, both hyposmolarity and ischemia rapidly enhance phosphorylation of p125^{FAK} and p38, without affecting ERK1/ERK2[7]. Kinase activation does not prove per se a link with the swelling-evoked osmolyte efflux. Both swelling and ischemia are complex phenomena resulting in a variety of events required for the cell to adapt to cell volume and to the stressful ischemic condition, evoking, in turn, many different signals which may or may not be connected with the activation of osmolytes fluxes. A more direct way to establish such a correlation is to examine the effect of tyrosine kinases on these fluxes. The MAP kinases ERK1/ERK2 appear not involved in the mechanism of osmolyte release, since either do not activate as in the present study [8,11,12] or, if activated, osmolytes fluxes remain unaffected when the activation is blocked[7,13,14]. Hyposmolarity also phosphorylates another MAP kinase, p38, in retina as well as in hepatocytes, cholangiocytes and hippocampal slices[8,12,13,14], but again, these events appear unlinked to osmosensitive taurine and Cl$^-$ permeation[8,13,14]. Therefore, even though the influence of tyrosine kinases on the osmolyte exit mechanism is well documented, the precise kinases involved remain largely unidentified. The closest approach to ascribe a specific tyrosine kynase to an osmolyte efflux pathway comes from studies on the influences of the src kinases p72^{syk} and p56^{lyn} on taurine efflux in skate blood cells[15] and of p56^{lyn} on Cl$^-$ currents in lymphocytes [16].

A tyrosine kinase target, the phosphoinositide-3 kinase (PI3-K), seems to be an important element in the signal transduction cascade connecting cell swelling to taurine and GABA efflux in the retina. This is indicated by its activation under both hyposmotic and ischemic conditions[7] as well as by the pronounced decrease of amino acid fluxes by PI3K blockers (Table 1). We have previously shown such a correlation for osmosensitive taurine efflux in cerebellar granule neurons[7] and hippocampal slices[11] but this is the first report of a similar connection for osmosensitive GABA efflux and for ischemia-evoked amino acid release. Besides our results in brain cells, PI3-K, activation by hyposmolarity has been observed so far in hepatocytes and the intestinal 407 cell line[12,17]. All these results, while stressing the importance of PI3-K in volume regulation processes provide no information about its link with specific components of the tyrosine kinase pathways, nor about its influence on the mechanism ultimately resulting in osmolyte efflux. Remarkably, the efflux of glutamate was much less affected by these blockers, again showing marked deviations from the typical amino acid osmolyte release pattern.

Amino acid release in the ischemic brain *in vivo* has been reported, but less is known about the possible modulation of this release by tyrosine kinase reactions. Ischemia activates phosphorylation of ERKs, p38, JNK and PI3-K in brain[18,19] and of PI3-K and MAPK in the retina[13], but, in most cases, phosphorylation occurs rather in connection with reperfusion, and in any case a link with amino acids fluxes has been established. An important point often raised, is the extent to which amino acid efflux in ischemia is a primary response to swelling concurrent with the ischemic episode. The similarities found between ischemia- and hyposmolarity-evoked GABA and taurine release suggest swelling as the primary stimulus for release. Also in ischemic models *in vivo*, taurine efflux shows properties essentially similar to those of osmosensitive fluxes described in retina and cultured cells, suggesting a main osmolyte-type response[21,3]. The release of glutamate is clearly different from those of taurine and GABA, notably in its lower sensitivity to tyrosine kinases and PI3-K blockers, as well as Cl to channel inhibitors. Noteworthy, however, even the hyposmolarity-evoked release of glutamate shows features different from those of taurine[8,13]. In any event, it is of interest to characterize the media for glutamate release during episodes which may contribute to exacerbate excitotoxicity.

ACKNOWLEDGEMENTS

This work was supported in part by grants No. 34886-M and 35806 from CONACYT and IN204900 from DGAPA, UNAM.

REFERENCES

1. Kirk ,K., 1997, Swelling-activated organic osmolytes channels. *J. Membr. Biol.* 158:1-16.
2. Lang, F., Busch, G.L., Ritter, M., Volkl, H., Waldegger, S., Gulbins, E., and Haussinge D.,1998, Functional significance of cell volume regulatory mechanism. *Physiol. Rev.* 8:247-306.
3. Pasantes-Morales, H., 1996, Volume regulation in brain cells: cellular and molecular mechanisms. *Metab. Brain Dis.*1 1:187-204.
4. Pasantes-Morales, H., Cardin, V., and Tuz, K., 2000, Signaling events during swelling and regulatory volume decrease. *Neurochem. Res.* 25:1301-1314.
5. Pasantes-Morales, H., Ochoa de la Paz, L.D., Sepúlveda, J., and Quesada, O., 1999, Amino acid as osmolytes in the retina. *Neurochem. Res.* 24:1339-1346.
6. Kimelberg, H.K., 1990, Chloride transport across glial membranes, in *Chloride Channels and Carriers in Nerve, Muscle, and Glial Cells.* Alvarez-Leefmans and Russell, Eds. Plenum Press, New York, pp.1-9.
7. Ochoa-de la Paz, L.D., Lezama, A., Torres-Marquez, M.E., and Pasantes-Morales, H., 2002, Tyrosine kinases and amino acids efflux under hyposmotic and ischaemic conditions in the chicken retina. *Pflüger Arch.* 445:87-96.
8. Morales-Mulia, S., Cardin, V., Torres-Marquez, M.E., Crevenna, A., and Pasantes-Morales, H., 2001, Influence of protein kinases on the osmosensitive release of taurine from cerebellar granule neurons. *Neurochem. Int.* 38: 153-161.
9. Franco, R., Torres-Marquez, M.E., and Pasantes-Morales, H., 2001, Evidence for two mechanisms of amino acid osmolyte release from hippocampal slices. *Pflugers Arch.* 442:791-800.
10. Pedersen, S.F., Beisner, K.H., Hougaard, C., Willumse, B.M., Lambert, I.H., and Hoffmann, E.K., 2002, Rho family GTP binding proteins are involved in the regulatory volume decrease process in NIH3T3 mouse fibroblasts. *J. Physiol.* 541: 7779-7796.
11. Feranchak, A.P., Berl, T., Capasso, J., Wojtaszek, P.A., Han, J., and Fitz, J.G., 2001, p38 MAP kinase modulates liver cell volume through inhibition of membrane Na+ permeability. *J. Clin. Invest.* 108:1495-1504.
12. Song, D., O'Regan, M.H., and Phillis, J.W., 1998, Protein kinase inhibitors attenuate cardiac swelling-induced amino acid release in the rat. *J. Pharm. Pharmacol.* 50:1280-1286.
13. vom Dahl, S., Schliess, F., Graf, D., and Haussinger, D., 200, Role of p38(MAPK) in cell volume regulation of perfused rat liver. *Cell Physiol. Biochem.*11:285-94.
14. van der Wijk, T., Tomassen, S.F., de Jonge, H.R., and Tilly, B.C., 2000, Signalling mechanisms involved in volume regulation of intestinal epithelial cells. *Cell Physiol. Biochem.*10:289-296.
15. Hubert, E.M., Musch, M.W., and Goldstein, L., 2000, Inhibition of volume-stimulated taurine efflux and tyrosine kinase activity in the skate red blood cell. *Pflugers Arch.* 440:132-139.
16. Lepple-Wienhues, A., Szabo, I., Wieland, U., Heil, L., Gulbins, E., and Lang, F., 2000, Tyrosine kinases open lymphocyte chloride channels. *Cell Physiol. Biochem.*10:307-312.
17. Feranchak, A.P, Roman, R.M., Schwiebert, E.M., and Fitz, J.G., 1998, Phosphatidylinositol 3-kinase contributes to cell volume regulation through effects on ATP release. *J. Biol. Chem.* 273:14906-14911.
18. Namura, S., Iihara, K., Takami, S., Nagata, I., Kikuchi, H., Matsushita, K., Moskowitz, M.A., Bonventre, J.V., and Alessandrini, A., 2001, Intravenous administration of MEK inhibitor U0126 affords brain protection against forebrain ischemia and focal cerebral ischemia. *Proc. Natl. Acad. Sci.* 98:11569-11574.
19. Park, K.M., Chen, A., and Bonventre, J.V., 2001, Prevention of kidney ischemia/reperfusion-induced functional injury and JNK, p38 and MAPK kinase activation by remote ischemic pretreatment. *J. Biol. Chem.* 276:11870-11876.
20. Hayashi, A., Kim, H.C., and de Juan, E. Jr., 1999, Alterations in protein tyrosine kinase pathways following retinal vein occlusion in the rat. *Curr. Eye Res.* 18:231-239.
21. Rutledge, E.M., Aschner, M., and Kimelberg, H.K., 1998, Pharmacological characterization of swelling-induced D-[3H]-aspartate release from primary astrocyte cultures. *Am. J. Physiol.* 274: C1511-C1520.

Effect of Taurine on Regulation of GABA and Acetylcholine Biosynthesis

DI SHA[*,#], JIANNING WEI[*], HONG JIN[*], HENG WU[*,#],
GREGORY L. OSTERHAUS[*], and JANG-YEN WU[*,#]
[*]Department of Molecular Biosciences, University of Kansas, Lawrence, KS, 66045
[#]Biomedical Sciences, Florida Atlantic University, Boca Raton, FL, 33431-0991 U.S.A.

1.   INTRODUCTION

Taurine is one of the most abundant amino acids in mammals[1] and is known to be involved in many important physiological functions, *e.g.*, as a trophic factor in the development of CNS[2], maintaining the structural integrity of the membrane[3], regulating calcium binding and transport[4-5], as an osmolyte[6], a neuromodulator[7], a neurotransmitter[8-10] [for review see reference 11] and a neuroprotector[12-14].

Recently, we have shown that taurine exerts its neuroprotection function through its inhibition of reverse mode of Na^+/Ca^{2+} exchanger and release of Ca^{2+} from the internal storage pools resulting in lowering the intracellular Ca^{2+}, $[Ca^{2+}]_i$[15]. Here we report that in addition to the numerous functions aforementioned, taurine has one more important function, namely, to regulate neurotransmitter synthesis presumably through its effect on Ca^{2+}-dependent protein kinase and phosphatase activity.

2.   EFFECT OF TAURINE ON PROTEIN KINASE C (PKC)-MEDIATED ACTIVATION OF HUMAN BRAIN GLUTAMATE DECARBOXYLASE$_{65}$ (HGAD$_{65}$)

HGAD$_{65}$ was found to be activated by PKC to an extent of about 50%. Furthermore, this activation by PKC was abolished in the presence of taurine in a dose-dependent manner (Fig.1). The activation of HGAD$_{65}$ by PKC was reduced from 50% to 20, 15, 5, and 0% by taurine at 0.5, 1, 2, and 10mM respectively. Taurine itself alone at the same concentration tested has no effect on HGAD$_{65}$ activity (data not shown).

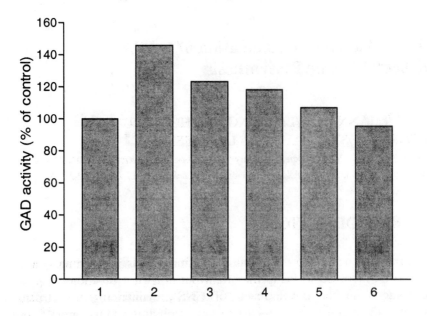

Figure 1. Effect of Taurine on the PKC activation of HGAD$_{65}$.

Lane 1, control: HGAD$_{65}$ in buffer alone; Lane 2, HGAD$_{65}$ treated with PKC; Lane 3, same as Lane 2, plus 0.5 mM taurine; Lane 4, same as Lane 2, plus 1 mM taurine; Lane 5, same as Lane 2, plus 2 mM taurine; Lane 6, same as Lane 2, plus 10 mM taurine.

For PKC treatment, HGAD$_{65}$ in standard GAD buffer (50mM KP, 1 mM AET, 0.2 mM PLP, pH = 7.2) was dialyzed extensively in dialysis buffer (15 mM Tris-Acetate, 1 mM AET, 0.2 mM PLP, pH = 7.2). Aliquot of HGAD$_{65}$ (~10-20 µg) was incubated at 30°C in 150 µl reaction mixture containing 200 µM ATP, 25 ng PKC, 103 µg/ml phosphatidylserine, 20 µg/ml diacylglycerol, 1 mM CaCl$_2$ in dialysis buffer for 30 min. No PKC was added in the control condition. For taurine treatment: HGAD$_{65}$ was incubated in the same reaction mixture as PKC treatment except that taurine was added to give a final concentration of 0.5 mM, 1 mM, 2 mM and 10 mM respectively. After incubation, the reaction mixture was divided into three aliquots and assayed for GAD activity. GAD activity was assayed by a radiometric method measuring the formation of $^{14}CO_2$ from [L-^{14}C] glutamic acid as described previously[16].

3. EFFECT OF TAURINE ON ATP-MEDIATED ACTIVATION OF CHOLINE ACETYLTRANSFERASE (ChAT) ACTIVITY.

Both membrane-associated ChAT (MChAT) and soluble ChAT (SChAT) were found to be activated by ATP. At 5mM, ATP activated MChAT and SChAT to 110 and 25% over the control values. (Fig. 2, Lane 2). In addition, ATP- mediated activation of both MChAT and SChAT was greatly inhibited by taurine. At 0.1mM, taurine reduced ATP activation of MChAT and SChAT from 110 and 25% to 25 and 0%, respectively (Fig.2, Lane 3). ATP activation of MChAT was completely abolished by taurine at 1mM concentration (Fig. 2, Lane 4). Taurine had no inhibitory effect on both MChAT and SChAT at the concentration tested.

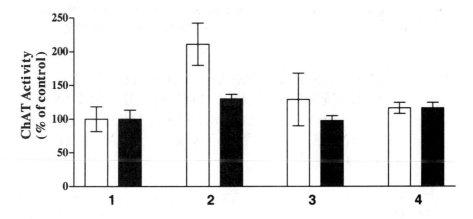

Figure 2. Effect of taurine on ATP-mediated activation of choline acetyltransferase (ChAT) activity.

Synaptosomes were prepared from fresh porcine brain as described [18]. Briefly, fresh porcine brains were homogenized in 0.32M sucrose, and the homogenate was centrifuged at 1,000 x g for 10 min. The supernatant solution was collected and centrifuged at 12,000 x g for 30 min. The resulting supernatant liquid was discarded and the pellet thus obtained was the crude synaptosome preparations. Synaptosome preparations were resuspended in 50mM potassium phosphate buffer (standard KP buffer, pH=7.4) and subjected to sonication 3 x 3 seconds. The suspension was centrifuged at 100,000 x g for 1 hour. The resulting supernatant liquid was referred as SChAT. The pellet thus obtained was resuspended in standard KP buffer and used as the source of MChAT.

The activity of both MChAT and SChAT was assayed by the radiometric method as previously described with some modifications[20]. In a typical assay, an aliquot of 50 µl of ChAT sample was incubated with 150 µl of standard reaction mixture (containing 50 mM choline chloride, 10 mM $(NH_4)_2SO_4$, 0.5 mg/ml bovine serum albumin (BSA), 0.1 mM acetyl-CoA, 37.5 µM neotigmine methyl sulfate, 1 mM dithiothreitol (DTT), 4 mM $MgCl_2$, 1.5 mM $CaCl_2$, 10 µCi/µmol [H^3] acetyl-CoA, 50 mM HEPES buffer, pH = 7.4) at 37°C for 30 min with constant shaking. In the case that ATP and taurine were tested, these compounds were added to the reaction mixture individually or together to give the final concentration as indicated. Lane 1, control; Lane 2, the same as Lane 1 except including 5 mM ATP; Lane 3, the same as Lane 2 except including 0.1 mM taurine; Lane 4, the same as Lane 2 except including 1 mM taurine. ChAT activity is expressed as percentage of activity using ChAT activity in control groups 100%. Data are expressed as the mean ± standard deviation. The bar indicates the standard deviation with n=3. Open column: MChAT activity; Solid column: SChAT activity.

4.   EFFECT OF TAURINE ON ATP/CALCIUM CALMODULIN KINASE II (CaM KII)-MEDIATED ACTIVATION OF MEMBRANE-ASSOCIATED CHOLINE ACETYLTRANSFERASE (MChAT) ACTIVITY.

Recently, we have reported that MChAT activity is greatly enhanced by ATP/CaM KII[16]. Here we found that MChAT activity was increased by almost 400% by a combination of ATP and CaM KII. Furthermore, this activation was markedly inhibited by taurine which reduced ATP/CaM KII activation of MChAT form 400% to only 35% at 5 mM (Fig. 3, Lane 5, 6). ATP alone also increased MChAT activity and ATP-mediated activation of MChAT was also inhibited by taurine similar to the results of Fig. 2 (Fig. 3, Lane 3, 4). Taurine alone at 5mM showed no significant effect on MChAT activity (Fig. 3, Lane 2).

MChAT samples were prepared as described in Fig. 2. ChAT activity was determined under following conditions: Lane 1: Control; Lane 2: 5 mM Taurine; Lane 3: 5 mM ATP; Lane 4: 5 mM Taurine + 5 mM ATP; Lane 5: 5 mM ATP + 1.5 mM Ca^{2+} + 1 µM Calmodulin; Lane 6: 5 mM Taurine + 5 mM ATP + 1.5 mM Ca^{2+} + 1 µM Calmodulin. ChAT activity is expressed as percentage of activity using ChAT activity in control group as reference, 100%. Data are expressed as the mean ± standard deviation. The bar indicates the standard deviation with n=3.

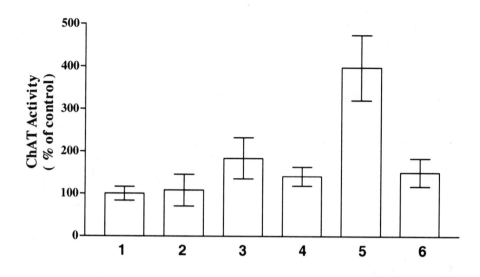

Figure 3. Effect of Taurine on ChAT Activity stimulated by ATP and CaM KII.

5.   CONCLUSION

Previously we have shown that the enzyme involved in the biosynthesis of neurotransmitter GABA and acetylcholine (ACh), GAD and ChAT, respectively are regulated by protein phosphorylation[17-19]. In this communication, we present evidence to show that taurine has one additional function, namely, to regulate the biosynthesis of neurotransmitter, e.g., GABA and ACh, in addition to its known functions as an osmolyte, inhibitory transmitter, Ca^{2+} regulator, trophic factor, membrane stabilizer and signaling molecule [for review, see reference 11]. Here we show that activation of both GAD65 and ChAT by ATP in the presence of PKC and CaM KII, respectively, is markedly reduced by taurine at 0.1-10mM.

This is physiologically significant since the concentration of taurine in the brain is in mM concentration[20]. In fact, it has been estimated in some neuronal types, such as neurons in the hippocampus, the concentration of taurine is around 20mM[21]. Therefore, at physiological concentration, taurine can exert its effect on neurotransmission through its effect on the synthesis of neurotransmitters, e.g. GABA and ACh. The mechanism of inhibition of ATP/Kinase-mediated activation of GAD and ChAT by taurine remains unclear. Since both PKC and CaM KII require Ca^{2+} for their activity, it is conceivable that taurine may inhibit kinase activity through its interaction with Ca^{2+} resulting in decrease in binding of Ca^{2+} to Ca^{2+}-dependent kinases.

Alternatively, taurine may interfere with Ca^{2+} binding to the kinases through its interaction with the Ca^{2+}-binding domains in those Ca^{2+}-dependent kinases. In summary, taurine is not only involved in regulation of neurotransmitters synthesis, but may also have a global effect on cellular functions since PKC and CaM KII are known to play important roles in many physiological and pathological processes.

ACKNOWLEDGEMENTS

The study was supported in part by grants from NSF, IBN-9723079, NIH, NS37851 and Schmidt Family Foundation at Florida Atlantic University.

REFERENCES

1. Jacobsen, J.G. and Smith, L.H. 1968. Biochemistry and physiology of taurine and taurine derivatives. *Physiol. Rev.* 48: 424-511.
2. Sturman, J. A. 1993. Taurine in development. *Physiol. Rev.* 73:119-147.
3. Moran, J., Salazar, P. and Pasantes-Morales, H. 1988. Effect of tocopherol and taurine on membrane fluidity of retinal rod outer segments. *Experimental Eye Research*. 45: 769-776.
4. Lazarewicz, J. W., Noremberg, K., Lehmann, A. and Hamberger, A. 1985. Effects of taurine on calcium binding and accumulation in rabbit hippocampal and cortical synaptosomes. *Neurochem. Int.* 7:421-428.
5. Lombardini, J. B. 1985. Effects of taurine on calcium ion uptake and protein phosphorylation in rat retinal membrane preparations. *J. Neurochem.* 45: 268-275.
6. Schaffer, S., Takahashi, K., and Azuma, J. 2000. Role of osmoregulation in the actions of taurine. *Amino Acids.* 19(3-4): 527-546.
7. Kuriyama, K. 1980. Taurine as a neuromodulator. *Fed. Proc.* 39:2680-2684.
8. Mandel, P., Pasantes-Morales, H. and Urban, P. F. 1976. Pages 89-105, in S. L. Bontig (ed.), *Transmitters in the Visual Process*, Pergamon, Oxford.
9. Lin, C.-T., Su, Y. Y. T., Song, G.-X. and Wu, J.-Y. 1983. Is taurine a neurotransmitter in rabbit retina? *Brain Res.* 337:293-298.
10. Okamoto, K., Kimura, H. and Sakai, Y. 1983. Evidence for taurine as an inhibitory neurotransmitter in cerebellar stellate interneurons: Selective antagonism by TAG (6-aminomethyl-3-methyl- ^{4}H, 1, 2, 4-benzothiadiazine-1,1-dioxide). *Brain Res.* 265(1):163-168.
11. Huxtable, R.J. Expanding the circle 1975-1999: sulfur biochemistry and insights on the biological functions of taurine. *Adv Exp Med Biol.* 2000;483:1-25.
12. Tang, X. W., Deupree, D. L., Sun, Y. and Wu, J.-Y. 1996. Biphasic effect of taurine on excitatory amino acid-induced neurotoxicity. Pages 499-506, in R. J. Huxtable, J. Azuma, M. Nakagawa, K. Kuriyama, and A. Bala, (eds.), *Taurine: Basic and Clinical Aspects*, Plenum Publishing Co.
13. El Edrissi, A. and Trenkner, E. 1999. Growth factors and taurine protect against excitotoxicity by stabilizing calcium homeostasis and energy metabolism. *J. Neurosci.* 19:9459-9468.

14. Saransari, P. and Oja, S. S. 2000. Taurine and neural cell damage. *Amino Acids.* 19(3-4): 509-526.

15. Chen, W. Q., Jin, H., Nguyen, M., Carr, J., Lee, Y. J., Hsu, C. C., Faiman, M. D., Schloss, J. V. and Wu, J.-Y. 2001 (a). Role of taurine in regulation of intracellular calcium level and neuroprotective function in cultured neurons. *J. Neurosci. Res.* 66: 612-619.

16. Wong, E., Schousboe, A., Saito, K., Wu, J. Y., Roberts, E. Immunochemical studies of brain glutamate decarboxylase and GABA-transaminase of six inbred strains of mice. *Brain Res.* 1974 Mar 15;68(1):133-42.

17. Sha, D., Jin, H., Kopke, R.D., Wu, J.Y. Choline Acetyltransferase: Regulation and Coupling with Protein Kinase and Vesicular Acetylcholine Transporter on Synaptic Vesicles (Submitted).

18. Bao, J., Nathan, B., Hsu, C. C., Zhang, Y., Wu, R. and Wu, J. Y. role of protein phosphorylation in regulation of brain L-glutamate decarboxylase activity. *J Biomed Sci.* 1(1994): 237-244.

19. Bao, J., Cheng, W.Y. and Wu, J. Y. Brain L-glutamate decarboxylase inhibition by phosphorylation and activation by dephosphorylatrion. *J. Biol. Chem.* 270(1995): 6464-6467.

20. Brandon, C., and Wu, J. Y. Purification and properties of choline acetyltransferase from *Torpedo Callifornia. J. Neurochem.* 30(1978): 791-797

21. Gaitonde, M.K. Sulfur Amino Acid. in *Handbook of Neurochemistry.* Edited by Abel Lajtha. 1970. PP. 262.

22. Olson JE, Li GZ. Osmotic sensitivity of taurine release from hippocampal neuronal and glial cells. *Adv Exp Med Biol* 2000;483:213-8.

Taurine Effect on Neuritic Outgrowth from Goldfish Retinal Explants in the Absence and Presence of Fetal Calf Serum

LUCIMEY LIMA, SUZANA CUBILLOS and FILI FAZZINO

Laboratorio de Neuroquímica, Centro de Biofísica y Bioquímica, Instituto Venezolano de Investigaciones Científicas, Apdo. 21827, Caracas 1020-A, Venezuela, llima@cbb.ivic.ve

Summary

Post-crush goldfish retinal explants were cultured in the absence of fetal calf serum and in the presence of calcium, albumin, glucose or taurine. There was an increase in length and density of the neurites in a dose-dependent manner when fetal calf serum was added to the medium. The addition of calcium to the basic medium, Leibovitz, produced a small increase in neuritic outgrowth at low concentrations, but inhibited outgrowth at higher levels. Albumin did not significantly modify neuritic outgrowth. Glucose, at variable concentrations, produced a bell-shaped increase in length and density of the neurites. Taurine, in the absence of fetal calf serum, produced a small increase in neuritic outgrowth at 10 mM. In the presence of glucose, taurine stimulated outgrowth was less efficient than in the presence of fetal calf serum. In order to investigate the mechanisms of action of taurine as a trophic agent, the culture medium in the absence of fetal calf serum will be supplemented with glucose.

1. INTRODUCTION

Neuritic outgrowth from goldfish retinal explants in the presence of fetal calf serum (FCS) is stimulated by the addition of taurine and is dramatically restricted in the absence of it[1]. A number of studies have been performed for investigating the influence of stimulating agents *per se* without the concomitant effect of several substances present in the serum[2-5]. Although the trophic effect of taurine is known to follow concentration-dependency, time-related modifications, variations with its own transport, modulation by calcium influx, relation to protein phosphorylation, among others[6], the requirements of specific molecules interacting with taurine during regeneration *in vitro* are unknown at the present. In previous reports it was found that the medium from optic tectum cultures and the coculture of optic tectum and retina differentially influence the patterns of neuritic outgrowth[7]. Thus, the definition of a medium in the absence of FCS will be useful for further studies concerning the mechanisms of action of taurine as a trophic agent in the retina.

2.   MATERIALS AND METHODS

2.1   Culture of Retinal Explants

Goldfish (*Carassius auratus*), 5-6 cm in length, were anesthetized in 0.05% tricaine. The retinas were dissected 10 days after a crush lesion of the optic nerve was made with fine forceps. Squares of retina (500 μm, 5-6 per retina) were prepared by sectioning the retina with a McIlwain tissue chopper. The squares were then placed, 10-14, in poly-L-lysine pre-coated tissue flasks (25 mm^2). FCS was used in some cultures in concentrations of 0.1, 0.5, 1, 2, 5 and 10%, but was omitted in those cultures in which the following conditions were tested: calcium 1.5, 3 and 6 mM; albumin, 0.01, 0.05, 0.1, 0.5, 1, 5 and 10%; or glucose, 0.01, 0.05, 0.5, 1, 2, 5 and 25 mM. Taurine, 0.2, 2, 4 and 10%[1], was added to some explants. The nutrient medium was Leibovitz, L-15, 3 ml per dish (Sigma) with 0.1 mg/ml of gentamicin and 10 or 20 mM of (*N*-[2-hydroethyl]piperazine-*N*'-[2-ethanesulfonic acid]) (HEPES). Sucrose was added in some experiments in concentrations of 0.5 to 10 mM. The density of the neurites was evaluated at 5 and 10 days after plating using a predetermined scale and the length was determined by using the program SigmaScanPro (Jandel). Results are expressed as mean ± SEM. Analysis of variance and the Student's *t* test for comparing results at 5 and 10 days were utilized. Statistical significance was accepted if P < 0.05.

3.   RESULTS

3.1   Neuritic Outgrowth in the Presence of Fetal Calf Serum, Calcium, Albumin or HEPES

Neuritic outgrowth (length and density) increased in the presence of FCS (0.1-10%) in a dose-dependent manner, and there was a significant increase in the length of the neurites between 5 and 10 days in culture. The length of the neurites was between 408.39 ± 31.63 for 0.1% and 1794.81 ± 62.46 for 10% FCS. The addition of calcium at several concentrations produced an increase in neuritic length from the retinal explants with the maximum increase at 3 mM, but reduced neuritic outgrowth at higher concentrations after 5 days in culture: 0, 3, 6 and 12 mM, 313.55 ± 17.49; 410.78 ± 26.23; 279.84 ± 27.10; and 122.45 ± 15.74 μm, respectively. A significant increase was observed at 10 days in culture. The addition of albumin did not increase the length of the neurites at 5 days in culture, but there was an increase in length with 1, 5 and 10 mM at 10 days in culture (Table 1). HEPES produced a slight increase in the length of the neurites at 20 mM, but a non significant change was observed with 20 mM, and there were no differences between 5 and 10 days in culture. The combination of albumin or HEPES with FCS did not result in any change with respect to FCS alone (data not shown).

Table 1. Effect of variable concentrations of albumin on neuritic length (μm) from goldfish retinal explants in the absence of fetal calf serum.

Albumin concentration (%)	5 days in culture	10 days in culture
0	426.66 ± 27.91	548.02 ± 42.94*
0.01	371.13 ± 23.77	440.24 ± 35.53*
0.05	363.14 ± 24.03	499.44 ± 35.60*
0.1	346.33 ± 24.03	633.60 ± 48.46*
0.5	365.72 ± 24.19	546.69 ± 34.24*
1	472.79 ± 30.24	913.85 ± 36.32*
2	425.14 ± 35.87	707.26 ± 32.57*
10	480.42 ± 26.55	846.52 ± 41.99*

5 days, $F_{(7,388)}$ = 3.98, $P < 0.001$; 10 days $F_{(7,327)}$ = 20.56, $P < 0.001$, * $P < 0.05$ with respect to 5 days in culture.

3.2 Neuritic Outgrowth in the Presence of Glucose

The length of the neurites was significantly increased by the addition of glucose to the medium in the absence of FCS (Fig. 1). This effect was evident at 0.5 mM glucose but was significantly less with 25 mM when compared to 0.5 mM although it was still higher than control or lower concentrations of glucose (Fig. 1). The presence of 10% FCS and increasing concentrations of glucose in the medium of outgrowing explants did not further elevate the length of the neurites already increased by FCS alone. The density of the neurites was modified in a similar manner (data not shown).

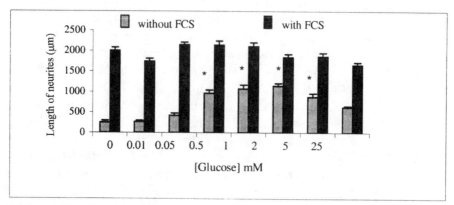

Figure 1. Effect of variable concentrations of glucose and glucose with 10% fetal calf serum on neuritic length from goldfish retinal explants at 5 days in culture. Without FCS, $F_{(7,224)}$ = 15.50, $P < 0.001$. With FCS, $F_{(7,549)}$ = 5.98, $P < 0.001$. * $P < 0.05$ with respect to 0.

3.3   Neuritic Outgrowth in the Presence of Taurine

The presence of 10 mM taurine in the medium without FCS produced a significant increase in the length of the neurites. In the presence of glucose there was a higher outgrowth, and the addition of taurine slightly increased the length of neurites at low concentrations but had no further effect at 2, 4 or 10 mM. In the presence of glucose and FCS there was an increase of outgrowth produced by taurine following a bell shaped concentration-dependency (Fig. 2). The presence of calcium, 3 and 6 mM, and 4 mM taurine did not produce any change with respect to calcium alone (data not shown). In the presence of HEPES, 10 and 20 mM, plus 4 mM taurine the length of the neurites was higher than in the absence of HEPES (data not shown).

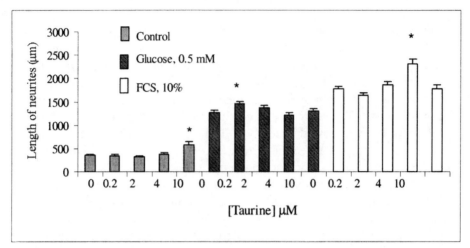

Figure 2. Effect of variable concentrations of taurine on neuritic length from goldfish retinal explants in the absence, $F_{(4,143)} = 4.37$, $P < 0.01$, or in the presence of fetal calf serum, $F_{(2,298)} = 13.11$, $P < 0.001$ and in the presence of glucose, $F_{(2,414)} = 3.10$, $P < 0.05$. * $P < 0.05$ with respect to 0.

4.   DISCUSSION

The interactions between glial Müller cells and neurons result in critical metabolic consequences. Müller cells have a low oxygen consumption and consequently they can spare oxygen to retinal neurons[5] and thus influence the ability of certain retinal neurons, such as ganglion cells, to regenerate. The presence of glucose in the medium was sufficient to preserve the outgrowth from retinal explants.   Taurine increased neuritic outgrowth in a more efficient manner than in the absence of glucose. Differences in glucose metabolism between glia and neurons have been reported, and a physiological interaction takes place between these cells[8]. For instance, the electrical function of cat retina is known to be preserved by glucose *in vitro*[9]. It is interesting to show that concentrations of glucose lower than those considered as normal, around 5 mM[8], were sufficient to elevate the outgrowth, but the stimulatory effect was lower with 25 mM. This observation and the effects with HEPES might be related to the regulation of the osmolarity[10]. However, sucrose up to 10 mM did not have any effect when added alone.

The above results indicate that, first of all, taurine can modulate retinal neuritic outgrowth by itself at high concentrations (10 mM). This effect is mainly evident for neuritic density, and that glucose, probably favoring the metabolic condition of the neurons, elevates the outgrowth as compared to Leibovitz medium alone. However, the stimulatory effect of taurine, evident by the increase in neurite length and density, was not dramatically observed in the presence of glucose. As previously described with dialysed FCS[1], the trophic effect of taurine was higher in the presence of FCS. This observation might be the consequence of a synergistic effect between components of the FCS and the amino acid taurine. One of the mechanisms of action of taurine as a trophic agent in the retina corresponds to the entrance of calcium[11] and calcium has been reported to stimulate neuritic outgrowth from retinal ganglion cells[12]. The relevance of calcium and protein phosphorylation in the retina has been reviewed recently[13]. The concomitant increase in calcium influx and the trophic effect of certain factors from FCS could be responsible for the elevation of neuritic outgrowth by taurine.

Chelators of extracellular or intracellular calcium produce a decrease in neuritic outgrowth from goldfish retinal explants[11]. Also calcium channel antagonists partially inhibit neurite outgrowth from adult rat olfactory cells[14]. However, supplementation with calcium only slightly increased neuritic outgrowth from retinal explants at 3 mM, and produced inhibition at 6 and 12 mM. Transferrin and albumin improve differentiation of neural tissue from the rat when used for *in vivo* transplantation[15] and increase survival of mouse cerebellar cells[16]. Albumin seems to participate in visual-cycle transport[17]. The presence of albumin in the medium did not favor the *in vitro* regeneration of the goldfish retina.

Although the evidence supports the importance of FCS for outgrowth at an elevated rate and magnitude, goldfish retinal explants outgrow in the absence of FCS in a modest manner. It has been recently demonstrated that neonatal mouse retinal explants can be cultured and remain viable for more than 4 weeks in a serum-free medium[2]. Characterization of minimal conditions for preserving the capacity of culturing nervous tissue are essential in the study of individual components or unidentified factors of different sources, such as constituents of the optic tectum, as it has been one of our aims with the godlfish retinal model[7].

Taurine displays trophic influences on the central nervous system, promotes proliferation and differentiation of human fetal neuronal cells[18], is related to migration of cells[19], increases differentiation of retinal cells[20], and stimulates outgrowth from retinal explants and ganglion cells[1,21]. The latter effect requires the presence of FCS, and only a small stimulatory effect is observed in the absence of FCS in the medium. For this and all above expressed reasons, the definition of a free-FCS medium is crucial for studying specific elements or target-derived preparations which contribute to the understanding of the mechanisms of action of taurine as trophic agent in the retina. The interaction of the optic tectum and taurine for modulating neuritic outgrowth of the retina will be performed in the absence of FCS and in the presence of 0.25 mM glucose.

ACKNOWLEDGEMENTS

This work was supported by the Grant S1-2001-903 from Fondo Nacional de Ciencia, Tecnología e Innovación (FONACIT). We appreciate the secretarial assistance of Mrs. Isabel Otaegui.

REFERENCES

1. Lima, L., Matus, P. and Drujan, B., 1988, Taurine effect on neuritic outgrowth from goldfish retinal explants. Int. J. Devl. Neuroscience, 6:417-420.
2. Caffe, A. R., Ahuja, P., Holmqvist, B., Azadi, S., Forsell, J., Holmqvist, I., Soderpalm, A. K. and van Veen, T., 2001, Mouse retina explants after long-term culture in serum free medium. *J. Clin. Neuroanat.*, 22:263-273.
3. Espinosa de los Monteros, A., Yuan J., McCartney, D., Madrid, R., Cole, R., Kanfer, J. N. and deVellis, J., 1997, Acceleration of the maturation of oligodendroblasts into oligodendrocytes and enhancement of their myelinogenic properties by a chemically defined medium. *Dev. Neurosci.* 19:297-311.
4. Sholl-Franco, A. and Araujo, E.G., 1997, Conditioned medium from activated spleen cells supports the survival of rat retinal cells in vitro. *Braz. J. Med. Biol.. Res.*, 30:1299-1303.
5. Winkler, B.S., Arnold, M.J., Brassell, M.A. and Puro, D.G., 2000, Energy metabolism in human retinal Müller cells. *Invest. Ophthalmol. Vis. Sci.* 41:3183-3190.
6. Lima, L., 1999, Taurine and its trophic effects in the retina. *Neurochem. Res.*, 24:13331338.
7. Cubillos, S., Urbina, M. and Lima, L., 2000, Differential taurine effect on outgrowth from goldfish retinal ganglion cells after optic crush or axotomy. Influence of the optic tectum. *Int. J. Devl. Neuroscience*, 18:843-853.
8. Lansel, N. and Niemeyer G., 1997, Effects of insulin under normal and low glucose on retinal electrophysiology in the perfused cat eye. *Invest. Ophthalmol. Vis. Sci.* 38: 792-799.
9. Onoe, S. and Niemeyer, G., 1992, Changing glucose concentration affects rod-mediated response in the perfused cat eye. *Nippon Ganka Gakkai Zasshi* 96:634-640.
10. Lleu, P.L. and Rebel, G., 1990, Effect of HEPES on the Na^+-Cl^-dependent uptake of taurine and beta-alanine by cultured glial cells. Modulation by composition and osmolarity of medium. *Neuropharmacology*, 29:719-725.
11. Lima L., Matus, P. And Drujan, B., 1993, Taurine-induced regeneration of goldfish retina in culture may involve a calcium-mediated mechanism. *J. Neurochem.* 60:2153-2157.
12. Ishida, A.T. and Cheng, M.-H., 1991, Cold inhibits neurite outgrowth from single retinal ganglion cells isolated from adult goldfish. *Exp. Eye Res.* 52:175-191.
13. Militante, J.D. and Lombardini, J.B. 2002, Taurine: evidence of physiological function in the retina. *Nutr. Neuroscience*, 5:75-88.
14. Sonigra, R.J., Brighton, P.C. Jacoby, J., Hall, S. and Wigley, C.B., 1999, Adult rat olfactory nerve ensheathing cells are effective promoters of adult central nervous system neurite outgrowth in coculture. *Glia* 25:256-269.
15. Belovari, T., Bulic-Jakus, F., Juric-Lekic, G., Maric, S., Jezek, D. and Vlahovic, M., 2001, Differentiation of rat neural tissue in a serum-free embryo culture model followed by in vivo transplantation. *Croat. Med. J.* 42:611-617.
16. Fischer, G., 1982, Cultivation of mouse cerebellar cells in serum free, hormonally defined media: survival of neurons. *Neurosci. Lett.* 28:325-329.
17. Adler, A.J. and Edwards, R.B., 2000, Human interphotoreceptor matrix contains serum albumin and retinal-binding protein. *Exp. Eye Res.*, 70:227-234.

18. Chen, X. C., Pan, Z.L., Liu, D.S. and Han, X., 1998, Effect of taurine on human fetal neuron cells: proliferation and differentiation. *Adv. Exp. Med. Biol.*, 442:397-403.
19. Maar, T., Moran, J., Schousboe, A. and Pasantes-Morales, H., 1995, Taurine deficiency in dissociated mouse cerebellar cultures affects neuronal migration. *Int. J. Devl. Neuroscience*, 13:491-502.
20. Altshuler, D., Lo Turco, J.J., Rush, J. and Cepko, C., 1993, Taurine promotes the differentiation of a vertebrate retinal cell type in vitro. *Development*, 119:1317-1328.
21. Matus, P., Cubillos, S. and Lima, L., 1997, Differential effect of taurine and serotonin on outgrowth from explants or isolated cells of the retina. *Int. J. Devl. Neuroscience*, 15:785-793.

Prevention of Epileptic Seizures by Taurine

ABDESLEM EL IDRISSI, JEFFREY MESSING, JASON SCALIA, and
EKKHART TRENKNER
*New York State Institute for Basic Research in Developmental Disabilities and The Center for
Developmental Neuroscience at The City University of New York, Staten Island, NY 10314*

1.   ABSTRACT

Parenteral injection of kainic acid (KA), a glutamate receptor agonist,
causes severe and stereotyped behavioral convulsions in mice and is used as
a rodent model for human temporal lobe epilepsy[1,3]. The goal of this study is
to examine the potential anti-convulsive effects of the neuro-active amino
acid taurine, in the mouse model of KA-induced limbic seizures.

We found that taurine (43 mg/Kg, s.c.) had a significant antiepileptic
effect when injected 10 min prior to KA. Acute injection of taurine increased
the onset latency and reduced the occurrence of tonic seizures. Taurine also
reduced the duration of tonic-clonic convulsions and mortality rate following
KA-induced seizures. Furthermore, taurine significantly reduced neuronal
cell death in the CA3 region of the hippocampus, the most susceptible region
to KA in the limbic system.

On the other hand, supplementation of taurine in drinking water (0.05 %)
for 4 continuous weeks failed to decrease the number or latency of partial or
tonic-clonic seizures. To the contrary, we found that taurine-fed mice showed
increased susceptibility to KA-induced seizures, as demonstrated by a
decreased latency for clonic seizures, an increased incidence and duration of
tonic-clonic seizures, increased neuronal death in the CA3 region of the
hippocampus and a higher post-seizure mortality of the animals.

We suggest that the reduced susceptibility to KA-induced seizures in
taurine-injected mice is due to an increase in GABA receptor function in the
brain which increases the inhibitory drive within the limbic system. This is
supported by our *in vitro* data obtained in primary neuronal cultures showing
that taurine acts as a low affinity agonist for $GABA_A$ receptors, protects

neurons against kainate excitotoxic insults and modulates calcium homeostasis. Therefore, taurine is potentially capable of treating seizure-associated brain damage.

2.   INTRODUCTION

Seizures are one of the most common pathological conditions encountered in humans and are more common in children than adults. The highest incidence of seizures occurs during the first year of life[7]. The increased excitability in the developing brain appears to be secondary to a developmental imbalance between maturation of excitatory and inhibitory circuits.

In the neonatal brain the main ionotropic receptors (GABA$_A$, NMDA, and AMPA/KA) display a sequential developmental pattern of participation in neuronal excitation[2]. GABA, the main inhibitory transmitter in the adult, provides the main excitatory drive to neurons at early stages of development because of a chloride gradient that leads to depolarization of young neurons rather than the hyperpolarization observed in adults. During this developmental stage, when GABA is excitatory, taurine might play a critical role as the major inhibitor of neuronal excitation by compensating for the lack of receptor-mediated neuronal inhibition.

The neonatal brain contains high levels of taurine[8]. As the brain matures its taurine content declines and reaches stable adult concentrations that are second to those of glutamate, the principal excitatory neurotransmitter in the brain. Taurine levels in the brain significantly increase under stressful conditions[10] (stroke, injuries, hypoxia), suggesting that taurine may play a vital role in neuroprotection. A possible mechanism of taurine's neuroprotection lies in its calcium modulatory effects and agonistic role on GABA$_A$.

The considerable pool of taurine in the brain must have a functional application. We have previously shown both *in vitro* and *in vivo* that taurine has a neuro-modulatory role on both excitatory and inhibitory signals. Taurine works concomitantly with GABA to activate GABA$_A$ receptors, thus enhancing neuronal inhibition. Taurine also acts downstream of glutamate receptor activation through the regulation of cytoplasmic and intra-mitochondrial calcium homeostasis[5], therefore preventing neuronal hyperexcitability. These two mechanisms indicate that taurine achieves the same goal: neuroprotection, through two independent means.

3. METHODS

3.1 Drug Administration

Kainic acid was dissolved in isotonic saline (at 3 mg/ml). Animals (FVB/N, two month old males) received either a single dose of kainic acid (20 mg/Kg) or saline, administered subcutaneously (s.c.). Taurine (43 mg/kg) was injected 10 min before KA injection or provided in drinking water (0.05%) for one month before seizure induction.

3.2 Tissue Fixation, Staining and Digital Microscopy

Seven days after KA injection, surviving mice were anesthetized with Nembutal and perfused transcardially with 3 ml PBS followed by 20 ml 2% paraformaldehyde. Coronal sections (8 μm) were stained with hematoxylin and eosin. Images of CA3 neurons were captured (20 sections/brain) with a Nikon microscope equipped with a digital camera. Digital images were analyzed, and live cells were scored by the presence of a visible nucleus and normal cell morphology by reviewers blind to the treatment of the mice. Presumptive live cells in each digital image were counted with Simple 32 software (Compix,Inc. Imaging system, Cranberry, PA). The number of live cells in all CA3 regions of KA-injected mice was averaged and normalized to the number of live cells obtained from the corresponding saline-injected animals.

3.3 ^{36}Cl$^-$ Influx Measurements

Cerebellar granule cells were cultured[12] for 5 days *in vitro*. Growth medium was poured off and cultures were washed three times with Earl's balanced salt solution (EBSS). After equilibration in 1ml of EBSS for 20 min at 37°C, ^{36}Cl$^-$ influx measurements were initiated by replacing this solution with an identical one containing 5 μCi/ml ^{36}Cl$^-$, GABA and taurine. Free radioligand was removed by three rapid (5 sec) rinses with EBSS. The radioactivity was extracted by 2 x 0.5 ml distilled water followed by 2 x 0.5 ml methanol and counted by standard liquid scintillation spectrometry.

3.4 Amino Acids Analysis

Brains were homogenized and blood samples were diluted in 5 % TCA then centrifuged at 12,500 rpm for 10 min. Supernatants were removed and filtered through 0.2 μm filter. Samples were then dried in a SpeedVac and derivatized in PITC. Derivatized samples were mixed in 2 ml of Sodium Acetate Trihydrate and Acetonitrile). 50 µl of this mixture were injected into a Beckman reverse phase C18 column (SpectraPhysics 8800 Ternary HPLC) and absorption maxima were determined with an LDC SpectorMonitor D at a wavelength of 254 nm.

3.5   Immunohistochemistry

Brain sections were incubated overnight at 4° C with a rabbit polyclonal antibody (1:1000 dilution, Chemicon) that recognizes Calbindin, and incubated with an HRP-conjugated goat anti-rabbit IgG antibody (1:2000, 1 hour at 25° C). The staining procedure was performed using the Vectastain Elite ABC kit and DAB substrates according to the supplier's instructions (Vector Laboratories, Burlingame, CA).

4.   RESULTS

4.1   Taurine Reduces the Severity of Limbic Seizures

Subcutaneous injection of KA (20 mg/kg) induces seizures that originate in the limbic system where the hippocampus, the dentate gyrus, and the entorhinal cortex are the structures most affected by KA[1,3]. KA can induce some or all of the following stereotypical behaviors: motionless stare, rearing and falling, clonic convulsions, tonic-clonic seizures (status epilepticus) and death. The occurrence of these behaviors, the time from injection to initiation of the behavior (latency) and the duration are measures of seizure severity (Table 1).

Table 1 shows that taurine-injected mice are less susceptible to KA seizures than littermate controls or taurine-fed. In the taurine-fed group, after KA (20mg/kg) injection, the latency to the onset of the first clonic seizure was significantly reduced ($p<0.05$) and the duration of the tonic convulsions was significantly increased ($p<0.05$) as compared to controls. Taurine injection did not affect the latency to the onset of the first tonic seizure but significantly reduced the duration of the status epilepticus ($p<0.05$) as compared to controls or the taurine-fed group. Latency to the onset and duration of seizure are expressed in minutes.

	(n)	Rearing & Falling	First Clonic Occurr	First Clonic Latency	Tonic-Clonic Occurr	Tonic-Clonic Duration	Mortality Rate	(n)
Control	6	83.3 %	83.3 %	24.6 ± 5.7	83.3 %	26.5 ± 9.5	33.3 %	2
Tau-fed	7	100 %	100 %	14.5 ± 1.9*	100 %	98.1 ± 29.8*	42.8 %	3
Tau-Inj	11	72.7 %	63.6 %	22.4 ± 4.4	63.3 %	13.9 ± 5.7*	18.2 %	2

4.2   Taurine and Kainic Acid in the Brain

To determine if the anticonvulsive effects of taurine were non-specific and due to interference with KA crossing the blood brain barrier, we measured the concentration of amino acids in the brain of controls, taurine-fed and taurine-injected mice. As shown in table 2, taurine concentration did not vary between experimental groups indicating that in the brain, taurine content is highly regulated. Taurine concentrations were the highest among the other measured amino acids. KA was not detected in the brain 45 min post injection, a time period during which limbic seizures have already been initiated. However, significant amounts of KA were detected in the blood.

Table 2 shows plasma and whole brain content of taurine, glutamate, Aspartate, GABA and kainic acid. Mice were injected with taurine 10 min before KA. 45 min after KA injection animals were perfused transcardially for 2 min with PBS to remove blood from the brain's vasculature. Before the perfusion was started, 100 μl of blood were removed after incision of the right atrium.

	Taurine	*Glutamate*	*Aspartate*	*GABA*	*Kainate*
Brain (μmol/g wet wt)					
Untreated	228 ± 4.3	155 ± 10.4	76 ± 16.6	67 ± 7.1	N.D.
KA-treated	227 ± 5.9	151 ± 7.9	77 ± 4.5	87 ± 3.3	N.D.
Tau-fed	224 ± 4.6	185 ± 16.3	99 ± 0.8	95 ± 2.5	N.D.
Tau-injected	216 ± 6.2	150 ± 9.9	105 ± 3.3	76 ± 7.8	N.D.
Blood (μmol/mL)					
Untreated	13 ± 0.7	5.6 ± 0.4	0.4 ± 0.01	1.2 ± 0.2	N.D.
KA-treated	16 ± 0.2	4.5 ± 0.7	0.6 ± 0.1	1.6 ± 0.1	45 ± 0.2
Tau-fed	10 ± 3.6	3.1 ± 1.3	0.4 ± 0.1	0.8 ± 0.3	27 ± 8.5
Tau-injected	14 ± 3.4	5.2 ± 1.6	0.9 ± 0.3	1.4 ± 0.2	21 ± 2.5

4.3   Taurine Prevents CA3 Cell Death After KA Seizures

A more sensitive measure of seizure severity is the quantification of neuronal degeneration in the CA3 region of the hippocampus 3-7 days after the seizure[3]. Cell death results from alterations in cell physiology initiated during seizure activity that are associated with glutamate-mediated excitotoxicity. This excitotoxicity is secondary to KA excitation and is regionally selective within the hippocampus[1,3]. Severe cell loss is reproducibly induced in the CA3 field with KA administration, whereas CA1 neurons and granule cells of the dentate gyrus are relatively well preserved (Fig. 1). This pattern of cell death is commonly described as Ammon's horn sclerosis[3].

As shown in figure 1, taurine injections significantly reduced the extent of KA-induced lesions in CA3 region of the hippocampus. Mice injected with taurine were also the most resistant to KA-induced seizures. Taurine-fed mice, on the other hand, exhibited the most severe seizures and more cell death in CA3

Detailed analysis of these regions captured in ~ 450 digital images from the sections of the 17 mice that survived the KA seizures and 4 untreated (saline injected) animals indicated that the KA-induced seizures were much more severe in the taurine-fed mice (Fig. 2). Taurine-injected mice presented approximately twice the number of surviving CA3 neurons than did taurine-fed mice. Furthermore, we found that CA3 cell death was correlated (r = 0.89) to seizure duration in individual animals. Thus, CA3 excitotoxic cell death is a highly informative measure of seizure severity.

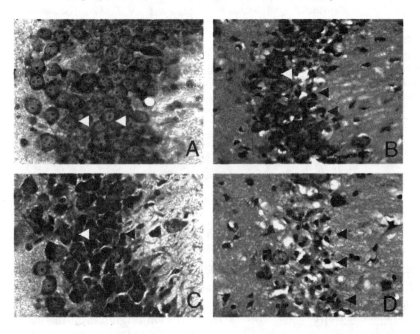

Figure 1. Histological evaluation of neurodegeneration in CA3 region of the hippocampus. Only the CA3 region was affected by KA treatment and a much higher level of neuronal degeneration is apparent in the control mouse than in the taurine-injected. These images are representative of the (A) saline-treated (n=4), (B) control (n=4), (C) taurine-injected (n=9) and taurine-fed (n=4) animals that survived seizure. Open arrowheads show healthy neurons, closed arrowheads point to the remaining nuclei of dead neurons

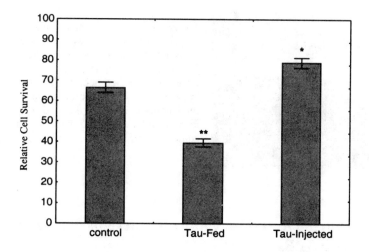

Figure 2. KA-induced cell death in CA3 region of the hippocampus. Presumptive live cells were identified by the presence of a visible nucleus and normal morphology. The number of live cells in all CA3 regions of KA-injected mice was averaged and normalized to the number of live cells obtained from the corresponding saline-injected controls. Data represent mean number of live cells (± SEM) in the CA3 region of the hippocampus in 4 control mice, 9 taurine-injected and 4 taurine-fed mice injected with KA. Cell death in CA3 regions of taurine-injected mice was significantly ($p<0.001$) lower than controls or taurine-fed mice (*). Cell death in CA3 regions of taurine-fed mice was significantly ($p<0.001$) higher than controls or taurine-injected mice (**).

4.4 CA3 Pyramidal Cells Are Most Susceptible to KA Induced Cell Death

CA3 pyramidal neurons are the target of excitatory afferents from the entorhinal cortex and the dentate gyrus (DG). In addition there are potent recurrent excitatory connections between CA3 pyramidal neurons that CA1 and DG neurons do not have[1,3]. Coupled with this, CA3 pyramidal cells have well-developed Ca^{2+} conductances that predispose them to sustained depolarizations and bursting activity[1,3]. Furthermore, we found that CA3 pyramidal cells do not express any calbindin (Fig. 3), a major calcium-binding protein, whereas other region of the limbic system heavily express this protein. These anatomical and functional characteristics of CA3 pyramidal cells make them more susceptible to sustained depolarizations triggered by KA.

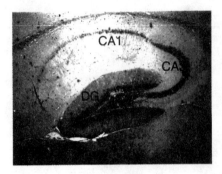

Figure 3. Calbindin immunoreactivity in the hippocampus. The expression of calbindin was analyzed by immunohistochemistry in brain sections. Calbindin is highly expressed in the CA1 and DG but not in CA3 region of the hippocampus. Elsewhere in the brain, calbindin immunoreactivity was detected in the frontal lobe, entorhinal cortex, diencephalon, superior colliculus, and Purkinje cells in the cerebellum.

4.5   Taurine Activates Chloride Currents Through GABA$_A$ Receptors

There is increasing evidence supporting the presence of functional interactions between GABA and taurine[8]. Taurine has been shown to increase plasma membrane chloride conductances by affecting bicuculine-sensitive chloride channels[4,9,13]. Taurine has also been shown to act as a partial agonist of GABA$_A$ receptors in synaptic membranes[11]. Here we show that taurine activates Cl⁻ influx through GABA$_A$ receptors in cerebellar granule cells *in vitro* (Fig. 4). These results suggest that the anti-convulsive effects of taurine might be mediated by direct interaction with the GABA$_A$ receptors *in vivo*.

5.   DISCUSSION

In this study, we examined the anti-convulsive effects taurine in a model of KA-induced seizure. We found that s.c. injection of taurine prior to seizure induction significantly reduced seizure severity, whereas long-term taurine intake though drinking water had an opposite effect, consistent with previous findings in rats[6]. We suggest that both effects are mediated through activation of GABA$_A$ receptors. The timing and duration of GABA$_A$ receptor activation determines when taurine has an inhibiting or promoting effect on seizure activity.

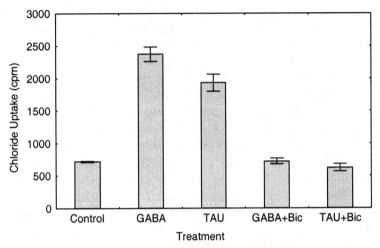

Figure 4. Taurine induces chloride uptake into cerebellar granule cells. Cells were treated with GABA (50 μM) or taurine (50 μM) and chloride uptake was initiated for 10 sec. The fact that bicuculline (10 μM), a GABA$_A$ receptor antagonist, blocked taurine-induced chloride influx indicates the GABA$_A$ receptor as the site of taurine action.

In the adult brain, inhibitory GABAergic interneurons modulate the activity of principal excitatory cells via their GABA$_A$ receptors, and thus adjust the excitatory output. The finding that taurine activates chloride currents through GABA$_A$ receptors (Fig. 4) suggests that taurine increases seizure threshold by potentiating the inhibitory drive in the brain and within the limbic system in particular. Thus taurine, by increasing excitatory input-resistance, renders the network that subserves KA seizure less excitable. However, this paradigm requires taurine to be injected shortly before KA. The question that arises then is why are taurine-fed mice more susceptible to KA-induced seizure?

The reasons for these discrepancies between acute and chronic administration of taurine are not yet clear, but it could be assumed that chronic administration leads to sensitization in structures that subserve seizure and/or desensitization of structures that protect from seizures. We speculate that both mechanisms might be operational. *In vitro* data indicate that chronic treatment with taurine reduces mitochondrial calcium concentrations[5]. Mitochondria are crucial organelles for both ATP production and calcium sequestration, especially under depolarizing conditions[5]. By reducing mitochondrial calcium content, the trans-membrane calcium gradient becomes greater in taurine treated-cells. Thus, glutamate depolarization induces a more rapid calcium influx. Because calcium is required to trigger seizure activity, taurine-fed mice would exhibit higher calcium currents that trigger more severe seizures than control mice. Alternatively, chronic taurine treatment may facilitate seizures through desensitization of the inhibitory system, primarily through GABA$_A$ receptors. Chronic elevated concentrations of taurine in the brain would lead to a desensitization of GABA$_A$ receptors or a down regulation of their expression. Because it has been established that GABA$_A$ receptors are involved in determining seizures and are activated by taurine (Fig. 4), it is legitimate to postulate a pharmacological link, through taurine, between the function of GABA$_A$ receptor and seizure threshold. Consistent with this, we found that taurine-fed mice have the highest GABA content in their brains as compared to the other experimental groups. On the other hand however, the taurine content of the brain was similar between treatment groups (table 2). These findings indicate that taurine levels in the brain are highly regulated but might be differentially compartmentalized between intracellular, extracellular, neuronal and non-neuronal cells. It is possible that taurine-fed mice have elevated extracellular taurine levels. This would lead to sustained activation or at least binding to GABA$_A$ receptors. Such a chronic interaction of taurine with GABA$_A$ receptors may lead to down-regulation of GABA$_A$ receptor function or expression. In response to these changes, there is increased synthesis of GABA by GABAergic neurons. Consistent with this, brains of taurine-fed mice have the highest GABA content.

In summary, this study shows that activation of GABA$_A$ receptors by taurine prior to KA treatment causes neuronal hyperpolarization which renders the brain more resistant to seizure induction. Whereas, sustained elevation of extracellular levels of taurine exposes GABA$_A$ receptors to high levels of the agonist and might have a desensitizing effect on these receptors, thus reducing the threshold for KA depolarization. We are currently investigating the mechanisms through which taurine affects the brain's threshold for seizures.

ACKNOWLEDGEMENTS

This work was supported by a grant from New York State Office of Mental Retardation and Developmental Disabilities and by The Center for Developmental Neuroscience at CSI of The City University of New York.

REFERENCES

1. Ben-Ari,Y., 1985, Limbic seizure and brain damage produced by kainic acid: mechanisms and relevance to human temporal lobe epilepsy. *Neuroscience*, 14:375-403.
2. Ben-Ari, Y., 2002, Excitatory actions of GABA during development: the nature of the nurture. *Nature Neurosci.*, 3:728-739.
3. Bruton, C.J., 1993, 'Status epilepticus. I: Pathogenesis'. *Dev. Med. Child. Neurol.*, 35:277-282.

4. del Olmo, N., Bustamante, J., del Rio, R.M., and Solis, J.M., 2000, Taurine activates GABA(A) but not GABA(B) receptors in rat hippocampal CA1 area. *Brain Res.*, 864:298-307.
5. El Idrissi, A., and Trenkner, E., 1999, Growth factors and taurine protect against excitotoxicity by stabilizing calcium homeostasis and energy metabolism. *J. Neurosci.*, 19:9459-9468.
6. Eppler, B., Patterson, T.A., Zhou, W., Millard, W.J., and Dawson, R. Jr., 1999, Kainic acid (KA)-induced seizures in Sprague-Dawley rats and the effect of dietary taurine (TAU) supplementation or deficiency. *Amino Acids.* 16:133-147.
7. Gregory, L., Ben-Ari, H., and Ben-Ari, Y., 1998, Seizures in the Developing Brain: Perhaps Not So Benign after All. *Neuron*, 21, 1231–1234.
8. Kuriyama, K., and Hashimoto, T., 1998, Interrelationship between taurine and GABA. *Adv. Exp. Med. Biol.*, 442:329-337.
9. Mellor, J.R., Gunthorpe, M.J., and Randall, A.D., 2000, The taurine uptake inhibitor guanidinoethyl sulphonate is an agonist at gamma-aminobutyric acid(A) receptors in cultured murine cerebellar granule cells. *Neurosci. Lett.*, 286:25-28.
10. O'Byrne, M.B., and Tipton, K.F., 2000, Taurine-induced attenuation of MPP+ neurotoxicity in vitro: a possible role for the GABA(A) subclass of GABA receptors. *J. Neurochem.*, 74:2087-2093.
11. Quinn, M.R., and Harris, C.L., 1995, Tautine allosterically inhibits binding of [^{35}S]-*t*-butylbicyclophosphorothionate (TBPS) to rat brain synaptic membranes. *Neuropharmacology*, 34:1607-1613.
12. Trenkner, E., 1991, Cerebellar cells in culture. In *Culturing Nerve Cells* (G. Banker and K. Goslin, eds.), MIT Press, pp. 283-307.
13. Wang, D.S., Xu, T.L., Pang, Z.P., Li, J.S., and Akaike, N., 1998, Taurine-activated chloride currents in the rat sacral dorsal commissural neurons. *Brain Res.*, 792:41-47.

Taurine Regulates Mitochondrial Calcium Homeostasis

ABDESLEM EL IDRISSI and EKKHART TRENKNER
New York State Institute for Basic Research in Developmental Disabilities and The Center for Developmental Neuroscience at The City University of New York, Staten Island, NY 10314

1.   ABSTRACT

We have investigated the protective role of taurine in glutamate-mediated cell death and the involvement of mitochondria in this process. In cultured cerebellar granule cells, glutamate induces a rapid and sustained elevation in cytoplasmic free calcium ($[Ca^{2+}]_i$), causing the collapse of the mitochondrial electrochemical gradient (MtECG) and subsequent cell death. We found that pre-treatment with taurine, did not affect the level of calcium uptake with glutamate but rather reduced its duration; the calcium increase was transient and returned to basal levels about 10 min after adding glutamate. Furthermore, taurine reduced mitochondrial calcium concentration under non-depolarizing conditions. Treatment of cerebellar granule cells with taurine enhanced mitochondrial activity as measured by rhodamine uptake, both in the presence or absence of glutamate. We conclude that taurine prevents or reduces glutamate excitotoxicity through both the enhancement of mitochondrial function and the regulation of intracellular (cytoplasmic and mitochondrial) calcium homeostasis.

2.   INTRODUCTION

In the CNS, calcium plays a key role in mediating glutamate excitotoxicity (for review see, [3,11,14,16,17]), therefore it is critical to understand how it is regulated. Glutamate-induced neuronal necrosis is preceded by a rapid increase in cytoplasmic-free calcium concentration[6,8,15,24]. There is increasing evidence that mitochondrial dysfunction plays a primary role in the initiation of both necrotic and apoptotic neuronal cell death[3,10,19,20,21,22]. Mitochondrial depolarization after exposure to glutamate has been reported to be an early event associated with neuronal calcium loading[1,13,4,12,28]. However, mechanisms involving mitochondria in glutamate excitotoxicity have been difficult to elucidate by virtue of the complexity of interactions of these organelles with the rest of the cell.

This study was designed to elucidate the mechanisms by which taurine regulates cytoplasmic and mitochondrial calcium homeostasis during glutamate excitotoxicity. We also evaluated mitochondrial activity in the presence of taurine in order to determine how taurine affect neuronal energy under excitotoxic conditions. And finally we will discuss possible strategies of how to prevent mitochondrial dysfunction in response to excitotoxic glutamate stimulation.

3.   METHODS

3.1 $^{45}Ca^{2+}$ - Accumulation

Cerebellar granule cells were prepared from 7-day-old mice as previously described[8,25,27]. After 4 days in culture cells were washed twice with Locke's solution [154 mM NaCl, 5.6 KCl, 3.6 mM NaHCO$_3$, 1.3 mM CaCl$_2$, 5.6 mM glucose, and 5 mM HEPES (pH 7.4)]. Additions were made to a final volume of 0.25 ml including 2 x 10^5 cpm of $^{45}CaCl_2$, which was added 10 seconds before the addition of the agonist. After 20 min (or as indicated) at room temperature, the cells were rapidly washed three times in 0.5 ml of Locke's solution containing 2 μM MK-801 to block NMDA receptor activity. Finally the total amount of $^{45}Ca^{2+}$ was determined in the lysate after the cells were dissolved in 0.5 ml of 0.1 M NaOH.

3.2   Calcium Imaging

Measurements of intracellular calcium concentration ($[Ca^{2+}]_i$) were performed in morphologically identified cerebellar granule cells grown on PDL-coated coverslips for 4 days. Confocal images of cellular fluorescence were obtained using a Nikon inverted epi-fluorescence microscope equipped with an oil immersion 60 x, 1.4 numerical aperture (NA) objective. Excitation wavelength used was 488 nm, the emission wavelength was 505 nm. All recordings were performed at room temperature (22-25 ^{0}C). The relative fluorescence values were averaged over all imaged cells, usually 80-100 neurons for each experiment. Every condition was evaluated in three or more experiments (for more descriptive details, see[8]).

3.3 Activity of the Mitochondrial Electrochemical Gradient

Quantitative determination of rhodamine 123 uptake was performed as described by[5] and modified by[8]. Cellular rhodamine content was normalized to total cellular protein. All treatments were done in triplicates and evaluated in three or more experiments.

4. RESULTS

4.1 Mitochondria Sequester Calcium

Cerebellar granule neurons accumulate calcium in an intracellular pool after glutamate depolarization. Calcium accumulation is inhibited by proton ionophores such as carbanyl cyanide p-(trifluoromethoxy)phenyl-hydrazone (FCCP) a compound that collapses the mitochondrial electrochemical gradient. We found that FCCP caused a rapid increase in cytoplasmic free calcium (Fig. 1), indicating that mitochondria sequester a significant amount of calcium ions. Mitochondrial inhibition by FCCP leads to elevated cytoplasmic calcium concentrations that can result in glutamate release from cells by exocytosis and/or transporter reversal, leading to secondary activation of glutamate receptors. Therefore, we have included in these experiments the N-methyl-D-aspartate (NMDA) receptor antagonist, MK801 (10 μM), in order to distinguish between a direct effect of FCCP on mitochondria and an indirect effect through the released glutamate on neurons.

The other cellular organelle that has been shown to sequester intracellular calcium is the endoplasmic reticulum (ER). Glutamate induced a dose-dependent accumulation of $^{45}Ca^{2+}$ into cerebellar granule cells (Fig. 2). Blocking calcium release from the ER with dantrolene (150 nM, 30 min) or calcium uptake into the ER with thapsigargin (1μM, 20 min) did not affect the total cellular $^{45}Ca^{2+}$ accumulation induced by glutamate (Fig. 2). These data indicate that, in cerebellar granule cells, mitochondria are the primary organelles that sequester calcium during glutamate stimulation.

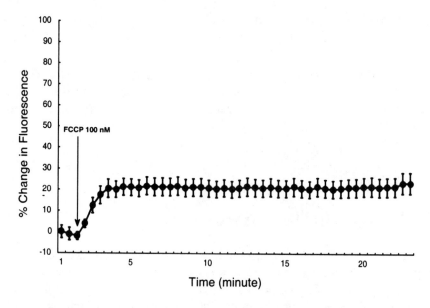

Figure 1. FCCP inhibits calcium sequestration into mitochondria. Relative changes in $[Ca^{2+}]_i$ were determined by confocal microscopy. Each data point represents mean ± SEM of the relative increases above baseline of $[Ca^{2+}]_i$ obtained from three separate experiments. Inhibition of the mtECG by FCCP (100nM) induced a rapid increase of $[Ca^{2+}]_i$ presumably released from mitochondria. MK801 (10 μM) was added to cells about 2 min before FCCP and was present throughout the experiment (see text). $[Ca^{2+}]_i$ remained at a sustained level throughout the recording.

4.2   Calcium Accumulation Into Mitochondria Depends on Glutamate Concentration

Consistent with calcium uptake studies (Fig. 2), Glutamate induced a dose-dependent increase in free-cytoplasmic calcium (Fig. 3). In order to determine the amount of calcium sequestered by mitochondria, we dissipated the mitochondrial membrane potential with FCCP to cause mitochondrial calcium release. We observed a rapid increase in $[Ca^{2+}]_i$, suggesting that accumulation of calcium into mitochondria is energy-dependent (Fig. 3). Furthermore, the amount of calcium released from mitochondria correlated with the initial concentration of glutamate used. Therefore mitochondria might play a critical role in clearing cytoplasmic calcium following glutamate depolarization thus preventing excessive increases in cytoplasmic calcium.

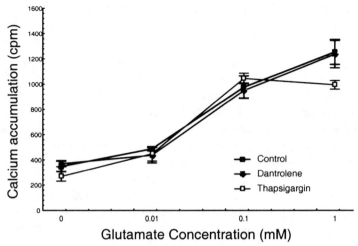

Figure 2. $^{45}Ca^{2+}$ accumulation into CGCs is not affected by dantrolene and thapsigargin. Cells were depolarized with various concentrations of glutamate as indicated. $^{45}Ca^{2+}$ accumulation was determined 15 min after depolarization. Each data point represents mean ± SEM of $^{45}Ca^{2+}$ accumulation from three experiments. At all glutamate concentration used, dantrolene and thapsigargin did not significantly affect the glutamate-induced $^{45}Ca^{2+}$ accumulation.

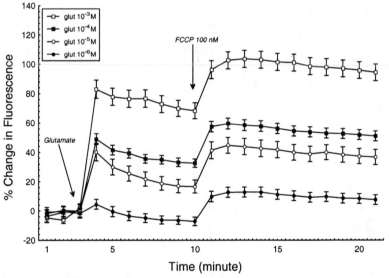

Figure 3. Glutamate induced a dose-dependent increase in cytoplasmic-free calcium and intra-mitochondrial calcium. Images were acquired at 60 sec intervals. Each data point represents mean ± SEM of the relative increases above baseline of $[Ca^{2+}]_i$ obtained from three separate experiments. Two cover slips were used for each glutamate concentration in each experiment. Each culture was treated with one concentration of glutamate for 10 min followed by FCCP (100 nM) for the remaining of the recording. Linear polynomial regression analysis using the last time point before addition of FCCP as a covariate, revealed a significant correlation between the initial glutamate concentration used and the relative increase in $[Ca^{2+}]_i$ induced by FCCP ; $r = 0.93$ [$F(1,198) = 1327.37$, $p < .0001$].

4.3   Regulation of Calcium Homeostasis by Taurine

We have previously shown that taurine protects cerebellar granule cells from glutamate excitotoxicity through regulation of intracellular calcium homeostasis[8]. Since mitochondria play a pivotal role in the regulation of intracellular calcium levels (Fig. 3), we evaluated the effects of taurine on both cytoplasmic and mitochondrial calcium homeostasis.

Glutamate depolarization induced a significant increase in $[Ca^{2+}]_i$ (82% above baseline) with no significant decrease over the period of recording (10 min, Fig. 4). Treatment of these cultures with 100 nM FCCP induced an additional significant and sustained increase in $[Ca^{2+}]_i$ (from 82% to 103 %) over the next 15 min (Fig. 4). However, in cells pre-treated with taurine (10 mM, 24h) glutamate elicited a rapid and significant increase (70 %) in $[Ca^{2+}]_i$ followed by a slow recovery towards baseline after 10 min. Addition of FCCP to taurine-treated cultures induced a significant increase of calcium (from 10 to 49 %) that was higher when compared to control cultures (from 82% to 103 %, Fig. 4).

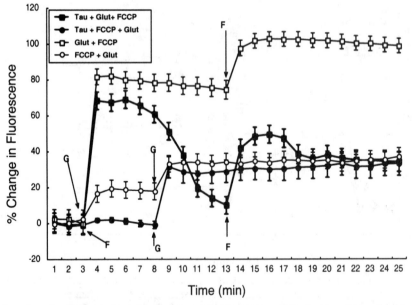

Figure 4. Regulation of calcium homeostasis by taurine. Cultures were treated with taurine (10mM) 24 h before the experiment. Each culture was treated either with glutamate (1mM) for 10 min followed by FCCP (100 nM) or with FCCP (100 nM) for 5 min followed by glutamate (1mM) as indicated. Pre-treatment with taurine is indicated by closed squares and circles. G: glutamate, F: FCCP.

FCCP induced a significant increase in $[Ca^{2+}]_i$ even when it was applied before glutamate (20 % above baseline; Fig. 1 & 4), indicating that mitochondria, in non-depolarized cerebellar granule cells, contain significant amounts of calcium. Interestingly, FCCP did not induce a significant release of mitochondrial calcium when cultures were pre-treated with taurine alone (~ 3 % above baseline, Fig. 4). These data strongly suggest that pre-treatment with taurine reduces mitochondrial calcium concentrations which explains the increased capacity of mitochondria to sequester calcium after depolarization with glutamate.

4.4   Taurine Enhances Mitochondrial Activity

Pretreatment of cultures with increasing concentrations of taurine counteracted the glutamate- and FCCP-induced decrease in rhodamine 123 accumulation (Fig. 5). These data argue for a causal relationship between the regulation of calcium homeostasis and the preservation of mitochondrial function. However, the finding that taurine enhanced mitochondrial activity, in the presence as well as absence of glutamate, suggests that mitochondrial protection by taurine is not merely a consequence of calcium regulation during excitotoxicity, but rather the function of taurine could be elicited even when calcium homeostasis was not perturbed (Fig.5). Therefore taurine has at least two distinct effects on cellular function: enhancement of mitochondrial activity, and regulation of intracellular calcium homeostasis.

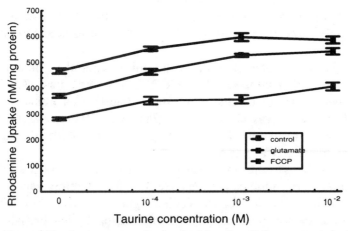

Figure 5. Taurine preserves mitochondrial function. Cells were treated with taurine 24h before addition of glutamate or FCCP. Each data point represents mean ± SEM of rhodamine accumulation normalized to total protein content of the cells in three separate experiments. Addition of glutamate or FCCP alone induced a statistically significant (p<.001) reduction in rhodamine accumulation. However, pretreatment (24h) with taurine significantly (p<.001) counteracted both glutamate- and FCCP-induced decrease in rhodamine uptake.

5.   DISCUSSION

Mitochondria provide two major functions: production of cellular energy and sequestration of calcium ions. Here, we investigate the involvement of taurine in the regulation of mitochondrial function and calcium homeostasis as a mechanism for taurine's role in the prevention of excitotoxic cell death.

Studies using radioactive $^{45}Ca^{2+}$ isotope to measure calcium uptake, suggest a direct relationship between calcium accumulation and subsequent death in neurons exposed to NMDA or glutamate[7,8,9]. This correlation was confirmed in this study (Fig. 2 & 3). Furthermore, we found that the amount of calcium sequestered by mitochondria was dependent on glutamate concentrations   (Fig. 3), indicating the importance of mitochondria in removing calcium from the cytoplasm.

Taurine did not affect the magnitude of $[Ca^{2+}]_i$, but rather down regulated $[Ca^{2+}]_i$ to basal levels approximately 10 min after glutamate-depolarization (Fig. 4). We have previously reported that taurine reduced glutamate-mediated   excitotoxicity,   mainly   through   regulation   of   calcium homeostasis[8,26]. Furthermore, our data extend the roles of taurine to regulate mitochondrial calcium homeostasis, since pre-treatment with taurine inhibited the FCCP-induced calcium release from mitochondria under non-depolarizing conditions (Fig. 4).

The exact mechanism as to how taurine influences calcium homeostasis in neuronal cells and their mitochondria is not yet clear. However, in other systems such as cardiomyocytes and retinal cells, taurine has been shown to modulate the activity of the voltage-dependent Ca^{2+} and Na^+ channels[18,23] and the Na^+/Ca^{2+} exchanger activity. Whether these mechanisms control calcium homeostasis in cerebellar granule cells and their mitochondria needs to be investigated.

As shown in figure 5, glutamate reduced the mitochondrial membrane potential which was further enhanced by FCCP. Our finding that FCCP caused an additional release of calcium from mitochondria after glutamate depolarization when $[Ca^{2+}]_i$ was at its plateau (Fig. 3 & 4) indicates that mitochondria partially maintain their bioenergetic competence during exposure to glutamate when $[Ca^{2+}]_i$ is elevated.

Finally, when cultures were pre-treated with taurine, mitochondrial activity was increased in the absence of glutamate (Fig. 5). Therefore, taurine may provide trophic support for cerebellar granule cells through enhancement of their bioenergetic capacity. Furthermore, taurine partially prevented the glutamate-induced decrease in mitochondrial activity (Fig. 5). Thus, we suggest a dual cellular function for taurine: increase of mitochondrial function and regulation of intracellular calcium homeostasis (cytoplasmic and mitochondrial). The culmination of these two functions, and possibly others, might be the mechanisms of promoting the survival and neuroprotection against glutamate excitotoxicity.

ACKNOWLEDGEMENTS

This work was supported by a grant from New York State Office of Mental Retardation and Developmental Disabilities and by The Center for Developmental Neuroscience at CSI of The City University of New York.

REFERENCES

1. Ankarcrona, M., Dypbukt, J.M., Orrenius, S., and Nicotera, P., 1996, Calcineurin and mitochondrial function in glutamate-induced neuronal cell death. *FEBS Lett.*, 394:321 324.
2. Beal, M.F., 1996, Mitochondria, free radicals, and neurodegeneration. *Curr. Opin. Neurobiol.*, 6:661-666.
3. Beal, M.F., Hyman, B.T., and Koroshetz, W., 1993, Do defects in mitochondrial energy metabolism underlie the pathology of neurodegenerative diseases? *Trends Neurosci.*, 6:125-131.
4. Castilho, R.F., Hansson, O., Ward, M.W., Budd, S.L., and Nicholls, D.G., 1998, Mitochondrial control of acute glutamate excitotoxicity in cultured cerebellar granule cells. *J. Neurosci.*, 18:10277-10286.
5. Chen, L.B., 1989, Fluorescent labeling of mitochondria. *Methods Cell Biol.*, 29:103-123.
6. Choi, D.W., 1990, The role of glutamate neurotoxicity in hypoxic-ischemic neuronal death. *Ann. Rev. Neurosci.*, 13:171-182.
7. Eimerl, S., and Schramm, M., 1994, The quantity of calcium that appears to induce neuronal death. *J. Neurochem.*, 62:1223-1226.
8. El Idrissi, A., and Trenkner, E., 1999, Growth factors and taurine protect against excitotoxicity by stabilizing calcium homeostasis and energy metabolism. *J. Neurosci.*, 19:9459-9468.
9. Hartley, D.M., Kurth, M.C., Bjerkness, L., Weiss, J.H., and Choi, D.W., 1993, Glutamate receptor-induced $^{45}Ca^{2+}$ accumulation in cortical cell culture correlates with subsequent neuronal degeneration. *J. Neurosci.*, 13:1993-2000.
10. Henneberry, R.C., 1989 The role of neuronal energy in the neurotoxicity of excitatory amino acids. *Neurobiol. Aging*, 10:611-613.
11. Kater, S.B., Mattson, M.P., Cohan, C., and Connor, J., 1988, Calcium regulation of the neuronal growth cone. *Trends Neurosci.*, 11:315-321.
12. Keelan. J., Vergun, O., and Duchen, M.R., 1999, Excitotoxic mitochondrial depolarisation requires both calcium and nitric oxide in rat hippocampal neurons. *J. Physiol.*, 3:797-813.
13. Khodorov, B., Pinelis, V., Vergun, O., Storozhevykh, T., and Vinskaya, N., 1996, Mitochondrial de-energization underlies neuronal calcium overload following a prolonged glutamate challenge. *FEBS Lett.*, 397:230-234.
14. Kristian, T., and Siesjo, B.K., 1996, Calcium-related damage in ischemia. *Life Sci.*, 59:357-367.
15. Lobner, D., and Lipton, P., 1990, Sigma-ligands and non-competitive NMDA antagonists inhibit glutamate release during cerebral ischemia. *Neurosci. Lett.*, 117:169-174.
16. Lynch, G., Larson, J., Kelso, S., Barrionuevo, G., and Schottler, F., 1983, Intracellular injections of EGTA block induction of hippocampal long-term potentiation. *Nature* 305:719-721.

17. Mattson, M.P., 1992, Calcium as sculptor and destroyer of neural circuitry. *Exp. Gerontol.*, 27: 29-49.

18. Militante, J.D., and Lombardini, J.B., 1998, Pharmacological characterization of the effects of taurine on calcium uptake in the rat retina. *Amino Acids* 15:99-108.

19. Nicholls, D.G., and Budd, S.L., 1998, Mitochondria and neuronal glutamate excitotoxicity. *Biochim. Biophys. Acta.*, 1366:197-112.

20. Nicotera, P., and Leist, M., 1997, Energy supply and the shape of death in neurons and lymphoid cells. *Cell Death Differ.*, 4: 435-442.

21. Petit, P.X., Susin, S.A., Zamzami, N., Mignotte, B., and Kroemer, G., 1996, Mitochondria and programmed cell death: back to the future. *FEBS Lett.*, 396:7-13.

22. Richter, C., Schweizer, M., and Ghafourifar, P., 1999, Mitochondria, nitric oxide, and peroxynitrite. *Methods Enzymol.*, 301:381-393.

23. Satoh, H., and Sperelakis, N., 1998, Review of some actions of taurine on ion channels of cardiac muscle cells and others. *Gen. Pharmacol.*, 30:451-463.

24. Siesjo, B.K., 1988 Acidosis and ischemic brain damage. *Neurochem. Pathol.*, 9:31-88.

25. Trenkner, E., 1991 Cerebellar cells in culture. In: *Culturing Nerve Cells* (G. Banker, and K. Goslin, eds.), MIT Press, pp.283-307.

26. Trenkner, E., El Idrissi, A., and Harris, C., 1996, Balanced interaction of growth factors and taurine regulate energy metabolism, neuronal survival and function of mouse cerebellar granule cells under depolarizing conditions. *Adv. Exp. Med. Biol.*, 403:507-517.

27. Trenkner, E., and Sidman, R.L., 1977, Histogenesis of mouse cerebellum in microwell cultures: cell reaggregation and migration, fiber and synapse formation. *J. Cell Biol.*, 75:915-940.

28. White, R.J., and Reynolds, I.J., 1996, Mitochondrial depolarization in glutamate-stimulated neurons: an early signal specific to excitotoxin exposure. *J. Neurosci.*, 16:5688-5697.

Taurine in Aging and Models of Neurodegeneration

RALPH DAWSON, JR.*

*Department of Pharmacodynamics, College of Pharmacy, University of Florida, Gainesville, Florida, USA

1.   INTRODUCTION

1.1   Brief Review of Recent Studies on Taurine and Aging

Taurine, the second most abundant amino acid in the CNS, is acknowledged to be important in cell volume regulation yet little is known about how aging might alter taurine homeostasis. Unfortunately, few investigators have focused on the study of taurine in aging and senescence. Previously we summarized much of the older work showing that taurine content declines in a number of tissues with advanced age in rodents[1]. I will briefly summarize some recent studies that have examined the effects of aging on taurine. I will also summarize our own findings that point to several potential mechanisms that may impact taurine homeostasis with advanced aging.

Both *in vivo*[2] and *in vitro*[3] studies have shown that basal taurine efflux in brain does not change substantially with age. These studies also examined the ability of glutamate receptor activation to enhance taurine efflux and again found that aging had no statistically significant effect to alter glutamate receptor-induced taurine release. Del Arco et al.[2] using microdialysis evaluated basal taurine efflux and taurine efflux after blocking glutamate uptake in the striatum and nucleus accumbens. An increase in basal taurine efflux from the nucleus accumbens was noted when young rats were compared to old rats, but there was a nonsignificant 36% drop in basal taurine efflux from the striatum when middle-aged rats were compared to aged rats. These authors also found glutamate efflux after uptake inhibition was lower in aged rats and a parallel trend was noted for taurine. We have found that taurine content in the striatum decreases in old rats with learning deficits while the decline was less marked for old rats that were not impaired in a spatial memory task[4]. Saransaari and Oja[3] found that potassium stimulated efflux rates of taurine increase in hippocampal slices from 12-18 month old mice when compared to 3 month old mice, but decline back to the 3 month level by 24 months of age. There was also a decline in the ability of NMDA to stimulate taurine efflux in 12-18 month old mice when compared to 3 month old mice, but the efflux rates in 24 month old mice were similar to the 3 month old controls. Overall, aging appears to have modest effects at best to alter basal taurine efflux in the brain of rodents.

Rivas-Arancibia et al.[5] examined the effects of acute taurine administration on memory and lipid peroxidation after ozone exposure in old (30 months) Wistar rats. Taurine was administered 5 minutes before or after acute ozone exposure. Ozone impaired both short-term and long-term memory in young rats (47 days old) and taurine given after ozone prevented short-term memory loss but had no effect on long-term memory. Old rats showed improvement in both short- and long-term memory when given taurine. Ozone impaired both short- and long-term memory in old rats and either mode of taurine treatment prevented an ozone-induced decrement in short-term learning. Long-term memory was more severely impaired when taurine was given prior to ozone treatment, but taurine administered after ozone treatment reduced memory impairment. Lipid peroxidation tended to increase in several brain regions when taurine was given prior to ozone treatment[5]. Taurine given after ozone treatment prevented lipid peroxidation in both young and old rats in most brain regions. Taurine was more effective in blocking lipid peroxidation in the striatum of old rats when administered prior to ozone treatment. This study suggests that taurine may have the ability to enhance cognitive function and reduce oxidative stress in old rats. We have found that taurine can reduce markers of oxidative protein damage in old rats[6].

Finally, a few studies have examined taurine or the effects of taurine in peripheral tissues of aged animals. Esquifino et al.[7] report that thymic concentrations of taurine, aspartate and glutamate decrease in aged rats and these decreases correlated with thymic ACTH levels. Age-related renal fibrosis was attenuated in aged F344 rats by 2% dietary taurine supplementation when given for up to 3 months[8]. Taurine treatment reduced renal mRNA for TGF-β1 and both type I and IV collagen[8]. Kishida et al.[9] examined the effects of 28 days of 5% taurine in young (5 weeks old) and middle aged rats (10 months old) fed a cholesterol free diet. These authors found aging increased plasma cholesterol in both male and female rats and that ovariectomy further increased cholesterol levels. Taurine supplementation was effective in lowering the age-related increase in plasma cholesterol in both intact and ovariectomized females, but was ineffective in male rats. We previously found a substantial age-related increase in plasma cholesterol in aged male F344 rats and in agreement with Kishida et al.[9] found no attenuation of this increase with taurine supplementation[10].

There is lack of information regarding taurine status in elderly humans, but disease, nutritional deficits and clinical problems with water balance may complicate a clear understanding of the effects of aging per se on taurine homeostasis in very old subjects. The elderly suffer from a high incidence of malnutrition associated with illness and in both hospital and nursing home environments poor nutritional status is often observed. Furthermore, a common clinical problem for the elderly is the regulation of fluid and electrolyte balance. Thus more research should be targeted to assess whether certain subpopulations of the elderly are at risk of being deficient in taurine.

1.2   Potential Mechanisms for an Age-Related Decline in Taurine

Absorption. The first bioavailability hurdle taurine must traverse is intestinal absorption. Taurine transport and tranporter regulation has been studied extensively. Unfortunately, little information is available in regard to how aging may affect intestinal absorption of taurine. We have examined the effects of taurine supplementation in several studies and middle-aged rats on these diets for 1 month or longer show serum levels of taurine around 700 μM while aged F344 rats given 1.5% taurine for 10 months had serum levels in excess of 1500 μM[10]. Taurine levels were measured in the small intestine and were not significantly altered by age[10]. Thus, at present we have no strong evidence to indicate a problem in intestinal uptake of taurine with aging.

Distribution. Once the gut absorbs taurine it must be distributed to tissues and cells. Skeletal muscle is the largest pool of taurine in the body, accounting for >70% of total body taurine content. The skeletal muscle pool of taurine turns over very slowly. Thus, skeletal muscle may serve as a sink for taurine after intestinal absorption or hepatic synthesis. Taurine biosynthesis appears to be minimal in skeletal muscle since knockout of the taurine transporter reduces muscle taurine content by more than 90%[11]. If aging muscle exhibited either an increase in taurine uptake or a decrease in taurine efflux, then other tissues could be deprived of access to circulating taurine. Taurine content has been reported to decline in the tibialis anterior muscle of 22 month old male Wistar rats[12] and the rectous femoris of C57Bl/J6 mice[13]. In contrast, other studies have found either no change or increases in muscle taurine content in aged rats and humans[14,15]. Type 1 muscle fibers contain higher concentrations of taurine[15] and factors such as denervation can increase muscle fiber taurine content. Thus, age-related changes in muscle could lead to enhanced sequestration of taurine by muscle leading to taurine deprivation in other tissues. We found that serum taurine content declines 30-40% in 3 different strains of aged rats and this would be consistent with a higher extraction efficiency of aged muscle for blood taurine. Clearly, further studies are needed to explore this hypothesis.

Excretion. The kidney is the major organ that governs whole body taurine homeostasis. Taurine transporter protein levels are adaptively regulated to allow for the efficient uptake or elimination of taurine from the plasma filtrate. Aging is associated with a decline in many renal functions. It would be logical to examine if diminished renal function would result in a greater loss of taurine via enhanced renal elimination. We[10] and others[16] have noted that aging does not result in a greater loss of taurine via the urine, but in fact results in lower levels of urinary taurine excretion. This would be consistent with a "perceived" need to renally conserve taurine and that advanced aging is a taurine deficient state.

Synthesis and/or Precursor Availability. We have previously shown that both cysteine dioxygenase (CDO) and cysteine sulfinic acid decarboxylase (CSAD) activity is substantially reduced in the liver of aged F344 rats[6,17]. This may be due to the severe age-related liver pathology normally seen in F344 rats. In contrast, tissue and serum taurine content declines in other strains of rats[1] without a loss of hepatic CSAD activity[14,17]. Hepatic CDO activity does appear to be reduced in all of the rat models of aging we have evaluated[14]. Sharing cysteine as a common precursor links taurine and glutathione (GSH) synthesis. Inhibiting γ-glutamylcysteine synthetase activity increases taurine production[18] and conversely supplementing cells with taurine boost GSH levels (unpublished findings). Paradoxically, both GSH and taurine appear to decline with age and γ-glutamylcysteine synthetase gene expression declines with age[19]. This points to age-related problems in regulating gene expression of the key enzymes necessary to maintain tissue levels of both taurine and GSH. Problems in precursor availability related directly to cysteine availability or alterations in the transsulfuration pathway from methionine to cysteine could also contribute to diminished synthesis of sulfur containing compounds. Homocysteine levels appear to rise as a function of age and could be one pathway draining cysteine away from both taurine and GSH production. The renal excretion of inorganic sulfate is decreased about 41% in aged F344 rats[20] suggesting some generalized deficit in sulfur metabolism with aging. Unfortunately, at the present time the lack of specific experimental evidence leaves much room for speculation as to the mechanisms involved in age-related taurine decline.

2.   TAURINE IN MODELS OF NEURODEGENERATION AND THERAPEUTIC IMPLICATIONS

Could an age-related decline in taurine contribute to the susceptibility of the aged nervous system to disease and injury? Taurine has been proven in many studies to have cytoprotective actions against a wide array of toxic agents and conditions[21]. Taurine has also been shown to play a key role in apoptosis and have antiapoptotic effects in certain models[22]. If cellular taurine content declines in aging, would this lower the threshold or promote cell death? Human neurodegenerative diseases have increasingly been associated with inflammatory mechanisms. Taurine is an important player in host defense mechanisms against self-inflicted immune-mediated tissue injury[22]. Would a lack of taurine exacerbate cellular injury by inflammatory mechanisms in age-related diseases? Poor nutrition, hyponatriemia, inflammation, oxidative stress etc. could all further promote age-related declines in taurine. I will briefly summarize data we have obtained to suggest that taurine can indeed protect against certain conditions thought to play a role in the pathogenic mechanisms in Alzheimer's disease. I will also suggest that chronic neuroinflammatory conditions may increase taurine utilization in free radical scavenging reactions or impair its biosynthesis.

Our first studies simply tested whether or not taurine could attenuate cellular injury caused by known neurotoxic fragments of β-amyloid. Incubation of human neuroblastomia cells (SK-N-SH) for 48 hours in the presence of β-amyloid$_{25-35}$ caused a significant reduction in cell viability that was blunted by co-administration of taurine (Figures 1 and 2). H_2O_2 production and/or calcium overload has been postulated to play a role in the mechanisms of Aβ toxicity. We tested whether taurine would be effective against H_2O_2 toxicity or calcium overload and found no protection by taurine against these direct modes of toxicity. PC12 cells grown as described for the SK-N-SH cells were treated for 48 hours with 500 μM H_2O_2 or 20 μM A23187, a calcium ionophore. These treatments caused a significant ($p<0.05$) loss of cell viability and taurine (1 mM) did not attenuate cell death as assessed by neutral red uptake (data not shown). While these experimental treatments may not exactly mimic the pathogenic events causing Aβ toxicity it does suggest taurine's mode of neuroprotection is rather complex. Taurine may exert some actions to mitigate neurotoxic injury by inhibiting metal catalyzed oxidation reactions[23]. Studies from our lab have shown that homotaurine can greatly increase iron-stimulated catecholamine oxidation reactions in a manner inhibited by the iron chleator, desferroxamine[23]. In contrast, taurine inhibits metal and nitric oxide stimulated oxidation of catecholamines[23].

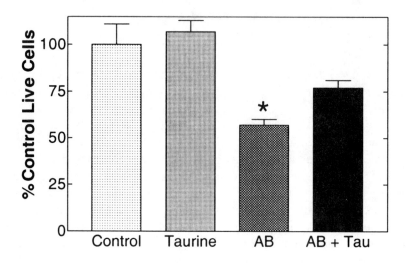

Figure 1. SK-N-SH cells grown in RPMI 1640 media with 10% FBS and antibiotics were treated for 48 hours with 20 μM β-amyloid$_{25-35}$ (AB), 500 μM taurine (Tau) or the combination of AB and Tau. Cell viability was determined by tryphan blue staining and cell counting. AB significantly (*p<0.05) reduced cell number and taurine significantly attenuated the effects of AB.

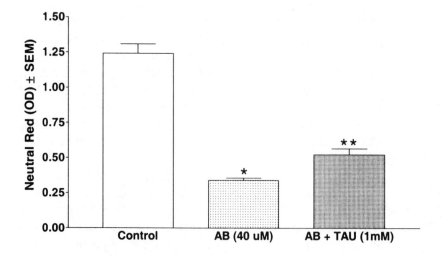

Figure 2. SK-N-SH cells grown as described in Figure 1 were treated for 48 hours with 40 μM β-amyloid$_{25-35}$ or a combination of AB and taurine (1 mM). Cell viability was determined by neutral red dye uptake. AB produced a significant (*p<0.05) reduction in cell viability and taurine attenuated AB's effects (**p<0.05).

Key players in neuroinflammatory reactions are microglia. Microglia can provide growth and support functions for neurons or release cytotoxic cytokines and free radicals (superoxide, HOCl). Evidence suggests that many human neurodegenerative diseases have many of the pathologic hallmarks of a chronic inflammatory condition. Some evidence suggests that polymorphisms in the gene for myeloperoxidase (MPO), the enzyme that catalyzes HOCl formation, may predict risk for Alzheimer's disease[24]. In fact β-amyloid can induce MPO expression in cultured microglia[24]. Previously we have shown that taurine can provide complete protection against HOCl toxicity to neurons in culture[25]. Almost nothing is known about the role of taurine in microglial function or how microglia regulate cellular levels of taurine. We have begun some basic characterization of microglial taurine status using primary cultures of microglia from the neonatal rat brain. These

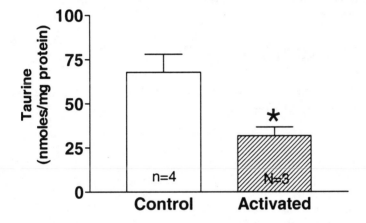

Figure 3. Primary cultures of rat microglia were activated by exposure to LPS (1 µg/ml) and γ-IFN (100 U/ml) for 24 hours. Taurine content declined significantly (*p<0.05) in these cells.

studies were conducted in collaboration with Dr. Wolfgang Striet at the University of Florida. Microglia cultured in DMEM supplemented with 10% FBS for 24 h have approximately 70 nmoles/mg protein of taurine. Trace amounts of hypotaurine were also detected in these cells, but it is uncertain if this is due to uptake from the surrounding medium or *de novo* synthesis. Evidence suggests that the former explanation is the most likely mechanism as preliminary studies have shown that these cells possess taurine transport activity. To determine if immunological activation would increase taurine utilization in cultured rat microglia, the cells were activated with a combination of lipopolysaccharide (LPS) and γ-interferon (γ-IFN). There was a marked drop in intracellular taurine (Figure 3) with no change in media taurine. This experiment suggests that immunological activation reduces intracellular pools of taurine, an effect not explained by enhanced efflux into the media. At present we do not know if this effect is a result of decreased biosynthesis or if taurine was consumed in scavenging reactions i.e. taurochloramine formation. LPS treatment for up to 48 hours does not alter taurine levels in C6 glioma cells, but does reduce hypotaurine levels significantly (see Dominy & Dawson chapter). Glia proliferation is a common consequence of aging and/or CNS injury and microglia are proposed to mediate some aspects of immune-mediated injury in Alzheimer's disease. An age-related decline in taurine could lead to enhanced vulnerability to immune-mediated tissue injury in human neurodegenerative diseases.

In summary, advanced aging may produce an imbalance between taurine availability and tissue-specific requirements for taurine. A failure by the elderly to maintain proper levels of nutrition could further diminish taurine availability and elevate the risk to certain neurodegenerative conditions.

ACKNOWLEDGEMENTS

The author gratefully acknowledges the technical assistance of John Dominy and Sean Kearns.

REFERENCES

1. Dawson, R., Eppler, B., Patterson, T.A., Shih, D., and Liu, S., 1996, The Effects of Taurine in a Rodent Model of Aging. In *Taurine 2: Basic and Clinical Aspects*, (R. Huxtable, J. Azuma, Nakagawa, K. Kuriyama, and A. Baba, eds.), Plenum, New York, pp. 37-50.
2. Del Arco, A., Segovia, G., Prieto, L., and Mora, F., 2001, Endogenous glutamate-taurine interaction in striatum and nucleus accumbens of the freely moving rat: studies druing the normal process of aging. *Mech. Ageing Dev.* 122: 401-414.
3. Saransaari, P. and Oja, S.S., 1997, Taurine release from the developing and ageing hippocampus: stimulation by agonist of inotropic glutamate receptors. *Mech. Ageing Dev.* 99: 219-232.
4. Dawson, R., Pelleymounter, M.A., Cullen, M.J., Gollub, M. and Liu, S., 1999, An age-related decline in striatal taurine is correlated with a loss of dopaminergic markers. *Brain Res. Bull.* 48: 319-324.
5. Rivas-Arancibia, S., Dorado-Martinez, C., Borgonio-Perez, G., Hiriart-Urdanivia, M., Verdugo-Diaz, L., Duran-Vazquez, A., Colin-Baranque, L., and Avila-Costa, M.R., 2000, Effects of taurine on ozone-induced memory deficits and lipid peroxidation levels in brains of young, mature, and old rats. *Environ. Res.* 82: 7-17.
6. Eppler, B. and Dawson, R., 2001, Dietary taurine manipulations in aged male Fischer 344 rat tissue: taurine concentration, taurine biosynthesis, and oxidative markers. *Biochem. Pharmacol.* 62: 29-39.

7. Esquifino, A.I., Garcia Bonacho, M., Arce, A., Cutrera, R.A., and Cardinali, D.P., 2001, Age-dependent changes in 24-hour rhythms of thymic and circulating growth hormone and adrenocorticotrophin in rats injected with Freund's adjuvant. *Neuroimmunomodulation*, 9: 237-246.

8. Iglesias-De La Cruz, C., Ruiz-Torres, P., Garcia Del Moral, R., Rodriguez-Puyol, M., and Rodriguez-Puyol, D., 2000, Age-related progressive renal fibrosis in rats with ACE inhibitors and taurine. *Am. J. Physiol.* 278: F122-F129.

9. Kishida, T., Akazawa, T., and Ebihara, K., 2001, Influence of age and ovariectomy on the hypocholesterolemic effects of dietary taurine in rats fed a cholesterol-free diet. *Nutri. Res.* 21: 1025-1033.

10. Dawson, R., Liu, S., Eppler, B., and Patterson, T., 1999, Effects of dietary taurine supplementation or deprivation in aged male Fischer 344 rats. *Mech. Ageing Dev.* 107: 73-91.

11. Heller-Stilb, B., van Roeyen C., Rascher, K., Hartwig, H-G., Huth, A. Seeliger, M.W., Warskulat, U. and Häussinger, D., 2001, Disruption of the taurine transporter gene *(taut)* leads to retinal degeneration in mice *FASEB J.* express article 10.1096/fj.01-0691fje. Published online December 28, 2001.

12. Pierno, S., De Luca, A., Camerino, C., Huxtable, R.J., and Conte Camerino, D., 1998, Chronic administration of taurine to aged rats improves the electrical and contractile properties of skeletal muscle fibers. *J. Pharmacol. Exp. Therapeut.* 286: 1183-1190.

13. Massie, H.R., Williams, T.R. and DeWolfe, L.K., 1989, Changes in taurine in aging fruit flies and mice. *Exp. Gerontol.* 24: 57-65.

14. Eppler, B. and Dawson, R., 1998, The effects of aging on taurine content and biosynthesis in different strains of rats. In: *Taurine 3. Cellular and Regulatory Mechanisms* (S. Schaffer, J.B. Lombardini, and R.J. Huxtable eds), Plenum Press, New York, pp.55-61.

15. Airaksinen, E.M., Paljarvi, L., Partanen, J., Collan, Y., Laakso, R. and Pentikainen, T., 1990, Taurine in normal and diseased human skeletal muscle. *Acta Neurol. Scand.* 81: 1-7.

16. Corman, B., Pratz, J., and Poujeol, P., 1985, Changes in anatomy, glomerular filtration, and solute excretion in aging rat kidney. *Am J. Physiol.* 248: R282-R287.

17. Eppler, B. and Dawson, R. 1999, Cysteine sulfinic acid decarboxylase and cysteine dioxygenase activities do not correlate with strain-specific changes in hepatic and cerebellar taurine content in aged rats. *Mech. Aging Dev.* 110: 57-72.

18. Beetsch, J.W. and Olson, J.E., 1998, Taurine synthesis and cysteine metabolism in cultured rat astrocytes: effects of hyperosmotic exposure. *Am. J. Physiol.* 274: C866-C874.

19. Liu, R-M. and Choi, J., 2000, Age-associated decline in γ-glutamylcysteine synthetase gene expression in rats. *Free Rad. Biol. Med.* 28: 566-574.

20. Bakhtian, S., Kimura, R.E. and Galinksy, R.E., 1993, Age-related changes in homeostasis of inorganic sulfate in male F-344 rats. *Mech. Aging Dev.* 66: 257-267.

21. Timbrell, J.A., Seabra, V., and Waterfield, C.J., 1995, The in vivo and in vitro protective properties of taurine. *Gen. Pharmac.* 26: 453-462.

22. Redmond, H.P., Stapleton, P.P., Neary, P., and Bouchier-Hayes, D., 1998, Immunonutrition: the role of taurine. *Nutrition* 14: 599-604.

23. Biasetti, M. and Dawson, R., 2002, Effects of sulfur containing amino acids on iron and nitric oxide stimulated catecholamine oxidation. *Amino Acids* 22: 351-368.

24. Reynolds, W.F., Rhees, J., Maciejewski, D., Paladino, T., Sieburg, H., Maki, R.A., and Masliah, E., 1999, Myeloperoxidasea polymorphism is associated with gender specific risk for Alzheimer's disease. *Exp. Neurol.* 155: 31-41.

25. Kearns, S. and Dawson, R., 2000, Cytoprotective effect of taurine against HOCl toxicity to PC12 cells. In: *Taurine 4. Taurine and Excitable Tissues.* (L. Della Corte, R.J. Huxtable, G. Sgaragli and K.F. Tipton eds.) Plenum Press, New York, pp. 563-570.

Taurine Stimulation of Calcium Uptake in the Retina
Mechanism of Action

JULIUS D. MILITANTE[*] and JOHN B. LOMBARDINI[*#]
[*]Department of Pharmacology, and [#]Department of Ophthalmology & Visual Sciences, Texas Tech University Health Sciences Center, Lubbock, Texas, USA

1.   INTRODUCTION

Taurine (2-aminoethanesulfonic acid) is a free amino acid found in high millimolar concentrations in animal tissues and various review articles and manuscripts have been written concerning its physiologic function[1-3]. Some suggested functions for taurine are related to neurotransmission, to osmoregulation and to protection against oxidative stress. Special attention has been paid to the mechanism of action of taurine in the retina, especially in the decades of the 70's and 80's, mainly because of experiments which suggest that taurine is most abundant in the retina compared to other tissue types and because of studies which linked taurine deficiency with visual dysfunction[4]. More importantly, the simple supplementation of the diet with taurine proved to be sufficient in alleviating the vision problem, theoretically through the replenishment of the depleted taurine levels.

Many studies have described the biphasic action of taurine on retinal calcium uptake[4]. At low concentrations of calcium (10-100 μM), ATP is a stimulatory agent, and taurine potentiates ATP-dependent calcium uptake. In contrast, ATP and taurine separately and together are inhibitory at higher calcium concentrations (>500 μM)[5]. The effects of taurine are observed both in whole retinal samples and in isolated rod outer segments (ROS)[5-8]. The mechanism behind these effects of taurine is not yet fully understood. More recent studies report that the inhibition of taurine uptake in retinal samples does not antagonize the stimulatory effects of taurine at low calcium concentrations[9,10]. The data suggested that the effects of taurine are not dependent on uptake, but rather on the binding of taurine on the retinal membrane. Moreover, taurine appears to stimulate calcium uptake through the activation of cGMP-gated cation channels[8], possibly by altering the membrane environment in which the channels are found.

The following experiments were performed to study the nature of the calcium uptake that taurine modulates, specifically to differentiate between calcium uptake through the retinal membrane and calcium binding to the retinal membrane. Calcium uptake was measured in retinal tissue by incubating retinal samples with radiolabeled calcium and measuring the amount that remained after filtering the sample through a glass fiber filter. In theory, the activity measured would be comprised of calcium taken up into the tissue and calcium that bound to the membrane, and is more correctly described as calcium accumulation. It is possible that taurine binds to the membrane and increases the binding of calcium to the membrane in addition to modulating calcium uptake. In these experiments, retinal tissue was osmotically lysed to denature active calcium transport. Calcium uptake was also measured with and without treatment with A23187, a calcium ionophore that makes membranes freely permeable to calcium. A23187 would allow the calcium gradient produced by uptake systems to dissipate when the tissue samples are washed and filtered, leaving only calcium bound to membranes. The effects of taurine on calcium binding can thus be studied. Calcium binding was also measured at 0°C in the presence of ATP and taurine. The use of lower incubation temperatures has been traditionally used to study binding as active uptake systems usually require a physiologic 37°C temperature to function adequately.

2.   METHODS

2.1   Preparation of tissue samples

Adult Sprague-Dawley or Wistar rats were anesthetized with CO_2 and killed by cervical dislocation or decapitation, after which the eyes were removed and either used immediately or frozen at -80°C to be thawed later for experimental use. Whole retinal tissue samples were isolated by cutting the cornea open and by gently teasing the tissue out of the eye cup into a 0.32 mM sucrose solution while on ice. All subsequent procedures were done on ice to maintain a 2°C temperature.

Krebs-Ringer-bicarbonate (KRB) buffer [118 mM NaCl, 25 mM $NaHCO_3$, 5 mM glucose, 1.2 KH_2PO_4, 4.7 KCl, 1.17 mM $MgSO_4$] with 10 μM $CaCl_2$ was prepared. KRB buffer was aerated with 5%/95% oxygen for 15 minutes and the pH of the solution adjusted to 7.4 with concentrated HCl.

To prepare osmotically-lysed membrane samples, retinal tissue was centrifuged for 15 minutes at 16,000 x g (4°C), resuspended in distilled water and gently homogenized with a glass-to-glass homogenizer. The homogenate was then recentrifuged and the pellet resuspended in the KRB buffer.

For the isolation of ROS, 0.3 mM mannitol was used instead of 0.32 mM sucrose. Retinal tissue was dissected out as before and the ROS were isolated by vortex-mixing the tissue for 10-20 seconds, allowing the tissue to settle, and the decanting the supernatant which contained the ROS in suspension. The procedure was sometimes repeated with the pellet to maximize ROS yield. The supernatant was then centrifuged for 15 minutes at 16,000 x g (4°C) and the remaining tissue components were discarded. Osmotically-lysed ROS samples were prepared by resuspending the isolated ROS in distilled water and by gentle homogenization was previously described. The tissue was then recentrifuged as before and resuspended in KRB buffer.

2.2 Calcium uptake assay

The assay was performed with lysed whole retinal homogenate and lysed ROS in KRB buffer. Reagents such as taurine and ATP were added to the reaction tube in the appropriate amounts and kept on ice until the start of the reaction. Identical amounts of $^{45}CaCl_2$ (400,000-5,000,000 dpm) were added to the tubes in the presence of 10 μM unlabeled $CaCl_2$. Then the reaction tubes were preincubated in a 37°C water bath for 2 minutes. Tissue samples were then added in equal amounts to start the reaction. The reaction was terminated by adding 3 ml of the chilled buffer and by immediate filtering through a glass-fiber filter in a Millipore apparatus. The filter was washed three times with 3 ml and then counted for radioactivity in a scintillation counter. The amount of $^{45}Ca^{++}$ taken up by the tissue sample was determined by subtracting the counts retained on the filter after a zero-time incubation with the retinal preparation.

2.3 Calcium binding assay

Calcium binding assays were done with either lysed whole retinal homogenate or lysed ROS in KRB buffer. Reagents such as taurine and ATP were added to the reaction tube in the appropriate amounts and kept on ice until the end of the reaction. Identical amounts of $^{45}CaCl_2$ (400,000-5,000,000 dpm) were added to the tubes in the presence of 10 μM $CaCl_2$. In preliminary binding experiments, calcium binding increased slowly and equilibrium was reached at around 60 minutes and maintained for 60 minutes more. Tissue samples were added in equal amounts to start the reaction, and the reaction was terminated after 90 minutes for whole retinal samples. Tissue samples were added in equal amounts to start the reaction, and the reaction was terminated after 90 and 60 minutes for whole retinal samples and for ROS samples, respectively. The reaction was terminated by adding 3 ml of the chilled buffer and by immediate filtering through a glass-fiber filter in a Millipore apparatus. The filter was washed three times with 3 ml and then counted for radioactivity in a scintillation counter. The amount of $^{45}Ca^{++}$ taken up by the tissue sample was determined by subtracting the counts from control reactions that were treated with 6N HCl.

2.4   Protein assay

The amount of protein used was assayed using the bincinchoninic acid (BCA) method. Briefly, standards and samples were mixed with a solution of BCA protein assay reagent and 4% copper II sulfate (50:1). The mixture was incubated in a 37°C water bath for 30 minutes and absorbance was read at 560 nm. Protein content was used to correct and standardize all the data measured. Commonly, 100-300 μg for the whole retinal homogenate or 50-150 μg for the ROS were used for each reaction.

2.5   Statistical analysis

Each data point was a measurement derived from an independent experiment. Statistical analyses were performed using the GraphPad Prism and Instat software. Data were analyzed using Student's t-test.

3.   RESULTS

The lysis of retinal membrane did not prevent the stimulation of calcium uptake in both whole retinal homogenates and in isolated ROS samples (Figures 1 and 2). ATP (1.2 mM) alone produced a significant increase in uptake while taurine (32 mM) alone did not. However, taurine potentiated the effects of ATP. The effects of ATP and taurine were completely eliminated by 10 μM A23187 treatment.

Calcium binding was measured in lysed whole retinal homogenates while on ice, and the effects of ATP and taurine were observed (Figures 3 and 4). The retinal samples were incubated with the appropriate reagents for 90 minutes and the samples were filtered through glass fiber filter paper. The radiolabeled $^{45}CaCl_2$ that was left bound to the retinal membrane was counted in a scintillation counter. Neither ATP nor taurine significantly changed calcium binding levels in the whole retinal homogenates.

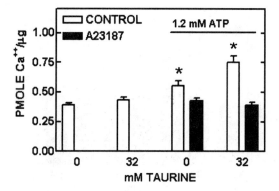

Figure 1. Calcium uptake in lysed retinal homogenate in the presence of 1.2 mM ATP and/or 32 mM taurine (mean ± SEM, N = 4-5). A23187 (10 μM) was used to inhibit the effects of ATP and taurine. An asterisk (*) indicates a significant difference compared to control (0 mM ATP, 0 mM taurine).

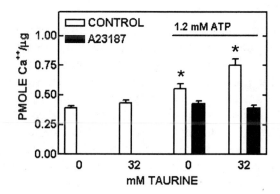

Figure 2. Calcium uptake in lysed ROS in the presence of 1.2 mM ATP and/or 32 mM taurine (mean ± SEM, N = 5). A23187 (10 μM) was used to inhibit the effects of ATP and taurine. An asterisk (*) indicates a significant difference compared to control (0 mM ATP, 0 mM taurine).

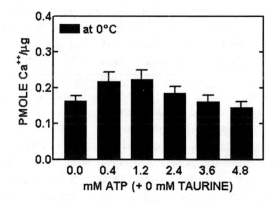

Figure 3. Calcium binding in lysed whole retinal homogenate in the presence of increasing concentrations of ATP (mean ± SEM, N = 5) and in the absence of taurine.

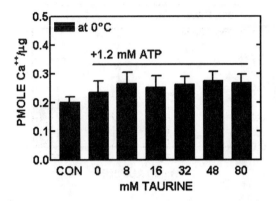

Figure 4. Calcium binding in lysed whole retinal homogenate in the presence of increasing concentrations of taurine (mean ± SEM, N = 3-4) and 1.2 mM ATP.

4.   DISCUSSION

Many similarities can be observed in cardiac sarcolemma calcium binding and retinal calcium uptake. Taurine has long been considered as a positive modulator of calcium binding in cardiac sarcolemma[11]. Specifically, taurine increased calcium binding in "intracellular" buffers (high sodium-low potassium) but did not affect low-affinity calcium binding in "extracellular" (low sodium-high potassium) buffers. Interestingly, when ATP was added, taurine increased calcium binding, regardless of the sodium concentration. The findings mirror the stimulatory effects of ATP and taurine observed in retinal calcium uptake. Two calcium binding sites were identified in the cardiac sarcolemma, one with dissociation constant (Kd) about a hundred fold greater than the other (~3.94 mM *vs* ~0.03 mM). Similarly, in the whole rat retina, two putative "uptake" sites have been described for calcium, exhibiting Km values of 2076 and 35μM[12]. Furthermore, previous studies with bovine ROS that used A23187 suggested that more than 50% of maximal calcium binding capacity is attained after only 1 minute incubation while active ATP-dependent calcium uptake continues for up to 30 minutes after that[13]. The current experiments measured uptake after only 2 minutes incubation and may in fact be composed mostly of calcium binding. It is thus a valid suspicion that the calcium uptake measured in retinal membranes in these esperiments may have a significant binding component, and that the stimulatory effects of taurine may also be directed toward this binding component, as it is with the cardiac sarcolemma.

However, differences should be noted between the methods behind sarcolemmal calcium binding and retinal calcium uptake. The calcium uptake components in the retina were observed with experiments performed at 37°C, a temperature that allows for the function of active calcium transport systems, with a very short incubation time. In contrast, the calcium binding experiments with cardiac sarcolemma were performed at 24°C for 45 minutes with a special equilibrium dialysis system that displayed an equilibrium time beginning at 30 minutes and lasting for at least 2 hours. The low temperature and delayed equilibrium are classic experimental parameters in binding experiments. In order to answer the question of whether calcium binding or uptake is involved in the effects of taurine in the retina, conditions were modified to allow for the elimination of active calcium uptake.

ATP stimulated calcium uptake in frog ROS[14]. Interestingly, the ROS membranes were found to be osmotically lysed, probably an artifact of tissue processing. While the osmotic lysis of bovine ROS membrane in water inhibited ATP stimulation of calcium uptake at 37°C temperatures, some level of stimulation could still be observed[13]. Similarly, the calcium uptake in both lysed whole retinal homogenate and ROS (Figures 1 and 2), also done at 37°C, exhibited some level ATP stimulation and in turn taurine potentiation. It is most interesting to note that the osmotic lysis of cell membranes does not completely obliterate the active transport of calcium, although clearly ATP-dependent uptake is compromised.

The calcium ionophore A23187 was employed to distinguish between calcium binding and calcium uptake. A23187 (0.9 µM) obliterated the stimulated effects of ATP on calcium uptake in both intact and lysed bovine ROS[13]. Calcium ionophores A23187 and X537A also inhibited ATP-dependent calcium uptake in frog ROS completely[14]. In the same manner, A23187 (10 µM) inhibited the effects of both ATP and taurine in rat retinal samples (Figure 1 and 2). It can be assumed that calcium uptake measured in the presence of the calcium ionophore corresponds to the calcium binding to membranes. Given that A23187 did not decrease uptake below that of control levels, the data suggests that in the absence of ATP and taurine, calcium uptake is almost totally composed of calcium binding. It is clear that ATP and taurine do not affect this calcium binding. Conversely, the stimulation of ATP and taurine corresponds to calcium uptake and not to binding. Experiments on calcium binding at 0°C support this notion in that ATP and taurine did not appear to have any stimulatory effect at all (Figures 3 and 4).

Previous studies suggest that the uptake or the diffusion of ATP into isolated bovine ROS is partly or completely necessary for the stimulatory effect of ATP on calcium uptake[13]. In contrast, other studies suggest that taurine uptake is not essential in the stimulation of calcium uptake and that taurine acts by binding to the membrane[9,10]. In addition, the effect of taurine on isolated ROS appears to be dependent on the activation of cGMP-gated channels. Thus, taurine appears to bind to the ROS membrane and stimulate the activation of cGMP-gated channels to allow calcium uptake through the membrane.

REFERENCES

1.   Wright, C.E., Tallan, H.H., Lin, Y.Y., and Gaull, G.E., 1986, Taurine: biological update. *Ann. Rev. Biochem.* 55:427-453.
2.   Huxtable, R.J., 1992, Physiological actions of taurine. *Physiol. Rev.* 72:101-163.
3.   Sturman, J.A., 1993, Taurine in development. *Physiol. Rev.* 73:119-147.
4.   Lombardini, J., 1991, Taurine: retinal function. *Brain Res. Rev.* 16:151-169.
5.   Militante, J.D. and Lombardini, J.B., 2000, Stabilization of calcium uptake in rat rod outer segments by taurine and ATP. *Amino Acids* 19:561-570.
6.   Liebowitz, S.M., Lombardini, J.B., and Allen, C.I., 1989, Sulfone analogues as modifiers of calcium uptake and protein phosphorylation in rat retina. *Biochem. Pharmacol.* 38:399-406.
7.   Militante, J.D. and Lombardini, J.B., 1998a, Effect of taurine on chelerythrine inhibition of calcium uptake and ATPase activity in the rat retina. *Biochem. Pharmacol.* 55:557-565.
8.   Militante, J.D. and Lombardini, J.B., 1998b Pharmacological characterization of the effects of taurine on calcium uptake in the rat retina. *Amino Acids* 15:99-108/
9.   Militante, J.D. and Lombardini, J.B., 1999a Taurine uptake activity in the rat retina: protein kinase C-independent inhibition by chelerythrine. *Brain Res.* 818:368-374.
10.   Militante, J.D. and Lombardini, J.B., 1999b), Stimulatory effect of taurine on calcium ion uptake in rod outer segments of rat retina is independent of taurine uptake. *J. Pharmacol. Exp.Ther.* 291:383-389.
11.   Sebring, L.S. and Huxtable, R.J., 1985, Taurine modulation of calcium binding to cardiac sarcolemma. *J. Pharmacol. Exp. Ther.* 232(2):445-451.
12.   Lombardini, J., 1983 Effects of ATP and taurine on calcium by membrane preparations of the rat retina. *J. Neurochem.* 40:402-406.
13.   Hemminki, K., 1975 Accumulation of calcium by retinal outer segments. *Acta Physiol. Scand.* 95:117-125.
14.   Pasantes-Morales, H., 1982, Taurine-calcium interactions in frog rod outer segments: taurine effects on an ATP-dependent calcium translocation process. *Vision Res.* 22:1487-1493.

The Nature of Taurine Binding in the Retina

JULIUS D. MILITANTE[*] and JOHN B. LOMBARDINI[*,#]
*Department of Pharmacology and #Department of Ophthalmology & Visual Sciences, Texas Tech University Health Sciences Center, Lubbock, Texas 79430, USA

1. METHODOLOGICAL IDIOSYNCRASIES

Lombardini and Prien[1] reported on the kinetics of taurine binding to rat retinal membranes and outlined the basic methodology behind binding experiments. Like most binding studies, the binding of radiolabeled ligand was counted in the presence of varying concentrations of unlabeled ligand under conditions which favor binding over transport. These conditions include low incubation temperatures and denaturation of tissue samples, both of which inhibit, in theory, active transport. Binding kinetics were then calculated using non-linear regression analysis of the binding curves through the range of total ligand concentrations, or alternately, linear regression analysis of the Scatchard plots. Binding kinetics include the dissociation constant (Kd) which is the total ligand concentration that produces half-maximal binding, and the estimated maximal binding level (Bmax). These values are used to specifically characterize the binding of a particular ligand.

While the above report[1] provided much information about taurine binding to the retina, the report also showed the idiosyncrasies of studying taurine binding to membranes. For example, kinetic estimates were shown to drastically shift with a change in the calculation of non-specific binding and in turn, specific binding levels. Non-specific binding is used as a correction factor in calculating specific binding and is routinely determined by competing out the labeled ligand with a concentration of unlabeled ligand that is ~1000 times greater than the Kd of the lowest-affinity component being resolved, i.e., if the Kd value is thought to be ~5 μM then non-specific binding is measured in the presence of 5 mM unlabeled ligand. Sodium-dependent taurine binding to retinal membranes exhibited 2 components, one with high affinity and one with low affinity. The amount of unlabeled taurine used to compete out the labeled ligand was increased from 2 to 80 mM when it was discovered that the estimated Kd of the low-affinity component (~334 µM) was only 5-15 times less than the 2 mM excess taurine used to determine non-specific binding. The estimate for the Kd and Bmax of the low-affinity component increased ~3.5 and ~5.5 with the use of 80 mM excess taurine. The various estimates of binding constants presented in the report can easily lead to confusion. Indeed, while the higher values are clearly more accurate, the lower values have been cited as the kinetic values for taurine binding in the rat retina[4].

Another problem that was observed was that binding constants could only be calculated within the range of unlabeled ligand used, in this case 0-1000 µM taurine[1], mainly because it is generally not acceptable to extrapolate beyond the binding curve generated by the actual data. Thus, the binding characteristics of taurine at the millimolar range could not be discerned with the data generated. The limitation is physiologically critical, as taurine is present within the cell in millimolar concentrations. In fact, in the retina, it is estimated that the taurine concentration is ~ 80 mM[2]. Binding studies with taurine have been performed in the presence of up to 100 mM unlabeled taurine with cardiac sarcolemma, and the Kd for low-affinity binding was estimated at ~ 19 mM[3]. In these studies, non-specific binding was determined in the presence of 250 mM taurine, as this was the highest concentration of taurine that could be dissolved in the buffer. However, 250 mM was in effect only 5-15 times greater than the Kd of the taurine binding measured, and thus is presented another problem in the study of taurine binding: the physiologic levels of taurine itself conflict with the traditional methods of binding experiments.

2.   FUTILE EXPERIMENTAL EXERCISE?

The estimation of taurine binding kinetics to retinal membranes was attempted in the face of the aforementioned difficulties. Briefly, adult Sprague-Dawley rats were sacrificed and retinae were isolated from the eyecups as previously described[1,7]. The retina was lysed in distilled water, homogenized and resuspended in Krebs-Tris HCl buffer, and used immediately for binding studies with radiolabeled [^{3}H]-taurine. Binding was performed at room temperature (22°C) for 60 minutes and terminated by filtration through a Millipore apparatus. Unlabeled taurine was used in the range 1-100 mM. Various alternate methods of calculating non-specific binding was tested and the denaturation of retinal tissue with 6N HCl was deemed as dependable, both in fact and in theory. Quite simply, this method produced the most consistent and predictable inhibition of specific taurine binding through the range of taurine concentrations used.

Not unexpectedly, the first major experimental challenge quickly became apparent as the data exhibited huge standard errors, especially at the higher levels of unlabeled taurine (Figure 1). Bound taurine is calculated both from the radioactivity counted and the total amount of taurine present, and so the same variations that produced little difference at lower taurine concentrations produced enormous differences at the higher levels. This inescapable methodological complication led directly to another analytical variation in the study. In the original report, several experiments were done to produce several independent estimates of Kd and Bmax, the means of which were then taken as the final values[1]. This is how kinetic constants are usually calculated and handled. However, in the present studies, the standard errors did not allow for the estimation of kinetic constants with either non-linear or linear regression analysis, at least at higher taurine levels. Only when raw binding data were pooled from all 4 independent experiments were data analyses possible. Non-linear regression analysis of the raw data and linear analysis of the Scatchard plot (Figure 2) were performed. The analyses of taurine binding in cardiac sarcolemma appears to have been done in the same manner[3].

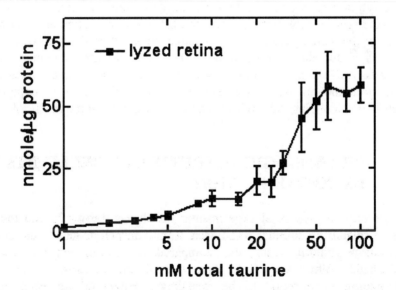

Figure 1. Taurine binding in lysed whole retina from rat (mean ± SEM). The data are pooled from 4 independent binding experiments.

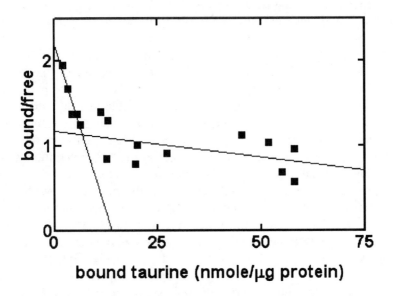

Figure 2. Scatchard analysis of the binding data presented in Figure 1.

Non-linear regression analysis of the binding curve revealed a 2-site binding system, with Kd values of 0.8476 ± 1.961 mM and 172.2 ± 120.5 mM. As if to continue the general theme of experimental aggravation, the dissociation constant for the lower affinity site did not follow the generally accepted requirement that the estimated Kd value be within the concentration range of the total ligand, *i.e.*, it was higher than 100 mM. Indeed, Scatchard analysis produced a regression line for the lower affinity component that was not statistically different from a horizontal line. Both observations point to a binding component that is essentially non-saturable, or to the experimental futility of calculating the binding constants for taurine at the millimolar level.

3.   PHYSIOLOGICAL TRUTH AND THE LIMITS OF EXPERIMENTATION

Before the proverbial experimental towel was thrown in and the white flag of scientific surrender was raised, a final desperate stand was made and the taurine binding data was compared to similar reports previously published. While very little data were found to exist, a few bits of information were found to be revealing. Huxtable[4] reviewed sodium-dependent taurine binding in neuronal-type membranes and showed Kd measurements with the corresponding range of taurine concentrations used in experiments with rat synaptosomal P_2B fractions, and rat glial, brain and synaptic membranes. In these studies that resolved a low-affinity type of taurine binding, the Kd was estimated to be higher than the highest concentration of taurine used. Had Huxtable[4] cited the correct kinetic constants in the original taurine binding studies of Lombardini and Prien[1], it would also have been observed that in the retina, the calculated Kd of the low-affinity site (~2738 μM) was higher than the highest taurine concentration used (1000 μM). By definition, Kd values determine the ligand concentration that results in half-maximal binding, and thus, in these experiments, the reported Kd values imply that maximal ligand binding is achieved at taurine concentrations that are significantly higher than taurine concentrations used and that, logically, maximal binding was not achieved. As half-maximal binding cannot be determined if maximal binding was not determined, these Kd values were probably significantly underestimated, suggesting that ligand binding affinity is much lower than expected. The apparent consistency represented either a brilliant hidden truth or some oddly shared experimental folly. If we choose the former, then we can move on to making some educated guesses as to the character to taurine binding. The data may simply represent a non-saturable type of binding that manifested itself as a dissociation constant that would float outside the range of total ligand used.

That the binding of taurine to neuronal type membranes is non-saturable is not unreasonable, given that taurine is for the most part assumed to interact with phospholipids in membranes[5]. High-affinity binding interactions are associated with minute amounts of endogenous ligands attaching to very specific protein molecules. If taurine were to act in this manner, then the vast amount of taurine actively sequestered by the cell would be practically useless. The massive binding of taurine to phospholipids was postulated to affect, among others, the binding of calcium to membranes, the operation of ion channels and protein phosphorylation processes. This notion is of great importance in the retina because of the high levels of taurine and the possible regulation of ion channels previously demonstrated[6]. Viewed from this perspective, it can be said that the physiologic action of taurine explains the difficulties encountered in the study of its binding to membranes.

REFERENCES

1. Lombardini, J.B. and Prien, S.D., 1983, Taurine binding by rat retinal membranes. *Exp. Eye Res.* 37:239-250.
2. Voaden, M.J., Lake, N., Marshall, J., Morjaria, B., 1977, Studies in the distribution of taurine and other neuroactive amino acids in the retina. *Exp. Eye Res.* 25:219-257.
3. Sebring, L.A. and Huxtable, R.J., 1986, Low affinity binding of taurine to phospholiposomes and cardiac sarcolemma. *Biochim. Biophys. Acta* 884:559-566.
4. Huxtable, R.J., 1989, Taurine in the central nervous system and the mammalian actions of taurine. *Prog. Neurobiol.* 32:471-533.

5.   Huxtable, R.J. and Sebring, L.A., 1986, Towards a unifying theory for the actions of taurine. *Trends Pharmacol. Sci.* 7:481-485.
6.   Militante, J.D. and Lombardini, J.B., 1998, Pharmacologic characterization of the effects of taurine on calcium uptake in the rat retina. *Amino Acids* 15:99-108.
7.   Militante, J.D. and Lombardini, J.B., 1999, Stimulatory effect of taurine on calcium ion uptake in rod outer segments of the rat retina is independent of taurine uptake. *J. Pharmacol. Exp. Ther.* 291:383-389.

INDEX

A7r5 cells, 6

Acamprosate, 481-484
   anti-craving taurine derivative, 481
   morphine interactions, 481
   treatment of alcoholism, 482

Acetaldehyde, 314

Acetaminophen toxicity, 314
   hepatic injury, 314
   N-acetyl-p-benzoquinone-imine, 314

Acetylcholine biosynthesis, 499-505

Actin filaments, 86

Acute myocardial infarction, 41-48

Adenosine, 424, 445
   A_1 receptor, 449

Aerobic exercise, 269-276

Aging, 537-545

Afterload pressure, 57-63

β-Alanine, 1, 28,29, 58, 61, 62, 255

Alcohol consumption, 246

Alcoholism
   therapies, 491

Alkaline phosphatase synthesis, 324

Alloxan, 97-104

Alveolar macrophages, 341-348, 349-356

Alzheimer's disease, 543, 357

Amino acids
   dabsylation of, 223
   derivatization of, 222
   determination, 299, 517
   efflux, 439, 494

Amiodarone, 310

Ammonia, 123-129, 463-470
   cerebral edema, 123
   excitotoxicity, 123
   swelling in brain slices, 125, 128

Angina pectoris, 42

Angiotensin II, 38, 39, 61

Anion channel blockers, 115

Anthracyclines, 414

Anti-calmodulin drugs, 116

Anticancer agents, 414
   Doxorubicin, 312, 411, 414

Antidepressant
   mirtazapine, 301

Antidiuretic hormone, 441

Anti-inflammatory activity, 329-339
   of Tau-Cl, 334

Anti-inflammatory factor, 13

Antioxidant nutrients, 272

Antioxidation, 213, 216, 255, 309, 411

Antineoplastic drugs, 411

Anti-pNCT antibody, 150

Anti-taurine antibody, 232

Aortic vascular smooth muscle cells, 14

Apolipoproteins, 263

Apoptosis, 314, 332

Arachidonic acid metabolites, 437

Arrhythmias, 20

Asparagine, 226

Asthma, 379, 403-410
   animal models, 408
   patients, 408

Astrocytes, 111, 368
   volume, 112

Atherogenic index, 81

Atherosclerosis, 5, 269

ATPase activity, 439

ATP production, 20

ATP-sensitive potassium ion channel, 315

ATP-stimulated calcium uptake, 547-554

Autoimmune Disease, 336

β-actin, 171

Bacterial endotoxin, 365

561

Balb/3T12 cells, 5
Benzodiazepine binding, 485
Beta-carotene, 43
Bicuculline, 126, 127
   sensitive chloride channels, 522
Bile acid pool, 261-267
   reabsorption, 266
   taurine effects on, 267
Bisphosphonates, 326
   therapy, 323
Bleomycin, 381-394, 395-402
   complex with iron, 310
   induced activation of nuclear factor-kB,
      381-394
   induced lung fibrosis, 381-394
   instillation in lungs, 395
Body mass index (BMI), 285-290
Bone, 323-328
8-Br-cAMP, 210
   taurine release, 210
Bronchoalveolar lavage fluid, 406
   taurine levels, 406
Butulated hydroxytoluene (BHT), 118
BV-2 cells, 358
   NO production, 359

C-6 cells, 365, 463-470, 471-479
CA3 pyramidal cells, 521
Caco-2 cells, 168, 214
Cadmium
   chloride channel blocker, 210
Calcium accumulation, 528, 530, 547-554
   assay, 549
   kinetics, 552
   mechanism of action, 547-554
Calcium binding, 549
   assay, 549
Calcium homeostasis, 21
   mitochondria, 527-536
Calcium imaging, 528
Calcium ionophore
   A23187, 553
   X537A, 553
Calcium metabolism, 315
Calcium overload, 20
Calcium paradox, 315
Calmodulin, 116
cAMP, 170

Canine taurine transporter, 26
Carbon tetrachloride toxicity, 310
Cardiac enzymes, 42
Cardiac fibroblasts, 28
Cardiac hypertrophy, 33
Cardiac myoblasts, 14
Cardiac taurine transporter, 33-41
Cardiomyopathy, 17
Cardioprotection, 19
Carnitine, 411
Catecholamines, 47
   oxidation, 541
   synthesis, 302
Cation transporters, 61
Cats, 1, 2
Cell proliferation, 412
Cell signal transduction, 116
Cell stretching, 39
Cell swelling, 123-129, 189
Cell volume disturbances, 183
Cell volume regulation, 183
Cerebral ischemia, 421-431, 433-444
Cerebrocortical slices, 123-129
cGMP levels, 456
Chelerythrine, 438
Chemo-attractants, 381
Chick embryo cardiomyocytes, 184
Chimpanzee spermatozoa, 308
Chloride channel blockers, 210
Chloride uptake, 202
Cholera toxin, 170
Cholesterol accumulation, 274
Cholesterol degradation, 261-267
Cholesterol diet, 263
Cholesterol intake, 271
Choline acetyltransferase activity, 501
Chronic hypernatremia, 309
Chronic Inflammataory disease, 336
Citrulline, 78
Clinical trials, 71
   limitations, 95
Coenzyme Q10, 41-48
   antioxidant effects, 42
   enhances cell membrane stabilization, 42
   free radical scavenger, 42
Colchicine, 86
Collagen induced arthritis (CIA), 336
Collagen synthesis, 324

Colon carcinoma cells, 411-417
Congestive heart failure, 33, 38
Contractile function, 60
Controlled inflammation, 216
Coronary flow rate, 61
Coronary heart disease, 5
Coronary thrombosis, 41
Cortisol, 47
COX 1 and 2, 53
Cultured cadiac myocytes, 25-31
Cyclooxygenase 1 and 2, 50, 333
CYP3A4 mRNA, 239
   induction by phenobarbital, 242
   induction by rifampicin, 239
Cysteine, 245
   catabolism, 475
Cysteine dioxygenase, 246
Cysteine sulfinate decarboxylase, 433
Cytochalasin D, 86
Cytochrome P450, 237
   induction of, 240, 242
   gene, 238
   mRNA, 237
Cytokines, 213-217, 385
Cytoprotection, 307-321
Cytoskeletal components, 116
Cytotoxic edema, 183

Dabsyl-hypotaurine, 226
   microdetermination, 227
Dabsyl-taurine, 221-228
D-aspartate, 191, 494
Depression, 297-304
   metabolic dysfunction, 302
Diabetes, 67-73, 83-90, 91-96, 97-104,
     205-212
   alloxan-treatment, 100
   human studies, 91-96
      limitations, 95
      subject criteria, 92
      study limitations, 95
      future studies, 95
   hyperosmotic component, 309
   polydipsia, 100
   polyphagia, 100
   polyurea, 100
   pre-diabetic state, 101
   retinopathy, 84

Diabetes (continued)
   streptozotocin-induced, 85
   type 1, 91
   type 2, 91
Dideoxyforskolin (DDF), 435
DIDS, 118, 440
Diethylnitrosamine-induced
hepatocarcinogenesis, 253-259
4,4-diiosothiocyanatostilbene-2,2'-disulfonic
     acid (DIDS), 118, 440
Diphenylene iodonium (DI), 119
Diisopropylfluorophosphate, 150
Dipyridamole, 446
Distribution of bile acids in the intestine, 264
DNA fragmentation, 379
Dot immunobinding assay, 230
Doxorubicin, 312, 411-417
Doxorubicin toxicity
   effect of antioxidants on, 413
Drug metabolizing enzymes, 241

Eadie-Hofstee plots, 29
EGTA, 21
Ehrlich cells, 190,415
   ascites tumour cells, 115
Embolic focal ischemia, 421-431
Embryonic chick cardiomyocytes, 18
EMT6 cells, 415
Endothelin, 61
Enzyme immunoassay, 50
Eosinophils
   peroidase, 379
Epigallocatechin-3-gallate, 392
Epileptic seizures, 515-525
   human temporal lobe, 515
   limbic seizures, 518
   status epilepticus, 518
Erα, 146
ERK signaling, 14
ERK phosphorylation, 14, 335
Estrogen receptors, 139-147
   in differential regulation of TauT promoter,
     144, 145
Estrogen response element (ERE), 141
Ethanol, 245, 485,-492
   acute adminisration, 247
   consumption, 489
   self-administration, 489, 491

Ethanolanime, 78
Experimental hepatic encephalopathy (HE),
    123
Expression of c-fos, c-jun and c-myc, 8

Fas-induced DNA fragmentation, 308
Fas-mediated apoptosis, 145, 332
Fibroblast-like synoviocytes (FLS), 332
Focal adhesion kinase (FAK), 120
Focal ischemia, 425
Forskolin, 438
Free radical scavenger, 42
Fundic glands, 232
    taurine immunoreactive cells, 232

GABA
    levels, 301
    synthesis, 499
GABA$_A$ receptors, 127, 522
Ganglion cells, 512
Gating of taurine transport, 149-157
Genistein, 495
Glia, 365
    Müller cells, 511
Global ischemia, 425
Glucose transporter, 85
Glutamate, 192
    accumulation, 458
    antagonists, 423
    calcium influx, 523
    depolarization, 523
    efflux, 439
    excitotoxicity, 519
    transport, 439
Glutamate decarboxylase (GAD), 499
Glutamine, 301
Glutathione, 245
    concentration, 246
Glutathione peroidase activity, 312
Glutathione-S-transferase, 50
Glycine, 81
Goldfish retinal explants, 507
Graft survival, 49
Guanidinoethanesulfonic acid (GES)
    dietary GES, 2
    toxicity, 8
    treatment of cells with GES, 8, 126, 127
Guanylyl cyclase, 455

Hamilton Scale for depression, 300
Hamster pancreatic insulinoma HIT-T15 cells,
    205
HbgA$_{1C}$, 93
HDL-cholesterol, 262
Heart failure, 41, 415
Heart rate, 59
Health drinks, 237
Heat shock protein, 55
HeLa cells, 115
Hepatic encephalopathy (HE), 123
Hepatocarcinogenesis, 253
High cholesterol diet, 261-267
High-affinity transport system, 37
Hill equation, 28
    coefficient, 201
Hippocampus, 107-114, 184, 445, 453
HIT-T15 cells, 206
HL60 cells, 377
HT-19 cells, 168
HT-29 cells, 30
Human aortic vascular smooth muscle cells
    (hVSMC), 13
Human brain glutamate decarboxylase, 499
Human breast cancer cells, 140
Human cervical cancer cells (SiHa), 161,
    176
Human colon carcinoma cell lines, 167-174
Human hepataic (HepG2) cells, 215,
    237-244
Human hepatocarcinoma cells (SK-Hep-1),
    161, 176
Human intestinal epithelial cells, 132
Human intestinal taurine transporter,
    131-138
    cloning of transporter, 134
    partial cDNA sequence, 136
    sequencing of transporter, 134
    production of polyclonal antibody to
        sequence, 135
Human leukemia cells (HL-60), 145
Human taurine transporter promoter,
    159-166
Hydrogen peroxide, 119, 373, 415
Hydroxyl radicals, 118
Hydroxyproline content in lung, 383
Hypercholesterolemia, 261
    effects of taurine, 261

Hyperglycemic ketoacidosis, 97
Hypertension, 43
Hypertriglyceridemia, 77
Hypertrophic cardiomyocytes, 36, 38
Hypochlorous acid, 13, 395
Hyponatremia, 107, 183
Hyposmolality, 107
Hypotaurine, 221-228, 250, 471-479
  antioxidant, 221
  conversion from taurine, 476
  depletion, 476
  effects of exogenous taurine exposure,
    473
  effects of LPS-induced activation, 473
  effects of LPS-induced depletion of, 476
  levels, 476
  production, 476
  synthesis, 477
Hypoxia, 445

IFN-γ signaling, 349-356
IκBα, 373-380
IL-1α, 383
  Effects of taurine and niacin on, 386
IL-1β, 215, 333
IL-6, 383
IL-8, 333
Immunohistochemistry, 518
Immunoregulatory activity of Tau-Cl
Infarct size, 43
Inflammatory bowel disease, 215
Inflammatory mediators, 341-348, 349,
    365-372
  assays, 367
Inflammatory response, 358
Inner medullary collecting duct, 191
iNOS, 53
  gene expression, 53, 341-348
  mRNA, 361
Insulin, 83-90
  binding, 85
  replacement therapy, 91
Intermittent claudication, 46
interstitial collage, 395
  overproduction, 395
Interstitial lung fibrosis, (ILF), 381
Intestinal epithelial cells, 215
Intestinal taurine transporter, 213-217

Intracellular calcium, 18
Intracellular taurine content, 35, 36
Intraepithelial macrophages, 232
Inulin, 124
Inward current, 17
Ionomycin, 210
  calciuim ionophore, 210
  taurine release, 210
Ion transport, 47
Ischemia, 445
  cerebral ischemia, 422
  forebrain ischemia, 423
  induced cardiac damage, 46
  neuronal injury, 422
  reperfusion injury, 49
  taurine protection, 422
Isolation of human eosinophils, 404
Isoproterenol, 312
Isovolumetric regulation in mammal cells,
    183-187
  mechanism, 184

JAK/Stat, 349-356
Jeju Island, 277
  women of, 277
Jurkat T-lymphocytes, 308

Kainic acid, 519
  induced seizures, 515
  receptors, 422
Ketamine, 49
Kidney, 50
Kinase inhibitors, 115
Korean college students, 291-295
  taurine intake levels, 291
Korean women, 269-276, 285-290
Kupffer cells, 216

Lactate dehydrogenase, 10, 44, 50
  cytotoxicity, 10
Langendorff preparation, 61
LDL-cholesterol, 81
Leukocytes, 401
  polymorphonuclear, 403
Leupeptin, 150
Light transmission, 111
Lineweaver-Burk plots, 29
5-Lipoxygenase, 116

Liver
  cholesterol, 81
  cytochrome P450, 241
  cysteine, 249
  ethanol, 249
  glutathione, 246
  hypotaurine, 226
  lipid analyses, 262
  measurements of enzyme activities, 246
  taurine, 249
LoVo cell lines, 412
  effect of oxygen pressure, 413
Low-affinity transport system, 37
L-type calcium current, 17
Luciferase activity, 179
  expression, 377
Luciferase-reporter constructs, 175-181
Lugol's iodine solution, 232
Lung fibrosis, 381
  animal models, 381
Lysophosphatidylcholine (LPC), 214

Macrophage-like (THP-1) cells, 215
Macrovascular complications, 67
Malondialdehyde (MDA), 44, 311
Maloni semialdehyde, 62
Mammary gland, 190
Mannitol, 125
MAP kinase/ERK cascade, 12
MCF-7 cells (human breast cancer), 139
  estrogen receptor-positive, 145
MDA-MB-231 cells, 143
Measurment of cytotoxicity, 7
Mechanical stretch of cultured myocytes, 38
Melatonin, 47
Membrane bilayers, 313
Memory
  long-term, 538
Met45, 378
Michaelis Menten constants, 37
Microdetermination of taurine/hypotaurine,
    221-227
Microdialysis
  amino acids, 422, 485
Microglial cells, 357-364
Mirtazapine, 297-304
Mitochondria, 527-536
  calcium content, 523

Mitochondria (continued)
  calcium homeostasis, 527
  calcium sequestering, 529
  depolarization, 528
  electrochemical gradient, 527
  respiratory activity, 42
  sequester calcium, 529
Mitogen-activated protein kinase (MAPK),
    335
Models of aging, 537
Models of neurodegeneration, 535-540
Mongolian gerbil, 425
Monocrotaline, 310
Multiple sclerosis, 357
Mutant R324G, 152
  kinetic analysis, 156
Myeloperoxidase system (MPO), 365
Myocyte hypertrophy, 38
Myo-inositol, 475

N-acetyl-cysteine, 392
Na^+/Cl^- - dependent taurine transporter,
    197-204
N-chloramine, 311
  bactericidal, 313
  fungicidal, 313
  inflammmation, 313
  neutrophils, 313
Necrosis, 46
Neonatal rat heart cells, 29
Neuritic outgrowth
  from goldfish retinal explants, 507-514
  in the presence of fetal calf serum, calcium,
    albumin, HEPES, 508
  in the presence of glucose, 509
  in the presence of taurine, 510
Neuronal cell death, 357
Neuronal dendarites, 112
Neurotransmitter release, 192
  amino acid, 192
Neutral amino acids, 115
Neutrophils, 47
NF-κB signaling, 341-348
NF-κB activation, 365-372
Niacin, 381-394
Niflumic acid, 186
NIH/3T3 cells, 5, 115
7-Nitroindazole, 456

N-methyl-D-aspartate, 422
Nocodazol inhibitor, 86
Nonmyocyte cultures, 25
Northern blot analysis of RNA, 26
Nitric oxide, 453-461
   formation, 457
   generators, 457
Nitric oxide production, 50, 357-364
NR8383 cells, 343
Nuclear factor κB, 216, 381-394
Nucleus accumbens, 487

Obesity
   deaths, 285
   In US, 285
Obesity Index, 285-290
Organic osmolytes, 189
Ornithine, 78
Ortho-vanadate, 438
Osmolar changes in mammalian cells, 183
Osmoregulation, 213, 308
Osmotic swelling, 49
Osteoblasts, 324
Osteoclasts, 324
Osteomalacia, 324
Osteoporosis, 324, 326
Ouabain, 433, 439
Ouabain-induced swelling in cortex, 439
Oxidation of hypotaurine, 223
Oxidized low density lipoprotein levels,
   269-276
Oxidoreductive mechanism, 334
Oxygen conditions
   ambient, 411
   hypoxia, 411
Oxygen consumption, 49
Oxygen radicals, 55
Ozone, 403

p53 tumor suppressor gene, 139
   wild type gene product, 146
Pancreatic β-cells, 205-212
Parietal cells, 232
Patch-clamp technique, 18
Penumbra, 366
Peripheral lymphocytes, 297
Peroxidase-anti-peroxidase complex, 231
Peroxidation, 270

Peroxynitrite, 357
Phagasomes, 373
Phenylalanine, 290, 376
8-Phenyltheophylline (8-PT), 448
Phorbol-myristate-acetate (PMA), 119
Phosphatidylinositol 3 kinase, 87
Phosphoethanolamine, 436
Phosphoinositide-3 kinse (PI3-K), 496
Phospholipase C
   inhibitor, 438
Phospholipase A2, 116
Phosphorylation, 150
Phosphoserine, 78
Plasma ammonia, 123
Plasma glucose levels, 99
Plasma triglyceride concentrations, 77
Plate aggregation, 46, 91
Platelet-derived growth factor (PDGF)
   development of atherosclerosis, 5
   proliferation of vascular smooth muscle
      cells, 5
Polyalcohols, 184
Polyamines, 184
Polymorphonuclear leukocytes, 349
Positive dP/dt, 60
Potassium current, 17
Preparation of cardiomyocytes, 34
Pre-steady-state currents, 199
Production of polyclonal antibody, 131-138
Proinflammatory cytokines, 216, 392
Proinflammatory mediators, 395
Proliferatiang cell nuclear antigen (PCNA),
   332
Prostaglandins, 50, 332
   E_2, 333
Protein kinase C, 119, 149, 167, 499
   phosphorylation, 149
Protein phosphatase, 116
Protein phosphorylation, 47
Protein synthesis, 86
Protein tyrosine kinase, 116
Pryamidal cells, 110
   diameter, 110
Pulmonary fibross, 381
Pulmonary inflammation, 341
   mediators, 341
Purkinje cell dendrites, 434
Pyrrolidine dithiocarbamate, 392

Rat stomach, 229
RAW264.7 cells, 132
Reactive aldehydes, 415
Reactive oxygen species, 118
Renal failure, 51
Renal (HEK293) cells, 215
Renal function, 51
Renal ischemia/reperfusion injury, 49-56
Renal tubular enzymes, 51
Reporter gene, 139
Retina, 83-90, 497, 507-514, 547-554,
     555-560
Retinal degeneration in cats
   decreased taurine levels, 1
Retinal membranes, 550
Retinal pigment epithelium, 83-90
   transport, 83
Retinol, 282
Reversed-phase high-performance liquid
chromatography, 221-228
Rheumatoid arthritis, 329-339
Rho
   constitutive, 115, 118
Rhodamine uptake, 527, 533
Rifampicin, 237-244
Rod outer segments, 309, 547

S-adenosylhomocysteine, 246
S-adenosylmethionine, 250
Scatchard plots, 555
Selenium, 70, 71
Serine, 78
Serotonin, 47
   agonists, 423
Serum creatinine, 50
Serum insulin levels, 99
SiHa cells, 164
Sino-atrial nodal cells, 17
Skate erythrocytes, 190
SK-Hep-1 genomic DNA, 177
Sodium/chloride transporters, 25
Steady-state currents, 199
Streptozotocin, 68, 76, 86, 205
Stretch-dependent ion channels, 61
Stroke, 357
Sulphoacetaldehyde
   concentration, 330
   cytotoxic, 330

Sulfur amino acid metabolism, 245-252
   alcohol consumption, 246
Superoxide anion generation, 404
Superoxide dismutase activity, 312
Synaptosomes, 501
Systemic hypertension, 57
Synovial hyperplasia, 334
Synoviocytes
   inhibition by Tau-Cl, 332

Tamoxifen, 435
Tapetum lucidum, 1
Taurine
   absorption, 539
   accumulation, 85
   activates chloride currents, 522
   activation of taurine-induced current, 201
   acute bronchoconstriction, 405
   adenosine release, 449
   administration, 41
   aging, 537-545
         decline in taurine levels, 539
   alleviates swelling due to ammonia, 125
   analogues, 323-328
   antagonist (TAG), 434
   anti-arrhythmic activity, 61
   anti-atherosclerotic effect, 13
   anti-convulsive effects, 522
   anti-dyslipidemic effect, 103
   antigen-induced acute bronchoconstriction,
         405
   antigen-induced increased cellularity, 407
   anti-inflammatory action, 373-380
   anti-inflammatory factor, 13
   anti-oxidant activity, 312
   asthma, 403-410
   autoradiographic detection, 84
   bicuculline-sensitivity, 128
   binding, 555-560
         kinetics, 556
   bleomycin, 395-402
   blood levels, 102
   bone formation, 323
   bone gain, 325
   bone metabolism marker, 326
   bone resorption, 323
   bone tissue formation, 323
   brain volume regulation, 107

Taurine (continued)
bronchoalveolar lavage fluid, 406
calcium binding, 548
calcium homeostasis, 21, 532
calcium metabolism, 315
calcium modulation, 27
calcium uptake, 547
cancer
taurine supplementation, 253-259
carbon tetrachloride toxicity, 311
cardiac pacemaker currents, 17-23
cell proliferation, 11,13
cellular injury induced by ischemia and
reperfusion, 54
cellular responses, 241
cell membrane stabilization, 261
cell swelling, 186
cell volume regulation, 107-114
cerebral ischemia, 421-431
chloramine, 13, 329-339
cholesterol catabolism, 5, 261-267
cholesterol diet, 264
cholesterol lowering effect, 5
chronotropic effect, 18
coenzyme Q10, 43
combinatin with niacin, 381-394
congestive hart failure, 33
conjugation with bile acids, 5, 307, 314
contractile force, 21, 61
coronary heart disease, 5
cytoprotective, 21, 41, 127, 197, 307-321
dabsyl-taurine, 223
decreases LDH leakage, 10
deficient heart, 60
degradation of cholesterol, 263
depletion, 57-63
depletion of hypotaurine by, 477
depression, 297-304
metabolic dysfunction, 302
derivative - acamprosate, 484
derivatization with dabsyl chloride,
221-228
determination, 221-228
microdetermination, 221
detoxification, 229
diabetes, 67-73, 75-82, 83-90, 91-96,
97-104
diarrhea, 99

Taurine (continued)
distribution in stomach, 232
dietary intake, 277-283
distribution, 539
dysrhythmias, 20
edema ameliorated by taurine, 123
effect on ATP/calcium calmodulin
kinase II, 502
effect on choline acetyltransferase activity,
501
effect on c-fos, c-jun, 12
effect on Goldfish retinal explants,
507-514
efflux, 433-444, 463-470
ethanol-induced taurine efflux,
485-492
melittin-induced, 117
efflux of taurine from hippocampal
slices, 185
efflux pathway, 115-122
chloride, 189-196
electron microcopic immunocytochemistry,
232
enhances induction of CYP3A4 by
rifampicin, 237-243
enhances mitochondrial activity, 533
epileptic seizures prevented by, 515-525
ERK phosphorylation, 12, 14
Evans Blue levels, 406
excitable tissues, 61
excretion, 539
exercise, 325
kidney, 540
fibrosis, 395-402
gene expression, 8
generation of tau-Cl, 330
growth, 77
heart failure, 46
heart rate, 61
health drinks, 237
high-affinity system, 29, 37
Hill equation/Hill coefficients, 203, 208
human blood peripheral lymphocytes,
297-304
hypertaurinuria, 335
hypocholesterolemic effect, 78
hypoglycemic effect, 92, 99, 103
hypolipidemic effect, 77

Taurine (continued)
  hypotaurine dynamics,   471-479
  immunocytochemistry,   231
  immunofluorescent double labelling,   231
  immunoreactivity,   84
  increase in coronary vascular resistance,
        61
  in drinking water,   395, 401
  induced inward current,   200
  induction by rifampicin,   237-244
  induction of iNOS,   55
  infarct size,   43, 46
  inflammation,   341, 478
  inhibitory nerve transmitter,   429
  insulin-induced taurine uptake,   85-87
  insulin receptor,   97
  intake in Korean college students,
        291-295
  intestinal absorption,   539
  intracellular osmolyte,   108
  intracellular taurine content of cardiocytes,
        26-27
  ionic channels,   19
  ionic currents,   20
  kainite-evoked excitotoxicity,   123
  kinetics of taurine uptake,   36, 37, 208,
        465
  LDL cholesterol,   78
  levels,   109
        of meat and fish,   291
        of myocytes,   27
  lipid levels,   263
  lipid lowering action,   13, 77
  low calcium concentrations,   548
  lowers cholesterol,   266
  lung,   403-410
        hydroxyproline content,   383
  lung fibrosis,   382
  lung inflammation,   395-402
  lung injury,   382
  lymphocytes,   300
  MDCK taurine transporter,   151
  mediated uptake of chloride,   202
  melittin-induced taurine efflux,   116
  metabolic bone diseases,   326
  metabolism of Tau-Cl,   330
  membrane stabilizer,   229, 309
  mitochondrial activity,   533

Taurine (continued)
  modulation by adenosine in normoxia,   446
  modulation by adenosine in ischemia,   448
  monochloramine,   330
  neuritic outgrowth of goldfish explants,   5
        10
        effect of calcium,   510
  neuromodulation,   229
  neuroprotection,   123
  neuroregulatory actions,   434
  neurotransmitter/neuromodulator,   235
  neutrophils,   235
  niacin,   382
  nitric oxide,   54
  NMDA receptor function,   127
  nutritional supplements,   237
        nutritional iimportance,   97
  obesity,   285-290
  oral supplementation,   102
  osmoregulation,   235, 434
  osmosensitive taurine release,   189-196,
        437
  osmotic agent,   71, 107
  osmotic stress,   307
  osteoporosis,   324
  oxidation of IκBα at Met45,   444
  pacemaking actitity,   20
  phosphorylation of ERK,   12
  pK2,   190
  plasma amino acid content,   78
  plasma cholesterol,   101
  plasma glucose,   77, 99
  plasma levels,   285-290
  plasma triglyceride concentrations,   77,
        101
  prevention of atherosclelrosis,   5
  prevention of CA3 cell death after kainic
        acid seizures,   519
  proliferation of vascular smooth muscle
        celis,   11
  prostaglandin synthesis,   54
  protein kinase C,   499
  reabsorption of bile acids,   267
  reduces cardiac dysfunction,   41
  reduces cell swelling,   128
  reduces coronary flow rate,   61
  reduces hyperglycemia and dyslipidemia,
        99

Taurine (continued)
  reduces mortality in diabetic rats, 67-75
  reduces severity of limbic seizures, 518
  regional differences in taurine intake,
      291-296
  regulates cholesterol metabolism, 266
  regulation of acetylcholine biosynthesis,
      499-505
  regulation of mitochondrial calcium
      homeostasis, 527-536
  regulation of GABA biosynthesis,
      499-505
  regulation of volume-sensitive taurine
      efflux pathway, 115-122
  relationship with various food intake, 280
  release, 120, 205-212, 453-461, 493-498
      assay, 207
      effects of cGMP, 456
      effects of nitric oxide-generating
          compounds, 454
      ionomycin, 211
      ischema-evoked release, 453-461
      nitric oxide-induced release and
          NMDA, 457
      tyrosine kinases, 493-498
  renal ischemia, 49-56
  retinal function, 83
  rheumatoid arthritis, 331
  serum insulin levels, 99,100
  serum taurine levels, 277-283
  sino-atrial nodal cells, 17-23
  specific fantibody, 229-236
  supplementation, 75, 78, 91-96, 99
  suppresses angiotensin II-mediated
      expression of c-fos and c-jun, 14
  swelling-activated amino acid release, 191
  swelling-activated taurine release, 191
  synthesis from cysteine, 245
  synthesis, 543
  taurine-bone interactions, 325
  therapeutic implications, 541
  thyroid taurine transporter, 137
  tissue distribution, 539
  transcriptional activity of taurine
      transporter, 175-181
  transport, 8, 149-157, 201, 213-217,
      463-470
      down-regulation of, 171

Taurine (transport continued)
    effect of ammonia, 463-470
    effect of glucose and insulin on, 85
    inhibitor (NPPB), 109-112
    in human colon carcinoma cell lines,
        167-174
    kinetics, 171, 197-204
    myocardial, 39
    mutant pNCT, 152
    PMA-induced down-regulation, 171
    regulated by cAMP, 167-174
    regulated by protein kinase C,
        167-174
    stoichiometry, 203
  transporter, 25-31, 33-40, 83-90,131-138,
        149-157,175-181, 197-204,
        207, 213-217
      intestinal, 137
      phosphorylation, 152
      regulation by cytokines, 213-217
      thyroid, 137
  transporter cDNA, 87
      cytoplasmic domains, 87
  transporter gene (TauT), 139
  transporter isoform, 87
  transporter mRNA, 28, 38
  transporter promoter, 159-166, 177
      sequence of human, 136, 159-166
      reporter plasmids, 179
  transporter protein, 26, 29
      transmembrane helices, 131
  treatment of animals, 51
  two transport systems, 84
  uptake assay, 87, 207, 464
  uptake inhibition, 208
  uptake in pancreatic β-cells, 205-212
  uptake in the retina, 83-90
      kinetics, 552
      regulation by protein kinases, 87
      rod outer segments, 550
      whole retina samples, 550
  uptake measurements, 27, 34, 150, 168
      effect of hyperosmolarity, 211
      effect of LPS-induced uptake, 477
      kinetics, 168
  urinary amino acid excretion, 81
  ventricular ectopics, 46
  volume regulator, 187

Taurine (continued)
  volume sensitive taurine efflux pathway,
      115-122, 211
  weight gain, 263
  whole blood taurine levels, 93-95,
      101,102
  whole retina samples, 549
Taurine bromamine, 378
  formation, 379
Taurine-Cl, 329-339, 341-348, 349-356,
      357-364, 365-372
  metabolism in rheumatoid arthritis,
      329-339
  Modification of IκBα, 375
  treatment of autoimmune disease, 336
  treatment of chronic inflammatory disease,
      336
Taurine-depleted heart, 57-63
Taut-1
  brain, 84
Taut-2
  associated with glial cells, 84
TauT, 139-147
  API-binding site, 163
  regulation by cytokines, 215
  regulation by food factors, 214
  regulation by hyperosmotic conditions,
      214
TauT gene, 143
  regulation of, 160
  transcriptional rate of, 171
TauT luciferase reporter gene, 180
TauT mRNA, 84, 173, 217
TauT promoter, 143-145, 176
  antisense promoter, 180
  human, 162, 180
  sequence analysis of promoter region, 180
TAUT-TRE luciferase plasmid, 163
TGF-β, 383
Thioacetamide-treated rats, 312
Thiobarbituric acid reactive substances, 44,
      94, 273
THP-1 cells
  macrophage-like, 215
Thrombolytic agents, 429
Thymidine incorporation, 7
  measurement, 7
TNF-α, 215-217, 343, 383

TNF-α (continued)
  effects of taurine and niacin on, 386
α-Tocopherol, 269, 282
Total cholesterol, 81, 262
Transactivation of TauT promoter by p53,
      143
Transducing signal, 494
  role of tyrosine kinases, 494
Transient forebrain ischemia, 421-431
Transsulfuration pathway, 245
Trolox, 414
T-type calcium current, 17
Tyrosine kinase, 493-498
  JAK1, 353
  JAK2, 353
Tyrosine phosphatases, 119
Tyrosine phosphorylation, 118, 324, 438
Tyrphostins, 436

UMR-106 cells, 324
Unstable angina, 43
Up-regulation of TauT transcription by E2,
      141
Uptake medium, 34
Urea, 50
Urinary free amino acids, 75-82
  excretion, 81

Vanadate, 119
Ventricular dysfunction, 46
Ventricular pressure, 59
Vinblastine, 411-417
Vitamin C, 44
Vitamin E, 44, 70, 272
Voltage-gated calcium channels, 423
Volume regulation, 107-114, 494
Volume sensitive anion pathway, 189
Volume sensitive organic osmolytes, 189

Whole blood taurine levels, 98
Women of Jeju Island, Korea, 277-283
Wortmannin, 88

Xenobiotics, 317
Xenopus laevis oocyte, 98,149,186, 207
  kinetic analysis of taurine transport, 155
  membrane isolation, 150
  microinjection, 150

Xenopus laevis oocyte (continued)
   mTauT-1 clone, 198
   phosphorylation of transporter, 154
   taurine uptake, 150
   western blot analysis, 150
Xylocain, 49

Zaprinast, 456